Aspartic Proteinases and Their Inhibitors
FEBS (Federation of European Biochemical Societies)
Advanced Course No. 84/07

Aspartic Proteinases and Their Inhibitors

Proceedings of the
FEBS Advanced Course No. 84/07
Prague, Czechoslovakia, August 20-24, 1984

Editor
Vladimír Kostka

Walter de Gruyter · Berlin · New York 1985

Editor

Vladimír Kostka, Dr., D. Sc.
Czechoslovak Academy of Sciences
Institute of Organic Chemistry and Biochemistry
Flemingovo náměstí 2
CS-166 10 Prague 6
Czechoslovakia

Library of Congress Cataloging in Publication Data

Aspartic proteinases and their inhibitors.
Includes bibliographies and indexes.
1. Aspartic proteinases--Congresses. 2. Aspartic proteinases--Inhibitors--Congresses. I. Kostka, Vladimír, 1930 – . II. Federation of European Biochemical Societies. [DNLM: 1. Peptide Peptidohydrolases--congresses. 2. Protease Inhibitors--congresses. QU 136 A838 1984]
QP609.A86A87 1985 574.19'256 85-13147
ISBN 0-89925-078-5 (U.S.)

CIP-Kurztitelaufnahme der Deutschen Bibliothek

Aspartic proteinases and their inhibitors : Prague, Czechoslovakia, August 20–24, 1984 / ed. Vladimír Kostka. - Berlin ; New York : de Gruyter, 1985.
(Proceedings of the FEBS advanced course ; No. 84,7)
ISBN 3-11-010179-3 (Berlin)
ISBN 0-89925-078-5 (New York)

NE: Kostka, Vladimír [Hrsg.]: Federation of European Biochemical Societies: Proceedings of the . . .

ISBN 3 11 010179 3 Walter de Gruyter · Berlin · New York
ISBN 0-89925-078-5 Walter de Gruyter, Inc., New York

 Printing: Gerike GmbH, Berlin. – Binding: Dieter Mikolai, Berlin. Printed in Germany.

PREFACE

The tradition, if it can be called that, of specialized meetings on aspartic proteinases is a short one. It is fifteen years since I attended my first proteinase meeting, the I.U.B. Symposium on Structure-Function Relationships of Proteolytic Enzymes in Copenhagen. If I remember correctly, the second morning of this Symposium was set apart for acid proteinases and it turned out that there was time enough to cover the whole field during a four-hour session. These were the days when studies on acid proteinases contained mostly specificity and mechanism investigations and when the structural knowledge of these enzymes was limited to a few fragmentary sequences in calf chymosin and hog pepsinogen, accounting together for less than 100 residues. The seven-year period following witnessed a rapid advance in our knowledge of acid proteinases, which came about mainly through the determination of primary and tertiary structures and through mechanism studies based on these structures. The Conference on Acid Proteases, organized by J. Tang in 1976 in Oklahoma City, Oklahoma, USA, clearly showed that acid proteases, although still bearing the old and vague name had "come of age". The Proceedings of this Conference, a collection of main lectures and titles of short and poster communications, became the first useful sourcebook for all further work in this field and has been frequently quoted ever since.

The increasing body of refined experimental data on acid proteases appearing at the end of the seventies (when the name of these enzymes was also changed to carboxylic or aspartic proteinases) gave evidence that a stage had been reached at which their physiological function not only could be understood but also modeled at the molecular level. This became evident in the workshop on the Structure and Activity of Aspartic Proteinases organized by J. Kay and T.L. Blundell in London in 1982. It was this excellent but short meeting that prompted my idea of organizing a "full size" meeting on this topic. Of the several possibilities, the form of a FEBS Advanced Course was chosen, mainly because of previous good experience with similar lecture courses. The original idea that overview lectures by invited speakers would be complemented by "problem-solving sessions" as in previous courses was abandoned, since such sessions give, as a rule, very little opportunity to less advanced or linguistically less well-equipped participants to express themselves. Instead the latter were offered the opportunity of presenting posters or short oral communications.

This book contains all lectures as given by invited speakers during the Course and the original version of the majority of the poster communications. While the final manuscript of the book was being prepared the authors of oral communications and of some posters made the suggestion of presenting the experimental data jointly as slightly longer articles. With the aim of making these proceedings a useful sourcebook rather than a verbatim record of the Course I accepted their proposal.

It became more than obvious at several points during the Course that the present nomenclature for aspartic proteinases remains on somewhat uncertain grounds. I am therefore grateful to B. Foltmann for having submitted his comments on our present situation in this respect even though they do not justify much optimism. Clearly, any proposal of a rational and consistent nomenclature for aspartic proteinases will have to await molecular characterization of enzymes and precursors from many organisms, preferably from those important phylogenetically and ontogenetically. These and other problems should be dealt with during one of the future meetings whose organization this book, I hope, may stimulate. If this turns out to be the case I will feel that my efforts were not in vain.

Vladimír Kostka

Prague, February 1985

ACKNOWLEDGEMENTS

For the Course I wish to acknowledge the generous financial support of the FEBS Advanced Courses Committee and to thank Prof. G. Bernardi, Chairman of this Committee, for his understanding, and Dr A. Kotyk, member of the same Committee, for the interest with which he followed the organization of the Course. My particular thanks are due to Dr J. Kay, University College Cardiff, Wales, UK, who gave me much from his personal store of experience with the organization of the London Workshop. I wish to acknowledge also the sponsorship of the Czechoslovak Biochemical Society and its Chairman, Prof. J. Škoda. The organization of the Course involved a great deal of effort for which my thanks are due to all members of the Organizing Committee and to the Course Secretariat, headed by Ms. Jabulková.

For this book my thanks go to all contributors, especially to those who submitted their manuscripts in due time and according to the Publisher's suggestions. It is a pleasure to acknowledge the help of my son, Mr. D. Kostka, and of Ms. J. Pelcová who undertook the preparation of the indexes. I am much indebted to Ms. E. Biskupová and to Ms. H. Pokorná for the typing work. Last but not least, I would like to extend my special thanks to Ms. Evelyne Glowka and to Ms. Monika Krumnow of Walter de Gruyter, Berlin, for their understanding and cooperation.

During the coffee break on the campus. Left to right: Meloun, Stepanov, Kostka, Wiederanders, Riechmann, Hallett, Pearl, Lah, Hausdorf, Turk, Takahashi, Andreeva, Kirchhübel, Oda, Olafsdōtir, Erkelens, Reid, Foltmann, Patestos, Blundell, Takahashi, Dreyer, Pfenninger, Mikeš, James, Harboe, Boger, Antonov, Samloff, Cherry, Breipohl, Schmitges, Kay, Raddatz, Safro, Dunn

The attentive listeners. Left to right (first row): Foltmann, Samloff, Antonov, Stepanov; (second row): Szécsi, Wiederanders, Lah, Turk, James

Discussing science over a cocktail. Left to right: Pohl, Kostka, Dunn

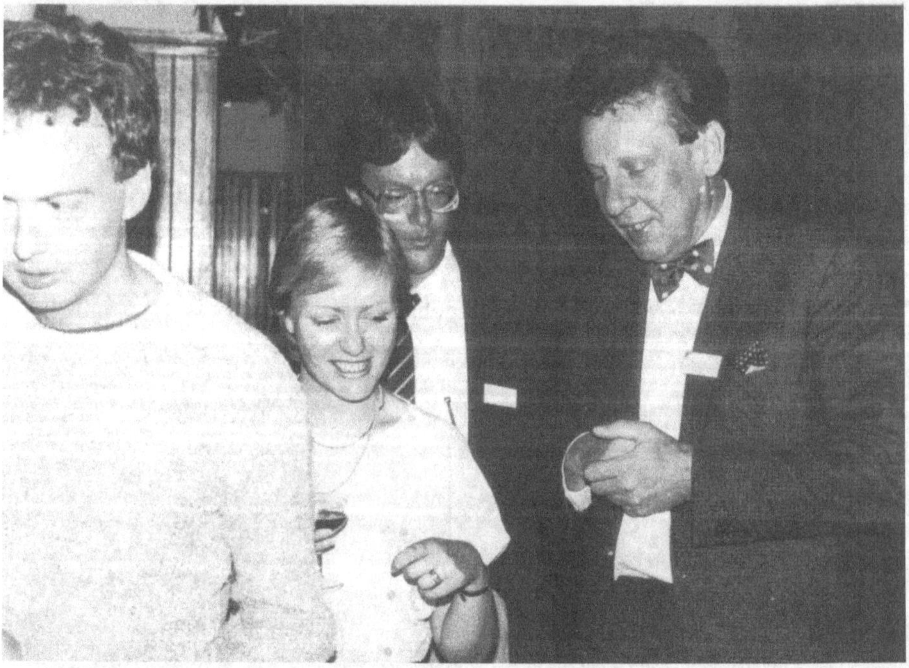

Get-together party. Left to right: Valler, Rolph, Dunn, Kostka
(All photographs taken by J. Plechatý)

Organizing Committee of FEBS Advanced Course No 84/07

V. Kostka (Chairman)

M. Baudyš

B. Meloun

O. Mikeš

I. Pichová

J. Pohl

J. Turková

Secretariat

D. Jabulková (Head)

E. Bulantová

J. Sedláčková

CONTENTS

INTRODUCTION

GENERAL ASPARTIC PROTEINASES

Isolation, molecular characteristics, and primary structures

Three-Dimensional Structures, Hydrolytic Mechanism and Specificity

ZYMOGEN ACTIVATION PATHWAYS

Renin

INHIBITORS OF ASPARTIC PROTEINASES

Analytical Methods

Occurrence and Role of Aspartic Proteinases in Biological Systems

Biotechnology Aspects of Aspartic Proteinases

ASPARTIC PROTEINASES AND THEIR INHIBITORS

John Kay
Department of Biochemistry, University College, P.O.Box 78,
Cardiff CF1 1XL, Wales, U.K.

Proteolytic enzymes are classified on the basis of their catalytic mechanism as belonging to one of four groups - the serine, cysteine, metallo and aspartic proteinases (1). In contrast to the detailed sequence information and 3-dimensional structures that have been produced for many enzymes belonging to the first three groups, relatively little is known about the aspartic proteinases.

These enzymes have been isolated from five major sources:

a) The stomachs of a number of species. Three different types of gastric enzyme have been resolved - the pepsins, chymosins and gastricsins - together with a less well-characterised component which has been termed slow-moving proteinase because of its electrophoretic mobility (2). Gastricsin is also produced by the prostate which is the source of the aspartic proteinase of seminal fluid (3).
b) the lysosomes of many cell types contain cathepsin D (and sometimes, cathepsin E).
c) tissues such as kidney and sub-maxillary gland produce renin.
d) microorganisms (but not bacteria) e.g. the proteinases from Endothia parasitica, penicillium janthinellum & mucor pusillus; yeast proteinase A. Scytalidium lignicolum also produces two acid proteinases (A & B) but it is uncertain at present whether these are genuine aspartic proteinases (4).
e) plants such as squash and cucumber (5).

In order to be classified as an aspartic proteinase,an enzyme must, by definition, be susceptible to inhibition by pepstatin and by the active-site directed affinity labels, diazoacetyl norleucine methyl ester and EPNP-epoxy (p-nitrophenoxy) propane.Each of the latter reacts specifically with the side chain carboxyl of a distinct aspartic acid residue (in positions 215 and 32 respectively in the linear sequence) to inactivate the enzyme. These residues must then be responsible for operating the catalytic mechanism in all of these enzymes (except Scytalidium proteinases) and so the subject matter of this volume is assigned its name.

Aspartic proteinases participate in a variety of physiological processes and alterations in the level of activity expressed may be associated with the onset of pathological conditions such as hypertension, gastric ulcers, muscular dystrophy and neoplastic diseases. In addition to their biological roles, aspartic proteinases have been used for centuries in two major areas of food processing-milk coagulation and the production of fermented foods from soya beans, rice and cereals. They are thus of enormous economic importance and the relative scarcity of the traditional milk clotting enzyme, calf chymosin, has led to the wide-spread introduction of rennet substitutes for cheese-making (6). Microbial enzymes, pig, cow and chicken pepsins have all been used industrially but since recombinant DNA technology has now been harnessed to introduce and express the gene for calf (pro) chymosin in E.coli (7, 8) and yeast (9) it seems likely that adequate supplies of recombinant chymosin will soon be available for cheese-manufacture-particularly with the incentive of a world-wide market for cheese that was estimated to be worth approx. £14,000 million in 1981!

All of the enzymes operate maximally in an acidic pH environment and all have molecular weights between 35-42,000 with the exception of cathepsin E, which appears to be a molecule of about 90,000 daltons, and the acid proteinase B from Scytalidium which is about half-sized at 22,000 daltons. The complete sequences of a number of the aspartic proteinases have been determined e.g. several microbial enzymes, pig and chicken pepsins (10), calf chymosin (11),mouse renin (12), pig cathepsin D (13) and corresponding data for human pepsinogen (14) and human renin (15) have been obtained from gene sequencing work. Extensive homology exists among all of these enzymes with particularly high conservation of residues around the catalytic aspartic acids, 32 and 215.

Only a few of these enzymes have yielded to X-Ray crystallographic analysis. The 3-dimensional structures of penicillopepsin (16), endothia proteinase (17), rhizopus chinensis proteinase (18) and pig pepsin (19) have been solved and refined to high resolution(1.8-2.1 A^{o}). All have broadly similar 3-dimensional structures (see 17,20) with dimensions of approx. 65x40 A^{o} and 30 A^{o} thick. This, together with the extensive sequence homologies that have been demonstrated, indicates that all of the aspartic proteinases are likely to have similar 3-dimensional conformations overall. However, each type of enzyme must have evolved subtle distinctions in structure, and therefore activity from the others in order to carry out its specific physiological function in its own environment. For example, cathepsin D is unstable at pH 2, the pH at which the pepsins are maximally active. At this pH, the pepsins readily degrade most proteins including haemoglobin, immunoglobulins and pancreatic ribonuclease whereas chymosin and the proteinase from Mucor pusillus have little general proteolytic activity towards proteins other than caseins. Similarly, renin has a very restricted specificity, directed to only one peptide bond in the circulating angiotensinogen and expressed at pH values between 5.5 and 8 depending on conditions. All of the enzymes appear to have an extended active site cleft (21) which can accommodate at least seven amino acids of a substrate in the S_4-S_3' sub-sites with a distinct preference for cleaving the bond between hydrophobic residues occupying the S_1-S_1' sites (22). The differences in specificity and activity may then be explained by discrete alterations in some or all of the sub-sites in the various enzymes.
Further distinctions among the enzymes are also to be found in other features. Pig pepsin is not a glycoprotein whereas chicken pepsin has a short oligosaccharide moiety (containing N-Acetyl-glucosamine and mannose) linked to Asn 113 and certain isoforms of human pepsin can be isolated with associated hyaluronic acid (23). Other enzymes such as the Mucor

miehei proteinase contain a higher percentage of carbohydrate and cathepsin D (24) is glycosylated in two locations (Asn67 and Asn183). Renin may be a glycoprotein depending on the species and tissue of origin.

An additional difference between mammalian and microbial aspartic proteinases is that, whereas the microbial enzymes do not appear to have to be synthesised in precursor form in order to facilitate folding, all of the mammalian enzymes are produced in the form of well-documented zymogens such as pepsinogen, prochymosin, progastricsin and prorenin (25). A precursor for cathepsin D has also been observed (26).

Activation of the zymogens to generate the physiologically-active enzyme form takes place upon acidification and approx. 45 residues are released in total in the activation segment (propart). For pig pepsinogen, for example, the activation has been shown to be predominantly an intramolecular reaction at pH values less than three and the sequence of residues removed is:-

				p5					p10
Leu -	Val -	Lys -	Val -	Pro -	Leu -	Val -	Arg -	Lys -	Lys -
				p15					p20
-Ser -	Leu -	Arg -	Gln -	Asn -	Leu -	Ile -	Lys -	Asp -	Gly -
				p25					p30
-Lys -	Leu -	Lys -	Asp -	Phe -	Leu -	Lys -	Thr -	His -	Lys -
				p35					p40
-His -	Asn -	Pro -	Ala -	Ser -	Lys -	Tyr -	Phe -	Pro -	Glu -
			p44	1					327
-Ala -	Ala -	Ala -	Leu -	Ile -	Gly ~		PEPSIN ~		Ala

Examination of the structure of pig pepsin indicates that the newly-generated NH_2-terminal Ile1 residue of the enzyme is located on the back face of the molecule, completely remote from the active site. Thus, it would be a physical impossibility for intramolecular cleavage of the

Leu p44 - Ile1 bond in the zymogen to occur without gross distortion of the entire molecule. Indeed, a number of investigations have indicated that under physiological conditions, the predominant activation process that occurs does not proceed by release of the intact propart but rather involves a sequential removal of shorter peptides in a stepwise fashion until all of the necessary amino acids have been cleaved off (25). In the case of pig pepsinogen, the first bond that is split in an intramolecular reaction is that between Leup16-Ilep17. Other zymogens undergo activation by slightly different mechanisms (27).

The objective of this process, of course, is to free the active site of encumbrances so that it may be fully operational under acidic conditions and available to interact with extraneous peptides to bring about their hydrolysis. As mentioned above, the various types of aspartic proteinase differ considerably in the selectivity that they display towards particular protein substrates. In most cases, it is difficult to gain any insight into these distinctions using polydisperse substrates such as naturally-occurring proteins. However, a wealth of information has been derived using synthetic peptide substrates (28) and recently the use of substrates such as

$$\text{Pro - Thr - Glu - Phe} \overset{\downarrow}{-} \text{Nph - Arg - Leu}$$

containing the chromophoric amino acid, p-nitro phenylalanine (Nph) in the P_1' position has permitted the cleavage of the scissile peptide bond to be followed spectrophotometrically (29) at a variety of pH values. The ability to interpret this type of data (30) on the basis of known or projected (modelled) protein structures has permitted logical deduction of the molecular architecture of some of the sub-sites in individual aspartic proteinases (see the chapters by Dunn et al, Foundling et al, James & Sielecki in this volume).

Interactions within all of the sub-sites serve to orient the substrate in such a way that the scissile peptide bond lies in close proximity to the catalytically-essential aspartic acid residues, 32 & 215. The carboxyl groups of these two aspartic acid residues appear to be sufficiently close to each other that they may share a proton (31). In addition, since these two residues are retained (in a very symmetric arrangement of two polypeptide strands) in all aspartic proteinases, it seems likely that one fundamental catalytic mechanism will operate in all of the enzymes. The consensus of current opinion is that no covalent intermediates are formed during the catalytic cycle (32). A general base type of mechanism is likely to be operative and the essential features of the reaction are depicted below. Electron density which might correspond to a hypothetical tightly-bound water molecule has been observed crystallographically (31).

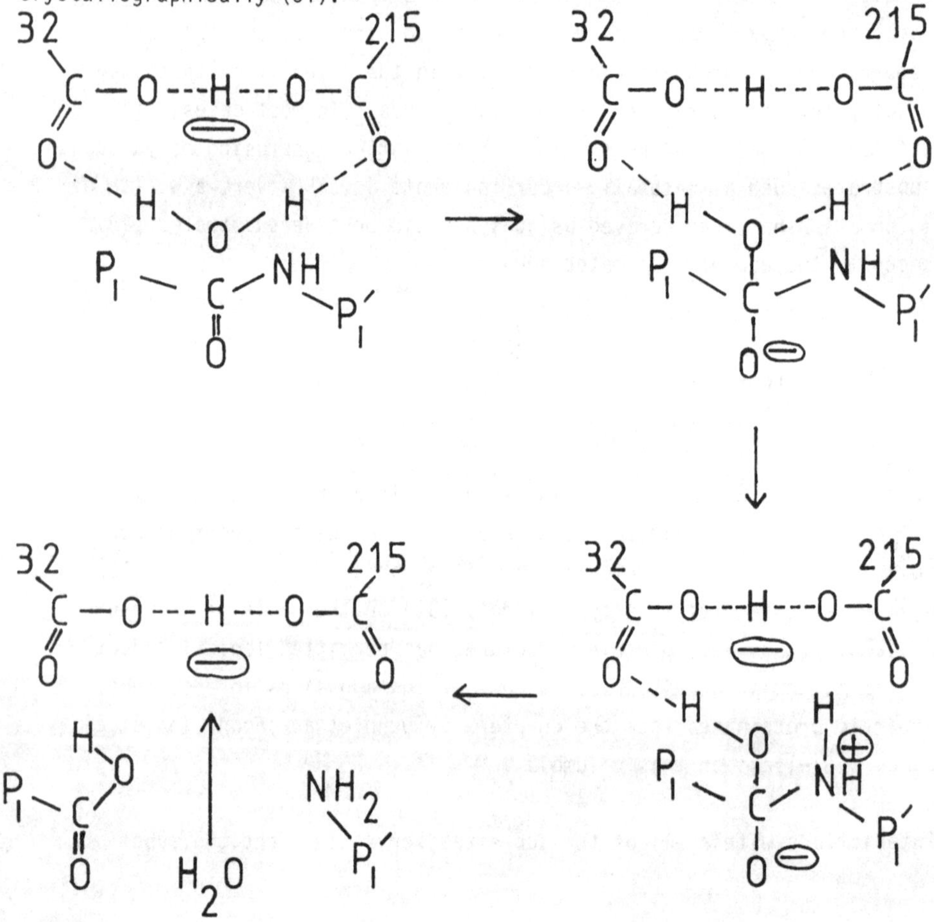

The carbonyl group of the scissile peptide bond becomes polarised thus increasing the susceptibility of the carbonyl carbon atom to nucleophilic attack and leading to the formation of a tetrahedral intermediate. This would be stabilised if the carboxyl anion could be accommodated in an "oxyanion hole". Transfer of a proton to the peptide nitrogen leads to the productive collapse of the tetrahedral intermediate and cleavage of the peptide bond.

Detailed investigations into the mechanism of the reaction pathway operated by these enzymes have been hampered by a number of practical difficulties but several complicating features have been observed and any proposed mechanism must also be able to account for these. For example, transpeptidation reactions may occur with short substrates under appropriate conditions and these enzymes can also catalyse ^{18}O exchange from $H_2\,^{18}O$ into the free carboxyl group of a short product such as acetylphenylalanine. Limitations on space will not permit consideration of these aspects here. However, for a further discussion of mechanistic proposals, the reader is referred to the chapters by James & Sielecki , and by Pearl, elsewhere in this volume.

The elucidation of the catalytic mechanism operative in other types of proteolytic enzyme (e.g. serine proteinases) has been facilitated by the ready availability of naturally-occurring inhibitors of these enzymes. Indeed, inhibitors of serine, cysteine and metalloproteinases are distributed ubiquitously throughout the biological world. In sharp contrast, however, naturally-occurring inhibitors of aspartic proteinases are relatively uncommon and are found in only certain specialised locations:

1) Proteins (Mol.Wt approx. 17,000) from Ascaris lumbricoides
2) The inhibitor peptide (containing 16/17 residues) released on activation of (pig/cow) pepsinogen(s)
3) Acylated pentapeptides (pepstatins) from various species of Actinomycetes
4) Renin-binding proteins. It is not certain whether the interaction of these proteins with renin actually blocks the activity of the enzyme.

5) The inhibitor of proteinase A in yeast.
With the exception of the last example, none of these naturally-occurring inhibitors can be considered to be of physiological importance to the host cell as regulators of the activity of its own aspartic proteinases (1). It is likely that modulation of this type of proteinase has to be achieved through regulation of the pH or of the rate of supply of substrate proteins.

Furthermore, the susceptibility of individual aspartic proteinases to inhibition by these compounds varies considerably. The Ascaris proteins, for example, are effective inhibitors of human,(33) pig and chicken pepsins, pig gastricsin and (rabbit) cathepsin E with a weaker influence on human gastricsin (Table 1) and little or no effect on the other aspartic proteinases tested.

This selectivity of inhibition is also observed with the pepsin inhibitor peptide obtained upon activation of pepsinogen. It has been known for almost 50 years that one of the peptides released on activation of (pig) pepsinogen binds to pepsin above pH 4.5 to stabilise the enzyme at higher pH values and to inhibit it (measured traditionally in a milk-clotting assay at pH 5.3). This inhibitor has been identified as the peptide released in the first step in the sequential activation of pig (and cow) pepsinogen(s) and is derived from the first 16/17 residues in the zymogens (34,35). There is considerable sequence homology in this region of the propart of many zymogens, including calf prochymosin, human and monkey progastricsins, human, chicken, cow, pig, monkey, bear and rabbit pepsinogens and even mouse and human prorenins. However, the naturally-occurring 1-16 propart from (pig) pepsinogen is able to inhibit only certain enzymes so that, given the extensive homologies in propart sequences, it seems likely that the feature(s) that determine susceptibility to inhibition must lie within the enzymes themselves. Detailed investigations have been carried out to examine this further (36,37) and the use of synthetic analogues containing only residues p1-p12 (13-16 are unnecessary for inhibitory potency) indicated that pig, cow and human pepsins were strongly inhibited whereas pig and

Table 1.

Enzyme	K_i (nM)
Pig pepsin	0.5*
Pig gastricsin	1.7*
Human gastricsin	26
Human cathepsin D	>700
Calf chymosin	>700
Endothia p.proteinase	>700
Mucor pus. proteinase	>1,200

Kinetic constants (K_i) for the inhibition of several aspartic proteinases by Ascaris lumbricoides inhibitors were determined at pH 3.1 and 37°. * = Data taken from reference 33.

human gastricsins were inhibited to a lesser extent and none of the other enzymes tested were susceptible to inhibition. The explanation for this is given in detail in the chapter by Dunn et al elsewhere in this volume but the results can be summarised briefly as follows: The binding of the inhibitor in the active site cleft depends on hydrophobic interactions centred around the region of Val p4 - Val p7 in the peptide (see above, for the sequence). This places the Lys (p3) residue into the S_3 sub-site. Examination of the protein structures indicates that one amino acid side chain in the enzyme approaches closely in this specificity pocket. This is identified as the residue at position 13 in the linear sequence. In all of the enzymes that are susceptible to inhibition, this residue is a glutamic acid. At pH values above about 4.5, this should be ionised and should form an electrostatic interaction with the (p3) lysine residue in the inhibitor peptide. In the non-susceptible enzymes, other residues (non-ionisable) are found to occur in position 13. Thus, the fifty year old observations on specificity and the pH dependence of inhibition can now be satisfactorily explained at the molecular level.

A further twist to this account has appeared recently. It seems that the equivalent peptide derived from the corresponding propart of mouse prorenin is an inhibitor of renin (38). Nothing is yet known of its effect on the other aspartic proteinases.

This differential susceptibility to inhibition by naturally-occurring compounds persists into what is perhaps the best-known category, the pepstatins. Two forms of the Acyl - Val - Val - Sta - Ala - Sta pentapeptide structure predominate in Actinomycetes and these differ only in the nature of their acyl substituent. These are the isovaleryl and acetyl derivatives and both isovaleryl and acetyl-pepstatins are very poorly soluble in aqueous solution. However, by modification of the naturally-occurring peptides, it is possible to introduce a hydrophilic lactyl residue as the acylating group and the resultant (shorter) Lactyl-Val-Sta-Ala-Sta = Lactyl- pepstatin is much more soluble in water (39).

The effects of these three inhibitors on several representative aspartic proteinases are shown in Table 2.

It can be seen that Lactyl-pepstatin is as effective as isovaleryl and acetyl-pepstatins in inhibiting pig pepsin and gastricsin but it interacts much less tightly than its hydrophobic counterparts with all of the other enzymes tested. Once again, this suggests that the residue occupying the S_3 sub-site of the enzymes plays a crucial role in determining the efficiency of inhibitor binding (37). Furthermore, whereas the isovaleryl and acetyl-pepstatins are equally potent towards all of the mammalian enzymes, the two microbial aspartic proteinases examined appear to be more susceptible (by approx. one order of magnitude) to isovaleryl-pepstatin. This suggests that the bulkier hydrophobic isovaleryl substituent occupies the S_4 sub-site in the microbial enzymes more efficiently than its smaller, acetyl counterpart.

Thus, once again, the different nature of the sub-sites of the active site cleft of individual enzymes is reflected by their distinctive susceptibility to inhibition. Nevertheless, some degree of inhibition is

Table 2.

	Isovaleryl -pepstatin	Acetyl -pepstatin	Lactyl -pepstatin
		K_i (nM)	
Pig pepsin	1	1	1
Pig gastricsin	20	20	70
Human gastricsin	100	120	5,800
Chicken pepsin	1	1	50
Calf chymosin	70	63	1,900
Cathepsin D	1	1	100
Endothia p.proteinase	0.5	9	∞
Mucor pus.proteinase	2	18	700

Kinetic constants (K_i) for the inhibition of several aspartic proteinases by isovaleryl, acetyl and lactyl-pepstatins were determined at pH 3.1 and 37°.

always observed with each pepstatin with every enzyme and this probably results from the energy of interaction derived from the first statine residue in the Acyl-Val-Val-Sta-Ala-Sta sequence being bound in close proximity to the two catalytic aspartic acid residues. The -CHOH-CH_2- structure intrinsic to statine has been suggested to be an analogue of the tetrahedral intermediate (or transition state) for the enzymic reaction and evidence in support of this has come from NMR and ESR studies and from data from X-Ray crystallographic analyses of complexes between Rhizopus proteinase (18) or penicillopepsin (40) with pepstatin (fragments). It must be borne in mind, however, that the 3S hydroxyl group of this statine residue may displace the bound water from the catalytic site (18,31) rather than reflecting the positioning of the carbonyl oxygen atom of the substrate (see above).

On the basis of these observations with naturally-occurring inhibitors, it would seem to be possible to design synthetic counterparts that should be specific for individual enzymes. The rationale behind such inhibitors has been to synthesise a peptide of an appropriate length and containing the amino acid residues known to be present in the naturally-occurring substrate for the enzyme but with the introduction of statine in place of the two residues contributing to the scissile bond of the substrate. Such an approach has led to the synthesis of highly potent inhibitors of human renin (41) and calf chymosin (42).

An analogous yet different strategy has been developed by Szelke and colleagues (43) whereby a chemical modification of the scissile peptide bond is used to introduce a non-hydrolysable analogue of the tetrahedral transition state formed during hydrolysis. Thus, instead of using the naturally-occurring statine residue as the centrepiece around which the inhibitor is constructed, the complete amino acid sequence of the substrate is retained in the inhibitor except that the scissile peptide bond - CONH - between residues P_1-P_1' of the substrate is replaced by, for example, the synthetic secondary amine - CH_2-NH- as the non-hydrolysable analogue of the transition state. Using this approach, a tight-binding inhibitor of human renin has been developed which is a synthetic analogue based on the sequence of residues known to occur on either side of the scissile -Leu-Val- peptide bond of human angiotensinogen but with a reduced -CH_2-NH- isostere in place of the -CONH- of the substrate (see the chapter by Szelke & Leckie elsewhere in this volume).

It is also possible, of course, to devise such inhibitors not on the sequence of residues found in naturally-occurring protein substrates but based on synthetic peptides that are known to be good substrates for certain enzymes. For example, two reduced isostere inhibitors have been synthesised as derivatives of the synthetic chromogenic substrate

Pro - Thr - Glu - Phe - Nph - Arg - Leu

that was described earlier in this chapter. The effects of these completely synthetic inhibitors on a number of aspartic proteinases are given in the chapter by Foundling et al elsewhere in this volume.

It would thus appear that while naturally-occurring inhibitors of aspartic proteinases may have little physiological significance in regulating their target enzymes in vivo, nevertheless such compounds and their synthetic counterparts have proved of inestimable value in facilitating distinction among the different types of aspartic proteinases. Such inhibitors can be utilised for diagnostic purposes to establish the nature of a 'newly-isolated' aspartic proteinase. As an illustration of this, consider the inhibition of human pepsin and gastricsin by isovaleryl and lactyl-pepstatins (Table 3).

It has been known for some time that human seminal plasma contains an aspartic proteinase and it is clear from the K_i values shown in Table 3 that this enzyme should be considered as a gastricsin rather than a pepsin. Confirmation of this diagnosis has been obtained since the seminal proteinase cross-reacts with an antiserum to human gastric gastricsin but not with anti-human pepsin (3). By applying this type of approach (for example with the Ascaris inhibitor), it may be possible to identify whether the slow-moving proteinase component of human gastric juice (2) is cathepsin D or E.

Thus, while considerable progress has been made since the publication of the previous book specifically concerned with the structure, activity and importance (biological and commercial) of the aspartic proteinases (44), much remains to be learned about the distinctions in molecular architecture and how these are reflected in the functions of the various enzymes. Inhibitors (naturally-occurring and synthetic) have permitted detailed biochemical and crystallographic investigations to be made but an understanding of the selectivity of such inhibitors may be of just as much importance for the design and synthesis of specific inhibitors for

Table 3

	Isovaleryl-pepstatin	Lactyl-pepstatin
	K_i (nM)	
Human pepsin	0.5	0.4
Human gastricsin	100	5,800
Seminal enzyme	160	4,000

Kinetic constants (K_i) for the inhibition of three human aspartic proteinases by Isovaleryl and Lactyl-pepstatins were measured at pH 3.1 and 37°.

use therapeutically in controlling individual aspartic proteinases (e.g. renin). These and other aspects will hopefully begin to unfold throughout the chapters which follow this introductory review.

Acknowledgements:
Research in the author's laboratory was sponsored by grants from the Medical Research Council, the Agricultural & Food Research Council, the Science and Engineering Research Council, by Celltech Ltd., Boots-Celltech Diagnostics Ltd and by The Wellcome Trust, Burroughs-Wellcome Fund and The Royal Society. I am very grateful to my many colleagues throughout the world who have contributed to this work by generously supplying samples of their purified proteins and inhibitors. In particular, however, I should like to express my appreciation of my colleagues Ben Dunn, Sandy Reid, Carole Rolph and Martin Valler, without whom none of it would have been possible. It is also a pleasure to acknowledge the superb secretarial and administrative contribution of Barbara Power.

References

1. Kay, J. Biochem.Soc.Trans.: 10, 277-280 (1982)
2. Samloff, I.M., Taggart, R.T.:this volume pp 79-95 (1984)
3. Reid, W.A., Vongsorasak, L., Svasti, J., Valler, M.J., Kay, J. : Cell Tissue Res. 236, 597-600 (1984).
4. Murao, S., Oda, K: this volume pp379-399 (1984).
5. Wilimowska-Pelc, A., Polanowski, A., Kolaczkowska, M.K., Wieczorek, M., Wilusz, T.:Acta Biochim. Pol. 30, 23-31 (1983).
6. Green, M.L., Valler, M.J., Kay, J.: J.Dairy Res. 51, 331-340 (1984).
7. Emtage, J.S., Angal, S., Doel, M.T., Harris, T.J.R., Jenkins,B., Lilley, G., Lowe, P.A. : Proc. Natl. Acad.Sci. (U.S.A.) 80, 3671-3675 (1983).
8. Marston, F.A.O., Lowe, P.A., Doel, M.T., Schoemaker, J.M., White, S., Angal, S: Biotechnology 2, 800-804 (1984).
9. Mellor, J., Dobson, M.J., Roberts, N.A., Tuite, M.F., Emtage, J.S., White, S., Lowe, P.A., Patel, T., Kingsman, A.J., Kingsman, S.M.: Gene 24, 1-14. (1983).
10. Baudys, M., Kostka, V.: Eur.J.Biochem. 136, 89-99 (1983).
11. Foltmann, B.: Essays Biochem. 17, 52-84 (1981).
12. Misono, K.S., Chang, J.J., Inagami, T.: Proc.Natl. Acad.Sci.(U.S.A.) 79, 4863-4867 (1982).
13. Shewale, J.G., Tang, J.J.N.: Proc.Natl.Acad.Sci. (U.S.A.) 81, 3703-3707 (1984).
14. Sogawa, K., Fujii-Kuriyama, Y., Mizukami, Y., Ichihara, Y., Takahashi, K.: J.Biol.Chem. 258, 5306-5311 (1983).
15. Hobart, P.M., Fogliana, P.M., O'Connor, B.A., Schaefer, I.M. & Chirgwim, J.M.: Proc.Natl.Acad.Sci (U.S.A.). 81, in press (1984).
16. James, M.N.G., Sielecki, A.: J.Mol.Biol. 163, 299-309 (1983).
17. Blundell, T.L., Foundling, S.I., Hemmings, A.M., Pearl, L.H., Sibanda, B.L., Watson, F.E.,Wood, S.P., Pedersen, V.B.: this vol. pp 151-161(1984).
18. Bott, R., Subramanian, E., Davies, D.R.: Biochemistry 21, 6956-6962 (1982).

19. Andreeva, N.S., Zdanov, A.S., Gustchina, A.E., Fedorov, A.A.: J.Biol.Chem. 259, in press (1984).
20. James, M.N.G., Sielecki, A.: this vol. pp 163-177 (1984).
21. Tang, J., James, M.N.G., Hsu, I-N., Jenkins, J.A., Blundell, T.L.: Nature 271, 618-620 (1978).
22. Powers, J.C., Harley, A.D., Myers, D.V.: Acid Proteinases, Structure, Function & Biology. (Tang.J. ed) pp 141-157. Plenum, N.Y. (1977).
23. Ryle, A.P., Foltmann, B.: this volume pp 97-100 (1984).
24. Takahashi, T., Schmidt, P.G., Tang, J.: J.Biol. Chem. 258, 2819-2830 (1983).
25. Kay, J.: The Enzymology of Post-Translational Modification of Proteins (R.B.Freedman & H.Hawkins, eds) pp 424-460. Academic Press. London (1980).
26. Puizdar, V., Turk, V.: FEBS Lett. 132, 299-301 (1981).
27. Kageyama, T., Takahashi, K.: Eur.J.Biochem. 141, 261-269 (1984).
28. Fruton, J.S.: Adv. Enzymol. 44, 1-36 (1976).
29. Dunn, B.M., Kammermann, B., McCurry, K.: Anal. Biochem. 138, 68-73 (1984).
30. Hofmann, T., Hodges, R.S., James, M.N.G.: Biochemistry 23, 635-642 (1984).
31. Pearl, L.H., Blundell, T.L.: FEBS Lett. 174, 96-101 (1984).
32. Dunn, B.M., Hofmann, T., Fink, A.L.: Biochemistry 23, in press (1984).
33. Abu-Erreish, G.M., Peanasky, R.J.: J.Biol.Chem. 249, 1566-1578 (1974).
34. Harboe, M., Andersen, P.M., Foltmann, B., Kay, J., Kassell, B.: J.Biol.Chem. 249, 4487-4494 (1984).
35. Dunn, B.M., Lewitt, M., Pham.C.: Biochem.J. 209, 355-362 (1983).
36. Dunn, B.M.: Arch.Biochem.Biophys. 214, 763-771 (1982).
37. Kay, J., Valler, M.J., Dunn, B.M.: Proteinase Inhibitors, Medical and Biological Aspects (N.Katunuma, H.Umezawa & H.Holzer, eds) pp 201-210. Springer-Verlag, Berlin (1983).

38. Evin, G., Devin, J., Castro, B., Menard, J., Corvol, P.: Proc. Natl. Acad.Sci (U.S.A.) 81, 48-52 (1984).

39. Kay, J., Afting, E-G., Aoyagi, T., Dunn, B.M.: Biochem.J. 203, 795-798 (1982).

40. James, M.N.G., Sielecki, A.R., Salituro, F., Rich, D.H., Hofmann, T.: Proc.Natl.Acad. Sci (U.S.A.) 79, 6137-6141(1984).

41. Boger, J.S., Lohr, N.S., Ulm, E.H., Poe, M., Blaine, E.H., Fanelli, G.M., Lin, T-Y., Payne, L-S., Schorn, T.W., Lamont, B.I., Vassil, T.C., Stabilito, I.I., Veber, D.F., Rich, D.H., Bopari, A.S.: Nature 303, 83-86 (1983).

42. Powell, M.J., Holdsworth, R.J., Baker, T.S., Titmas, R.C., Bose, C.C., Phipps, A., Eaton, M., Rolph, C.E., Valler, M.J., Kay, J.: this vol. pp 479-483 (1984).

43. Szelke, M., Leckie, B., Hallett, A., Jones, D.M., Sueiras, J., Atrash, B., Lever, A.F.: Nature 299, 555-557 (1982).

44. Acid Proteases, Structure, Function & Biology (J.Tang, ed) Plenum Press, New York (1977).

COMMENTS ON THE NOMENCLATURE OF ASPARTIC PROTEINASES

Bent Foltmann
Institute of Biochemical Genetics, University of Copenhagen,
Ø. Farimagsgade 2 A, (DK) 1353 Copenhagen K, Denmark.

Introduction

The questions of nomenclature and classification are shortly dealt with in other chapters of this volume. The aim of these comments is to present a general overview of the problems, and to present a discussion of the recommended IUB nomenclature (1) relative to our present knowledge about the structure, function and biology of aspartic proteinases. The nomenclature and classification of enzymes in many cases represent a compromise between historical tradition and a rational approach. But even in attempts of a rational approach it may be difficult to obtain complete consensus among all biochemists. The present comments give a personal point of view, and they contain a few changes relative to previous suggestions for classification of gastric proteinases (2,3).

Definitions

The following definitions (4) may be useful for a discussion on the relationships among the enzymes:

Homologous enzymes show such great similarities in their structures that we must assume a common ancestry.

Paralogous enzymes are homologous enzymes that have arisen by gene duplications and subsequently evolved side by side in a single line of descent. Such enzymes generally perform different functions within one organism.

Orthologous enzymes are homologous enzymes in which the structural differences arise from speciation. This means that we find phyletic correspondance between the history of their genes and the history of the taxa from

Aspartic Proteinases and their Inhibitors

which they derive. Such enzymes generally perform the same functions within different species.

Analogous enzymes have a similarity in their catalytic functions, but the differences in their structures are so great, that we must assume that the similarities are due to a convergent evolution.

Furthermore the term isoenzyme is used for enzymes that catalyse the same reaction, and occur as genetically determined variants of the amino acid sequences within one species.

In order to have a separate designation for posttranslational modifications of enzymes with a common amino acid sequence, such multiple forms may be called isoforms.

The aspartic proteinases of the vertebrates are synthesized and secreted as proenzymes (zymogens). The proenzymes are classified in the same way as their corresponding enzymes. The suffix -ogen is well established for pepsinogen. Other proenzymes are characterized by the prefix pro-, e.g. prochymosin. The prefix pre- is now commonly used for proteins that still carry the signal peptide, which is necessesary for crossing the membrane of the endoplasmatic reticulum, e.g. preprochymosin (5).

Irreversible activation takes place through a limited proteolysis that finally removes about 45 amino acid residues from the N-terminus of the proenzyme. The segment that is removed, is often called the activation peptide. Here the designation propart is suggested. The logic is, that the enzyme is left when the propart is removed from the proenzyme. This way, one also omits the linguistic inconsistency that activation peptides may have inhibitory effects.

Criteria for classification.

In the IUB nomenclature, enzymes are primarily classified after the reaction which they catalyse, and subsequently divided in sub- and sub-subgroups according to their specificities against substrates. Due to overlapping specificities, the classification of proteinases is an exception to the general rule. These enzymes are devided into sub-subgroups on the basis of the chemistry of their active sites. This is a step toward classification on the basis of structure, though structure is not adopted as a criterion for classification.

From the contributions to this book it appears that all enzymes which are numbered in sub-subgroup EC 3.4.23 also are homologous. Proteinase B from Scytalidium lignicolum has not yet been given an EC number, and it is seen from the paper by Murao and Oda that its structure is not homologous to the aspartic proteinases in general. In my opinion this enzyme may be regarded as analogous to the aspartic proteinases, and it should tentatively be classified with the number EC 3.4.99. When more is known about this and related enzymes, it may be an advantage to create a new sub-subgroup.

Inhibitors are apparently suitable for classification of proteinases into sub-subgroups. Though to a varying degree, all aspartic proteinases are inhibited by pepstatin, diazoacetyl norleucine methyl ester (DAN), and 1,2-epoxy-3-(p-nitrophenoxy)propane (EPNP). These types of inhibitions reflect a homology of the active centres. Thus, if a new proteinase is inhibited by all three types of inhibitors, there is a high degree of probability, that such an enzyme should be classified in group EC 3.4.23.

As regards the specificities of the individual aspartic proteinases, examples are given below. With the extended binding site and a possible cooperative effect of binding in the subsites, it is, on the basis of our present knowledge, difficult to suggest a definitive classification into individual EC numbers from the substrate specifities.

The allocation of individual numbers for the aspartic proteinases carry the stamp of the evolution of biochemistry. Due to their easy accessibility the gastric proteinases from mammals, and especially from pig, have been subject for many investigations. Three gastric proteinases from pig had been purified when the first IUB nomenclature was published, and in addition calf chymosin was well known. We therefore have four numbers for gastric proteinases, whereas all the aspartic proteinases from fungi are collected with one number (EC 3.4.23.6).

It is an advantage for clinical biochemists and for dairy technologists to have a number system at their disposal for characterization of the gastric enzymes, but interspecific comparison raises some problems. The available information on amino acid sequences of proenzymes from man, pig and cattle indicates that separate genes for pepsinogen A, - C and prochymosin were present before the divergence of mammals. Hence these may be regarded as 3 groups of paralogous enzymes.

Sequencing still requires a great deal of effort, and for comparison of gastric proteinases from different species advantage has been taken that proteins which have at least 70 to 80% of identity in their primary structures generally also have a partial immunochemical identity when tested with polyclonal antisera. Proteins with less than 70% of identity normally show no immunochemical cross reactivity. We have found that pepsin A from goat, zebra, horse, dog, cat, and seal may by precipitated by antisera raised against pepsin A from man, pig or cattle (6). This is consistent with results from amino acid sequencing that indicate about 75% of identity between pepsinogen A from different species. Conversely, pepsinogen A from one species does not show a partial immunochemical identity with pepsinogen C from the same species. Thus we may conclude that pepsinogen A from different species form a group of orthologous enzymes, that are more related to each other than to the paralogous counterparts of their own species. It should be added, however, that recent results from this institute have shown, that the cross reactivity among different types of pepsinogen A varies considerably with antisera from different rabbits.

Among pepsinogens from non-mammalian species we have detailed information about chicken pepsinogen only. The sequence indicates that it is slightly more similar to pepsinogen A than to other mammalian pepsinogens (7). But the differences are small only, and one can not draw definitive conclusions on the basis of the structure of a single non-mammalian pepsinogen.

Genetically determined isoenzymes and posttranslationally modfified isoforms are found among many aspartic proteinases. Without structural analyses, it is very difficult to distinguish between the two types of variants. Both types are numbered and characterized by their electrophoretic mobilities, (no. 1 having the greatest mobility towards the anode).

Due to their clinical interest, the human pepsinogens have been subject to many investigations, and these pepsinogens show a very complex picture, (cfr. Samloff et al., this book). Because of poor resolution in the first electrophoretic analyses, a group of fast moving components of human pepsinogens and pepsins are all collected with the electrophoretic number 1. Until further information is available it is advisable to maintain the traditional numbering of the electrophoretic components of human pepsinogen A. But a consecutive numbering including both pepsinogen A and -C must be dissuaded.

Comments on the individual enzymes.

(For a schematic comparison of older designations of gastric proteinases and references to these see (3)).

3.4.23.1: Pepsin A.

The predominant gastric proteinase produced in the fundus of adult mammals. Small amounts are found in other tissues, in bloodserum and in urine. (Foltmann et al., Samloff et al., this book). In most papers the name pepsin refers to pepsin A. Until a definitive test for the classification of non-mammalian gastric proteinases is established, these may be designated as pepsins without further qualifications.

3.4.23.2: Pepsin B.

Isolated as a minor component from pig gastric juice only. Characterized by a low activity towards hemoglobin and considerable activity toward acetyl-phenylalanyl-diiodotyrosine (8). The previous suggestion (3) that pepsin B should be related to the electrophoretically slow moving proteinase from human gastric mucosa is not supported.

3.4.23.3: Pepsin C or gastricsin.

In the gastroentestinal tract of mammals the proenzyme is secreted in fundus, antrum and in duodenum. Small amounts are found in other tissues, especially in the seminal vesicles and prostatic glands. These give rise to the zymogen in the seminal fluid (Foltmann et al., this book).

The present trend is that biochemists prefer the designation gastricsin (in order to emphasize the difference from pepsin A), whereas clinicians prefer the designations pepsin A and -C for the major proteinases of the gastric juice. If the EC number is added, misunderstandings should be avoided. Gastricsins from pig and man have virtually no activity toward acetyl-phenylalanyl-diiodotyrosine (9) whereas that from cattle has some activity toward this substrate (10).

3.4.23.4: Chymosin.

Neonatal gastric proteinases that are found in the gastric juice of mammals with postnatal uptake of immunoglobulins. The enzymes a have high milk-clotting and a low general proteolytic activity (3). The name rennin has been abandoned in order to avoid mistakes with renin.

3.4.23.5: Cathepsin D.

Intracellular aspartic proteinases with a wide distribution (11). That from pig spleen is well characterized (Tang, this volume). A membrane bound proteinase from erythrocytes has previously been designated cathepsin D, this enzyme is immunologically related to the electrophoretically slow moving proteinase from gastric juice. But none of the two show immunological cross reactivity with liver cathepsin D (Foltmann et al., this book). It is up to a future decision, if no. 3.4.23.5 shall cover several types of aspartic proteinases with the only common feature that they have an intracellular distribution.

3.4.23.6: Microbial aspartic proteinases.

This entry includes a large number of proteinases from fungi and amoebae. Comparison among these show considerable variations both in specificities and in structure.

3.4.23.7 - 10: Open numbers in the present version of the nomenclature.

Previously used for different microbial proteinases.

3.4.23.11: Thyroid aspartic proteinase.

Isolated from pig thyroid glands. Poorly characterized, may be related to cathepsin D.

3.4.23.12: Nepentes aspartic proteinase.
3.4.23.13: Lotus aspartic proteinase.
3.4.23.14: Sorghum aspartic proteinase.

In addition to the enzymes that are numbered above (references in (1)), aspartic proteinases have been isolated from several other plants e.g. cucumber seeds (Polanowski et al., this book), rice seeds (12), and wheat leaves (13).

3.4.23.15: Renin.

Isolated from kidneys and submaxillary glands. Transferred to group 3.4.23 in supplement 2 (1981) of the recommended Enzyme Nomenclature. The enzyme has a high degree of specificity towards angiotensinogen, (cfr. contributions to this book).

Aspartic proteinases from invertebrates are not numbered in the Enzyme Nomenclature. Only few reports about proteinases from invertebrates have been published, and their chemistry is not well characterized. But inhibition studies have shown the presence of aspartic proteinases in antarctic krill (14) and nematodes (15).

Concluding remarks:

With the information that is available to-day, a rational nomenclature and classification of the aspartic proteinases present great problems. Future investigations may serve to elucidate if it is possible to find general principles for the classification of the proteinases from plants, fungi, invertebrates and for the different types of intracellular cathepsins.

References:

1. Enzyme Nomenclature: Recommendations of the Nomenclature Committee of the International Union of Biochemistry (1978).
2. Foltmann, B., Pedersen, V.B.: In "Acid proteases" (Tang, J., ed.) 3-22, Plenum Press, New York (1977).
3. Foltmann, B.: Essays in Biochem. 17, 52-84 (1981).
4. Fitch, W.M.: In "Molecular Evolution" (Ayala, F.J., ed.) 160-178, Sinauer Ass.Inc.Sunderland, Ma., USA (1976).
5. Harris, T.J.R. et al.: Nucleic Acids Res. 10, 2177-2187 (1981).
6. Foltmann, B., Axelsen, N.H.: FEBS Proc. 60, 271-280 (1980).

7. Baudys, M., Kostka, V.: Eur.J.Biochem. 136, 89-99 (1983).

8. Ryle, A.P.: Meth.Enzymol. 19, 316-336 (1970).

9. Tang, J.: Meth.Enzymol. 19, 406-421 (1970).

10. Martin, P., Trieu-Cuot, P., Collin, J.C., Ribadeau Dumas, B.: Eur.J.Biochem. 122, 31-39 (1982).

11. Barrett, A.J.: In "Proteinases in Mammalian Cells and Tissues" (Barrett, A.J., ed.) 209-247, Elsevier-North Holland, Amsterdam-London (1977).

12. Doi, E., Shibata, D., Matoba, T., Yonezawa, D.: Agric.Biol.Chem. (Tokyo) 44, 741-747 (1980).

13. Frith, G.J.T., Peoples, M.B., Dalling, M.J.: Plant Cell Physiol. 19, 819-824 (1978).

14. Kimoto, K., Thanh, V.V., Murakami, K.: J.Food Sci. 46, 1881-1884 (1981).

15. Gertler, A., Madar, Z., Haas, Y.: Comp.Biochem.Physiol. 64B, 357-362 (1979).

FUNGAL ASPARTYL PROTEINASES

Valentin M. Stepanov

Institute of Genetics and Selection of Industrial Microorganisms,

113545, Moscow, USSR

Introduction

Two main reasons explain the stable interest towards fungal aspartyl proteinases. These enzymes being readily available have found practical application as the logical substitutes for mammalian proteins. On the other hand, it appears that the fungal enzymes represent one of the most ancient forms of aspartyl proteinases. Therefore in the absence of sufficient data on plant and Protista acid proteinases the comparative study of fungal enzymes open a promising way to better understanding of aspartyl proteinases evolution.

In this paper we discuss the results gained in the course of research carried out at the Institute of Genetics and Selection of Industrial Microorganisms and Chemistry Department of Moscow State University.

Isolation and Purification of Fungal Aspartyl Proteinases.

Among various approaches used to isolate these enzymes, the tendency to exploit their anionic properties might be clearly traced. To this end the use of anionites appears to be method of choice, but the presence of carbohydrases ought to be taken in consideration, which is especially important, when the surface cultures of microscopic fungi

Aspartic Proteinases and their Inhibitors

are to be treated. The latter enzymes destroy cellulose or agarose matrixes thus forbidding the application of conventional ion exchangers, specifically on the early phases of the procedure.
One of the possible solutions of this problem consists in the use of the materials stable towards the enzymatic attack, e.g., of Biogels or Acrylexes for gel filtration. In our hands an anion exchanger on the basis of macroporous silica - aminosilochrom turned to be efficient for the initial steps of fungal aspartyl proteinases isolation, e.g., of those produced by Aspergillus awamori (1) or Aspergillus foetidus (2).
Affinity chromatography was found to be an exceptionally efficient method of aspartyl proteinases purification. Several convenient and easily available sorbents have been suggested for this purpose in our laboratory. These sorbents were prepared by the attachment of cyclopeptide antibiotics -- gramicidin S or bacitracin to agarose derivatives or to aminosilochrom (for review see 3,4). Both antibiotics can be bound at the active sites of various proteinases, but remain resistant to further hydrolysis as a consequence of high D-amino acids content and, possibly, of specific conformational traits characteristic for these cyclopeptides. The cyclopeptides attachment to the matrixes was achieved by conventional CNBr activation procedure (used to prepare bacitracin- of gramicidin S- Sepharose) or via the condensation with aminosilochrom by the reaction with p-benzoquinone. It ought to be stressed, although, that the Sepharose-based sorbents are not suitable for the first steps of fungal enzymes purification for the evident reasons discussed above. On the contrary, bacitracin -silochrom being completely stable towards hydrolases turned to be effective on the early steps of purification procedures. In several cases the successive use of bacitracin-silochrom and bacitracin-Sepharose (or respective sorbents that contained gramicidin S as the ligand) gave us satisfactory results. Thus, gramicidin S and bacitracin, being affinity ligands of general type, have proved their efficiency for aspartyl proteinases isolation.
Table 1 summarizes the isolation procedure applied by us for puri-

fication of Trichoderma viride aspartyl proteinase (5).

Table 1.Purification of T.viride aspartyl proteinase.

Step	Total protein A_{280}	Activity total units	Activity specific units/A_{280}	Yield %	Purification times
Culture extract	21000	100000	4.8	=100	=1
$(NH_4)_2SO_4$ precipitation	18600	96000	5.2	96	1.1
Acrylex P-10	3800	93000	10.5	93	2.2
Aminosilochrom	444	41000	92.5	41	19.2
Bacitracin-sepharose	26	40000	1540	40	320
Sephadex G-25	18.5	40000	2160	40	450

The table 2 illustrate the results of purification of various fungal proteinases. It might be seen that affinity chromatography steps were of decisive importance for the success of the whole procedures.
This approach turned to be also efficient for purification of aspartyl proteinase produced by edible mushroom - basidiomycete Russula decolorans (7). Recently the same principles have been extended on the isolation of plant proteinases, e.g., carboxyl proteinase from buckwheat (8).

Affinity sorbents that contain gramicidin S or bacitracin as the ligands show some extent of selectivity - there is certain difference in specificity of both ligands towards various enzymes.

Table 2. Purification of fungal aspartic proteinases.

Enzyme source	Sorbents used	Yield %	Purification times	Ref.
A.awamori	aminosilochrom gramicidin S-sepharose ECTEOLA-cellulose	35	97	(1)
A.awamori	aminosilochrom bacitracin-sepharose	52	360	(6)
A.foetidus	aminosilochrom bacitracin-sepharose (twice)	76	240	(2)
T. viride	aminosilochrom bacitracin-sepharose	40	450	(5)
T.lignorum	aminosilochrom bacitracin-sepharose	12	163	(5)
Russula decolorans	gramicidin S-silochrom DEAE-sephadex bacitracin-sepharose	43	136	(7)

Comparative Biochemistry of Fungal Aspartyl Proteinases.

All fungal aspartyl proteinases studied in our laboratory belong to the pepsin family which is in accordance with the whole complex of their biochemical characteristics. One of the most important reactions used for the attribution of newly isolated enzymes - their complete inactivation by diazoacetyl reagents in the presence of Cu^{++} ions. For this reaction - at least as important as the popular inhibition by pepstatin-we suggested to use the colored reagent - N-diazoacetyl-N'-2,4-dinitrophenyl ethylenediamine that blocks specifically Asp-215 or the corresponding residue of the active site:

$$N_2CHCO\text{-}NHCH_2CH_2NH\text{-}C_6H_3(NO_2)_2 \quad + \quad HOOC\text{-}CH_2\text{-}(Asp\text{-}215) \quad \xrightarrow{Cu^{++}}$$

$$(NO_2)_2C_6H_3NHCH_2CH_2NH\text{-}COCH_2\text{-}(Asp\text{-}215)$$

Its incorporation into the enzyme molecule might be easily assayed by the UV absorption of dinitrophenylamino moiety (for review see 9).

The comparison of the data collected on fungal aspartic proteinases indicates rather high extent of similarities of the enzymes produced by the fungi that belong to the same genus. Thus , the amino acid composition of aspartyl proteinases produced by Aspergilli is very similar (Table 3). The same refers also to their N-terminal sequences traced for more than 30 residues - at least 90% of homology was observed (1,2) The aspartyl proteinases produced by two Trichoderma species also possess rather similar amino acid composition (Table 3), although few meaningful differences might be indicated. Apparently, these differences lead to noticeable distinctions in the functional properties of the enzymes. Thus. the specific milk-clotting activity was found to be equal to 2100 units /A_{280} and 5400 units/A_{280} for the proteinases from T.viride and T.lignorum respectively, their specific activities against hemoglobin were equal to 7 and 21 units/A_{280}, optimal pH values - 2.3 and 2.8.

The differences between the enzymes produced by the fungi species that belong to the distinct genera appears to be substantially larger. The most striking seems to be very pronounced difference in the primary structures of Mucor mieheiand other fungal aspartic proteinases. The proteinases from Penicillium janthinellum and Aspergillus awamori show relatively high homology extent - 58% of coinciding residues

Table 3. Amino acid composition of fungal proteinases.

Amino acid	Trichoderma viride	Trichoderma lignorum	Aspergillus awamori	Aspergillus foetidus	Aspergillus oryzae
Lys	5	9	15	15	15
His	2	4	3	3	4
Arg	6	6	2	1	1
Asp	27	36	38	33	36
Thr	39	26	31	29	23
Ser	38	46	39	39	27
Glu	27	25	30	31	27
Pro	13	14	18	17	17
Gly	41	35	35	33	35
Ala	28	23	24	23	26
Cys	2	2	2	2	3
Val	37	28	24	22	24
Met	0	0	0	0	1
Ile	11	16	14	13	14
Leu	17	19	19	20	19
Tyr	13	12	16	17	15
Phe	11	14	16	16	15
Trp	3	4	3	4	5
Sum	319	319	329	318	307

whereas further improvement of this figure is likely so far as few uncertainties in the latter sequence would be eliminated.

The differences in the amino acid sequences of fungal and mammalian aspartyl proteinases are as a rule substantially larger. Thus, the share of identical residues in swine pepsin and Aspergillus awamori aspartyl proteinase (10,11) corresponds to 28%, whereas for the pair swine pepsin - penicillopepsin this value is equal to 29%.

On the other hand, the differences between the amino acid sequences of fungal aspartyl proteinases might be much more pronounced than that indicated above for the pair Aspergillus awamori versus Penicillium janthinellum aspartyl proteinases. Thus, homology extent between Mucor miehei and Penicillium janthinellum enzyme attains the marginal value of 21%, whereas rather long extensions and hardly traceble homology for several stretches induce suspiction that the tertiary structure of M.miehei enzyme might be rather different from those found for other fungal proteinases. Hence, we prefer, to consider M.miehei proteinase as a special case, probably, the result of strong divergence both in primary and tertiary structures from typical ones. Bearing in mind substantial divergence of the primary structures among fungal aspartyl proteinases, one is compelled to aknowledge that it would be hard to indicate the principal differences between the sequences of fungal and mammalian aspartyl proteinases. This reflects in the conservatism of the tertiary structures (12-15).

The use of chemical modifications also did not reveal any dramatic differences between these two groups of aspartyl proteinases. The modification of the aspartyl residues that belong to the catalytic site of the enzymes, e.g., with diazoacetyl reagents proceeds equally well both with mammalian and fungal enzymes. Therefore, even marginal differences in the enzymes behaviour towards modifying reagents might be of interest. One indication on the possible difference in the active sites reactivity was observed by Lysogorskaya et al. in our laboratory, who had found that water soluble carbodiimide of the following structure:

$$C_6H_5\text{-}N{=}N\text{-}C_6H_4\text{-}N{=}C{=}N\text{-}CH_2CH_2CH_2\overset{+}{N}(CH_3)_3$$

inhibited Aspergillus awamori aspartyl proteinase by incorporation of one reagent residue into enzyme molecule. The reagent was found to be distributed between Asp-215 and

an aspartyl residue within the N-terminal sequence, presumably Asp-11. On the contrary, only slow and non specific inactivation was observed when the same reagent was applied for swine pepsin modification (16). Comparable results were found for Aspergillus awamori proteinase inactivation by dicyclohexyl carbodiimide.
Tarasova et al. in our laboratory have compared the behaviour of swine pepsin and Aspergillus awamori pepsin toward aromatic diazonium reagents. (17). It turned out that the bulk of p-nitrophenyl diazonium reacts with Tyr-189 - the residue localized within the active site cleft and apparently involved in the substrate binding. It should be mentioned that this position is always occupied by tyrosine or phenylalanine. The modification of Tyr-75 was also observed but to substantially lesser extent. Pepsin modification by aromatic diazonium lead to more extended modification and, in addition to Tyr-189 several tyrosines located on the surface of the molecule were substituted. It appears that these tyrosines participate in formation of hydrophobic clusters on the perifery of the molecule.
It would be premature to interpret these rather subtle differences in the terms of functional distinction between two groups of aspartyl proteinases. Nevertheless, it is tempting to discuss the eventual participation of such clusters on mammalian pepsin surface as the points of the activation peptide attachment. Usually, the interaction between the latter and the core of pepsinogen molecule would be described in the terms of electrostatic interactions, but simple inspection of the activation peptides sequences shows that these are enriched in hydrophobic residues - ca. 10 positions are invariably occupied by Leu, Phe, Ile, Met or Tyr residues. Hence, one may avait the presence of counterparts for these hydrophobic structures on the surface of pepsin. It might happen that the hydrophobic clusters on the surface of mammalian pepsin correspond to the imprint of the hydro-

phobic patches of the activation "tail".

Problems Concerning Aspartyl Proteinases Evolution.

Aspartyl proteinases are not distributed as ubiquitously as other classes of proteolytic enzymes - serine, metallo- or thiol proteinases. No data are available to us on their occurence in bacteria, especially in relatively well studied bacilli or streptomycetes. Archebacteria also do not appear to possess these enzymes. Hence, microscopic fungi are the most ancient organisms known to produce aspartyl proteinases. Presumably, relatively late appearance of these enzymes might be explained by their specific adaptation to acidic environment that is a prerequisite for denaturation of the protein substrate. It is understandable, that the maintenance of acidic pH might be attained by formation of specific morphological structures, like lysosomes, digestive vacuoles etc, characteristic for more advanced biological taxones. Anyway, relatively recent advent of aspartyl proteinases might explain well preserved in their tertiary structures evidences of internal duplication (12,13). One might assume that the same factor explains comparatively low extent of the functional differentiation of aspartyl proteinases. Essentially, the whole family of these enzymes embrace the proteinases of the same structural (and functional) type - that of pepsin. In this sense, aspartyl proteinases that belong to pepsin family. differ sharply from serine proteinases that form strikingly diverse chymotrypsin family that includes highly sophisticated enzymes of blood-clotting system, complement etc. Even subtilisin family, apparently less specialized, contains dimeric enzymes, ATP-dependent proteinases . One has to take in account that our knowledge of aspartic proteinases is rather limited, especially scarce is the information on plant enzymes. but nevertheless one can hardly escape the conclusion that aspartyl proteinases are rather weakly specialized.

It appears that the most crucial step in their functional evolution on the way from fungal to mammalian proteinases consisted in the aquirement of the zymogenes. Remarkably, this apparently important step left almost no traces in the chemistry of mammalian enzymes . Indeed, the occurence of hydrophobic clusters on the surface of mammalian enzymes seems to be the only known distinction that deserves to be mentioned, but its importance would be cleared only after the tertiary structure of pepsinogen will be established.

Two alternative explanation might be suggested for this rather strange fact. First, it is possible, that the way to block the activity used in pepsinogens - the attachment of rather bulky "screen" - is rather simple one. The activation of pepsinogen consists essentially in the removal of this screen and,therefore it does not require complicated adjustment of the enzyme structure to highly ordered activation pathway as in the case of serine proteinases. It appears that similar principle is used in the formation of procarboxypeptidase structures. It would be interesting to learn, if carboxypeptidase T recently found in Thermoactinomycetes (18) possess a proenzyme. Characteristically enough, the activation of procarboxypeptidases might proceeds ambiguously as in the case of pepsinogen (19,20).

Second, it is not yet unambiguously shown that fungal aspartyl proteinases do not have precursors. Indeed, the evidences presented are of negative nature. It should be mentioned. that bacterial serine proteinases - subtilisins - that had been assumed to be devoid of precursors, were recently shown to possess precursor structure identified only by the gene sequencing. Apparently, the prosubtilisins exist only as transient entities and comparable situation might exist in the case of fungal aspartyl proteinases.

Table 4. Amino acid sequences of swine pepsin (SP),Penicillium janthinellum (PP) and Aspergillus awamori (AP) aspartyl proteinases

25
SP A L I G D E P L E N Y L - D T E Y F G T I G I G T P A Q
PP A A S C V A T N T P T A N D E E Y I T P V T I G G T T -
AP S K G S A V T T P Q N N D E E Y L T P V T V G K S T -

53
SP D F T V I F D T G S S N L W V P S V Y C S S L A C S D H
PP - L N L N F D T G S A D L W V F S T E L P A S Q Q S G H
AP - L H L B F D T G S A B L W V F S D E L P S S E R T G H

80
SP N Q F N P D D S S T F E A T S Q E L S I T Y G T G S M -
PP S V Y N P S A T G - K E L S G Y T W S I S Y G D G S S A
AP N V Y T P S S S A T K - L S G Y T W N I S Y G N G S S A

108
SP T G I L G Y D T V Q V G G I S D T N Q I F G L S E T E P
PP S G N V F T D S V T V G G V T A H G Q A V Q A A Q Q I S
AP S G D V Y R D T V T V G G V T N T K Z A V(S,A,A,Q)K I(S,

136
SP G S F L Y Y A P F D G I L G L A Y P S I S A S G A T P V
PP A Q F Q Q D T N N D G L L G L A F S S I N - T V Q P Q S
AP S,Z)F Z Z V B G G Z B S G(A,Z.A)Y S S I N - T V Q P K A

164
SP F D N L W D Q G L V S Q D L F S V Y L S S N D D S G S V
PP Q T T F F D T V K S S L A O P L F A V A L K H Q Q P G V
AP Q T T F F D T V K S Q L N S P L F A V Q L K H B A P G V

191
SP V L L G G I D S S Y Y T G S L N W V P V - S V E G Y W Q
PP Y D F G F I D S S K Y T G S L T Y T G V N N S Q G F W S
AP Y B F G Y I B B S K Y T G S I T Y T D A D S S E G Y W G

219
SP I T L D S I T M D G E T I A C S G G C Q A I V D T G T S
PP F N V N S Y T A G S Q S G D G F - - - S G I A D T G T T
SP F N P N G Y S I G D S S S S G F - - - S A I A D T G T T

Table 4. (cont.)

```
                                                            245
SP  L L T G P T S A I A I N I Q S D I G A - S E N S D G E
PP  L L L L B D S V V S Q Y Y S Q V S G A Q Q D S N A G G
AP  L I L L D D E I V L - D G S Z V S G(A,Z,B)E E A D G G

                                                            272
SP  M V I S C S S I D Q L P D I V F T I D G V Q Y P L S P
PP  Y V F Z C S S S L P B F S V S I S G Y T A T V P G S L
AP  Y V F X C S T T P P B F T X X I G D Y K A V G P K - Y

                                                            298
SP  S A Y I L Q D D D S - C T S G F E G M D V P T S S G E
PP  I N Y G P S G N G S T C L G G I N S N - - - - S G I F
AP  I N Y A P G F B I(Z,S,C,Y)G G I N S N - - - - S G L G

                                                            325
SP  L W I L G D V F I R Q Y Y T V F D R A N N K V G L A P
PP  Y S I F G D I F L K S Q Y V V F D S D G P Q L G F A P
AP  L S L L G B V F L K S Q Y V V F B S Z G P K L G F A A

SP  V A
PP QQ A
AP  Q A
```

References

1. Kovaleva G.G., Ostoslavskaya V.I., Surova I.A., Revina L.P., Kotlova E.K., Nemtsova E.R., Levin E.D., Timokhina E.A., Baratova L.A., Belyanova L.P. , Stepanov V.M. : Bioorganicheskaya Khimiva 6, 1765-1777 (1980).

2. Ostoslavskaya V.I., Kotlova E.K., Stepanov V.M., Rudenskava G.N., Baratova L.A., Belyanova L.P.: Bioorganicheskaya Khimiya 5, 595-603 (1979).

3. Stepanov V.M., Rudenskava G.N. : J.Appl.Biochemistry 5, 420-428 (1983).

4. Stepanov V.M., Rudenskaya G.N., Gaida A.V., Osterman A.L. : J.Biochem.Biophys.Methods 5,177-186 (1981).

5. Gaida A.V., Osterman A.L., Rudenskaya G.N., Stepanov V.M. Biokhimiva 46, 181-190 (1981).

6. Kovaleva G.G., Yusupova M.P.,Lysogorskaya E.N., Balandina G.N., Stepanov V.M. : Biokhimiya 42, 534-539 (1977).

7. Rudenskaya G.N., Gaida A.V., Stepanov V.M.: Biokhimiya 45,561-567 (1980).

8. Belozersky M.A., Dunaevsky Ya.E., Rudenskaya G.N., Stepanov V.M.: Biokhimiya 49.479-485 (1984).

9. Lysogorskaya E.N., Stepanov V.M. : Khimiva Prirodnykh soedinenii 1980,591-607.

10. Kotlova E.K., Ostoslavskaya V.I., Revina L.P.,Kovaleva G.G.,Surova I.A., Levin E.D., Timokhina E.A..Stepanov V.M.: Bioorganicheskaya Khimiya 7,58-67 (1981).

11. Lobareva L.S., Stepanov V.M. : Uspekhi Sov Biologicheskoi Khimii 19, 33-105 (1978).

12. Andreeva N.S., Gustchina A.E. : Biochem.Biophys.Res.Commun. 87, 32-42 (1979).

13. Tang J., James M.N.G., Hsu I.-N., Jenkins J.A.,Blundell T.L.: Nature 271, 613-621 (1978).

14. Subramanian E., Swan I.D.A., Liu M., Davies D.R., Jenkins J.A., Tickle I.J., Blundell T.L. : Proc.Nat.Acad.Sci.USA 74,556-559 (1977).

15. Hsu I.-N., Delbaere L.T.J., James M.N.G.,Hofmann T. : Nature 266,

140-145 (1977).

16. Lysogorskaya E.N., Balandina G.N., Stepanov V.M.: Bioorganicheskaya Khimiya 3, 537-541 (1977).
17. Tarasova N.I., Lavrenova G.I., Stepanov V.M. : Biokhimiya 46, 369-375 (1981).
18. Osterman A.L., Stepanov V.M., Rudenskaya G.N., Khodova O.M., Tsaplina I.A., Yakovleva M.B., Loginova L.G. : Biokhimiya 49, 292-301 (1984).
19. Stepanov V.M., Lavrenova G.I., Adly K., Belyanova L.P., Baratova L.A. : Bioorganicheskaya Khimiya 3,673-675 (1977).
20. Stepanov V.M., Timokhina E.A., Baratova L.A., Belyanova L.P., Korzhenko V.P., Zhukova I.G. : Biochem.Biophys.Res.Commun. 45, 1482-1487 (1971).

STRUCTURE AND PROPERTIES OF PROTEINASE A FROM SACCHAROMYCES CARLSBERGENSIS AND SACCHAROMYCES CEREVISIAE.

Thomas Dreyer, Barbara Halkjær, Ib Svendsen and Martin Ottesen.
Carlsberg Laboratory, Department of Chemistry, Gamle Carlsberg-vej 10, DK-2500 Copenhagen Valby.

Introduction.

The vacuoles of Saccharomyces cerevisiae contain an aspartic proteinase (1,2). A similar proteinase has been isolated from Saccharomyces carlsbergensis (3,4). This enzyme has now been isolated in larger scale and compared with proteinase A. Partial amino acid sequences are presented for both enzymes.

Results.

The aspartic proteinase from S.carlsbergensis was purified to homogeneity according to the scheme in Table 1. Some of its properties are, in Table 2, compared with the properties of proteinase A, pepsin and renin. The results indicate no differences between proteinase A from S.carlsbergensis and S.cerevisiae. Further evidence for similarity was obtained by N-terminal amino acid sequence determination, using a Beckman Sequencer 890C. Of 36 investigated N-terminal residues all were identical in the two enzyme preparations.
Proteinase A from S.cerevisiae has been chosen for further sequence studies. Fragments obtained from cyanogen bromide cleavage and from digestion with trypsin, were purified using gel filtration, reverse-phase-HPLC and ion-exchange-FPLC before sequencing. So far 38% of the sequence has been determined, revealing a high degree of homology with mammalian gastric and lyso-

Aspartic Proteinases and their Inhibitors

somal aspartic proteinases, at present 44% homology with pepsin and with renin, see Fig. 1 where residues are numbered according to the pepsin sequence.

The catalytically active aspartic acids, Asp32 and Asp215, in aspartic proteinases (6), are conserved in proteinase A, and so are residues in positions 30, 33, 39, 216, and 222, which correspond to S_1 and S_1' positions in e.g. pepsin (6).

The ion pair between an aspartic acid in position 11 and a lysine in position 308, is an important initial part of the proposed charge relay mechanism, aimed at increasing the electrophilicity of a proton shared by Asp32 and Asp215 (6). Position 11 in proteinase A, and in the lysosomal angiotensinogen converting enzyme, renin (7), is an asparagine. Renin operates at pH near neutral, and hence there is no need for lowering the pK of Asp32 and Asp215 (7), while proteinase A is catalytically active down to pH 2.0, which should require the proposed involvement of the aspartic acid in position 11.

Clearly the role of Asp11 in the mechanism of aspartic proteinases should be reevaluated, based on the above finding.

REFERENCES.

1. Hata,T.,Hayashi,R.,Doi,E.: Agric.Biol.Chem.31,357-367 (1967).
2. Meussdoerffer,F.,Tortora,P.,Holzer,H.: J.Biol.Chem.255,12087-12093 (1980).
3. Maddox,I.S.,Hough,J.S.: Biochem.J.117,843-852 (1970).
4. Woods,F.C.,Kinsella,J.E.:J.Food Biochem.4,79-98 (1980).
5. Saheki,T.,Matsuda,Y.,Holzer,H.:Eur.J.Biochem.47,325-32(1974).
6. James,M.N.G.,Hsu,I.-N.,Hofmann,T.,Sielecki,A.R.:in "Structural Studies on Molecules of Biological Interest",ed Dodson,G. Claredon,Oxford 350-389 (1981).
7. Blundell,T.,Sibanda,B.L.,Pearl,L.: Nature 304,273-275 (1983).

Table 1.

Procedure	Total mg protein	Proteinase A S.cer. mg	Spec.act.
Supernatant after acid activation of ether plasmolysate	85,000	1,608	19
Batchwise DE-52 chromatography	9,000	774	86
Gradient DE-52 chromatography	1,300	620	477
Phenyl Sepharose chromatography	490	456	930
Sephadex G-100 chromatography	428	428	1000

Purification scheme for proteinase A from S.cerevisiae. Proteinase A was assayed using acid denatured hemoglobin at pH 3.2.

Table 2.

	Proteinase A		Pepsin	Renin
	S.cer.	S.carlsb.		
Molecular weight	41,000	41,000	35,500	40,000
Carbohydrate cont.	8.5%	8.5%	0	+
Hexosamine cont.	3.2%	3.2%	0	?
Isoelectric point	3.8-4.1	3.8-4.1	1.0	5.7
pH optimum	3.2	3.2	1.8-2.0	5.5-6.5
pH stability	2.0-7.5	3-7	1-6	?
Inhib.by DAN	+	+	+	+
Inhib.by EPNP	+	+	+	+
Inhib.by pepstatin	+	+	+	+
Inhib.by I^A	+	+	-	?
Milk clotting act.	-	-	+	?

Data for pepsin: Ryle,A.P. in "Methods in Enzymology", eds. Perlmann,G.E.,Lorand,L.,19,316-337 (1970).

Data for renin: Inagami,T.,Chang,J.J.,Dykes,C.W.,Takii,Y., Kisaragi,M.,Misono,K.S.,Fed.Proc.42,2729-2734 (1983).

Fig.1

```
                          1         10        20        30
Prot.A S.carlsberg.    GGHDVPLTNYLNAQYYTDITLGTPPQNPKVILDTGSSNL
Prot.A S.cerevisiae    GGHDVPLTNYLNAQYYTDITLGTPPQNPKVILDTGSSNL
Pepsin (hog)            IGDEPLENYLDTEYFGTIGIGTPAQDFTVIFDTGSSNL
Mouse subm.renin  SSLTDLISPVVLTNYLNSQYYGEIGIGTPPQTFKVIFDTGSANL

               40     AB  50        60        70        80
Prot.A S.cer WVPSNEC---------------------------------------
Pepsin       WVPSVYCS__SLACSDHNQFNPDDSSTFEATSQELSITYGTGSMTGILG
Renin        WVPSTKCSRLYLACGIHSLYESSDSSSYMENGDDFTIHYGSGRVKGFLS

                90        100       110       120       130
Prot.A S.cer -------------------------------------------------
Pepsin       YDTVQVGGISDTNQIFGLSETEPGSFLYYAPFDGILGLAYPSISASGAT
Renin        QDSVTVGGITVT_QTFGEVTELPLIPFMLAQFDGVLGMGFPAQAVGGVT

                 140       150       160B        170       180
Prot.A S.cer -------------------------------------------------
Pepsin       PVFDNLWDQGLVSQDLFSVYLSSNDD__SGSVVLLGGIDSSYYTGSLNW
Renin        PVFDHILSQGVLKEKVFSVYYNRGPHLLGGEVV_LGGSDPEHYQGDFHY

                    190       200       210       220
Prot.A S.cer -----KAYWELKFEGIGLGDEYAELESHG_AAIDTGTSLITLP_SGLAE
Pepsin       VPVSVEGYWQITLDSITMDGETIACSGGCQAIVDTGTSLLTGPTSAIAI
Renin        VSLSKTDSWQITMKGVSVGSSTLLCEEGCEVVVDTGSSFISAPTSSLKL

                     240       250       260       270      A
Prot.A S.cer MINAEIGAKKGWTGQYTLDCNTRDNLPDLIFNINGY-------------
Pepsin       NIQSDIGASENSDGEMVISCSSIDSLPDIVFTINGVQYPLSPSAYILQ_
Renin        IMQA_LGAKEKRLHEYVVSCSQVPTLPDISFNLGGRAYTLSSTDYVLQY

            B       290       300       310       320
Prot.A S.cer-------------------------------------------------
Pepsin      _DDDSCTSGFEGMDVPTSSGELWILGDVFIRQYYTVFDRANNKVGLAPVA
Renin       PNDKLCTVALHAMDIPPPTGPVWVLGATFIRKFYTEFDRHNNRIGPALAR
```

PEPSTATIN-SENSITIVE PROTEINASE FROM *CHLAMYDOMONAS REINHARDTII* CELLS

Tadeusz Wilusz, Antoni Polanowski
Institute of Biochemistry, University of Wrocław, Poland

Raymond F. Jones
Department of Biology, SUNY at Stony Brook, U.S.A.

Introduction

It has been established that an unicellular organism of *Chlamydomonas reinhardtii* contains several enzymes which are capable of hydrolyzing hemoglobin, dipeptides, specific substrates for amino- and carboxypeptidases and cathepsin A (1, 2). A proteinase degrading hemoglobin under acidic condition was examined during all stages of growth, development and differentiation (3). The present communication describes the purification and some properties of this proteinase.

Results and discussion

Plant material. Large-scale culture of *Chlamydomonas reinhardtii* (strain 137$^-$) were grown on high salt media (3) under continuous illumination to a cell concentration of approximately 9 x 10^6 per ml. The cells were harvested and washed twice with cold distilled water before being frozen at -20^{0}C.

Purification and properties of the enzyme. The frozen cells were thawed and extracted for half an hour with two volumes of 0.01 M acetate buffer, pH 5.8 and then centrifuged at 10 000 x g for 30 min. The pellet was re-extracted and centrifuged as above. Only some 35-40% of entire activity

Aspartic Proteinases and their Inhibitors

detected in sonicated cells was extractable under applied conditions. However, we have failed to extract the remaining activity even with 1 M NaCl and Triton X100. That is why we have limited our procedure to the leaking form of the enzyme. The cell extract was stirred for 15 min with charcoal (1 g per 100 ml) then clarified by centrifugation. To the supernatant NaCl was added to the concentration of 1 M followed by acidification to pH 5.1. The decreasing of pH below 5.1 led to the co-precipitation about 35% of the enzyme with balast proteins. To separate the proteinase from the solution an affinity chromatography was employed (Fig. 1). It has been established that the enzyme was more effectively adsorbed to immobilized pepstatin at room temperature but its desorption was much more efficient at lower temperature. Also it was found that the use of such an amount of immobilized pepstatin which bounded not more than 30% of the total activity of the extract markedly improved the recovery of the enzyme. Thus the small amount of immobilized pepstatin was used repeatedly to gain a higher yield. The enzyme solution was mixed with a pepstatin--AH Sepharosa (4) in a ratio of 3 g of sucked off slurry to one liter of

Fig. 1. Chromatography of proteinase on immobilized pepstatin. Arrows indicate a change of the buffers.

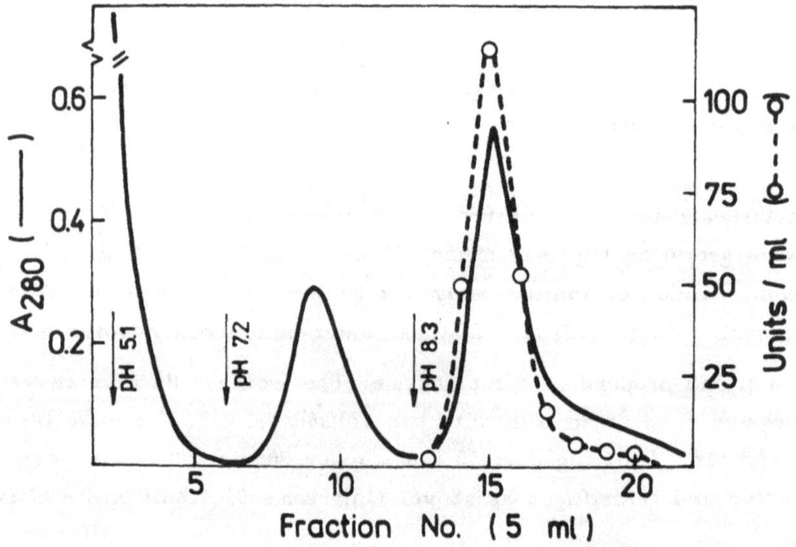

Table 1. Purification of the pepstatin-sensitive proteinase from 300 g of *Chlamydomonas reinhardtii* cells.

Purification step	Protein mg	Activity total units	Activity specific units/mg	Yield (%)
Extract	3445	4932	1.43	100
Charcoal treatment	2392	4508	1.88	91.4
Affinity chromatography	15.2	2621	172.7	53.1
Ultrafiltration	2.1	2053	977.6	41.6
DEAE-cellulose	1.8	1955	1086	39.6
Preparative PAGE	1.5	1779	1186	36.1

The activity was determined in Folin-Lowry reaction according to Barrett (7) using hemoglobin as substrate

extract. After 30 min stirring the resin was transferred into a column, cooled down to 6°C and washed out with a cold 0.05 M acetate buffer, pH 5.1 followed by 0.05 M Tris-HCl buffer, pH 7.2. The enzyme was eluted with 0.05 M Tris-HCl buffer, pH 8.3. All used buffers contained 1 M NaCl. The obtained preparation was concentrated in dialysis tubing under reduced pressure to remove low mol. wt. contaminations. The final purification was achieved with chromatography on DEAE-cellulose at pH 8.2 followed by preparative PAGE at the same pH. The results of purification are summarized in Table 1. From 300 g of cells, 1.5 mg of the enzyme were obtained with about 830-fold purification and 36% yield. In the analytical gel electrophoresis at pH 4.5 the enzyme preparation migrated as a single band but at pH 8.3 it resolved into three closely spaced bands with a predominance of the midle one. The enzyme had a molecular mass of 35 500 as determined on Sephadex G 75 column and showed the maximal activity on acid-denatured hemoglobin in pH range 3.6 to 4.2. Some properties of the *Chlamydomonas* proteinase were compared with those of cathepsin D isolated from calf thymus (5). For both enzymes hemoglobin was better substrate than bovine serum albumin, although, relatively the last mentioned protein was more susceptible to *Chlamydomonas* proteinase. The alga enzyme upon incubation for 24 hrs at pH 5.6 and room temperature preserved full its activity while the activity of cathepsin D dropped by 80%. A month period of storage at both $4^{o}C$ and $-20^{o}C$ did not affect

Chlamydomonas proteinase activity. Both enzymes were linearly inhibited by pepstatin while the cathepsin D inhibitor from potatoes (6) was active only to thymus cathepsin D.

Since O'Kane (2) has localized the hemoglobin digesting acid proteinase in alga vacuoles, a counter part of animal lisosomes, therefore, one can consider the pepstatin-sensitive proteinase from *Chlamydomonas* as a plant cathepsin D-like enzyme, what already was suggested by Jones (3).

References

1. Lang, W.C., Blatt, D., Plapp, R.: Plant Cell Physiol. *30*, 657-665 (1979).
2. O'Kane, D.J.: Ph. D. Thesis. SUNY at Stony Brook, N.Y., U.S.A. (1978).
3. Jones, R.F.: Ann. N. Y. Acad. Sci. *175*, 648-659 (1970).
4. Kregar, J., Urh, J., Smith, R., Umezawa, H., Turk, V. In: Intracellular Protein Catabolism II (V. Turk, Marks, H. eds.). Plenum Press. New York and London. 250-254 (1977).
5. Smith, R., Turk, V.: Eur. J. Biochem. *48*, 245-254 (1974).
6. Keilova, H., Tomasek, V.: Collection Czechoslov. Chem. Commun. *41*, 489-497 (1979).
7. Barrett, A.J., In: Lysosomes. A. Laboratory Handbook (J.T. Dingle, ed.). North-Holland. Amsterdam. 46-135 (1972).

This work was supported by Research Project of the MSzWiT No. R. 1.9 and by NSF Grant No. INT76-22205.

PURIFICATION AND CHARACTERIZATION OF ASPARTIC PROTEINASES FROM *CUCUMIS SATIVUS* AND *CUCURBITA MAXIMA* SEEDS

Antoni Polanowski, Tadeusz Wilusz, Maria K. Kołaczkowska,
Maciej Wieczorek, Anna Wilimowska-Pelc
Institute of Biochemistry, University of Wrocław, Poland

Marian Kuczek
Institute of Physiology and Biochemistry, Medical School, Wrocław, Poland

Introduction

In contrast to the extensively studied aspartic proteinases from animal tissues relatively little is known about these enzymes from plant material. Recently we have isolated an aspartic proteinase from the seeds of cucumber *(Cucumis sativus)* (1). The present communication deals with a comparison of cucumber proteinase with an aspartic enzyme separated from the seeds of squash *(Cucurbita maxima)*.

Results and discussion.

1) Purification of the enzymes. Ground seeds were extracted with 0.05 M acetate buffer, pH 4.5. The clarified supernatant was brought to 80% saturation with ammonium sulfate and the precipitate was dissolved in water. After dialysis the solutions were acidified to pH 3.6, NaCl was added at the concentration of 0.5 M and the proteinases were adsorbed on 10-20 ml of a pepstain-AH Sepharose (2). The enzymes were eluted with 0.1 M Tris-HCl buffer, pH 8.3, containing 0.5 M NaCl. The procedure was repeated 3-5 times to gain a higher yield. The enzyme from cucumber after dialysis and concentration in a stream of cold air was finaly purified on

Aspartic Proteinases and their Inhibitors

Sephacryl S-200 column. The squash proteinase after concentration in dialysis tubing under reduced pressure was subjected to DEAE A-50 (0.05 M Tris-HCl buffer, pH 8.3) followed by Sephadex G 75 chromatography. From 1 kg of seeds 13.6 and 10.4 mg enzyme preparations from cucumber and squash, respectively, were obtained (Table 1). The proteinases were homogeneous on polyacrylamide gel electrophoresis at pH 8.3 and 4.5.

2) Some properties of the enzymes. Both proteinases in spite of their origin from different botanical genera resembled each other in many respects. Their molecular weights estimated on Sephadex G-75 column were found to be 40-42 kDa. They differed only slightly in their isoelectric points and optimal pH's. The isoelectric point determined by isoelectric focusing for cucumber proteinase was 5.2 and its maximal activity against acid-denatu-

Table 1. Purification of the proteinases from 1 kg of cucumber and squash seeds.

Purification step	Cucumber		Squash	
	Protein mg	Specific activity units/mg	Protein mg	Specific activity units/mg
Extract	40800	n.d.	39600	3.0
Precipitation with $(NH_4)_2SO_4$	17630	2.1	5358	9.3
Pepstatin-Sepharose 4B	24.8	847.8	34.3	952.0
Sephacryl S-200	13.6	1159	-	-
DEAE-Sephadex	-	-	12.2	1654
Sephadex G-75	-	-	10.4	1770

Protein in the extract was determined by the method of Goa (5) and at further steps according to Lowry (6). A unit of activity was defined as an increase of 0.1 absorbance unit at 280 nm per 30 min in TCA-soluble products of hemoglobin digestion (1).

Fig. 1. Effect of pepstatin (A) and diazoacetyl-DL-norleucine methyl ester (B) on the activity of the proteinases from cucumber (circle) and squash (square). The incubation mixture in B contained 880 molar excess each of DAN and cupric ions (full symbols); in the control the samples of the enzymes were incubated with curpic ions without DAN (open symbols).

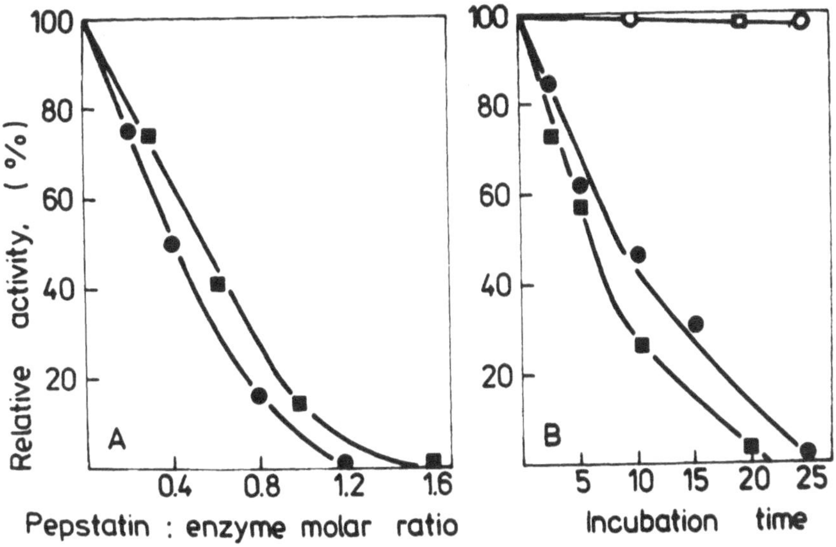

red hemoglobin was at pH 3.2, whereas those values for squash enzyme were 5.6 and 3.6, respectively. Changes in ionic strenght of the buffer over the range 0.05-0.25 μ did not affect the activities of the enzymes at optimal pH's. In addition to hemoglobin which appeared to be the best substrate, the enzymes are able, to a lesser extent, to hydrolyze serum and egg albumin, edestin from hemp seeds and the endogenous cucumber and squash seeds globulins. It is worthy to note that the last mentioned proteins besides hemoglobin were the most susceptible substrates for the enzymes (determined ty the increase in free amino groups with trinitrobenzenesulphonic acid). Moreover, both proteinases are able to coagulate milk casein at pH 6.3. The proteinases stored in the frozen state (pH 8-9) for at least 3 months retained full activity. When kept hovewer at 5°C their stability depends on pH. Within 6 days at pH 8-9 either in the presence or without of 0.5 M NaCl their activity remained almost unchanged. At pH below 5 they gradually lost the activity, especially in the presence of 0.5 M NaCl. Divalent cations (Ca, Mg, Zn, Cu, Mn, Fe, Hg) EDTA, the -SH-group reagent (pCMPSA, DTT, iodoacetamide), synthetic inhibitors of serine proteinases (PMSF, DFP) and endogenous trypsin inhibitors from resting seeds of cucumber and squash (3) had no effect on the activity of both proteinases. Only group-specific inhibitors of aspartic proteinases: pepstatin, diazoacetyl-DL-norleucine methyl ester (Fig. 1) and 1,2,3-epoxy(*p*-nitrophenoxy)-propane effectively inhibited these enzymes.

3) Specificity of the squash seeds proteinase. Using oxidized B-chain of bovine insulin as a substrate it was established under conditions described in (4) that the enzyme is able to split the peptide bond between Phe1--Val2, Leu15-Tyr16, Tyr16-Leu17, Phe24-Phe25 and Phe25-Tyr26. Studying the rate of hydrolysis it is apparent that squash proteinase preferencially hydrolyzes those linkages involving the amino group of tyrosine.

References

1. Wilamowska-Pelc, A., Polanowski, A., Kołaczkowska, M.K., Wieczorek, M., Wilusz, T.: Acta Biochim. Polon. *30*, 23-31 (1983).

2. Kregar, J., Urh, J., Smith, R., Umezawa, H., Turk, V.; In Intracellular Protein Catabolism II (V. Turk, H. Marks eds.) Plenum Press, New York and London. 250-254. (1977).

3. Leluk, J., Otlewski, J., Wieczorek, M., Polanowski, A., Wilusz, T.: Acta Biochim. Polon. *30*, 127-138 (1983).

4. Kuczek, M., Nowak, K., Kołaczkowska, M.K.: Acta Biochim. Polon. *30*, 3-10 (1983).

5. Goa, J.: Scand. J. Clin. Lab. Invest. *5*, 218-222 (1953).

6. Lowry, O.H., Rosebrough, N.J., Farr, A.L., Randall, R.J.: J. Biol. Chem. *193*, 265-275 (1951).

This work was supported by Research Project of the MSzWiT no. R.1.9.

ISOLATION AND MOLECULAR CHARACTERISTICS OF AVIAN PEPSINS

Vladimír Kostka, Iva Pichová, Miroslav Baudyš
Institute of Organic Chemistry and Biochemistry, Czechoslovak Academy of Sciences, 166 10 Prague 6, Czechoslovakia

Introduction

The rapid advance in studies on the molecular characteristics and hydrolytic mechanism of aspartic proteinases during the past fifteen years has extended our knowledge of mainly mammalian pepsins and proteinases of the pepsin type from fungi. In contrast information on enzymes from other organisms is meagre and usually limited to fragmentary data on a sole representative of an individual species at the most. Surprisingly enough this holds also for aspartic proteinases of birds, even though early knowledge (1) of peptic activity in chicken forestomachs dates back to 1908 and the enzyme itself and its precursor have been isolated and partly characterized (2) in 1938, probably as the second proteinase and its zymogen at all. Studies in this field continued, however, only some 25 years later and concentrated mainly on the purification and resolution of the individual isoforms, first of the enzyme (3) and later also of the zymogen (4). While furnishing important data, such as the pH-profile of activity of the enzyme, its electrophoretic behavior and also the important finding of the inability of chicken pepsin to cleave low molecular substrates this work was not based, however, on a structure knowledge of the protein. It was not until 1969 that the first sound study in this field appeared (5) providing molecular characteristics of the enzyme and its zymogen. Investigations by other authors which followed soon thereafter (6,7) and aimed to draw a parallel between chicken pepsin and mammalian pepsins as regards

Aspartic Proteinases and their Inhibitors

the multiplicity of forms (8,9) yielded controversial information, especially as regards the carbohydrate moiety and the size of the zymogen propart. These facts, together with the interesting catalytic characteristics of chicken pepsin, which can be modulated by labeling its free SH-group (10), its other unusual molecular features, and also the practical importance of the enzyme as a rennet substitute (11) led us to engage in a complex study of chicken pepsin(ogen) as a model of gastric, vertebrate nonmammalian aspartic proteinase.

In this study we will report on the preparation of chicken pepsin and its zymogen in forms suitable for more exacting studies, such as of the zymogen activation pathway. The molecular characteristics of these preparations will be compared with those of other pepsins and of duck pepsin(ogen). Later in this volume (cf. pp. 301 and pp.245) the activation of chicken pepsinogen and the catalytic properties of the enzyme will be treated.

Results

Preparation of avian pepsins and pepsinogens

Isolation of chicken and duck pepsin. The procedures based on acid extraction of the forestomach mucosa (e.g. 2,12) with subsequent precipitation steps yield the enzyme contaminated with partly autolyzed products because of prolonged exposure of the material to acid conditions. Rapid methods of one-step isolation from crude mucosal extracts are recommendable. One of these based on the affinity of chicken pepsin for mercurial supports (13) was developed in the author´s laboratory (14). The use of affinity chromatography methods is limited by the availability of suitable inhibitors. Immobilized pepstatin A, usually employed for this purpose, gives a very tight complex with chicken pepsin from which the enzyme can be displaced by 6M guanidine only with simultaneous denaturation.

The preparation of the naturally occurring pepsin inhibitor from Ascaris lumbricoides (15) is troublesome and involves health hazard because of the presence of an allergen in the biological material. Affinity chromatography methods using this inhibitor (e.g. 16) are therefore likely to receive little application only. One of the novel affinity methods lacking these disadvantages makes use of a synthetic inhibitor of chicken pepsin, peptide Val-D-Leu-Pro-Phe-Phe-Val-D-Leu, a substrate analog of the enzyme. The binding capacity of the inhibitor immobilized via its amino group to 6-aminohexanoic acid-Sepharose 4B with a six carbon spacer is more than 1 mg/ml over the pH-range 2-5. Affinity chromatography of crude forestomach extracts on these supports (17) gives a 70-80% yield of chicken pepsin with specific activity increased 10-20 times. The enzyme displaced by buffers at pH 6 and higher is native and homogeneous. A similar treatment of duck forestomach extracts is not equally efficient because the enzyme is less inhibited by the peptide. The method is of general applicability and has been used for the isolation of bovine spleen cathepsin D (17) and two mammalian gastric pepsins (18).

Isolation of chicken pepsinogen. Unlike hog pepsinogen the chicken zymogen shows no affinity for polylysine-Sepharose and there is no simple method of its isolation from alkaline mucosal extracts. A modification of the procedure (5) using repeated acetone precipitation and dual anion chromatography followed by gel filtration gives a 71% yield of the zymogen (Table 1.). Attempts to use this material for zymogen activation studies (19) revealed 3 % of peptic activity detectable at pH 5.5-6.0 without activation of the zymogen by the assay (20) using <Glu-His-Nph-Phe-Ala-Leu.NH_2 as substrate. The contamination with the active enzyme is readily removed by chromatography of the pepsinogen on a Pepstatin-Sepharose 4B column (Figure 1). The completely pepsin-free zymogen thus obtained is, however, inhomogeneous when examined by SDS-PAGE.

Table 1

Isolation and purification of chicken pepsinogen

Step	Weight mg	Activity total $U \times 10^5$	Activity specific U/mg prot.	Yield %	Purification x
Chicken forestomachs	43 400				
Crude extract	21 620				
Dialyzed extract of 2nd acetone powder	4 524	14.50	321	100	1
Preparation after DEAE-cellulose chromatography	474	12.25	2 582	84.50	8.1
rechromatography on DEAE-cel.	444	11.78	2 654	81.24	8.3
gel filtration on G-100	339	10.30	3 039	71.03	9.5
chromatography on Pepstatin-Sepharose	289	9.65	3 337	66.55	10.4
chromatography on Phenyl-Sepharose : NC	99	2.51	2 539	17.31	7.9
CPG	106	3.31	3 136	22.82	9.8

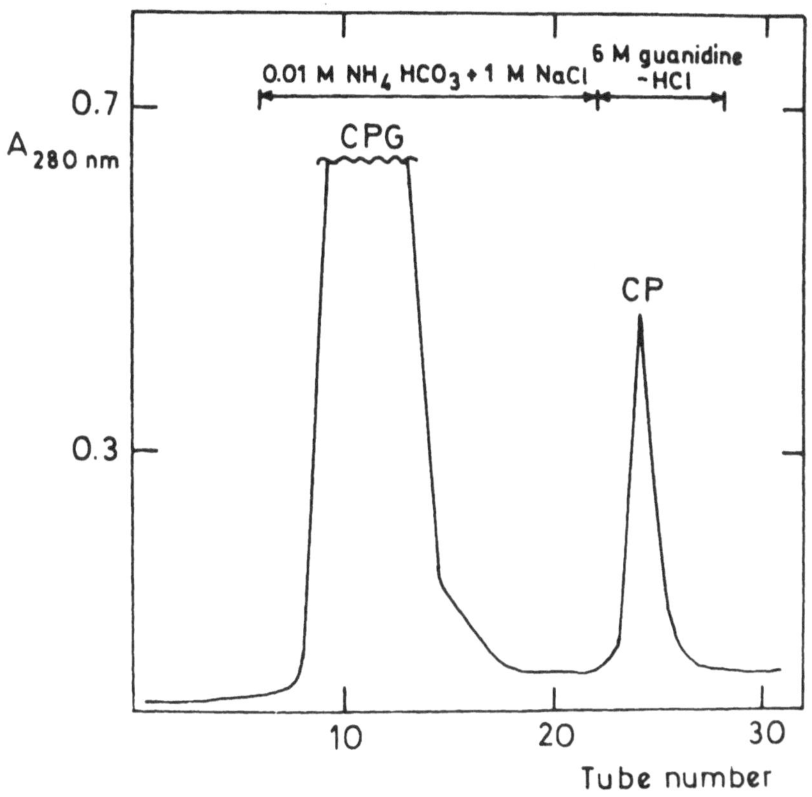

Figure 1. Chromatography of chicken pepsinogen on Pepstatin-Sepharose 4B. Column 1 x 10 cm, flow rate 20 ml/h. The horizontal bars indicate the application of the individual eluents.

Chromatography on Phenyl-Sepharose CL-4B (Figure 2) yields homogeneous chicken pepsinogen (CPG) and one additional high molecular weight component (NC) whose identity was examined as described below. The final yield of chicken pepsinogen thus prepared is 22.8 % (Table 1).

Isolation of duck pepsinogen. The same procedure as that de-

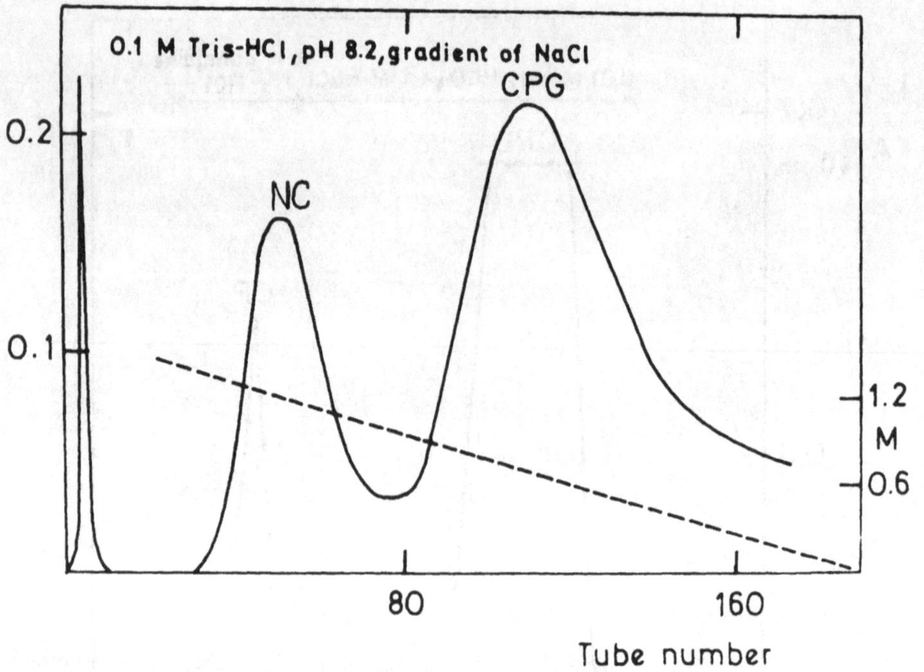

Figure 2. Hydrophobic chromatography of chicken pepsinogen on Phenyl-Sepharose CL-4B. Column 2.4 x 50 cm, flow rate 60 ml/h. The salt gradient is marked by a dashed line and NaCl concentration shown on the right-hand ordinate. Left-hand ordinate shows absorbance at 280 nm.

veloped for the chicken zymogen was used. The protocol is shown in Table 2. The preparation freed of contamination with the active enzyme is homogeneous when assayed by SDS-PAGE and the hydrophobic chromatography step can be omitted. The molecular characteristics of duck pepsinogen and of the pepsin obtained by its activation are summarized in Table 3.

Table 2

Isolation and purification of duck pepsinogen

Step	Weight	Activity		Yield	Purification
		total	specific		
	mg	$U \times 10^5$	U/mg prot.	%	x
Crude extract	39 705				
Dialyzed extract of 2nd acetone powder	26 428	1.930	7.32	100	1
Preparation obtained after DEAE-cellulose chromatography	1 782	0.858	48.17	44.45	6.6
rechromatography on DEAE-cellulose	1 052	0.679	64.58	35.18	8.8
gel filtration on G-100	160	0.587	366.87	30.41	50.1
chromatography on Pepstatin-Sepharose	126	0.484	384.10	25.10	52.5
chromatography on Phenyl-Sepharose	120	0.404	336.67	20.93	46.0

Table 3

Molecular characteristics of duck pepsin and pepsinogen

	zymogen	enzyme
Amino acid composition[+]	Lys_{16} His_5 Arg_{11} Asp_{31} Thr_{26} Ser_{39} Glu_{37} Pro_{26} Gly_{32} Ala_{25} Cys_6 Val_{25} Met_6 Ile_{21} Leu_{31} Tyr_{15} Phe_{18} Trp_4	Lys_{11} His_1 Arg_7 Asp_{29} Thr_{26} Ser_{36} Glu_{28} Pro_{16} Gly_{31} Ala_{24} Cys_6 Val_{23} Met_6 Ile_{20} Leu_{23} Tyr_{15} Phe_{18} Trp_4
Carbohydrate moiety	$GlcNAc_8$ Man_3	$GlcNAc_8$ Man_3
N-terminal amino acid	Leu	Ala
pH-stability [a,b]	pH 6-10 100 % pH 11.5 50 % pH 12.1 0 %	pH 6-7.5 100 % pH 9.2 50 % pH 9.6 0 %
activation, halflife [c]	20 s (pH 2, 14 °C)	-
inhibition		
EPNP [b]	-	complete (25 °C, 75 h)
heptapeptide [b,d] IC_{50}	-	0.1 mM
DAN [b]		complete (25 °C, 5 min)
pepstatin [e]	-	$k_i \sim 10^{-10}$ M
pH-optimum [f]	-	4.0

[a] incubation at the given pH, 20 min at 25 °C, [b] activity toward hemoglobin, [c] <Glu-His-Phe-Phe-Ala-Leu.OMe as substrate, fluorescamine detection, [d] Val-D-Leu-Pro-Phe-Phe-Val-D-Leu, [e] <Glu-His-Nph-Phe-Ala-Leu.NH_2 as substrate, [f] Lys-Pro-Ala-Glu-Phe-Nph-Ala-Leu as substrate

+ For the calculation of the amino acid composition pepsinogen and pepsin were assumed to contain 374 and 324 amino acid residues, respectively.

Evidence for the presence of the chicken pepsin-activation peptide (p1-p42) complex in pepsinogen preparations

Hydrophobic chromatography on Phenyl-Sepharose CL-4B of pepsin-free chicken pepsinogen yields two components (Figure 2) of which the second one is the zymogen (CPG). The material in the first peak (NC) has the same amino acid composition as CPG corresponding to 367 residues determined by sequence analysis (23). NC and CPG differ in carbohydrate content (5 GlcNAc and 3 Man in NC compared to 7 GlcNAc and 4 Man in CPG) and in mobility during SDS-PAGE. Whereas CPG is homogeneous, NC gives a high molecular weight band, slightly faster than that of CPG and corresponding in mobility to pepsin, and a fast, low molecular weight band. Sequence analysis of NC revealed the presence of N-terminal sequences of pepsinogen and pepsin at a 1:1 ratio. The two components of NC were resolved by gel filtration on Sephadex G-50 in 0.1 M acetate, pH 5.3, containing 8M urea. The high molecular weight component has the same amino acid composition as pepsin and is N-terminated with threonine, the small component with N-terminal serine corresponds to the amino acid composition of the chicken pepsinogen propart (p1-p42). Hence, NC is the complex of chicken pepsin and its activation peptide, first reported as chicken "pepsin inhibitor" in 1941 (21). Since the whole process of preparation of the pepsinogen is carried out at a pH excluding its conversion to the active enzyme (22), the complex is obviously present in the mucosa. To check on this possibility crude pepsinogen after the first acetone precipitation (Table 1) was chromatographed on a column of Phenyl-Sepharose CL-4B under the conditions described in Figure 2. The material eluted in the same volume as NC was purified by DEAE-cellulose chromatography at pH 7.6 and by gel filtration on Sephadex G-100 in 0.04 M NH_4 HCO_3. The process of preparation was checked by SDS-PAGE of the individual products (Figure 3). The amino acid composition of the final product was the same as that of the complex (NC) isolated from

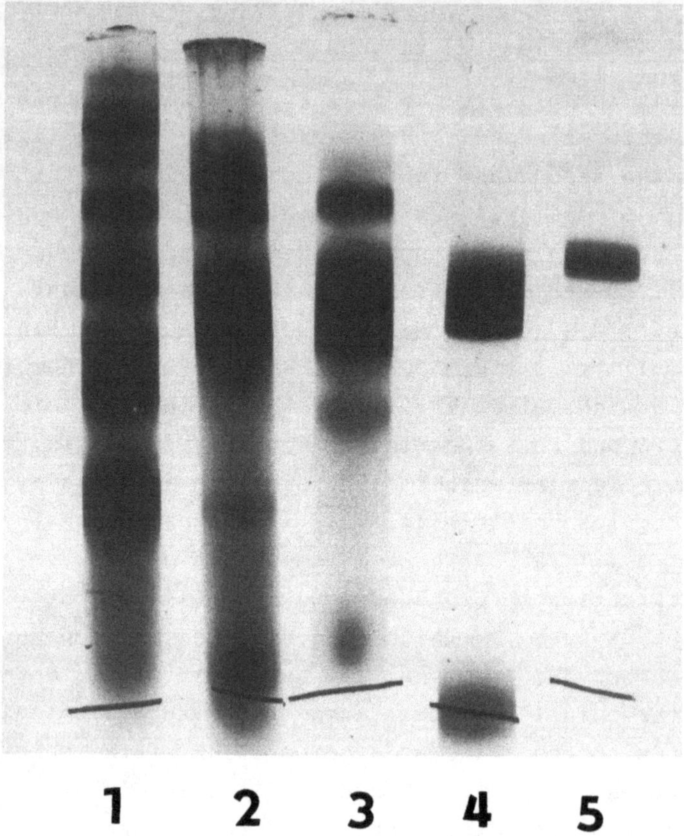

Figure 3. Isolation of chicken pepsin-activation peptide (p1-p42) complex from chicken forestomachs. SDS-PAGE pattern of: 1 - crude extract of mucosa, 2 - pepsin-activation peptide (p1-p42) complex (NC) after chromatography on Phenyl-Sepharose CL-4B, 3 - NC after chromatography on DEAE-cellulose, 4 - NC, 5 - chicken pepsinogen (CPG). The position of the marker (bromophenol blue) is marked by horizontal bars.

pure pepsinogen preparations (Figure 2). The SDS-PAGE pattern of the complex gave two bands (Figure 3) one of pepsin and the other one of the propart peptide (p1-p42). These results show that the pepsin-activation peptide (p1-p42) complex is present in the mucosa of the forestomachs and is extracted together with the pepsinogen.

Minor components of chicken pepsin and pepsinogen

The existence of a minor component accounting for 10-20 % of the pepsinogen content of chicken forestomach mucosa was reported first by Levchuk and Orekhovich (4) yet the pepsin obtained by its activation had the same properties as the main component. Essentially the same observation was made by Bohak (5) who moreover was able to obtain evidence showing that the same zymogen was contained in both the minor and the major fraction whose different chromatographic behavior was probably affected by contaminants. The existence of at least five different pepsinogens was postulated by Green and Llewellin (7). An effort to draw a parallel between the components of hog and chicken pepsinogens was made by Donta and Van Vunakis (6) who isolated a minor pepsinogen, allegedly analogous to hog pepsinogen C, which, however, they failed to activate to one single enzyme. In order to get insight into this controversial information we repeated their experiments under identical conditions starting with a pH 6.9 phosphate extract of the foretomachs to avoid the possible loss of the component of high negative charge suspected by some authors (7). The neutral extract was resolved by anion exchange chromatography on DE-32 cellulose. The material eluted by 0.3-0.8 M NaCl was pooled and rechromatographed to yield a major fraction containing pepsinogen and a minor peak. Efforts to separate the latter on DE-52 cellulose were unsuccessful and yielded heterogeneous samples. Chicken pepsin shows a considerably high alkali stability reported long ago (1). Extraction of the mucosa with neutral buffers and repeated chromatography at neutral pH is likely to cause proteolytic digestion of pepsinogen by pepsin thus giving rise to products which yield inhomogeneous, ill-definable pepsins upon activation. In contrast no minor components are observed if an alkaline buffer (pH 8.6, cf. 5, 22) is used for extraction and any pepsin present thus inactivated. From the results of our experiments the existence of a zymogen in chicken forestomach mucosa corresponding to hog or human progastricsin (pepsinogen C) seems unlikely.

Covalent structures

Amino acid sequences. The covalent structure of chicken pepsinogen determined in our laboratory (23) shows that the zymogen consists of 367 amino acid residues in a single polypeptide chain cross-linked by 3 disulfide bonds (Cys 45-Cys 50, Cys 206-Cys 210, and Cys 250-Cys 283, pepsin numbering, cf. 24). There is a cysteine residue in position 115 which finds no counterpart in any other aspartic proteinase including duck and Japanese quail (25) pepsin.

Our knowledge of the amino acid sequence of duck pepsin(ogen) limited to short N-terminal parts of the chain. Their alignment with the corresponding parts of chicken pepsinogen and pepsin is demonstrated in Figure 4.

Oligosaccharide units and glycosylation site. Chicken pepsin contains one oligosaccharide unit which is the only post-translational modification of the protein. The heterogeneity of this unit (26) may be the cause of the microheterogeneity of the protein. The glycosylation site is asparagine 68 localized in the position occupied by phosphoserine in hog pepsin. Mention deserves also the fact that one of the oligosaccharide units of hog spleen cathepsin D is linked to the asparagine residue adjacent to the glycosylation site of chicken pepsin (Asn 67, cf. 27). The complete sequence of the oligosaccharide unit is not known. Preliminary results of proton magnetic resonance measurements performed by Dr. P. Schmidt in Prof. Tang´s laboratory are best compatible with the structure shown in Figure 5. The structure of the oligosaccharide unit is in agreement with the present state of knowledge of N-linked oligosaccharides, biosynthesized first as a larger unit via

```
                                            p10                                               p20
DPG   Leu-Pro-Ala-Lys-Ile-PRO-LEU-Arg-LYS-Met-LYS- X -LEU-ARG-Gln-Arg-LEU-Glu-Glu- X -

CPG   Ser-Ile-His-Arg-Val-PRO-LEU-Lys-LYS-Gly-LYS-Ser-LEU-ARG-Lys-Gln-LEU-Lys-Asp-His-

                                            P30                                               p40
                         Phe     Phe
DPG   GLY-LEU- X -GLU- X -   -LEU-   - X -HIS- X -His-ASN-...
                         Lys     Lys

CPG   GLY-LEU-Leu-GLU-Asp-Phe-LEU-Lys-Lys-HIS-Pro-Tyr-ASN-Pro-Ala-Ser-Lys-Tyr-His-Pro-


              1                                 10
                              Ala
DP            Ala-Gly-THR- X - X -   -GLU-PRO-Leu-Gln-ASN-TYR-MET-Asn-Asn-Glu-TYR-TYR
                              Ser

CPG   Val-Leu-Thr-Ala-THR-Glu-Ser-Tyr-GLU-PRO-Met-Thr-ASN-TYR-MET-Asp-Ala-Ser-TYR-TYR


         20
                  Ala
DP    GLY-THR-Thr-   -ILE-GLY- X - ...
                  Ser

CPG   GLY-THR-Ile-Ser-ILE-GLY-Thr- ...
```

Figure 4. Alignment of N-terminal amino acid sequences of duck and chicken pepsinogen. Identities in duck (DPG) and chicken (CPG) pepsinogen are marked by capital letters, identities with other gastric aspartic proteinases are underlined. DP - N-terminal amino acid sequence of duck pepsin.

a dolicholphosphate intermediate (28). The remaining five GlcNAc residues are sulfated and the NMR data show the sulfation to be (at least predominantly) on the position 6. The presence of sulfated sugars in glycoproteins was originally believed to be a characteristic feature of mucopolysaccharides only. However, during the past years acid labile sulfate has been reported to exist in many glycoproteins and polypeptides (30,31). While the oligosaccharide units of cathepsin D, an intracellular aspartic proteinase are essential for lysozome targeting of this enzyme, the biological role of the carbohydrate moiety of avian pepsins as of many other proteins remains unknown.

Homology of chicken pepsin with other aspartic proteinases

A sequence alignment of chicken pepsinogen, hog pepsinogen (32-35) and calf prochymosin (36) is shown in Figure 6. Homologies among all three zymogens, which have 44 % residues in common, are observed throughout the length of the whole chain especially around Asp 32 and Asp 215, Tyr 75 and Gly 303, in the proparts and in the carboxy-terminal region. Chicken pepsinogen has a 62% sequence identity with hog pepsinogen and a 57% identity with calf prochymosin whereas the corresponding values for the active enzymes are 63 % and 56 %. The high degree of sequence homology with other aspartic proteinases around the catalytically important aspartic acid residues seems to indicate a similarity in three-dimensional structure.

Concluding remarks

The present stage of our knowledge of aspartic proteinases is still not sufficient enough to explain the structure-function relationship of these enzymes, to trace their evolution yet even to classify the representatives of the individual species. Chicken pepsin may serve as an example of an enzyme which cannot be unambiguously identified with a correspond-

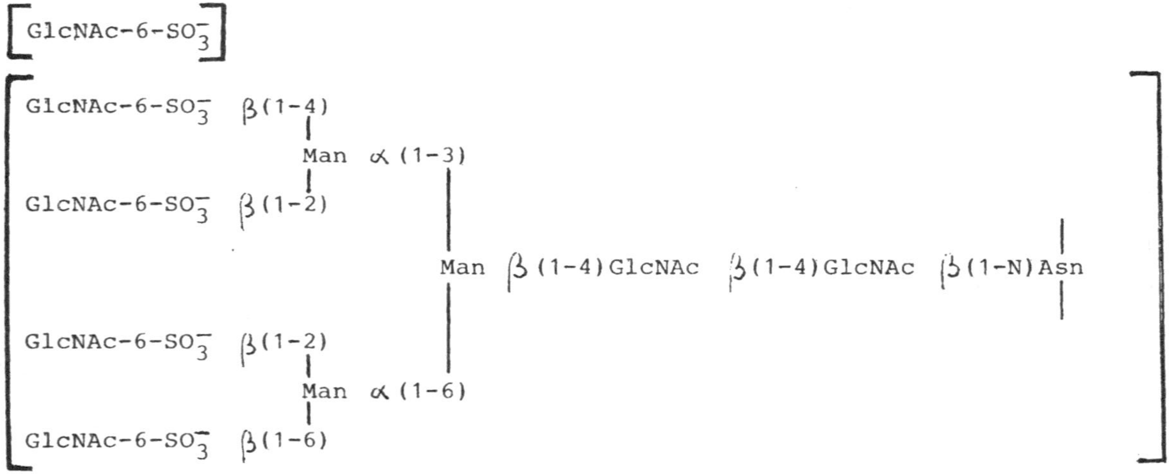

Figure 5. Partial structure of the carbohydrate moiety of chicken pepsin-(ogen) (29). The site of attachment of the remaining [GlcNAc] residue in the oligosaccharide unit is not known as yet.

```
                      p10                                        p20
HPG         Leu-Val-Lys-Val-PRO-LEU-Val-Arg-Lys-LYS-SER-LEU-ARG-Gln-Asn-LEU-Ile-Lys-Asp-GLY-Lys-LEU-Lys-
CPG     Ser-Ile-His-Arg-Val-PRO-LEU-Lys-Lys-Gly-LYS-SER-LEU-ARG-Lys-Gln-LEU-Lys-Asp-His-GLY-Leu-LEU-Glu-
CPC Ala-Glu-Ile-Thr-Arg-Ile-PRO-LEU-Tyr-Lys-Gly-LYS-SER-LEU-ARG-Lys-Ala-LEU-Lys-Glu-His-GLY-Leu-LEU-Glu-

                p30                             p40                     1
HPG ASP-PHE-LEU-Lys-Thr-His-Lys-His-Asn-Pro-Ala-SER-LYS-TYR-Phe-Pro-Glu-Ala-Ala-Ala-Leu-Ile-Gly-Asp-Glu-
CPG ASP-PHE-LEU-Lys-Lys-His-Pro-Tyr-Asn-Pro-Ala-SER-LYS-TYR-His-Pro-Val-Leu-Thr-Ala-Thr-Glu-Ser-Tyr-Glu-
CPC ASP-PHE-LEU-Gln-Lys-Gln-Gln-Tyr-Gly-Ile-Ser-SER-LYS-TYR-Ser- - - - -Gly-Phe-Gly-Glu-Val-Ala-Ser-Val-

                        10                                  20
HPG PRO-Leu-Glu-ASN-TYR-Leu-ASP-Thr-Glu-TYR-Phe-GLY-Thr-ILE-Gly-Ile-GLY-THR-PRO-Ala-GLN-Asp-PHE-Thr-VAL-
CPG PRO-Met-Thr-ASN-TYR-Met-ASP-Ala-Ser-TYR-Tyr-GLY-Thr-ILE-Ser-Ile-GLY-THR-PRO-Gln-GLN-Asp-PHE-Ser-VAL-
CPC PRO-Leu-Thr-ASN-TYR-Leu-ASP-Ser-Gln-TYR-Phe-GLY-Lys-ILE-Tyr-Leu-GLY-THR-PRO-Pro-GLN-Glu-PHE-Thr-VAL-

30      +                           40                                    50
HPG Ile-PHE-ASP-THR-GLY-SER-SER-Asn-Leu-TRP-VAL-PRO-SER-Val-TYR-CYS-Ser-SER-Leu-ALA-CYS-Ser-Asp-HIS-Asn-
CPG Ile-PHE-ASP-THR-GLY-SER-SER-Asn-Leu-TRP-VAL-PRO-SER-Ile-TYR-CYS-Lys-SER-Ser-ALA-CYS-Ser-Asn-HIS-Lys-
CPC Leu-PHE-ASP-THR-GLY-SER-SER-Asp-Phe-TRP-VAL-PRO-SER-Ile-TYR-CYS-Lys-SER-Asn-ALA-CYS-Lys-Asn-HIS-Gln-

                        60                                  70
HPG Gln-PHE-Asn-PRO-Asp-Asp-SER-SER-THR-Phe-Glu-Ala-Thr-Ser-Gln-Glu-Leu-Ser-ILE-Thr-TYR-GLY-THR-GLY-SER-
CPG Arg-PHE-Asp-PRO-Ser-Lys-SER-SER-THR-Tyr-Val-Ser-Thr-Asn-Glu-Thr-Val-Tyr-ILE-Ala-TYR-GLY-THR-GLY-SER-
CPC Arg-PHE-Asp-PRO-Arg-Lys-SER-SER-THR-Phe-Gln-Asn-Leu-Gly-Lys-Pro-Leu-Ser-ILE-His-TYR-GLY-THR-GLY-SER-

80                                      90                                      100
HPG MET-Thr-GLY-ILE-LEU-GLY-TYR-ASP-THR-VAL-Gln-VAL-Gly-Gly-ILE-Ser-Asp-Thr-Asn-GLN-Ile-Phe-GLY-LEU-SER-
CPG MET-Ser-GLY-ILE-LEU-GLY-TYR-ASP-THR-VAL-Ala-VAL-Ser-Ser-ILE-Asp-Val-Gln-Asn-GLN-Ile-Phe-GLY-LEU-SER-
CPC MET-Gln-GLY-ILE-LEU-GLY-TYR-ASP-THR-VAL-Thr-VAL-Ser-Asn-ILE-Val-Asp-Ile-Gln-GLN-Thr-Val-GLY-LEU-SER-

                        110                                 120
HPG Glu-Thr-GLU-PRO-GLY-Ser-Phe-Leu-Tyr-TYR-Ala-Pro-PHE-ASP-GLY-ILE-LEU-GLY-Leu-ALA-Tyr-PRO-SER-Ile-Ser-
CPG Glu-Thr-GLU-PRO-GLY-Ser-Phe-Phe-Tyr-TYR-Cys-Asn-PHE-ASP-GLY-ILE-LEU-GLY-Leu-ALA-Phe-PRO-SER-Ile-Ser-
CPC Thr-Gln-GLU-PRO-GLY-Asp-Val-Phe-Thr-TYR-Ala-Glu-PHE-ASP-GLY-ILE-LEU-GLY-Met-ALA-Tyr-PRO-SER-Leu-Ala-

130                                     140                                     150
HPG Ala-Ser-Gly-Ala-Thr-PRO-VAL-PHE-ASP-ASN-Leu-Trp-Asp-Gln-Gly-LEU-VAL-Ser-GLN-ASP-LEU-PHE-SER-VAL-TYR-
CPG Ser-Ser-Gly-Ala-Thr-PRO-VAL-PHE-ASP-ASN-Met-Met-Ser-Gln-His-LEU-VAL-Ala-GLN-ASP-LEU-PHE-SER-VAL-TYR-
CPC Ser-Glu-Tyr-Ser-Ile-PRO-VAL-PHE-ASP-ASN-Met-Met-Asn-Arg-His-LEU-VAL-Ala-GLN-ASP-LEU-PHE-SER-VAL-TYR-
```

160 170

HPG Leu-Ser-Ser-Asn-Asp-Asp-Ser-Gly-SER-Val-Val-Leu-Leu-GLY-Gly-ILE-ASP-Ser-Ser-TYR-Tyr-THR-Gly-Ser-Leu-
CPG Leu-Ser-Lys-Asp-Gly-Glu-Thr-Gly-SER-Phe-Val-Leu-Phe-GLY-Gly-ILE-ASP-Pro-Asn-TYR-Thr-THR-Lys-Gly-Ile-
CPC Met-Asp-Arg-Asp-Gly-Gln-Glu- - -SER-Met-Leu-Thr-Leu-GLY-Ala-ILE-ASP-Pro-Ser-TYR-Tyr-THR-Gly-Ser-Leu-

180 190 200

HPG Asn-TRP-VAL-PRO-Val-Ser-Val-Glu-Gly-TYR-TRP-GLN-Ile-THR-Leu-ASP-Ser-Ile-THR-Met-Asp-Gly-Glu-Thr-Ile-
CPG Tyr-TRP-VAL-PRO-Leu-Ser-Ala-Glu-Thr-TYR-TRP-GLN-Ile-THR-Met-ASP-Arg-Val-THR-Val-Gly-Asn-Lys-Tyr-Val-
CPC His-TRP-VAL-PRO-Val-Thr-Val-Gln-Gln-TYR-TRP-GLN-Phe-THR-Val-ASP-Ser-Val-THR-Ile-Ser-Gly-Val-Val-Val-

210 + 220

HPG ALA-CYS-Ser-Gly-Gly-CYS-GLN-ALA-ILE-Val-ASP-THR-GLY-THR-SER-Leu-LEU-Thr-Gly-PRO-Thr-Ser-Ala-Ile-Ala-
CPG ALA-CYS-Phe-Phe-Thr-CYS-GLN-ALA-ILE-Val-ASP-THR-GLY-THR-SER-Leu-LEU-Val-Met-PRO-Gln-Gly-Ala-Tyr-Asn-
CPC ALA-CYS-Glu-Gly-Gly-CYS-GLN-ALA-ILE-Leu-ASP-THR-GLY-THR-SER-Lys-LEU-Val-Gly-PRO-Ser-Ser-Asp-Ile-Leu-

230 240 250

HPG Ile-Asn-ILE-Gln-Ser-Asp-Ile-GLY-Ala-Ser-Glu-Asn-Ser-Asp-Gly-GLU-Met-Val-ILE-Ser-CYS-Ser-Ser-Ile-Asp-
CPG - -Arg-ILE-Ile-Lys-Asp-Leu-GLY-Val-Ser- - - - -Ser-Asp-Gly-GLU- - - - -ILE-Ser-CYS-Asp-Asp-Ile-Ser-
CPC - -Asn-ILE-Gln-Gln-Ala-Ile-GLY-Ala-Thr-Gln-Asn-Gln-Tyr-Asp Gly-GLU-Phe-Asp-ILE-Asp-CYS-Asp-Asn-Leu-Ser-

260 270

HPG Ser-Leu-PRO-Asp-Ile-Val-PHE-Thr-ILE-ASN-GLY-Val-Gln-Tyr-Pro-LEU-Ser-Pro-SER-ALA-TYR-Ile-Leu-Gln-Asp-
CPG Lys-Leu-PRO-Asp-Val-Thr-PHE-His-ILE-ASN-GLY-His-Ala-Phe-Thr-LEU-Pro-Ala-SER-ALA-TYR-Val-Leu-Asn-Glu-
CPC Tyr-Met-PRO-Thr-Val-Val-PHE-Glu-ILE-ASN-GLY-Lys-Met-Tyr-Pro-LEU-Thr-Pro-SER-ALA-TYR-Thr-Ser-Gly-Asp-

280 290 300

HPG Asp-Asp-Ser-CYS-Thr-Ser-GLY-PHE-Glu-Gly-Met-Asp-Val-Pro-Thr-Ser-Ser-Gly-Glu-Leu-TRP-ILE-LEU-GLY-ASP-
CPG Asp-Gly-Ser-CYS-Met-Leu-GLY-PHE-Glu-Asn-Met-Gly-Thr-Pro-Thr-Glu-Leu-Gly-Glu-Gln-TRP-ILE-LEU-GLY-ASP-
CPC Gln-Gly-Phe-CYS-Thr-Ser-GLY-PHE-Gln-Ser-Glu-Asn- - - - - - -His-Ser- - -Gln-Lys-TRP-ILE-LEU-GLY-ASP-

310 320

HPG VAL-PHE-ILE-ARG-Gln-TYR-TYR-Thr-Val-PHE-ASP-ARG-ALA-ASN-ASN-Lys-VAL-GLY-LEU-Ala-Pro-Val-Ala
CPG VAL-PHE-ILE-ARG-Glu-TYR-TYR-Val-Ile-PHE-ASP-ARG-ALA-ASN-ASN-Lys-VAL-GLY-LEU-Ser-Pro-Leu-Ser
CPC VAL-PHE-ILE-ARG-Glu-TYR-TYR-Ser-Val-PHE-ASP-ARG-ALA-ASN-ASN-Leu-VAL-GLY-LEU-Ala-Lys-Ala-Ile

Figure 6. Alignment of amino acid sequences of hog pepsinogen (HPG, ref. 32-35), chicken pepsinogen (CPG, ref. 23), and calf prochymosin (CPC, ref. 36). Residues identical in all three zymogens are set in CAPITAL letters. The pepsin numbering system (cf. 24) is used.

ing mammalian counterpart. While being different in several respects, such as e.g. pH-optimum of activity and stability, activity towards substrates, from the best known representative of the mammalian gastric enzymes, hog pepsin, its structure features (Asx/Glx and Ile/Leu ratio) resemble more those of pepsin A type proteinases rather than gastricsins. The sequence homology of chicken pepsin(ogen) with hog pepsinogen and with calf prochymosin is almost the same and thus sequence resemblance may hardly serve as a classification criterion. From their studies the Japanese authors (25) concluded that the main pepsinogen of Japanese quail is similar to chicken pepsinogen C (6). In the present study we failed, however, to find any "pepsinogen C" in our material.

From the results of this study the existence of a chicken pepsinogen component equivalent to mammalian progastricsin (pepsinogen C) is dubious. Any conclusions suggesting a close similarity between avian pepsins and gastricsin (25) rest therefore on uncertain grounds. Clearly, more work will be needed to cast light on some of the peculiar features of the molecules of avian pepsins, such as the free SH-group in chicken pepsin and the oligosaccharide moieties in chicken and duck pepsin.

References

1. Hammarsten, O.: Z. Physiol. Chem. 56, 18-80 (1908).
2. Herriott, R.M., Bartz, Q.R., Northrop, J.H.: J. Gen. Physiol. 24, 213-246 (1938).
3. Ginodman, L.M.: Akad. Med. Nauk SSSR, Inst. Biol. i. Med. Khim. 2, 54-70 (1962).
4. Levchuk, T.P. Orekhovich, V.N.: Biokhimiya 28, 1004-1010 (1963).
5. Bohak, Z.: J. Biol. Chem. 244, 4638-4648 (1969).
6. Donta, S.T., Van Vunakis, H.: Biochemistry, 9, 2791-2797 (1970).

7. Green, M.L., Llewellin, J.M.: Biochem. J. 133, 105-115 (1973).

8. Ryle, A.P.: Biochem. J. 96, 6-16 (1965).

9. Etherington, D.J., Taylor, W.H.: Nature 216, 279-280 (1967).

10. Becker, R., Schechter, Y., Bohak, Z.: FEBS Lett. 36, 49-52 (1973).

11. Gutfeld, M., Rosenfeld, P.P.: Dairy Industries 40, 52-55 (1975).

12. Bohak,Z.: Methods Enzymol. 19, 347-358 (1970).

13. Sluyterman, L.A.Ae., Wijdenes, J.: Biochim. Biophys. Acta 200, 593-595 (1970).

14. Keilová, H., Kostka, V.: Collect. Czech. Chem. Commun. 40, 574-579 (1975).

15. Peanasky, R.J., Abu-Erreish, G.M.: In Proc. Int. Res. Conf. Proteinase Inhibitors (Fritz, H., Ed.), pp. 281-293, W. de Gruyter, Berlin-New York 1971.

16. Keilová, H., Salvetová, A., Kostka, V.: Collect. Czech. Chem. Commun. 40, 580-584 (1975).

17. Pohl, J., Zaoral, M., Jindra Jr., A., Kostka, V.: Anal. Biochem. 139, 265-271 (1984).

18. Tomášek, V., Pohl, J., Kostka, V., Gan-Erdene, T., Parevsuren, B., Dorzhpalam, B.: this volume, pp. 73-78.

19. Pichová, I., Pohl, J., Štrop, P., Kostka, V.: this volume, pp. 301-308.

20. Pohl, J., Baudyš, M., Kostka, V.: Anal. Biochem. 133, 104-109 (1983).

21. Herriott, R.M.: J. Gen. Physiol. 24, 325-328 (1941).

22. Kostka, V., Keilová, H., Grüner, K., Zbrožek, J.: Collect. Czech. Chem. Commun. 42, 3691-3704 (1977).

23. Baudyš, M., Kostka, V.: Eur. J. Biochem. 136, 89-99 (1983).

24. Foltmann, B.: Essays in Biochem. 17, 52-84 (1981).

25. Esumi, H., Yasugi, S., Mizuno, T., Fujiki, H.: Biochim. Biophys. Acta 611, 363-370 (1980).

26. Baudyš, M., Kostka, V., Grüner, K., Pohl, J.: Collect. Czech. Chem. Commun. 47, 709-718 (1982).

27. Shewale, J.G., Takahashi, T., Tang, J.: this volume, pp. 101-116.

28. Chapman, A., Li, E., Kornfeld, S.: J.Biol. Chem. 254, 10243-10249 (1979).

29. Schmidt, P.G., Tang, J.: personal communication.

30. Pierce, J.G., Parsons, T.F.: Ann. Rev. Biochem. 50, 465-495 (1981).

31. Neufeld, E.F., Ashwell, G.: In The Biochemistry of Glycoproteins and Proteoglycans (Lennarz, W.J., ed.) pp. 241-266, Plenum Press, New York 1981.

32. Ong , E.B., Perlmann, G.E.: J.Biol. Chem. 243, 6104-6109 (1968).

33. Pedersen, V.B., Foltmann, B.: FEBS Lett. 35, 255-256 (1973).

34. Morávek, L., Kostka, V.: FEBS Lett. 43, 207-211 (1974).

35. Sepulveda, P., Marciniszyn Jr., J., Liu, D., Tang, J.: J. Biol. Chem. 250, 5082-5088 (1975).

36. Foltmann, B., Pedersen, V.B., Jacobsen, H., Kaufmann, D., Wybrandt, G.: Proc. Natl. Acad. Sci. USA 74, 2321-2324 (1977).

PEPSINS OF YAK AND CAMEL. ISOLATION AND MOLECULAR CHARACTERISTICS

Vladimír Tomášek, Jan Pohl, Vladimír Kostka

Department of Biochemistry, Institute of Organic Chemistry and Biochemistry, Czechoslovak Academy of Sciences
166 10 Prague 6, Czechoslovakia

Tudevin Gan-Erdene, Badra Parevsuren, Banzaryn Dorzhpalam

Institute of Chemistry, Mongolian Academy of Science,
Ulan Baatar, People's Republic of Mongolia

Introduction

The pepsins of numerous species have been isolated and characterized (cf. (1) and this volume) yet there is no information on the gastric proteinases of less common mammals such as the two Mongolian endemic species, yak and two-humped camel. This study was undertaken in an effort to isolate the pepsins from these species and also to check the suitability of the affinity chromatography procedure developed in this laboratory and using a specific synthetic pepsin inhibitor as affinant for rapid isolation of pepsins from crude extracts.

Results

Material and Methods

Preparation of crude extracts. The stomach mucosa was cut off, washed with 0.9% NaCl, ground in a meat grinder, and homogenized in a blender. The homogenate was extracted with a 2.5-fold volume of 10 mM sodium phosphate buffer, pH 7, at 5 °C twice for 24 h. The suspension was centrifuged, the pooled

Aspartic Proteinases and their Inhibitors

supernatants were saturated to 70 % with ammonium sulfate and the precipitate was centrifuged off.

Isolation of pepsins by affinity chromatography. The crude precipitate was dissolved in dilute ammonia, pH 8, and dialyzed against the same solution. Before chromatography the pepsinogens were activated by adjusting the pH of the solution with 2M HCl to 2 (25 °C , 10 min). The insoluble part of the activation mixture was centrifuged off and the supernatant was diluted with an equal volume of 0.03 M sodium acetate buffer at pH 3.3, 0.5 M in NaCl. The diluted sample solution was applied to a 1 x 9.3 cm column of Sepharose 4B with immobilized synthetic inhibitor of aspartic proteinases of the pepsin type, Val-D-Leu-Pro-Phe-Phe-Val-D-Leu as affinant (VLPFFVL-Sepharose) (2). The proteins which had not been adsorbed were washed off with the same buffer. The pepsins bound to the column were eluted by 0.05 M sodium phosphate buffer, pH 6.43, containing 20 % (v/v) of dioxane. The affinity chromatography of yak and camel pepsin is shown in Figure 1 and 2, respectively. The active fractions were pooled and lyophilized. In one single affinity chromatography step the pepsins were enriched more than 180 times.

Molecular characteristics of yak and camel pepsin
SDS-PAGE of both pepsins is shown in Figure 3. The molecular weights determined by this procedure are 31 000 for yak pepsin and 33 000 for camel pepsin. Both enzymes behaved as homogeneous products on SDS-PAGE. Their amino acid composition is shown in Table 1. The most striking difference is in the content of basic amino acid residues which is much higher in camel pepsin. The isoelectric focusing pattern in a thin polyacrylamide gel layer over the range pH 2-5 (Pharmalyte) is shown in Figure 4. Both pepsins give 3 components during isoelectric focusing which rather than in the mobility (pI) differ in the intensity of the individual bands. The strongest

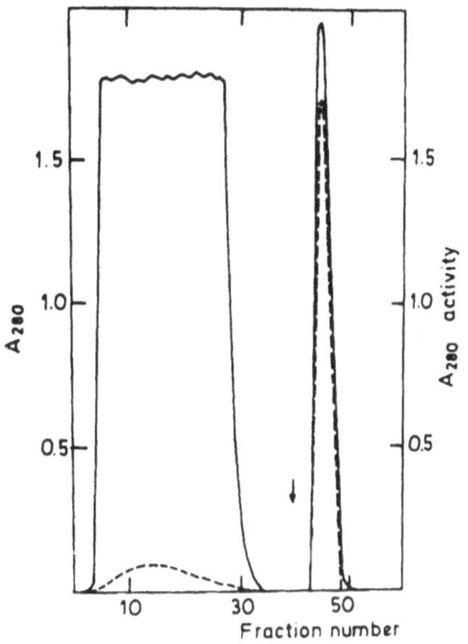

Figure 1. Chromatography of yak pepsin on VLPFFVL-Sepharose 4B. Column 1 x 9.3 cm, flow rate 20 ml h^{-1}. One-ml fractions were collected into 0.1 M acetic acid. —— $A_{280\ nm}$, ---- proteolytic activity (hemoglobin).

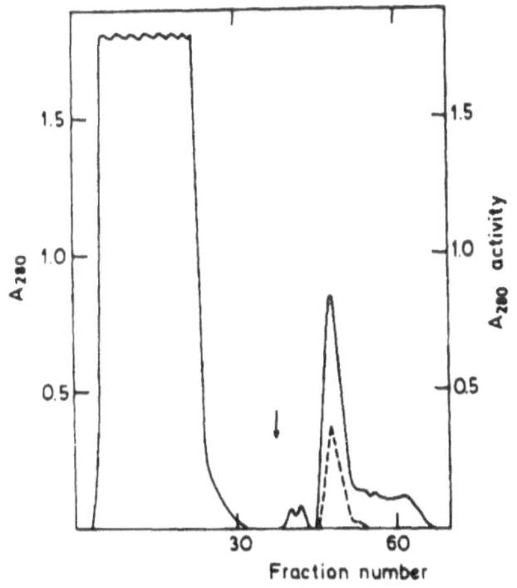

Figure 2. Chromatography of camel pepsin on VLPFFVL-Sepharose 4B. Conditions see legend to Figure 1.

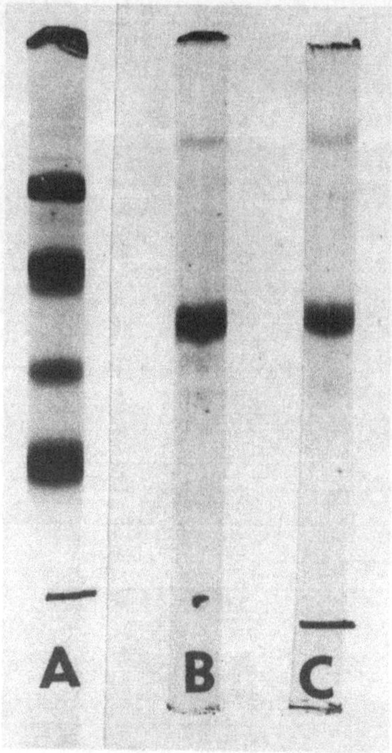

Figure 3. SDS-PAGE of yak and camel pepsin. A - reference mixture of proteins (from top to bottom) serum albumin, ovalbumin, chymotrypsinogen, ribonuclease, B - yak pepsin, C - camel pepsin

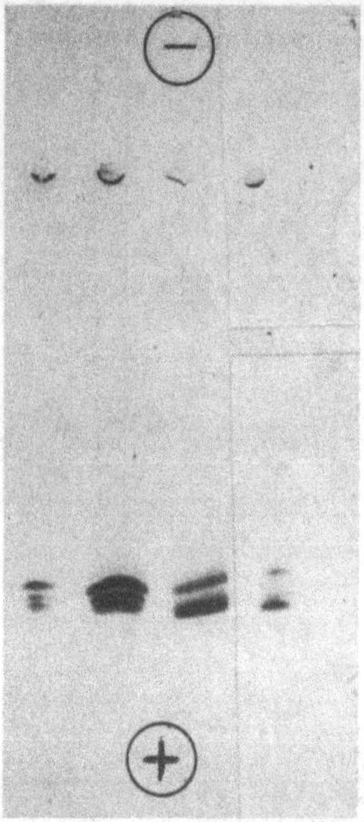

Figure 4. Isoelectric focusing of yak and camel pepsin (Pharmalyte pH 2-5), A,B - yak pepsin, C,D - camel pepsin

band in yak pepsin is the most acidic one whereas the corresponding band in camel pepsin is less acidic.

<u>Kinetic constants.</u> The K_m- and k_{cat}-values were determined by the method of Lineweaver and Burke from the initial reaction rates measured with the water-soluble chromophoric substrate <Glu-His-Nph-Phe-Ala-Leu.NH_2 (3). The hydrolysis was carried out in 0.1 M sodium acetate buffer, 1 mM in EDTA, at pH 4.9 and 30 °C. The concentration of the enzymes was 93.9 nM in each case, the substrate concentration varied between 0.071 mM and 0.495 mM in assays of yak pepsin, and between 0.094 mM-

0.472 mM in assays of camel pepsin.

Table 1

Amino acid composition of yak and camel pepsin

	residue/mol pepsin	
	yak	camel
Lysine	1	7
Histidine	1	2
Arginine	3	8
Aspartic acid	33	34
Threonine	22	21
Serine	40	39
Glutamic acid	30	42
Proline	14	10
Glycine	31	29
Alanine	14	20
Half-cystine	6	6
Valine	22	17
Methionine	2	4
Isoleucine	24	20
Leucine	19	25
Tyrosine	15	13
Phenylalanine	13	10
Tryptophan	4	3
Total	294	310

Table 2

Hydrolysis of <Glu-His-Nph-Phe-Ala-Leu.NH_2 by yak and camel pepsin

pepsin	K_m mM	k_{cat} s^{-1}	K_m/k_{cat} mM/s^{-1}
yak	0.81	14.4	17.5
camel	0.31	3.2	10.2

Conclusions

The pepsins were isolated from the gastric mucosa of two mammalian species so far not studied from yak and two-humped camel. Both enzymes were characterized by determination of molecular weight by SDS-PAGE, isoelectric focusing pattern, amino acid composition and kinetic constants of hydrolysis of a synthetic substrate, <Glu-His-Nph-Phe-Ala-Leu.NH_2.

Affinity chromatography on Sepharose 4B with immobilized synthetic inhibitor of aspartic proteinases of the pepsin type, Val-D-Leu-Pro-Phe-Phe-Val-D-Leu, is an efficient method of isolation of pepsins from crude mucosa extracts.

References

1. Pedersen, V.B., Foltmann, B.: In Acid Proteases, Structure, Function, and Biology (Tang, J., ed.), pp. 1-18, Plenum Press, New York 1977.
2. Pohl, J., Zaoral, M., Jindra, Jr., A., Kostka, V.: Anal. Biochem. 139, 265-271 (1984).
3. Pohl, J., Baudyš, M., Kostka, V.: Anal. Biochem. 133, 104-109 (1983).

MOLECULAR VARIANTS OF HUMAN ASPARTIC PROTEINASES

I. Michael Samloff, R. Thomas Taggart and Klaus J. Hengels

V.A. Medical Center, Research Service (151), Sepulveda, California 91343, U.S.A., and UCLA School of Medicine, Los Angeles, California

Introduction

Our current understanding about the heterogenity of the aspartic proteinases in human gastric mucosa has developed mainly during the past two decades. During this time, the results of studies by several groups of investigators have established a) that the mucosal lining of the human stomach contains three immunochemically distinct types of acidic proteinases (1, 2, 3), b) that each immunochemically distinct type consists of variants that differ in net ionic charge and/or molecular weight (2, 3, 4, 5, 6), c) that the three types differ to varying degrees in their biochemical characteristics (7, 8, 9), topology in gastroduodenal mucosa (2, 4, 5), cellular origins (10, 11, 12), and distribution in body fluids (13, 14, 15, 16, 17, 18), and d) that there is variation among individuals in both the number and concentration of the molecular variants due to genetic and nongenetic factors (19, 20, 21, 22, 23).

The intent of this paper is to review current information on the heterogenity of the aspartic proteinases in man and to present recent data indicating that the degree of heterogeneity is more extensive than previously thought.

Aspartic Proteinases and their Inhibitors

Overview of the Human Aspartic Proteinases

Nomenclature

At present, there is not complete agreement on the nomenclature for the aspartic proteinases in human gastric mucosa. The terminology for the three immunologically distinguishable types that has gained widest acceptance in the clinical literature is pepsinogen I (PG I), pepsinogen II (PG II) and slow moving protease (SMP) (24). The latter, as implied by its designation, is a non-pepsinogen proteinase. Two of its characteristics are that is not inactivated by sequential acidification and alkalinization (2, 5), and that it is less acidic than the pepsinogens. Consequently, the SMP has the slowest rate of anodal migration in agar gel, but is the first to elute from DEAE ion exchange media. The SMP has also been referred to as P IV (2), and as gastric cathepsin (8). PG I, which is homologous with pepsinogen A (Pg A) of other species, is normally the major acidic proteinase in gastric mucosal extracts. PG II is less acidic than PG I and is homologous with pepsinogen C (Pg C) of other species. It has also been referred to as gastricsinogen (25) and as progastricsin (26). The terms Pg A, Pg C and SMP will be used in this paper.

Early studies using ion exchange chromatography and agar gel electrophoresis suggested that Pg A consisted of two isozymogens that differed in net ionic charge (2, 3, 4), but later studies revealed additional heterogeneity. In 1969, Samloff and Townes (5) developed an improved electrophoretic technique and identified five isozymogens of Pg A in extracts of gastric mucosa (27). These were designated Pg 1 through Pg 5 in order of decreasing electrophoretic mobility (Figure 1). A sixth isozymogen of Pg A, which migrated cathodal to Pg 5, was subsequently identified in urine and was termed Pg 5s (28). These six isozymogens of Pg A will henceforth be referred to as Pg A1, A2, A3, A4, A5 and A6, respectively.

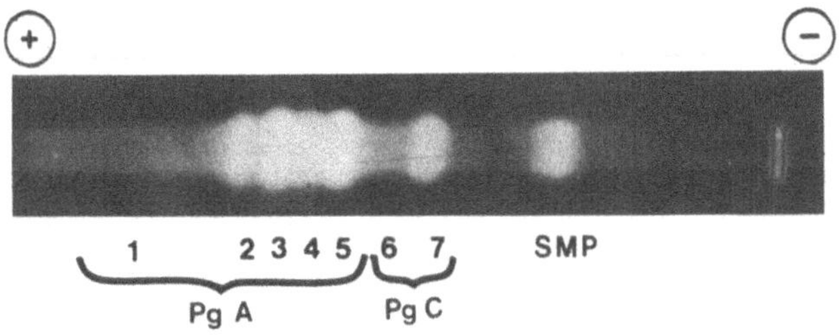

Figure 1. Zymogram of acidic proteinases in an extract of human gastric mucosa obtained by electrophoretic analysis in agar gel. Note that while Pg A2 - A5 appear as discrete bands, Pg A1 is characterized by a diffuse zone of proteolytic activity.

In early studies, Pg C appeared to be chromatographically and electrophoretically homogeneous (2, 4), but it was later resolved by agar gel electrophoresis into two isozymogens, designated Pg 6 and Pg 7 (5). These will henceforth be referred to as Pg C6 and C7, respectively (Figure 1). Recent information on additional electrophoretic and chromatographic heterogeneity of Pg A and Pg C will be presented below.

Cellular origins in gastroduodenal mucosa

Immunohistochemical studies using antisera specific for Pg A and Pg C have shown that Pg A is produced by the chief and mucus neck cells in fundic gland mucosa (10), whereas Pg C is produced by these cells and by three types of clear staining glands. These are the cardiac glands in the gastric cardia, the pyloric glands in the gastric antrum, and Brunner's glands in the proximal duodenum (11, 12). The SMP appears to be produced by the surface epithelial cells (29). Thus the aspartic proteinases in human gastroduodenal mucosa are derived from six cell types.

Distribution in body fluids

The mechanism by which pepsinogens are translocated from the peptic cells into the circulation is not known with certainty, but both Pg A and Pg C are present in serum (14, 30, 31). In contrast, urine normally contains only Pg A (13, 14, 15) and seminal fluid contains only Pg C (3, 16, 17). Pg C in seminal fluid appears to arise from the prostate and the seminal vesicles (26), but it is immunologically, electrophoretically, and chromatographically indistinguishable from Pg C in gastric mucosa (3, 16, 17).

Genetic polymorphism

Early studies revealed that some individuals lack Pg A5 in their urine, and that the absence of Pg A5 in urine is associated with an absence of this isozymogen in gastric mucosa (14). Conversely, individuals with Pg A5 in their urine also had Pg A5 in their mucosa. A subsequent population and family study demonstrated that about 14% of the Caucasian population in the United States lack Pg A5, and showed that this deficiency, termed phenotype B, is inherited as an autosomal recessive trait (19). The pepsinogen B phenotype appears to be present in about 20% in United States Blacks (32). In contrast, populations originating in East Asia are all phenotype A, i.e. none lack Pg A5 (33).

In addition to the absence of Pg A5, there are also differences among individuals with respect to the relative concentrations of the Pg A isozymogens in both urine and gastric mucosa (14, 20, 34). This variation has also been shown to be a genetic trait (21, 35). Taggart et al. (20) identified nine phenotypes in a United States population, all of which could be explained by the codominant expression of four alleles at a single genetic locus (Figure 2). However, additional

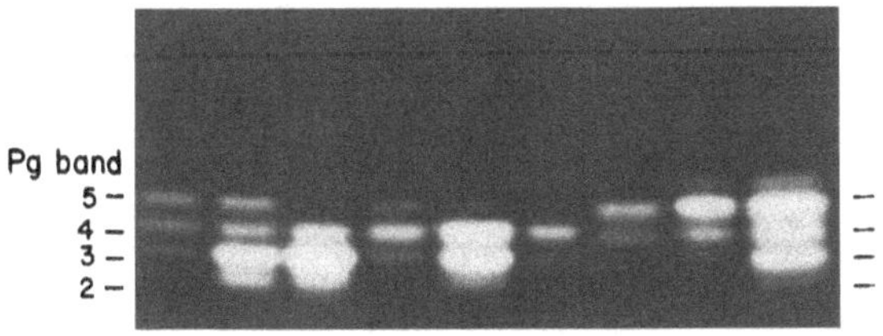

Figure 2. Zymogram of acidic proteinases in human urine obtained by electrophoretic analysis in polyacrylamide gel. Nine pepsinogen phenotypes are shown.

phenotypes were found in European populations (21) that could not be explained by this or any previous genetic model, and Frants et al. (36) have recently proposed that Pg A3, A4, and A5 are coded by separate genes and that the concentration of each is determined principally by the number of gene copies.

There has been little work on pepsinogen polymorphism in other species, but such has been shown in the mouse (37) and in the Wistar rat, but not in several other strains of rat (38).

Recent studies have also shown that there is not always perfect concordance between the Pg A phenotype pattern in urine and gastric mucosa (34). The most common discrepancy is a difference in the apparent concentration of Pg A5 relative to that of the other isozymogens. The reason for the discordance has not been established. Genetic polymorphism of the Pg C isozymogens has not been reported.

There is limited information on the association of the pepsinogen phenotypes with clinical disorders in man. There are two reports of an inreased frequency of phenotype A in patients with duodenal ulcer (39, 40), and one of an increased prevalence of phenotype A in patients with gastric cancer (41). An increased prevalence of phenotypes characterized by a strong Pg A5 band has also been reported recently in patients with established gastric cancer and in those with premalignant disorders of the gastric mucosa (42).

Other genetic and nongenetic factors

The concentrations of Pg A and Pg C in both gastric mucosa and serum have been reported to be higher than normal in patients with duodenal ulcer (43). Furthermore, in some families with a prominent history of duodenal ulcer, an elevated serum Pg A level has been found to be inherited as an autosomal dominant trait (22).

In pernicious anemia, a disorder characterized by a loss of the Pg A-producing chief cells and an increase of Pg C-producing pyloric glands, there is a marked decrease in serum Pg A levels, but serum Pg C levels are within the normal range (30). Lesser degrees of gastritis are also associated with nonparallel changes in serum Pg A and Pg C levels, and with a decrease in the ratio of Pg A to Pg C (23).

Additional Variants of the Human Aspartic Proteinases

Gel filtration and electrophoretic studies

Most of our information about the heterogeneity of the aspartic proteinases in gastric mucosa is based on the results of studies in which agar gel electrophoresis was used to resolve the molecular variants. Because of the large pore size of agar gel, the rates of migration of the aspartic proteinases in this medium is dependent primarily on their net ionic charge. Thus, the six isozymogens of Pg A and the two isozymogens of Pg C that have been demonstrated by this analytic technique can be thought of as "charge" isomers.

There is, however, evidence for the existence of both "charge" and "size" isomers of the aspartic proteinases. Agar gel electrophoresis of eluate fractions of gastric mucosal extracts obtained by gel filtration on both Sephadex G-100 and Bio-Gel A-0.5m has shown that Pg A1 elutes ahead of Pg A2 - A5 (5, 30). Furthermore, the elution profile of Pg A1 is characterized by the stepwise emergence of proteolytic zones that exhibit a progressive decrease in anodal mobility with decreasing molecular weight (Figure 3). Thus, Pg A1 appears to be extraordinarily heterogeneous, and _in toto_, to have both a higher molecular weight and a higher acidic charge than the other isozymogens of its group. In addition to Pg A1, the zymograms of the eluate fractions also revealed two zones of proteolytic activity that emerged from the column

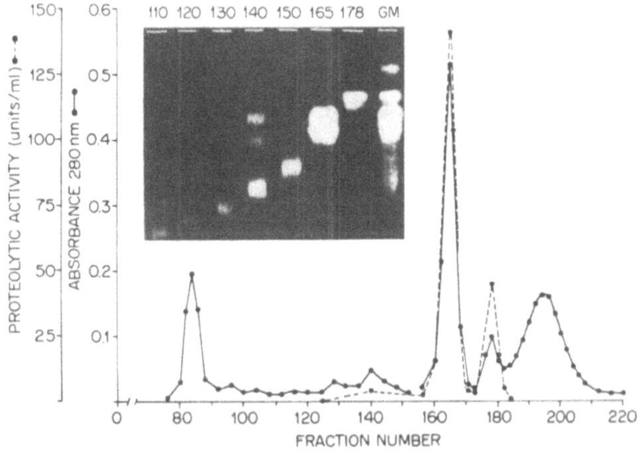

Figure 3. Gel exclusion chromatography of a partially purified preparation of acidic proteinases on Bio-Gel A-0.5m. Inset. Zymograms of selected eluate fractions and of the original extract of gastric mucosa (GM). Pg A1 is seen in eluate fractions 110 through 150 as a series of proteolytic zones that exhibit a progressive decrease in anodal mobility with decreasing molecular weight. There is a small peak of acid proteinase activity in eluate fraction 140 which contains, in addition to a component of Pg A1, two zones of proteolytic activity with electrophoretic mobilities similar to Pg A2 - A5.

ahead of Pg A2 - A5, but which migrated toward the anode at essentially the same rates as Pg A2 - A5 (Figure 3).

To further investigate the latter two proteolytic zones and the possible existence of additional molecular variants of the aspartic proteinases, we analyzed an extract of gastric mucosa, urine and seminal fluid by both agar gel and polyacrylamide gel electrophoresis. The latter was performed as described previously by Taggart et al. (44). The zymograms produced by electrophoresis of the samples in agar gel and in polyacrylamide gel are shown in Figures 4 and 5, respectively.

The results obtained by the two analytic techniques differed in several respects. One difference is that by polyacrylamide gel electrophoresis, the mucosal extract is seen to contain seven proteolytic bands that migrate cathodal to Pg C7 (Figure 5). These are made up of two subgroups; a more anodal subgroup, which contains four bands, and a more cathodal subgroup, which consists of three bands. The aspartic proteinases represented by these seven proteolytic bands have been shown to correspond to the two zones of proteolytic activity in gastric mucosa that were previously demonstrated to elute from Bio-Gel A ahead of Pg A2 - A5, but to have, in agar gel, similar rates of anodal migration as Pg A2 -A5 (see Figure 3). The finding that they migrate cathodal to Pg A2 - A5 in polyacrylamide gel is in keeping with the fact they have higher molecular weights than Pg A2 - A5 and the sieving effect of this medium.

Conversely, the apparent absence of these proteolytic zones in the zymogram of the mucosal extract analyzed by agar gel electrophoresis can be explained by the lack of a sieving effect of this medium, and the fact that their net ionic charge is similar to that of Pg A2 - A5. Thus, they comigrate with Pg A2 - A5 in agar gel and are obscured by the latter isozymogens. Since this group of aspartic proteinases has also been shown to react with antiserum to Pg A, but not with antiserum to Pg C, it has been referred to as big-Pg A (6).

Several other differences in the zymograms obtained by agar gel and polyacrylamide gel electrophoresis can also be explained by interactions between the characteristics of each gel and the known differences in the molecular weights and acidic charges of the aspartic proteinases. One such difference is that Pg A1 is evident in the zymogram of the gastric mucosal extract obtained by agar gel electrophoresis (Figure 4), but not in the zymogram obtained by polyacrylamide gel electrophoresis (Figure 5). This has been shown, in

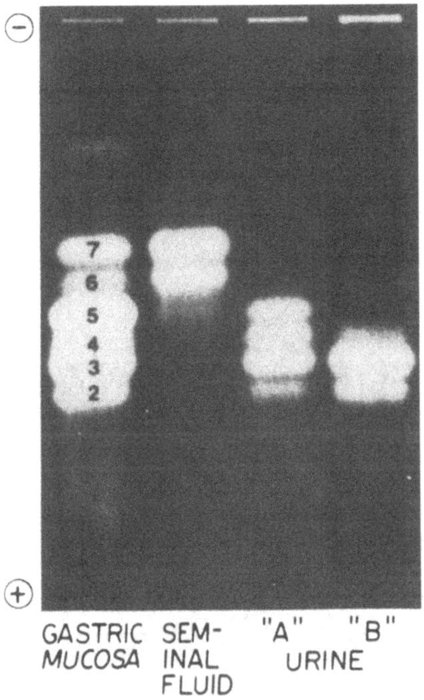

Figure 4. Zymograms of acidic proteinases in an extract of gastric mucosa, phenotype A and B urines, and seminal fluid obtained by agar gel electrophoresis.

separate studies, to result from retardation of the anodal mobility of Pg A1 in polyacrylamide gel. In this medium, Pg A1 migrates as a broad zone that coincides with the anodal mobilities of Pg A2 - A5 and Pg C6 and C7. It is therefore obscured by these proteinases.

Another difference between the two sets of zymograms is that in agar gel, Pg C6 in seminal fluid migrates cathodal to Pg A5 in urine (Figure 4), whereas in polyacrylamide gel Pg C6 and Pg A5 have similar rates of anodal migration (Figure 5). Thus, in the zymogram of the gastric mucosal extract obtained by polyacrylamide gel electrophoresis, Pg A5 and Pg C6 appear to be represented by a single proteolytic band, which is labelled "5,6" (Figure 5). That this interpretation is

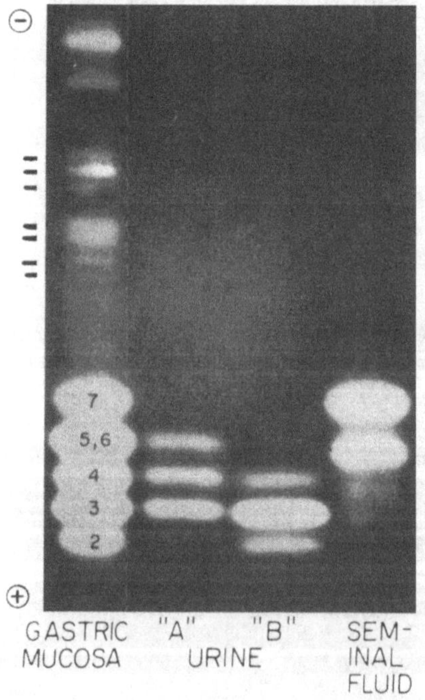

Figure 5. Zymograms obtained by polyacrylamide gel electrophoresis of the same samples shown in Figure 4.

correct has been shown by studies using antisera to Pg A and Pg C to identify each component within the single proteolytic band. The increase in anodal mobility of Pg C6 relative to that of Pg A5 in polyacrylamide gel can be explained by the fact that Pg C6 has a lower molecular weight than Pg A5 (30). It is appropriate to note at this point that Pg A6 and Pg C6 comigrate in agar gel and produce a single proteolytic band. The two components of this band can also be distinguished by immunochemical techniques.

Yet another difference between the zymograms is that in polyacrylamide gel, the specimen of seminal fluid is seen to contain, in addition to Pg C6 and C7, two proteolytic bands that migrate anodal to Pg C6 (Figure 5). These have been shown to react with antiserum to Pg C, but not with antiserum to Pg A, indicating that they represent additional isozymogens of Pg C. In agar gel, these bands appear as a diffuse, poorly demarcated zone of proteolytic activity. The difference between the two zymograms, in this instance, is probaly due to the better resolution that can be obtained in polyacrylamide gel using a discontinuous buffer system and a stacking gel.

Lastly, the SMP appeared to consist of two proteolytic bands by both agar gel and polyacrylamide gel electrophoresis (Figures 4 and 5). In purified preparations of SMP, however, four discrete bands have been resolved by polyacrylamide gel electrophoresis.

Chromatographic and immunochemical studies

The results of recent studies in our laboratory have also shown that there are two molecular variants of Pg A3 and of Pg A4. The initial evidence came from ion exchange chromatography of a purified sample of Pg A. Gradient elution of the sample resulted in five protein peaks (Figure 6).

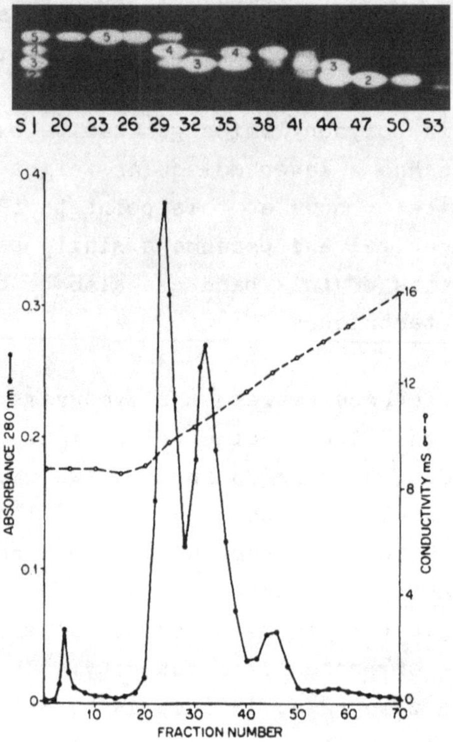

Figure 6. Gradient elution of a partially purified preparation of Pg A. Inset. Zymograms of selected eluate fractions. See text for details.

Polyacrylamide gel electrophoresis of the eluate fractions revealed that the first peak was devoid of proteolytic activity, that the second peak (Fractions 20, 23 and 26) contained mainly Pg A5, that the trough between the second and third peak contained mainly Pg A4 (Fraction 29), and that the ascending limb of the third peak contained mainly Pg A3 (Fraction 32). However, the descending limb of the third peak (Fractions 35 and 38) was found to contain a second Pg A4 band, while the ascending limb of the fourth peak (Fractions 41 and 44) was found to contain a second Pg A3 band. Pg A2 eluted mainly in the descending limb of the fourth peak and continued into the small fifth peak (Fractions 44, 47, 50 and 53).

The second line of evidence for the existence of two variants of Pg A3 and A4 came from studies with murine monoclonal antibodies. The monoclonal antibodies were made against a preparation containing Pg A2 - A5 (45), and were tested by immunoabsorption to determine if any might be specific for only some of the Pg A isozymogens. The test procedure consisted of incubating each antibody with a sample of Pg A and then subjecting the mixture to polyacrylamide gel electrophoresis. In the presence of antibody excess, the disappearance of an isozymogen from the zymogram indicates that it has reacted with the antibody. Whereas most of the monoclonal antibodies resulted in the complete removal of all the Pg A isozymogens, a few, exemplified by antibody II-219, produced a pattern characterized by the complete loss of Pg A2 and Pg A5, but only a partial loss of Pg A3 and Pg A4 (46). The residual isozymogen pattern was identical to that seen in individuals with pepsinogen phenotype CC (Figure 2).

Further immunabsorption experiments were then carried out to examine the possible relationship between the two variants of Pg A3 and A4 that had been identified by ion exchange chromatography (see Figure 6) and the two variants that were identified by immunabsorption. These studies revealed that the II-219 monoclonal antibody is specific for the first Pg A3 variant and the second Pg A4 variant that elutes from the column. Thus, the Pg A isozymogens can now be divided into two immunologically distinct subgroups, which we have termed α-Pg A and β-Pg A (47). The α-Pg A isozymogens react with the II-219 antibody and comprise Pg A2, A3α, A4α, and A5. The β-Pg A izozymogens consist of Pg A3β and A4β. In addition, Pg A6 has also been found to belong to the β-subgroup.

Summary and Conclusions

A large body of information is now available on the heterogeneity of the aspartic proteinases in human gastric mucosa. Immunochemical studies using polyclonal antisera have shown that there are three immunologically distinguishable types, while chromatographic and electrophoretic studies have demonstrated that each type consists of multiple molecular variants that differ in net ionic charge and/or molecular weight. It has also been shown that no single analytic technique is able to resolve all of the variants. Moreover, Pg A, Pg C and SMP have been shown to differ in their cellular origins, and Pg A and Pg C have been found to differ in their distribution in body fluids. In addition, population studies have shown that there is marked variability among persons in both the number and the concentration of the Pg A isozymogens, and family studies have shown that this variability is due, in large part, to genetic factors. Genetic and nongenetic factors have also been found to influence Pg A and Pg C levels in gastric mucosa and in the circulation. Recent studies using monoclonal antibodies, have uncovered additional heterogeneity by showing a single epitope difference between two of the isozymogens of Pg A. It is likely that additional molecular variants remain to be discovered and that future studies will delineate the molecular bases for the extensive variation of the aspartic proteinases in human gastric mucosa.

Acknowledgement

This work was supported in part by Research Grant AM 32015 from the National Institutes of Health, Bethesda, Maryland.

References

1. Rapp, W., Aronson, S.B., Burtin, P., Grabar, P.: J. Immunol. 92, 579-595 (1964).
2. Kushner, I., Rapp, W., Burtin, P.: J. Clin. Invest. 43, 1983-2002 (1964).
3. Hirsch-Marie, H.: Biol. Gastroenterol. 2, 109-122 (1968).
4. Seijffers, M.J., Segal, H.L., Miller, L.L.: Am. J. Physiol. 205, 1099-1105 (1963).
5. Samloff, I.M.: Gastroenterology 57, 659-669 (1969).
6. Hengels, K.J., Taggart, R.T., Samloff, I.M.: Gastroenterology 82, 1083 (1982).
7. Tang, J., Mills, J., Chiang, L., de Chiang, L.: Ann. N.Y. Acad. Sci. 140, 688-696 (1967).
8. Mangla, J.C., Guarasci, G., Turner, M.D.: Biochem. Med. 10, 83-96 (1974).
9. Roberts, N.B., Taylor, W.H.: Biochem. J. 169, 617-624 (1978).
10. Samloff, I.M.: Gastroenterology 61, 185-188 (1971).
11. Samloff, I.M., Liebman, W.M.: Gastroenterology 65, 36-42 (1973).
12. Weinstein, W., Lechago, J., Samloff, I.M., Bowes, K.L.: Clin Res 25, 690A (1977).
13. Seijffers, M.J., Segal, H.L., Miller, L.L.: Am. J. Physiol. 206, 1106-1110 (1964).
14. Samloff, I.M., Townes, P.L.: Gastroenterology 58, 462-469 (1970).
15. Hirsch-Marie, H., Conte, M.: Rev. Franc. Etudes Clin. Biol. 14, 977-983 (1969).
16. Seijffers, M.J., Miller, L.L., Segal, H.L.: Proc. Soc. Exp. Biol. Med. 118, 405-409 (1965).
17. Samloff, I.M., Liebman, W.M.: Clin. Exp. Immunol. 11, 405-414 (1972).
18. Liebman, W.M., Samloff, I.M.: Biol. Neonat. 33, 174-176 (1978).
19. Samloff, I.M., Townes, P.L.: Science 168, 144-145 (1970).
20. Taggart, R.T., Karn, R.C., Merritt, A.D., Yu, P.L., Conneally, P.M.: Hum. Genet. 52, 227-238 (1979).
21. Korsnes, L., Gedde-Dahl, T.Jr.: Ann. Hum. Genet. 43, 199-212 (1980).

22. Rotter, J.I., Sones, J.W., Samloff, I.M., Richardson, C.T., Gursky, J.M., Walsh, J.H., Rimoin, D.L.: N. Engl. J. Med. 300, 63-65 (1979).

23. Samloff, I.M., Varis, K., Ihamaki, T., Siurala, M., Rotter, J.I.: Gastroenterology 83, 204-219 (1982).

24. Samloff, I.M.: Gastroenterology 60, 586-604 (1971).

25. Foltmann, B., Jensen, A.L.: Eur. J. Biochem. 128, 63-70 (1982).

26. Chiang, L., Contreras, L., Chiang, J., Ward, P.H.: Arch. Biochem. Biophys. 210, 14-20 (1981).

27. Samloff, I.M.: J. Immunol. 106, 962-968 (1971).

28. Gedde-Dahl, T.Jr., Korsnes, L., Thorsby, E., Olaisen, B., Bratlie, A, Siverts, A.: Cytogenet. Cell Genet. 22, 301-303 (1978).

29. Hirsch-Marie, H., Loisillier, F., Touboul, J.P., Burtin, P.: Lab. Invest. 6, 623-632 (1976).

30. Samloff, I.M.: Gastroenterology 82, 26-33 (1982).

31. Ichinose, M., Miki, K., Furihata, C., Kageyama, T., Hayashi, R., Niwa, H., Oka, H., Matsushima, T., Takahashi, K.: Clin. Chim. Acta 126, 183-191 (1982).

32. Townes, P.L., White, M.R.: Am. J. Hum. Genet. 26, 252-254 (1974).

33. Samloff, I.M., Liebman, W.M., Glober, G.A.: Am. J. Hum. Genet. 25, 178-180 (1973).

34. Petersen, G.M., Rotter, J.I., Samloff, I.M.: In: Pepsinogens in Man. Clinical and Genetic Advances. Editors: J. Kreuning, I.M. Samloff, J.I. Rotter, A.W. Eriksson. Alan R. Liss, New York (in press).

35. Taggart, R.T., Yu, P.L., Karn, R.C., Conneally, P.M., Merritt, A.D.: Cytogenet. Cell Genet. 22, 335-340 (1978).

36. Frants, R.R., Pronk, J.C., Pals, G., Defize, J., Westerveld, B.D., Meuwissen, S.G.M., Kreuning, J., Eriksson, A.W.: Hum. Genet. 65, 385-390 (1984).

37. Szymura, J.M., Taylor, B.A., Klein, J.: Biochem. Genet. 20, 1211-1219 (1982).

38. Muto, N., Tani, S.: Biochem. Genet. 20, 1189-1193 (1982).

39. Samloff, I.M., Cole, D.: Gastroenterology 68, 980 (1975).

40. Ellis, A., McConnell, R.B.: Gastroenterology 83, 1261-1263 (1982).

41. Ellis, A., Hughes, S., McConnell, R.B.: Br. J. Cancer 46, 289-290 (1982).

42. Westerveld, B.D., Pals, G., Defize, J., Pronk, J., Frants, R.R., Kreuning, J., Eriksson, A.W., Meuwissen, S.G.M.: In: Pepsinogens in Man. Clinical and Genetic Advances. Editors: J. Kreuning, I.M. Samloff, J.I. Rotter, A.W. Eriksson. Alan R. Liss, New York (in press).

43. Libman, L.J., Samloff, I.M.: Gut 19, A998 (1978).

44. Taggart, R.T., Miller, R.B., Karn, R.C., Trimble, J.A., Craft, M., Ripberger, J., Merritt, A.D.: In: Proceedings of Electrophoresis '78. Editor: N. Catsimpoolas. Elsevier, New York 1978, 231-242.

45. Taggart, R.T., Samloff, I.M.: Science 219, 1228-1230 (1983).

46. Taggart, R.T., Samloff, I.M.: Am. J. Human Genet. 34, 64A (1982).

47. Taggart, R.T., Samloff, I.M.: Gastroenterology 86, 1272 (1984).

HUMAN PEPSINS 1 AND 2 ('FAST PEPSINS'): HETEROGENEITY AND CARBOHYDRATE CONTENT

Andrew P. Ryle
Department of Biochemistry, University of Edinburgh

Bent Foltmann
Institute for Biochemical Genetics, Copenhagen

Introduction

Human pepsins 1 and 2 have been described (1) as two minor proteolytic components of human gastric juice that migrate faster than the major enzyme, pepsin 3, on electrophoresis in agar gel at pH 5. Pepsin 1 is of particular interest because of a correlation between its concentration in gastric juice and peptic ulceration (2) and because of its high activity in digesting collagen (3) and gastric mucoprotein (4). Pepsins 1 and 2 have been isolated from gastric juice by chromatography on DEAE-cellulose (5); their enzymic activities are very similar to that of pepsin 3 (5,6).

It is known that human pepsins and their zymogens fall into two immunologically distinct groups (7) and that pepsins 3 and 5 have different, but related, amino acid sequences (8). Little else is known of the chemical differences between the human pepsins. We report that a fraction of human gastric juice containing pepsins 1 and 2 comprises at least six proteolytic components associated with acidic carbohydrate.

Results

On chromatography of gastric juice on DEAE-Sephadex A-50, or DEAE-

Aspartic Proteinases and their Inhibitors

Sepharose CL6B at pH 4 with a NaCl gradient pepsins 5, 4, 3 and 3a are eluted before 0.25 M-NaCl. At about 0.3 M-NaCl a small irregular peak of enzymic activity is eluted. This report is mainly concerned with such material obtained from a chromatography of four pooled gastric juices from patients undergoing acid-output tests, though similar material has been obtained from several other chromatograms. After exclusion chromatography on Sephadex G-100 133 mg of active material (here called 'fast pepsins') was recovered and had $A_{280} = 0.22\ cm^2.mg^{-1}$.

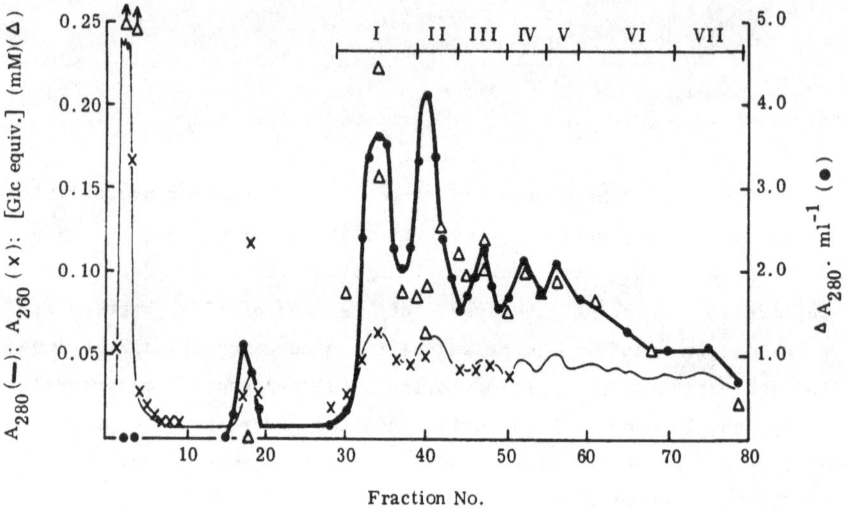

Fig. 1 Rechromatography of 'fast pepsins' at pH 3.1. DEAE-Sepharose 4B, 0.9 x 12 cm; 25 mM-citrate; 5.5 ml fractions, 22 ml/h; linear gradient from 0 to 0.55 M-NaCl. The bars indicate fractions pooled as 'fast pepsin' subfractions FPI to FPVII. Yield: a few mg of each.

Rechromatography of 100 mg of this material (Fig. 1) gave peaks of activity accompanied by peaks of carbohydrate (9), although much carbohydrate was eluted before the enzymes. Successive subfractions contained enzymes of increasing mobility at pH 3, but only FPVI and FPVII, which were identical, contained a single enzyme. Altogether six enzymes were detectable.

The 'fast pepsin' subfractions showed reactions of identity with pepsin 3, whether tested by immunodiffusion or by rocket electrophoresis with antiserum raised against pepsin 3 or against the whole 'fast pepsin' fraction.

Fig. 2 Effect of treatment with hyaluronidase.
Two different preparations of 'fast pepsins' and normal gastric juice were electrophoresed at pH 4 without (-) and with (+) previous incubation with testicular hyaluronidase.

Treatment of the whole 'fast pepsin' fraction with testicular or Streptococcal hyaluronidase markedly decreased its mobility (Fig. 2); neuraminidase and potato phosphatase had no effect. The figure also shows the electrophoresis of a second preparation of 'fast pepsins' from three different juices. It lacks the fastest-moving components of the first preparation, but after treatment with hyaluronidase the two preparations become indistinguishable. When examined by agar gel electrophoresis at pH 5 the whole first 'fast pepsin' preparation and FPVII behaved as pepsin 1, FPII as pepsin 2, and the second preparation as mainly pepsin 2 with some pepsin 1. (We are grateful to Dr. W.H. Taylor who examined some samples for us and confirmed these findings).

Table 1. Some analytical data

Carbohydrate contents are expressed as moles per mole protein, assuming that the protein portions have M_r 37000 and $A_{280} = 1.3\ cm^2.mg^{-1}$.

	Total hexose (9)	Uronic Acid (11)	Amino-sugar (12)	Molecular Weight no hy'ase	after hy'ase
FPII	62	26	79	n.d.	n.d.
FPIII	41	20	24	44000	41000
FPIV	43	22	29	45500	41000
FPV	46	46	n.d.	n.d.	n.d.

The presence of uronic acids and amino sugars, in some fractions in nearly equimolar ratios, (Table 1) is characteristic of hyaluronic acid. FPIV, tested with Azure A (10) gave a distinct metachromasia, as does hyaluronic acid; pepsin 3 gave no such effect. Fraction FPIV, after electrophoresis at pH 3 and electro-elution from the gel retained both its normal mobility and its sensitivity to treatment with hyaluronidase.

Conclusion

Pepsins 1 and 2 together comprise at least six different proteolytic components, all of the same immunological group as pepsin 3. These proteinases are strongly associated with material having some of the properties of hyaluronic acid. We do not know whether the binding of the polysaccharide and protein is covalent.

References

1. Etherington, D.J., Taylor, W.H.: Nature 216, 279-280 (1967).
2. Walker, V., Taylor, W.H.: Gut 21, 766-771 (1980).
3. Etherington, D.J., Roberts, N.B., Taylor, W.H.: Clin. Sci. 58, 30P (1980).
4. Pearson, J.P., Roberts, N.B., Taylor, W.H.: Clin. Sci. 66, 61P (1984).
5. Roberts, N.B., Taylor, W.H.: Biochem. J. 169, 607-615 (1978).
6. Roberts, N.B., Taylor, W.H.: Biochem. J. 179, 183-190 (1979).
7. Samloff, I.M.: J. Immunol. 106, 962-968 (1971).
8. Foltmann, B., Pedersen, V.B.: in: "Acid Proteases: Structure, Function & Biology" ed. Tang, J. Exp. Med. & Biol. 95, 3-22 (1977).
9. Hirs, C.H.W.: Meth. Enzymol. 11, 411-413 (1967).
10. Szirmai, J.A., Balasz, E.A.: Histochimica Acta, Suppl. 1, 56-59 (1958).
11. Bitter, T., Muir, H.: Analyt. Biochem. 4, 330-334 (1962).
12. Gatt, R., Berman, B.R.: Analyt. Biochem. 15, 167-171 (1966).

THE PRIMARY STRUCTURE OF CATHEPSIN D AND THE IMPLICATIONS FOR ITS BIOLOGICAL FUNCTIONS

Jaiprakash G. Shewale, Takayuki Takahashi, Jordan Tang

Laboratory of Protein Studies, Oklahoma Medical Research Foundation and The Department of Biochemistry and Molecular Biology, University of Oklahoma Health Sciences Center, Oklahoma City, OK 73104

Among the better studied aspartic proteases, cathepsin D is unique in its intracellular function. It is located in the lysosomes of all mammalian cells. The main physiological role of cathepsin D is the break down of tissue proteins. This function differentiates cathepsin D from most other aspartic proteases, such as pepsin, renin, and penicillopepsin, which are secreted and hydrolyze proteins outside the cells. Therefore, a structure-function comparison of cathepsin D with other aspartic proteases may provide insights into the structural features which are responsible for the functions and regulation of an intracellular enzyme. For this reason, cathepsin D has been actively studied by many investigators (for review, see Ref. 1,2,3). During the past few years, we have undertaken a study of the structure and function relationships of cathepsin D and our findings are summarized in this report.

The Amino Acid Sequence of Cathepsin D

In order to determine the amino acid sequence of cathepsin D, we first devised a large-scale purification scheme of this enzyme from porcine spleen and carried out preliminary characterizations (4,5). Porcine spleen cathepsin D is mainly a two-chain protein, although a small fraction of the single

Aspartic Proteinases and their Inhibitors

chain species is always present (4). The molecular weight is about 45,000. (Because cathepsin D is a glycoprotein, its molecular weight is often over estimated due to its lower mobility in SDS-gel electrophoresis.) The chains, which are noncovalently associated, were separated (5) and the amino acid sequences separately determined (6,7). The entire amino acid sequence of the two-chain cathepsin D is shown in Figure 1.

Porcine spleen cathepsin D contains 339 amino acid residues in two peptide chains with a molecular weight of 36,779, not counting the oligosaccharide units. The light chain has 97 residues and a molecular weight of 10,548 while the heavy chain contains 242 residues and a molecular weight of 26,231. In Figure 1, the light chain is aligned before the heavy chain because the light chain is known to occupy the amino-terminal portion of the single-chain cathepsin D (4). There are seven half-cystine residues. Six of these are located near the corresponding positions as in pepsin and other aspartic proteases (see homology alignment below). The seventh half-cystine, located at residue 27, is likely to be a cysteine in the native protein (6). We observed two positions with structural microheterogeneity. Lysine and serine were both found at position 228 while glutamine and glycine were seen at position 241.

Homology of Cathepsin D with Other Aspartic Proteases

Cathepsin D is strongly homologous to other aspartic proteases. An example of the sequence alignment against pepsin and renin is shown in Figure 2. Among all three proteases, near 33% of the residues are identical. Although the homology exists through the length of the molecule, it is particularly strong within several regions. This includes the regions near the active-site Asp-32 and Asp-215, and carboxyl-terminal region (residues 300-327). These findings suggest that cathepsin D

1 (1) 10 (10) 20
Gly-Pro-Ile-Pro-Glu-Val-Leu-Lys-Asn-Tyr-Met-Asp-Ala-Gln-Tyr-Tyr-Gly-Glu-Ile-Gly-

(20) 30 (30) 40
Ile-Gly-Thr-Pro-Pro-Gln-Cys-Phe-Thr-Val-Val-Phe-Asp-Thr-Gly-Ser-Ser-Asn-Leu-Trp-

(40) 50 (50) 60
Val-Pro-Ser-Ile-His-Cys-Lys-Leu-Leu-Asp-Ile-Ala-Cys-Trp-Ile-His-His-Lys-Tyr-Asn-

(60) 70* (70) 80
Ser-Gly-Lys-Ser-Ser-Thr-Tyr-Val-Lys-Asn-Gly-Thr-Thr-Phe-Ala-Ile-His-Tyr-Gly-Ser-

(80) 90 (90) 100
Gly-Ser-Leu-Ser-Gly-Tyr-Leu-Ser-Gln-Asp-Thr-Val-Ser-Val-Pro-Ser-Asn Val-Gly-Gly-
LC ← → HC

(100) 110 (110) 120
Ile-Lys-Val-Glu-Arg-Gln-Thr-Phe-Gly-Glu-Ala-Thr-Lys-Gln-Pro-Gly-Leu-Thr-Phe-Ile-

(120) 130 (130) 140
Ala-Ala-Lys-Phe-Asp-Gly-Ile-Leu-Gly-Met-Ala-Tyr-Pro-Arg-Ile-Ser-Val-Asn-Asn-Val-

(140) 150 (150) 160
Val-Pro-Val-Phe-Asp-Asn-Leu-Met-Gln-Gln-Lys-Leu-Val-Asp-Lys-Asp-Ile-Phe-Ser-Phe-

(160) 170 (170) 180
Tyr-Leu-Asn-Arg-Asp-Pro-Gly-Ala-Gln-Pro-Gly-Gly-Glu-Leu-Met-Leu-Gly-Gly-Ile-Asp-

(180) 190 * (190) 200
Ser-Lys-Tyr-Tyr-Lys-Gly-Ser-Leu-Asp-Tyr-His-Asn-Val-Thr-Arg-Lys-Ala-Tyr-Trp-Gln-

(200) 210 (210) 220
Ile-His-Met-Asn-Gln-Val-Ala-Val-Gly-Ser-Ser-Leu-Thr-Leu-Cys-Lys-Gly-Gly-Cys-Glu-

Lys (220) 230 240
Ala-Ile-Val-Asp-Thr-Gly-Thr-Ser-Leu-Ile-Val-Gly-Gln-Pro-Glu-Glu-Val-Arg-Glu-Leu-

Gln (240) 250 (250) 260
Gly-Lys-Ala-Ile-Gly-Ala-Val-Pro-Leu-Ile-Gln-Gly-Glu-Tyr-Met-Ile-Pro-Cys-Glu-Lys-

(260) 270 (270) 280
Val-Pro-Ser-Leu-Pro-Asp-Val-Thr-Val-Thr-Leu-Gly-Gly-Lys-Lys-Tyr-Lys-Leu-Ser-Ser-

(280) 290 300
Glu-Asn-Tyr-Thr-Leu-Lys-Val-Ser-Gln-Ala-Gly-Gln-Thr-Ile-Cys-Leu-Ser-Gly-Phe-Met-

(290) 310 (300) 320
Gly-Met-Asp-Ile-Pro-Pro-Pro-Gly-Gly-Pro-Leu-Trp-Ile-Leu-Gly-Asp-Val-Phe-Ile-Gly-

(310) 330 (320) 339 (327)
Arg-Tyr-Tyr-Thr-Val-Phe-Asp-Arg-Asp-Leu-Asn-Arg-Val-Gly-Leu-Ala-Glu-Ala-Ala

Figure 1. Amino acid sequence of porcine spleen cathepsin D. Glycosylated aspargines are indicated by *. Residues are numbered starting from NH_2-terminus of cathepsin D. Numbers in parenthesis indicate the pepsin residue numbers (16). Positions 228 and 241 are structurally heterogeneous. The light chain (LC) ends at residue 97 and the heavy chain (HC) starts at residue 98.

```
                                1                   10                  20
Cathepsin D                     G P I P E V L K N Y M D A Q Y Y G E I G I G T P P Q C
Renin           S S L T D L I S P V V L T N Y L N S Q Y Y G E I G I G T P P Q T
Pepsin                          I G D E P L E N Y L D T E Y F G T I G I G T P A Q D

      30                  40                  47A B     50                  60
F T V V F D T G S S N L W V P S I H C K L L D I A C W I H H K Y N S G K S S T Y
F K V I F D T G S A N L W V P S T K C S R L Y L A C G I H S L Y E S S D S S S Y
F T V I F D T G S S N L W V P S V Y C S S - - L A C S D H N Q F N P D D S S T F

    CH2O
    |       70                  80                  9091A B C D                 100
V K N G T T F A I H Y G S G S L S G Y L S Q D T V S V P S N V G G I K V E R Q T
M E N G D D F T I H Y G S G R V K G F L S Q D S V T V - - - - G G I T V T - Q T
E A T S Q E L S I T Y G T G S M T G I L G Y D T V Q V - - - - G G I S D T N Q I

                  110                 120                 130                 140
F G E A T K Q P G L T F I A A K F D G I L G M A Y P R I S V N N V V P V F D N L
F G E V T E L P L I P F M L A Q F D G V L G M G F P A Q A V G G V T P V F D H I
F G L S E T E P G S F L Y Y A P F D G I L G L A Y P S I S A S G A T P V F D N L

                  150       155A B      160A                      170
M Q Q K L V D K D I F S F Y L N R D P G A Q P G G E L M - L G G I D S K Y Y K G
L S Q G V L K E K V F S V Y Y N R G P H L L - G G E V V - L G G S D P E H Y Q G
W D Q G L V S Q D L F S V Y L - - S S N D D - S G S V V L L G G I D S S Y Y T G

          CH2O
          |
  180                     190                 200                 210
S L D Y H N V T R K A Y W Q I H M N Q V A V G S S L T L C K G G C E A I V D T G
D F H Y V S L S K T D S W Q I T M K G V S V G S S T L L C E E G C E V V V D T G
S L N W V P V S V E G Y W Q I T L D S I T M D G E T I A C S G G C Q A I V D T G

  220                     230                 240                 250
T S L I V G Q P E E V R - E L G K A I G A V P L I Q G E Y M I P C E K V P S L P
S S F I S A P T S S L K - L I M Q A L G A K E K R L H E Y V V S C S Q V P T L P
T S L L T G P T S A I A(I)N I Q S D I G A S E N S D G E M V I S C S S I D S L P

  260                     270                 280A B C D                290
D V T V T L G G K K Y K L S S E N Y T L K V S Q A G Q T I C L S G F M G M D I P
D I S F N L G G R A Y T L S S T D Y V L Q Y P N R R D K L C T V A L H A M D I P
D I V F T I D G V Q Y P L S P S A Y I L Q D D - - - - D S C T S G F E G M D V P

          300                 310                 320                 327
P P G G P L W I L G D V F I G R Y Y T V F D R D L N R V G L A E A A
P P T G P V W V L G A T F I R K F Y T E F D R H N N R I G F A L A R
T S S G E L W I L G D V F I R Q Y Y T V F D R A N N K V G L A P V A
```

Figure 2. Comparison of amino acid sequences of cathepsin D, renin and pepsin. Gaps have been placed to maximize the homology. Boxed areas contain residues that are identical in all three proteins. Residues are numbered according to the numbering system of pepsin (16).

has a similar three-dimensional structure to other aspartic proteases (8-11). They also support the thesis that the active-site structure and catalytic mechanism of cathepsin D is very similar to other aspartic proteases. When compared individually, cathepsin D has 49% identical residues with mouse submaxillary renin (12,13), 46% with human renin (14), 48% with porcine pepsin (15,16), 45% with chymosin (17) and 26% with penicillopepsin (9). Cathepsin D may be most closely related to renin in its structure. This is supported by the high percent of identical residues and other structural evidence. For example, the disulfide bridges between half-cystine residues 45 and 50 in cathepsin D and renin both form an eight-residue loop (Figure 2), while pepsin and other aspartic proteases have a six-residue loop. The alignment in Figure 2 also shows that cathepsin D and renin share insertions at residues 155A, B and residues 280A-D.

Oligosaccharide Units and Glycosylation Sites in Cathepsin D

Cathepsin D contains two oligosaccharide units, one each in the light and heavy chains (4). We have determined the amino acid sequence near the two glycosylation sites (18), thus the location of these carbohydrates in cathepsin D sequence are known. Both oligosaccharides are N-linked asparagines at positions 67 and 183 (Figure 2). Several different oligosaccharides from both chains were isolated and their structures determined using proton nuclear magnetic resonance (18). A total of eight different oligosaccharides were found (Figure 3). These are predominantly high mannose-type oligosaccharides, known to be present on lysosomal enzymes (19). However, minor complex types of oligosaccharides are also present. Five different oligosaccharides were found on Asn-67 of the light chain. Four of these are high mannose-type having 3, 5, 6, and 7 mannoses, respectively (Figure 3) and the fifth structure contains a third GlcNAc. Two out of three oligosaccharides from the heavy chain

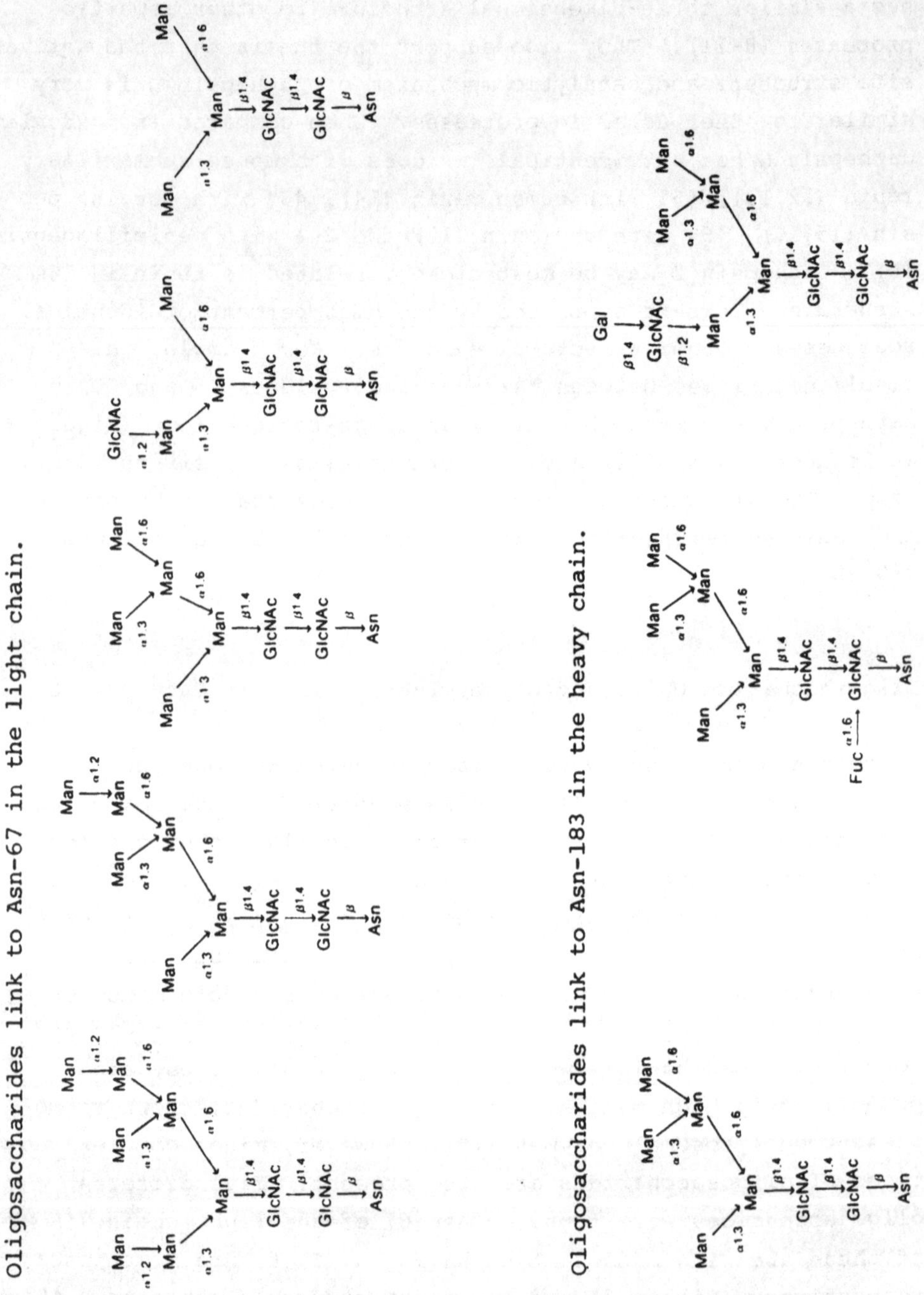

Figure 3. Oligosaccharide structures found in cathepsin D.

have 5 mannoses. However, one of the two, which represents 1/3 of the total, contains an additional fucose (Figure 3).

The biological function of the high mannose oligosaccharide on cathepsin D and other lysosomal enzymes is well known. These oligosaccharides are phosphorylated to form terminal mannose-6-phosphates which serve as markers for lysosome targeting after the biosynthesis of these enzymes (20,21). Hasilik and Neufeld (22) have shown that cathepsin D is phosphorylated at oligosaccharides in both chains during biosynthesis. However, the lysosome targeting role of mannose-6-phosphate does not explain the presence of large numbers of structures observed for the oligosaccharides on each glycosylation site in cathepsin D. One possibility for the origin of these oligosaccharides is that they are the products of random hydrolysis by acid glycosidases in the lysosome. The more complex carbohydrates with extra GlcNAc and fucose, however, probably do not participate in the lysosome targeting, thus their biological function is not known.

Since the tertiary structure of cathepsin D is likely to be highly homologous to that of other aspartic proteases, it is possible to use the crystal structures of the latter to approximate for the glycosylation positions in the former. As expected, the glycosylation positions, Asn-67 and Asn-183 are found on the surface of the tertiary structure away from the active site (Figure 4). It is interesting that the structure near one of the glycosylation sites, Asn-67, appears to be the region of frequent post-translational modifications in many other aspartic proteases. Porcine pepsin is phosphorylated at Ser-68 (15,16). Chicken pepsin is glycosylated at an asparagine equivalent to residue 68 (23). In human renin (14), the glycosylation site is an aspargine at either residue 67 or 68, depending on the alignment of sequences. Also, the amino acid sequences around Asn-67, residues 63-72, are highly variable among all aspartic proteases. It seems possible that the

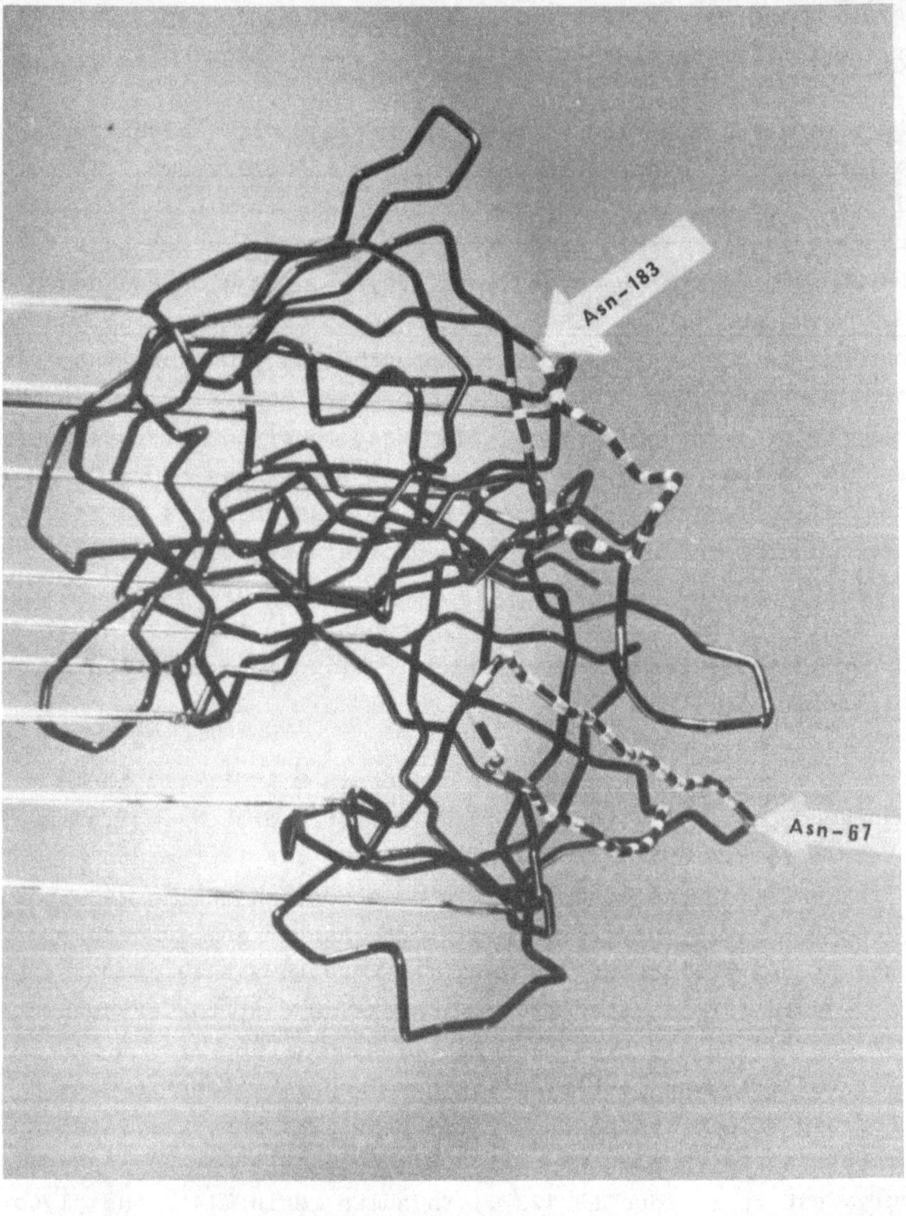

Figure 4. Glycosylation positions of cathepsin D as shown on a wire model of penicillopepsin.

function of this region of structure in aspartic proteases may be that of recognition markers similar to the lysosomal targeting of mannose-6-phosphate in cathepsin D.

Cathepsin D apparently contains another site which is recognized for mannose phosphorylation. The enzymic system responsible for the phosphorylation of mannoses on the lysosomal enzymes has been described (24-29). N-acetylglucosamine-1-phosphate is first transferred from UDP-N-acetylglucosamine to an acceptor mannose on a lysosomal enzyme. A phosphodiesterase then removes the N-acetylglucosamine leaving a mannose-6-phosphate on the glycoprotein. Reitman and Kornfeld (29) have found that lysosomal enzymes are preferentially phosphorylated by N-acetylglucosaminyl phosphotransferase, the enzyme that catalyzes the first step in this pathway. Using cathepsin D as acceptor of GlcNAc-phosphate, the apparent Km is about 200 times lower than the Km value for ribonuclease B, a nonlysosomal glycoprotein (30). These observations suggest cathepsin D, as well as other lysosomal enzymes, contains a recognition site for the transfer of GlcNAc-phosphate by the transferase. This hypothesis was substantiated by the inhibition of the transferase reaction by deglycosylated cathepsin D (30,31), which had an apparent Ki value similar to the Km value for cathepsin D. Interestingly, neither the isolated light or heavy chain of cathepsin D is a good acceptor for the phosphotransferase catalyzed reaction, indicating that the transferase recognition of this site is dependent on the native tertiary structure of cathepsin D. The nature and location of this phosphotransferase recognition site on cathepsin D molecule have not been determined.

The Processing of Cathepsin D Precursors

From the biosynthetic studies, cathepsin D is known to be synthesized as precursor and ultimately processed to the two-chain structure (22,32). The processing of cathepsin D precursors is interesting since all lysosomal enzymes studied for biosynthesis have been found to be proteolytically processed from larger precursors (33-39). Thus the processing of cathepsin D may serve as a model for the structural features involved in the processing of this group of enzymes.

Erickson, Conner and Blobel (32) have established that cathepsin D is synthesized as prepro-form, with a "leader" peptide (pre) and an activation peptide (pro). The leader sequence appears to contain 20 amino acids rich in leucine residues, much like the leader sequences of the secretory proteins. The function and the proteolytic processing of the cathepsin D "leader" sequence are likely the same as those for the secretory proteins, namely the translocation of the synthesized polypeptide across the endoplasmic reticulum membrane and a cleavage of leader peptide almost immediately thereafter. The partial structure of the 44-residue "pro" sequence obtained by Erickson *et al.* (32) appears to be homologous to the corresponding sequence in pepsinogen. Additionally, procathepsin D has been shown to gain enzymic activity in an acid environment (40). This evidence suggests that procathepsin D may be activated by an intramolecular mechanism in an acid environment as in the case of pepsinogen (41-43). However, firm support for acid activation must be established by the study of isolated procathepsin D especially in view of the fact that prorenin is apparently activated by the proteolysis of another protease at the carboxyl side of a Arg-Arg sequence (13,14). Another very interesting observation by Erickson and Blobel is the possibility that proteolytic processing of cathepsin D precursor may take place at the carboxyl terminal region (44). If substantiated, this carboxyl terminal processing would be unique among the aspartic proteases.

The processing of the single chain to the two chain cathepsin D takes place between residues Asn-91C and Val-91D (Figure 2). Some cathepsin D molecules also have two amino acids (Val and

Gly) removed from the NH_2-terminus of the heavy chain (unpublished results), resulting in the previously observed sequence heterogeniety of the heavy chain (4). From the homology alignment in Figure 2, it is clear that a four-residue addition with the sequence Pro-Ser-Asn-Val, residues 91A - 91D, is present in cathepsin D but not in other aspartic proteases. Rat liver cathepsin H is proteolytically processed between residues Asn-167C and Gly-167D (papain numbers, Ref. 45, see Figure 2). Porcine spleen cathepsin B, a structurally homologous enzyme, is partially processed at this same site (46). In both cathepsins B and H, an addition of 4 residues is seen around this processing site, when these structures are aligned with papain (45). This addition is located on the surface of papain crystal structure (47). Therefore, there may be a general structural feature in lysosomal enzyme processing.

Concluding Remarks

From the comparison of the primary structure of cathepsin D to other aspartic proteases, it is apparent that cathepsin D has a very similar three-dimensional structure to other enzymes. The major structural difference is the presence of two oligosaccharides. There is still a structurally unidentified site on cathepsin D which is recognized for enzymic mannose phosphorylation. These structural features are, as discussed above, essential for the lysosomal targeting of cathepsin D. The catalytic apparatus of cathepsin D should be structurally very similar to other aspartic proteases, since both active site residues are found at the expected locations in the sequence. There are unusual aspects of cathepsin D catalysis, however, that are not adequately explained by the structural information. First, as compared to pepsin, cathepsin D hydrolyzes most protein substrates poorly (3), even though the specificities of the two enzymes are similar (1). Second, unlike other aspartic proteases, small synthetic peptides are poor cathepsin D substrates.

Thus, there are few effective synthetic substrates for cathepsin D assay. A possible explanation for these points is that the cathepsin D may have a more stringent structural requirement for its side chain binding pockets, much like that for renin. A more stringent specificity for cathepsin D, however, would suggest possible specific proteolytic functions, perhaps in addition to lysosomal digestions.

Many different physiological and pathological functions outside of lysosomes have been suggested for cathepsin D (for review, see Ref. 3). There are two specific proteolytic functions which seem interesting for consideration here. First is the renin-like activity of cathepsin D. Not only is cathepsin D structurally most closely related to renin (see above), it is also the only other aspartic protease capable of producing angiotensin I from renin substrates (48). The question of whether cathepsin D ever releases angiotensin under physiological or pathological conditions is still unanswered. It seems unlikely, however, that cathepsin D can hydrolyze antiotensinogen in the plasma where not only is the pH too high, but also the cathepsin D activity is apparently inhibited by plasma (3). The second possible function of cathepsin D outside of lysosomes is in the processing of protein precursors. Of particular interest is the recent study of Helseth and Veis (49), who proposed that cathepsin D is responsible for the processing of procollagen to collagen via a pathway of secretion and endocytosis. If cathepsin D is indeed involved in protein processing, then a more stringent specificity would seem logical.

Acknowledgements

The authors would like to thank Drs. Jean Hartsuck and Robert Delaney for their help during the preparation of this manuscript. This work was supported by NIH grants GM-20212 and AM-01107.

References

1. Keilova, H.: In: Tissue Proteinases (Barrett, A.J. and Dingle, J.T., eds.), p.45-64 Amsterdam:North Holland 1971.
2. Barrett, A.J.: In: Tissue Proteinases (Barrett, A.J. and Dingle, J.T., eds.), p.109-127 Amsterdam:North Holland 1971.
3. Barrett, A.J.: In: Proteinases in Mammalian Cells and Tissues (Barrett, A.J., ed.), p.209-248 Amsterdam:North Holland 1977.
4. Huang, J.S., Huang, S.S., Tang, J.: J. Biol. Chem. 254, 11405-11417 (1978).
5. Takahashi, T., Tang, J.: Methods Enzymol. 80,565-581 (1981).
6. Takahashi, T., Tang, J.: J. Biol. Chem. 258,6435-6443 (1983).
7. Shewale, J.G., Tang, J.: Proc. Natl. Acad. Sci. USA 81, 3703-3707 (1984).
8. Andreeva, N.S., Fedorov, A.A., Gustchina, A.E., Riskulov, R.R., Safro, M.G., Shutzkever, N.E.: Mol. Biol. Moscow (Engl Transl.) 12,704-707 (1978).
9. Hus, I.-N., Delbaere, L.T.J., James, M.N.G., Hofmann, T.: Nature 266,140-145 (1977).
10. Jenkins, J., Tickle, I., Sewell, T., Ungaretti, L., Wollmer, A., Blundell, T.: In: Acid Proteases (Tang, J., ed.), New York:Plenum Press 1977.
11. Subramanian, E., Swan, I.D.A., Liu, M., Davies, D.R., Jenkins, J.A., Tickle, I.J., Blundell, T.L.: Proc. Natl. Acad. Sci. USA 74,556-559 (1977).
12. Misono, K.S., Chang, J.-J., Inagami, T.: Proc. Natl. Acad. Sci. USA 79,4863-4867 (1982).
13. Panthier, J.J., Foote, S., Cambraud, B., Strosberg, A.D., Covol, P., Rougeon, F.: Nature 298,90-92 (1982).
14. Imai, T., Miyazaki, H., Hirose, S., Hori, H., Hayashi, T., Kageyama, R., Ohkubo, H., Nakamishi, S., Murakami, K.: Proc. Natl. Acad. Sci. USA 80,7405-7409 (1983).

15. Tang, J., Sepulveda, P., Marciniszyn, J., Jr., Chen, K.C.S., Huang, W.-Y., Tao, N., Liu, D., Lanier, J.P.: Proc. Natl. Acad. Sci. USA 70,3437-3439 (1973).
16. Sepulveda, P., Marciniszyn, J., Jr., Liu, D., Tang, J.: J. Biol. Chem. 250,5082-5088 (1975).
17. Foltmann, B., Pedersen, V.B., Kauffman, D., Wybrandt, G.: J. Biol. Chem. 254,8447-8456 (1979).
18. Takahashi, T., Schmidt, P.G., Tang, J.: J. Biol. Chem. 258,2819-2830 (1983).
19. Kornfeld, R., Kornfeld, S.: In: The Biochemistry of Glycoproteins and Proteoglycans (Lennarz, W.J., ed.), p. 1-34 New York:Plenum Press 1980.
20. Neufeld, E.F., Ashwell, G.: In: The Biochemistry of Glycoproteins and Proteoglycans (Lennarz, W.J., ed.), p. 241-266 New York:Plenum Press 1980.
21. Sly, W.S., Fischer, H.D.: J. Cell. Biochem. 18,67-85 (1982).
22. Hasilik, A., Neufeld, E.F.: J. Biol. Chem. 255,4946-4950 (1980).
23. Baudyś, M., Kostka, V., Grüner, K., Pohl, J.: Collection Czechoslovak Chem. Commun. 47,709-718 (1982).
24. Reitman, M.L., Kornfeld, S.: J. Biol. Chem. 256,4275-4281 (1981).
25. Hasilik, A., Waheed, A., vonFigura, K.: Biochem. Biophys. Res. Commun. 98,761-767 (1981).
26. Varki, A., Kornfeld, S.: J. Biol. Chem. 255,8398-8401 (1980).
27. Waheed, A., Hasilik, A., vonFigura, K.: J. Biol. Chem. 256,5717-5721 (1981).
28. Varki, A., Kornfeld, S.: J. Biol. Chem. 256,9937-9943 (1981).
29. Reitman, M.L., Kornfeld, S.: J. Biol. Chem. 256,11977-11980 (1981).
30. Lang, L., Tang, J., Roberts, M., Kornfeld, S.: Fed. Proceedings 43,1350 (1984).
31. Lang, L., Reitman, M., Tang, J., Roberts, M., Kornfeld, S.:

Manuscript in preparation (1984).

32. Erickson, A.H., Conner, G.E., Blobel, G.: J. Biol. Chem. 256,11224-11231 (1981).
33. Hasilik, A., Tanner, W.: Eur. J. Biochem. 85,599-608 (1978).
34. Skudlarek, M.D., Swank, R.T.: J. Biol. Chem. 254,9939-9942 (1979).
35. Myerowitz, R., Neufeld, E.F.: J. Biol. Chem. 256,3044-3048 (1981).
36. vonFigura, K., Klein, U., Hasilik, A.: In: Biological Chemistry of Organelle Formation (Bücker, T. *et al.*, ed.), p.207-219 Berlin:Springer.
37. Brown, J.A., Johreis, G.P., Swank, R.T.: Biochem. Biophys. Res. Commun. 99,691-699 (1981).
38. Waheed, A., Hasilik, A., vonFigura, K.: Eur. J. Biochem. 123,317-321 (1982).
39. Hasilik, A., Neufeld, E.F.: J. Biol. Chem. 255,4937-4945 (1980).
40. Hasilik, A., vonFigura, K., Conzelmann, E., Nehrkorn, H., Sandhoff, K.: Eur. J. Biochem. 125,317-321 (1982).
41. Bustin, M., Conway-Jacobs, A.: J. Biol. Chem. 246,615-620 (1971).
42. Al-Janabi, J., Hartsuck, J.A., Tang, J.: J. Biol. Chem. 247,4628-4632 (1972).
43. McPhie, P.: J. Biol. Chem. 247,4277-4281 (1972).
44. Erickson, A.H., Blobel, G.: Biochemistry 22,5201-5205 (1983).
45. Takio, K., Towatari, T., Katunuma, N., Teller, D.C., Titani, K.: Proc. Natl. Acad. Sci. USA 80,3666-3670 (1983).
46. Takahashi, T., Dehdarani, A., Schmidt, P.G., Tang, J.: J. Biol. Chem. 259, in press (1984).
47. Drenth, J., Jansonius, J.N., Koekoek, R., Sluyterman, L.A.A., Wolthers, B.G.: Phil. Trans. Roy. Soc. Lond. B. 257,231-236 (1970).
48. Dorer, F.E., Lentz, K.E., Kahn, J.R., Levine, M., Skeggs, L.T.: J. Biol. Chem. 253,3140-3142 (1978).

49. Helseth, D.L., Jr., Veis, A.: Proc. Nat. Acad. Sci. USA 81,3302-3306 (1984).

SOME UNEXPECTED PROPERTIES OF CATHEPSIN D

Bernd Wiederanders, Heidrun Kirschke, Susanne Schaper
Physiologisch-chemisches Institut der Universität Halle,
DDR-4020 Halle/Saale

Martin J. Valler, John Kay
Department of Biochemistry, University College, Cardiff, U.K.

ANSON's method [1] for assaying "acid" proteinases using denatured hemoglobin as substrate has two disadvantages: 1) the relationship of v vs. [E] is not linear [1,2], and 2) the method is not very sensitive (see Fig. 1).

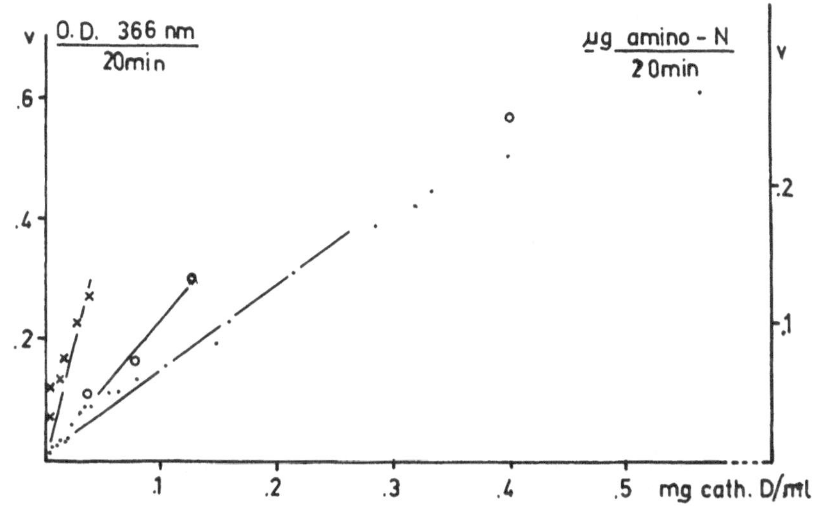

Fig.1
Apparent reaction velocities for the degradation of various proteins at different concentrations of cathepsin D at 37° C. o——o Hb at pH 3.2; x——x azocasein in 3 M urea at pH 5.0; •——• azo-Hb in 3 M urea at pH 4.5. The activities are given either as increase of O.D. at 366 nm (azocasein and azo-Hb) or of α-amino-nitrogen (Hb) in the TCA supernatants after 20 min incubation. Cathepsin D from rat liver.

Aspartic Proteinases and their Inhibitors

Nevertheless, since protein substrates are much less expensive than synthetic peptides for general use, the suitability of diazotised derivatives of Hb and casein [3] as possible substrates for cathepsin D was examined. Urea was also included (at 3 M) in order to render the proteins more susceptible to digestion.

It was found that azocasein in the presence of 3 M urea is a more sensitive substrate than denatured hemoglobin, even if the hydrolysis of the latter is measured by the increase of α-amino-nitrogen in the TCA supernatant instead of following its difference in extinction at 280 nm. In the presence of urea, the pH optimum for the hydrolysis of azocasein by cathepsin D appeared to be 5.0 (Fig. 2) and the activities measured at pH values between 3.5 and 6.5 were apparently greater than those obtained in the absence of urea. Preincubation of cathepsin D solutions (0.5 mg/ml for 24 h in 3 M urea at 4° C) did not alter this effect.

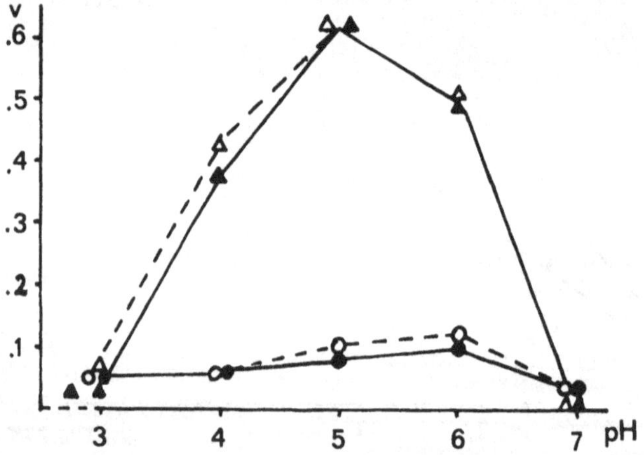

Fig. 2
The effect of pH and urea on the apparent reaction velocity of cathepsin D. △ ○ without preincubation; ▲ ● after 24 h preincubation of the enzyme in 3 M urea at the pH values indicated; △ ▲ in the presence of 3 M urea; ○ ● without urea. The activity of the enzyme was determined under the conditions used in the preincubation. Substrate was azocasein (0.5 g/ 100 ml), 20 min at 37° C. Cathepsin D from rat liver.

In order to distinguish whether urea was exerting its influence predominantly on azocasein or on the enzyme, cathepsin D activity was monitored using a synthetic peptide substrate, Lys-Pro-Ala-Glu-Phe-(NO_2)Phe-Arg-Leu [4].
At pH 5.3, cathepsin D (bovine spleen) had very little activity towards this peptide (ΔE_{300}/min = 8 x 10^{-5}) and 3 M urea did not increase this (ΔE/min = 3 x 10^{-5}). Indeed, at pH 3.1 the native enzyme (ΔE/min = 1 x 10^{-2}) was unable to hydrolyse the substrate in the presence of 3 M urea (ΔE/min = 4 x 10^{-4}).
The effect of 3 M urea on the activity of cathepsin D at several pH values was also followed by means of this spectrophotometric method. At pH 6.6 and 25° C, bovine spleen cathepsin D was completely stable for 2 hours in the presence of 3 M urea. At pH 3.1, the denaturing effect of urea was dependent on the temperature of incubation. At 4° C, approximately 50 % of the activity remained after 24 h whereas the enzyme was completely inactivated within 15 min of adding 3 M urea at 37° C. Thus, it would appear that in the azocasein assay, the enzyme undergoes a rapid inactivation below about pH 4.0 during the first few minutes of the incubation at 37° C. Thus, the activity that is observed at lower pH values is due to the diminishing fraction of the enzyme that can express its catalytic affect before becoming denatured.
To examine this further, a sample of cathepsin D was partially inactivated by incubation for 10 min at 37° C in 3 M urea at pH 3.1 (19% of the original activity remained). This was then diluted ten-fold into pH 3.1 buffer and the change in expressed activity was followed with time of incubation at pH 3.1 and 37° C. The activity increased, until after 15 min it had returned to a level of 41 % of the original value. Thereafter, it decreased steadily with prolonged time of incubation.
Urea may then be imparing some effect on the conformation of the cathepsin D molecule and the nature of this influence varies with pH. Lah et al. [5,6] have recently reported on

the changes in the spatial arrangement of cathepsin D in the presence of urea as well as guanidinium•HCl. They postulated a highly labile but more active transition state of the enzyme in the presence of these chaotropic agents. SDS-gel electrophoresis (± ß-mercaptoethanol) of rat liver cathepsin D used with azocasein in our studies indicated that it was a single chain species (Fig. 3). Thus, the change in specific activity obseved by Lah et al.[6] on the basis of conversion of a single chain enzyme into a two chain protein, cannot account for our observation.

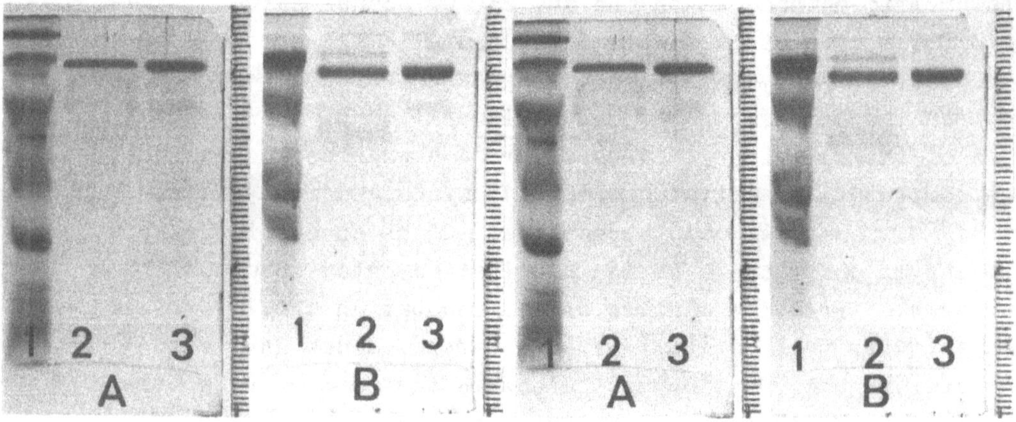

Fig. 3
SDS-PAGE of 13 /ug rat liver cathepsin D in 15 % gels containing 6 M urea. Denaturation was done under (A) reducing or (B) nonreducing conditions. Lane 1: MW marker proteins (at the top and the 2nd positions are bovine serum albumin and ovalbumin, resp.). Lanes 2 and 3: cathepsin D preparations of different purity.

Urea thus has complex effects on the conformation and activity of cathepsin D: It inactivates the enzyme very rapidly at low pH values whereas it has much less inactivating effect as the pH nears neutrality. It increases the susceptibility of azocasein to attack by the enzyme and it improves the solubility of the split products in TCA as reported in [7]. Nevertheless, the combination of these factors result in a sensitive, convenient assay method which can be used to measure cathepsin D activity.

References

1. Anson, M.L.: J. Gen. Physiol. 20, 565-574 (1937).
2. Takahashi, T., Tang, J.: Methods Enzymol.(Lorand, L., Ed.) 80 C, 565-580, Academic Press, New York 1981.
3. Charney, J., Tomarelli, R.M.: J. Biol. Chem. 171, 501-505 (1947).
4. Dunn, B.M., Parten, B., Jimenez, M., Rolph, C.E., Valler, M.J., Kay, J.: this volume,221-243(1985).
5. Lah, T., Turk, V., Pain, R.P.: Period. Biol. 85, 95-100 (1983).
6. Lah, T., Drobnič-Kosorok, M., Turk, V., Pain, R.P.: Biochem. J. 218, 601-608 (1984).
7. Wojtowicz, A.T., Odense, P.: Can. J. Biochem. 48, 1050-1053 (1970).

NEW CHARACTERISTICS OF A HIGH MOLECULAR WEIGHT ASPARTIC PROTEINASE FROM BOVINE BRAIN

Anahit Azaryan, Nina Barkhudaryan, Armen Galoyan
Institute of Biochemistry
Yerevan 375044, USSR

Bernd Wiederanders
Institute of Physiological Chemistry
Halle 4020, GDR

Introduction

During the last decade a great deal of data have accumulated indicating the existence of a high molecular weight (HMW) aspartic proteinase in normal and pathological tissues, in addition to the major lysosomal aspartic proteinase cathepsin D (EC 3.4.23.5). This enzyme has been classified by one group of authors as a cathepsin E-like enzyme (1-3), by the others as a dimer or precursor of cathepsin D (4-8).

Results

The purification and some characteristics of a HMW aspartic proteinase from human and bovine brain have been reported by us in 1983 (9, 10). Human and bovine brain cortex contain a HMW aspartic proteinase (M_r about 90 K) which has been purified to electrophoretical homogeneity by a procedure involving ammonium sulfate fractionation (30-80 %), gel filtration on Sephadex G-100, affinity chromatography on pepstatin-Sepharose, and isoelectric focusing. The enzyme was assayed fluorimetrically with pyridoxyl-hemoglobin (Hb) at pH 3.2 and spectro-

Aspartic Proteinases and their Inhibitors

photometrically with azohemoglobin and azocasein at pH 5.0. The enzyme is highly sensitive to pepstatin ($K_I = 10^{-8}$M), the $K_{M\ (app.)}$ for pyridoxyl-Hb is 10^{-5}M. The pH-optimum of hydrolysis of pyridoxyl-Hb is 3.2 and the pI of the enzyme is 5.6. When subjected to polyacrylamide gel electrophoresis in sodium dodecyl sulfate the enzyme moves as one band corresponding to proteins with M_r of about 90 K; under reducing conditions (2.5 % β-mercaptoethanol) this band disappears and a new band corresponding to proteins with M_r of about 45-50 K can be observed.

The specific activity of the HMW proteinase is 1/10 of the activity of cathepsin D from the same source (pyridoxyl-Hb as substrate) and 1/7 of the activity of cathepsin D assayed with azosubstrates. In the presence of 3M urea in the assay mixture the specific activity of both the HMW proteinase and of cathepsin D increase 5 times (determined with azocasein as substrate) and 3 times (with azoHb as substrate) at pH 5.0 (Table 1). The HMW proteinase hydrolyzes bovine serum albumin and γ-globulin (10 and 5 %, respectively of the hydrolysis of Hb); the same results were obtained with cathepsin D.

Table 1

Hydrolysis of azosubstrates by brain aspartic proteinases

Substrate	Enzyme activity (ΔE_{366}/min/mg)	
	Cathepsin D	HMW proteinase
Azocasein	3.7	0.49
Azocasein in 3M urea	20.35	2.68
AzoHb	2.4	0.34
AzoHb in 3M urea	6.7	0.88

The final concentration of the azosubstrates was 1 %.

In order to compare the immunological properties of cathepsin D and the HMW proteinase the antiserum to bovine brain cathepsin D was raised in rabbits. Anti-bovine brain cathepsin D IgG gave a single precipitin line with bovine brain cathepsin D as well as with the HMW aspartic proteinase (native and urea-treated) but there was no identity reaction with cathepsin D (Fig. 1).

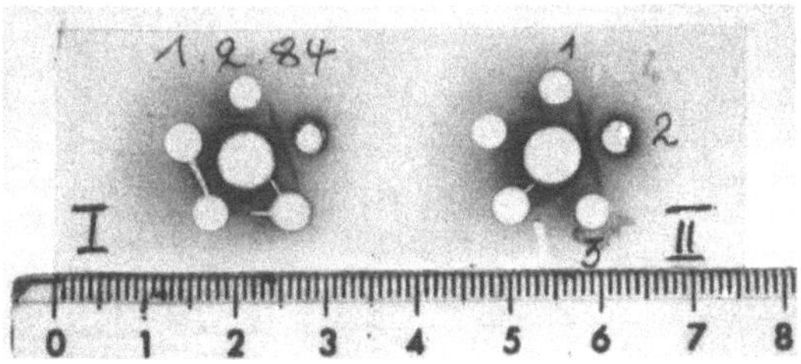

Fig. 1 Double immunodiffusion analysis of the HMW proteinase using anti-bovine brain cathepsin D IgG. I - HMW proteinase not treated with urea, II - urea-treated HMW proteinase. The center well contained anti-bovine brain cathepsin D IgG, the outer wells 1 and 3 - bovine brain cathepsin D, well 2 - HMW aspartic proteinase.

The data obtained in this study provide evidence showing that the HMW enzyme, whose existence in mammalian brains is reported for the first time, is an aspartic proteinase with properties similar to those of cathepsin D from the same source. It is not identical, however, to cathepsin E (EC 3.4.23.-) because a) cathepsin E is more active against serum albumin than against Hb; b) cathepsin E shows a lower pH-optimum (2.5) of hydrolysis of both albumin and Hb; c) cathepsin E is a very acidic protein with a pI = 4.1-4.4; d) cathepsin E is not precipitated by anti(cathepsin D) serum (11, 12), (see

Table 2

Comparison of properties of cathepsin D, E and HMW aspartic proteinase

	Cathepsin D (EC 3.4.23.5)	Cathepsin E (EC 3.4.23.-)	HMW proteinase (EC 3.4.23.-)
Recommended substrate	Hemoglobin	BSA[a], HSA[b]	Hemoglobin
pH-optimum	3.2-3.5	2.5	3.2
Molecular weight	42-50 K	100 K	90 K
Multiple forms	Yes	No	No
pI-value	4.4-6.8	4.1-4.4	5.6
K_I (pepstatin)	$0.6 \cdot 10^{-8}-10^{-8}M$	N.D.	$10^{-8}M$
$K_{M(app)}$ (hemoglobin)	$0.5 \cdot 10^{-5}-0.8 \cdot 10^{-5}M$	N.D.	$10^{-5}M$
Low molecular weight substrates (smaller than pentapeptides)	Not split	Not split	Not split
Endopeptidase activity	High	Moderate	Specific activity 1/10 and 1/7 of that of cathepsin D with Hb and azocasein, respectively.
Endogenous protein substrates	MBP, neurofilament proteins, aldolase	N.D.	N.D.
Specificity of cleavage	Phe-Phe, Phe-Tyr, Tyr-Phe, Leu-Tyr	Same as cathepsin D (insulin B chain as substrate)	N.D.
Immunological properties		No cross-reaction with anti(cathepsin D) serum	Cross reacts with anti-bovine brain cathepsin D IgG, but no cross reaction with cathepsin D

a bovine serum albumin, b human serum albumin

Table 2). It remains to be demonstrated whether the high molecular weight aspartic proteinase isolated by us is an individual enzyme or rather a complex of cathepsin D and an inhibitor.

References

1. Yamamoto, K., Katsuda, N., Kato, K.: Eur. J. Biochem. 92, 499-508 (1978).
2. Yamamoto, K., Katsuda, N., Nimeno, N., Kato, K.: Eur. J. Biochem. 95, 459-467 (1979).
3. Turk, V., Kregar, J., Gubenšek, F., Popovič, T., Lah, T.: Proc. FEBS Meet. 60, 317-330 (1980).
4. Yago, N., Bowers, W.: J. Biol. Chem. 250, 4749-4754 (1975).
5. Bowers, W., Beyer, C., Yago, N.: Biochim. Biophys. Acta 497, 272-280 (1977).
6. Yasani, B., Yasani, M., Talbot, M.: Biochem. J. 169, 287-295 (1978).
7. Kazakova, O., Orekhovitch, V.: Biokhimiya 44, 1762-1767 1979).
8. Huang, J., Huang, S., Tang, J.: J. Biol. Chem. 254, 11405-11417 (1979).
9. Barkhudaryan, N., Azaryan, A.: Neurokhimiya 2, 280-287 (1983).
10. Azaryan, A., Barkhudaryan, N.: 4th All-Union Symposium on Medical Enzymology, Alma-Ata, Abstracts pp. 10-11, 1983.
11. Barrett, A.J.: Res. Monogr. Cell Tissue Physiol. 2, 209-248 (1977).
12. Lapresle, C.: Tissue Proteinases (Barrett, A.J., Dingle, J., eds.), pp. 135-155, North Holland, Amsterdam 1971.

ISOLATION AND PROPERTIES OF AN ASPARTIC PROTEINASE FROM PIG INTESTINAL MUCOSA

V.K. Antonov, M.I. Zilberman, T.I. Vorotyntseva
Shemyakin Institute of Bioorganic Chemistry, USSR Academy of Sciences, 117988 Moscow

Introduction

During the past years relatively little attention has been paid to intestinal aspartic proteinases (1-2). Kregar et al. (1) found that the enterocytes of pig small intestine contain an intracellular acid proteinase. The latter was partially purified and characterized as cathepsin D with a molecular weight of about 40,000 which is typical of this group of enzymes. In the course of a study on transport processes in the mucosa of the small intestine of pig acid proteolytic activity was observed in the mucosa homogenate. After centrifugation this activity was localized in lysosomes. This paper deals with the isolation and purification of the proteinase responsible for this activity, describes its molecular characteristics, substrate specificity, and its interaction with inhibitors. The acid proteinase which we isolated in homogeneous state from pig intestinal mucosa belongs to the group of aspartic proteinases. Its molecular weight is 60,000. The enzyme molecule consists of two subunits 30,000 each.

Results

The purified enzyme was isolated from the homogenate of the small intestinal mucosa by precipitation with ammonium sulfate (30-70 % saturation), acid precipitation (pH 4.8), affinity chromatography on pepstatin-Sepharose 4B, and gel filtration on Sephadex G-100. The experimental protocol is given in Table 1.

Table 1. Purification of aspartic proteinase from the mucosa of the small intestine of pig

Purification steps	Protein mg	Specific activity unit/mg	Activity units	Purification -fold	Yield %
Homogenate	5,900	0.24	1,440	1	100
Precipitation with $(NH_4)_2SO_4$	1,837	0.27	500	1	36
Acid precipitation	582	0.71	415	3	30
Pepstatin-Sepharose 4B	1.66	134.0	222	558	16
Pepstatin-Sepharose 4B	0.55	322.0	177	1,342	13
Sephadex G-100	0.26	391.0	103	1,630	7

The specific activity of the pure preparation increased 1,630times with respect to the activity of the homogenate. The molecular weight of the enzyme (60,000 $\pm$ 5,000) was determined by gel filtration on a calibrated Sephadex G-100 column and by titration with pepstatin (3). The homogeneity of the preparation was demonstrated by electrophoresis in 7.5 % polyacrylamide gel (Fig. 1 a, b). In 12.0 % gel containing 0.1 % of SDS the enzyme migrated as a single band of molecular weight 30,000, both in the presence and absence of 2-mercaptoethanol (2 %). This shows that two noncovalently bound subunits are present in the enzyme molecule (Fig. 2). Upon isoelectric

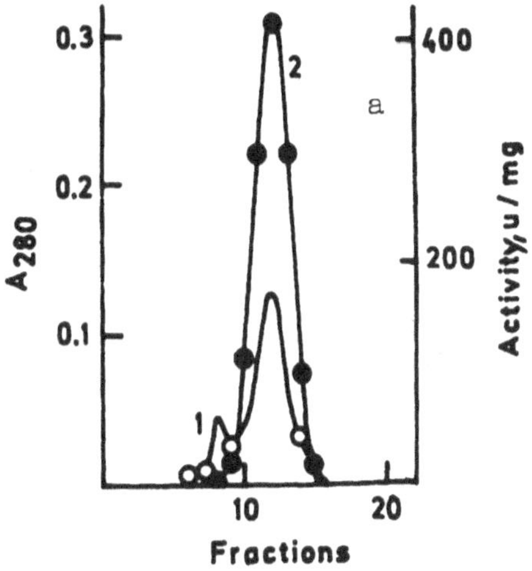

Fig. 1a Gel filtration on Sephadex G-100 of enzyme fraction obtained by chromatography on pepstatin-Sepharose 4B. About 180 units of the enzyme were applied to 1.1x50 cm column of Sephadex G-100 equilibrated with 0.02 M sodium phosphate buffer, pH 7.4 and 0.2 M sodium chloride. The flow rate was 10 ml/h and fractions of 2 ml were collected. 1 - protein, 2 - aspartic proteinase activity. b Polyacrylamide gel electrophoresis of the purified enzyme. The electrophoresis was carried out in 7.5% gel using Tris-glycine buffer (pH 8.3). The electrophoresis was run with a 30 ug sample of the protein and the gels were stained with Coomassie brilliant blue for proteins.

focusing in the presence of a mixture of ampholynes at pH 4-6 and 5-8 multiple enzyme forms were found which showed pI-values of 5.1, 5.25, 5.6, 6.9, 7.5, 8.0 (Fig. 3).

The protein content of the enzyme preparations was determined spectrophotometrically from the difference in absorbance at 235 nm and 280 nm (6) (extinction coefficient $A_{280}^{1\ mg/ml}$ - 1.65, at pH 7.4).

The amino acid composition of the purified enzyme is shown in Table 2.

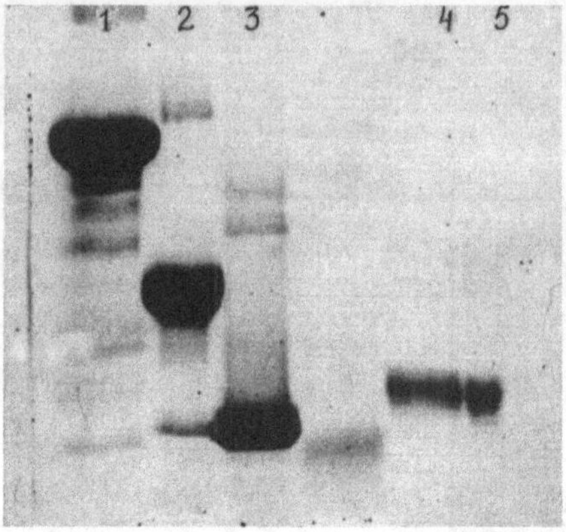

Fig. 2 Sodium dodecyl sulfate polyacrylamide gel electrophoresis of aspartic proteinase. The proteins (30 ug) were denatured 3 min at 100 °C in 0.1 % SDS/2 % mercaptoethanol before the electrophoresis. 1 - bovine serum albumin, 2 - ovalbumin, 3 - chymotrypsinogen, 4 - aspartic proteinase, 5 - aspartic proteinase without mercaptoethanol.

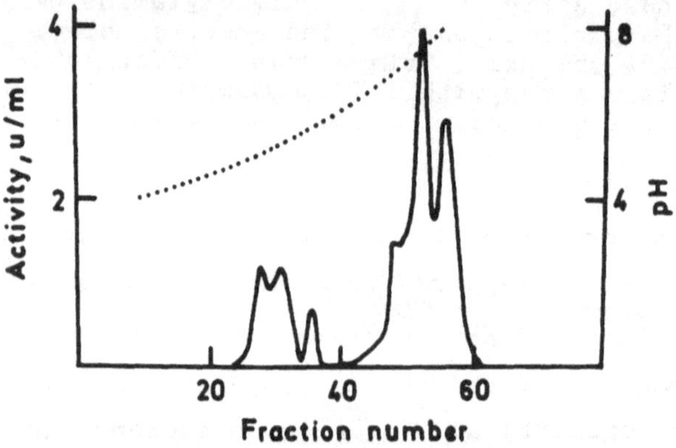

Fig. 3 Isoelectric focusing of aspartic proteinase. The enzyme solution was subjected to column electrophoresis at 2 °C with a sucrose density gradient containing 2 % of the pH 4-6 and 2 % of the pH 5-8 ampholyte. Fractions (2 ml) were collected after isoelectric focusing for 46 h at a constant voltage of 400 V. The enzyme activity (——) and pH (·) of each fraction were determined.

Table 2 Amino acid composition of aspartic proteinase from pig intestinal mucosa, pig spleen cathepsin D, and rat spleen cathepsin E-like proteinase

Amino acid	Aspartic proteinase from pig intestinal mucosa		Cathepsin D from pig spleen (7)	Cathepsin E-like acid proteinase from rat spleen (8)
	residues/			
Lys	28	0.46	0.51	0.24
His	8	0.13	0.14	0.16
Arg	13	0.22	0.31	0.11
Asp	42	0.70	0.83	0.72
Thr	31	0.51	0.51	0.46
Ser	47	0.78	0.91	0.93
Glu	55	0.92	0.91	0.67
Pro	36	0.60	0.60	0.44
Gly	66	1.10	0.89	0.97
Ala	31	0.51	0.71	0.46
Cys/2	n.d.	n.d.	0.23	0.08
Val	28	0.46	0.66	0.56
Met	7	0.11	0.20	0.12
Ile	25	0.42	0.43	0.36
Leu	39	0.65	0.77	0.50
Tyr	24	0.40	0.37	0.13
Phe	20	0.33	0.29	0.33
Trp	12	0.20	0.11	-
Total	512			

Treatment of the protein in the gel slab with the Schiff reagent according to Zacharius (4) showed that the enzyme is a glycoprotein containing according to quantitative analysis (5) 5.9 % of sugars, namely mannose and galactose at a ratio 1:3.

The pH dependence of the hydrolysis of hemoglobin is shown in Fig. 4. The pH optimum is 3.2 in 0.1 M sodium citrate buffer.

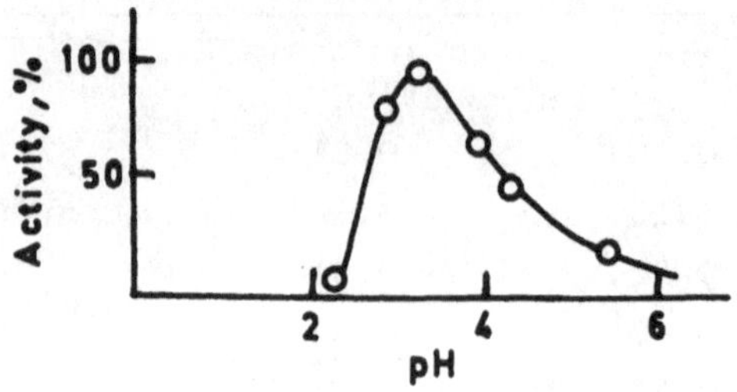

Fig. 4 pH-activity curve of the purified enzyme. The enzyme was incubated with 1 % hemoglobin at 37 °C for 30 min. The absorbance of the trichloroacetic acid soluble products was measured at 280 nm. Maximum activity is taken to represent 100 %.

The kinetic parameters of hydrolysis of the Phe(4-NO_2)-Nle bond in hexapeptide TFA HLeu-Ser-Phe(NO_2)-Nle-Ala-Leu-OMe (K_m=4.5·10^{-5} M; k_{cat}=1.7 s^{-1}) were determined. The isolated enzyme cleaves the bond Leu_{15}-Tyr_{16} in the B-chain of oxidized insulin. Pepstatin inhibits the hydrolysis of both hemoglobin and TFA HLeu-Ser-Phe(NO_2)-Nle-Ala-Leu-OMe by the aspartic proteinase from pig intestinal mucosa (K_i=2.2·10^{-9} M and 1.2·10^{-9} M, respectively). N-diazoacetyl-N′(2,4-dinitrophenyl)ethylenediamine inhibits the hydrolysis of hemoglobin by the enzyme by 80 %.

The isolated enzyme belongs to acid proteases with the pH-optimum below 6.0. As follows from the character of inhibition by pepstatin and N-diazoacetyl-N′(2,4-dinitrophenyl)ethylenediamine the enzyme is an aspartic proteinase. The amino acid composition of the enzyme (Table 2) does not differ any significantly from the data on cathepsin D and E. The pH-optimum, kinetic parameters of hydrolysis of substrates by the proteinase and its inhibition seem to indicate its similarity to cathepsin D. However, the molecular weight of the aspartic proteinase is different from the typical value observed with cathepsin D (35,000-45,000); neither are the two identical subunits characteristic of this group of enzymes. The isolated enzyme selectively cleaves the bond Leu_{15}-Tyr_{16} in the B-chain of oxidized insulin whereas cathepsins D cleave several bonds of this substrate.

References

1. Kregar, I., Turk, V., Lebez, D.: Enzymologia 33, 80-88 (1967).
2. Davies, P.H., Messer M.: Biol. Neonate 45, 197-202 (1984).
3. Knight, C.G., Barrett, A.J.: Biochem. J. 155, 117-125 (1976).
4. Zacharius, R.M., Zell, T.E., Morrison, J.H., Woodlock, J.J.: Anal.Biochem. 30, 148-152 (1969).
5. Khorlin, A.Ya., Gorodilova, V.V., Mandrik, E.V., Mirsoyanova, M.N., Shiyan, S.D., Markin, V.A., Alekhina, T.V., Yermoshina, N.V.: Eksp. Onkol. 2, 34-39 (1980).
6. Whitaker, J.R., Granum, P.E.: Anal. Biochem. 109, 156-159 (1980).
7. Cunningham, M., Tang, J.: J. Biol. Chem. 251, 4528-4536 (1976).
8. Yamamoto, K., Katsuda, N., Kato, K.: Eur. J. Biochem. 92, 499-508 (1978).

X-RAY DIFFRACTION ANALYSIS OF PORCINE PEPSIN STRUCTURE

Natalia Andreeva, Alexander Zdanov, Alla Gustchina, Alexander Fedorov
Institute of Molecular Biology, USSR Academy of Sciences, Moscow, USSR.

Introduction

Pepsin is the most important member of a group of enzymes which are called "aspartyl proteases". Many biochemical and physico-chemical properties of these enzymes were discovered during investigations of porcine pepsin. The three dimensional structure of porcine pepsin was established several years ago in our laboratory by x-ray diffraction studies of monoclinic crystals obtained from water-ethanol solutions (1,2). The structure was found to be homologous to that of mould aspartyl proteases which were studied simultaneously (3,4).
At present we have 2 Å resolution data on the refined coordinates of 2436 atoms of a porcine pepsin molecule (5,6). As the results of the refinement show the actual object of studies is ethanol-inhibited enzyme. The ethanol molecules bound to the active carboxyls of Asp-32 and Asp-215 are clearly visible at 2 Å resolution difference Fourier maps (6,7). Their presence is obviously the consequence of the crystallization procedure performed by using ethanol as precipitating reagent.
The crystallographic work with monoclinic porcine pepsin crystals is not yet completely finalized: the atomic parameters of some outer loops need to be refined better, and all water molecules have to be localized in the unit cell. Only several water molecules at the region of the active

Aspartic Proteinases and their Inhibitors

site are detected at present.

The general structural properties of pepsin

The three dimensional structure of all aspartyl proteases which contain more than 2,000 atoms is rather complicated. We attempted to describe this structure in terms simple as possible to make understandable its main properties. Such description was facilitated by the presence of the intramolecular symmetry. The stereoview of pepsin main-chain conformation is presented in Fig.1

Fig. 1. Stereoview of pepsin main-chain conformation.

The molecule consists of two domains separated by a large cleft. The N-terminal domain contains the first half of the polypeptide chain, including residues 1-175 (approximately), the rest of the molecule, containing residues 176-327, forms the C-terminal domain. However, C-terminal β-loop, 309-327, located between domains,forms hydrophobic contacts with both and can be considered as a part of either domain. The two domains of pepsin, as well as of other aspartyl proteases, have topologically similar folding of the chain (2,8). Superposition of C-αatoms of the pepsin domains shows the presence of 62 equivalent pairs with root mean square deviation of coordinates equal to 3.08 Å. Together with data on the internal sequence homology of porcine pepsin in the region of the two active aspartic acid residues (9) this observation suggested that the evolution of these enzymes included the duplication of a gene corresponding to some precusor protein with molecular weight half of that of pepsin molecule, followed by gene fusion (8).

The similarity of the domains makes it possible to describe their structures in terms common for both of them and to use the unique system of labeling for chain segments. We propose here such system of labeling which seems to be convenient for the practical work.

The domains of pepsin and other aspartyl proteases possess also some kind of symmetry and contain topologically similar regions. This property of aspartyl protease structure was found during the investigation of pepsin electron density maps (2), it was also observed during special studies of fungi enzyme "endothiapepsin"(10). The presence of the intradomain symmetry implies the presence of the repetition of certain structural motif along the chain. Such repetition really exists (See Fig.2.), and can be described as follows: the repeating motif contains about 80 amino acid residues and includes the β-hairpin, or A-element, the loop with the arms far enough to incorporate the additional extended segment between them (B-element), the helical type segment, or C-element, and the second β-hairpin, or D-element. Between loops and helices some irregular segments are observed, therefore the repeating units are not completely identical.

The N-terminal domain is started from the segment containing an extended part, then after the turn the two repeating units follow one after the other. The C-terminal domain is started from a rather long β-strand, then the sequence of the two repeating motifs completes the molecule. After the refinement of pepsin atomic coordinates this description is not changed in general, although some helical type turns are revealed in irregular regions.

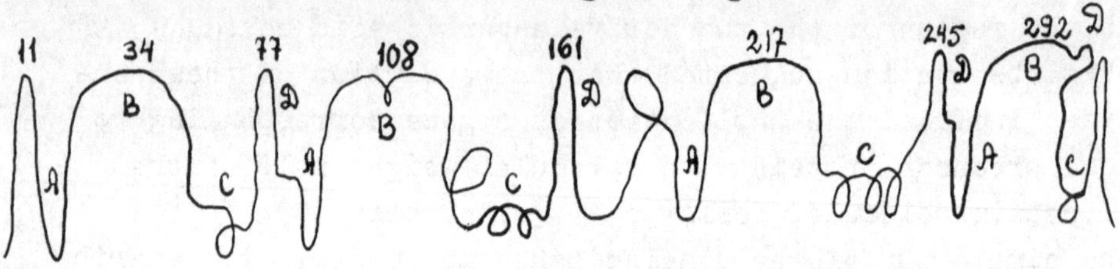

Fig.2. The linear sequence of loops and helices along pepsin chain.

How are the repeating motifs arranged in the three dimensional structure of the domain? The general scheme of this structure is presented in Fig.3.

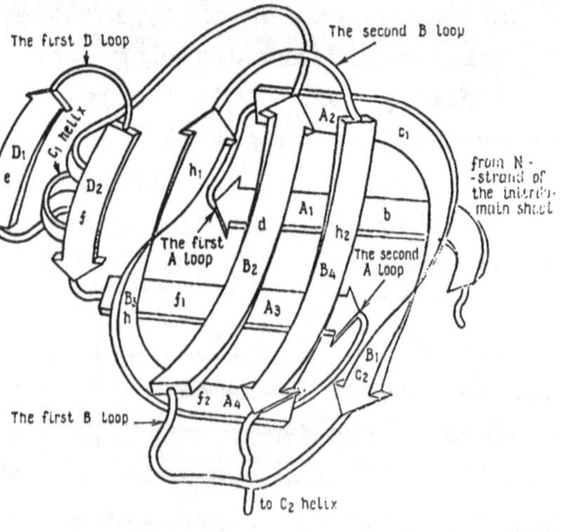

Fig.3a. The scheme of the domain structure, showing the intradomain β-sheet. Capital letters correspond to the proposed labeling of strands; lower case letters correspond to the labeling proposed in Ref. (8). The D-loop of the second repeating motif is involved in the formation of the interdomain sheet.

As can be seen from Fig.3a the domain structure has some kind of a symmetry: the central part is formed from the symmetrically related strands of the A and the B-loops. This scheme is essentially the same as the part of the scheme presented in Ref. (2). After partial refinement of pepsin structure, the symmetrically related fragments of the repeating motifs were superimposed on each other and root mean square deviations of the C-α coordinates were calculated. The calculations showed the presence of 25 equivalent pairs of C-α atoms in the A and the B-loops of N-terminal domain with root mean square deviation equal to 1.58 Å. For 12 equivalent pairs of C-α atoms in the B-loops of the C-terminal domain the root mean square deviation of coordinates is equal to 1,47 Å. These figures support the conclusion about the presence of the intradomain symmetry in pepsin molecule, which can be detected by the visual inspection of the structure. The calculations show that the C- and D-elements are not related by the same translational symmetry as the A- and B-loops.

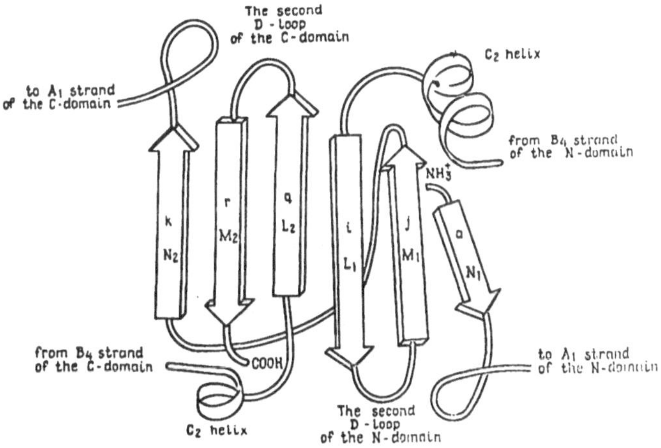

Fig. 3b. The interdomain β-sheet of the aspartyl proteases

N-terminal strands of the domains and their second D-loops are involved in the formation of the interdomain β-sheet. This antiparallel β-sheet is clearly symmetrical (Fig.3b). The presence of the intradomain symmetry and the repeating units suggested to us the possibility that aspartyl protease structure evolved by double duplication or multiplication of a gene corresponding to a smaller substructure or a fragment. However, one should keep in mind another possibility: the intradomain symmetry could be the consequence of a folding process, as one of the way to get a compact structure is to form it from similar asymmetrical objects arranged in a symmetrical manner.(A good example is the paintings of M.C.Escher (11)). It is obvious that every repeating unit is asymmetrical, it has not the compact structure, while mutual arrangement of the two units provides the formation of the compact hydrophobic nuclei. In that event the repetition and the intradomain symmetry would not be relevant to multiplication of the genetic material.

This consideration does not apply to the interdomain symmetry. The resemblance of the domain structures seems to be clear proof of gene duplication as the domains fold separately and form independently similar compact structure. There is a very low probability to obtain such similarity by the covergent evolution.

The general scheme showing the combination of the interdomain sheet and the domain structures in a molecule is presented in Fig.4. As can be seen from this scheme the dominating secondary structure of pepsin and related enzymes is the β--structure. We describe it in terms of three sheets - one inter- and the two intradomain.

The intradomain sheet consists of two layers A and B. The bottom A layer is formed from the two antiparallel A-hairpins and contains A_1, A_2, A_3 and A_4 strands. The upper B-layer is formed from the two B-loops and contains B_1, B_2, B_3 and B_4 strands. The first D-loop of the domain combines with the B-layer by the antiparallel interaction.

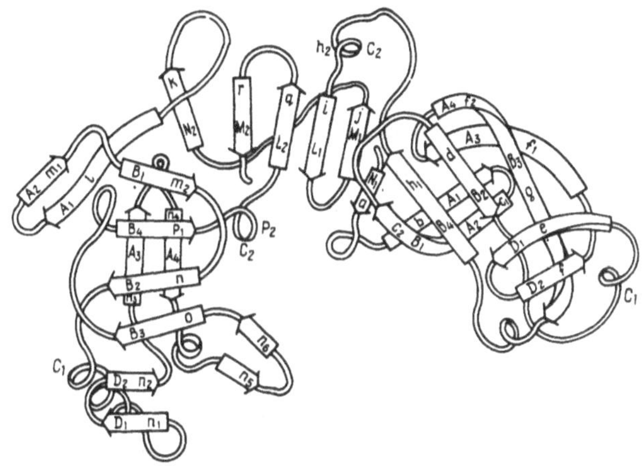

Fig. 4. General scheme of pepsin structure which shows the mutual arrangement of the intra- and interdomain sheets. The intradomain symmetry can be shown only partly in this figure. Capital letters correspond to the proposed labeling of strands; lower case letters correspond to the labeling proposed in Ref.(8).

The mutual arrangement of the B-loops (Fig.3a) is a very specific one. It seems that the loops penetrate into each other forming the compact structure. The simplified scheme of their arrangement is presented in Fig.5.

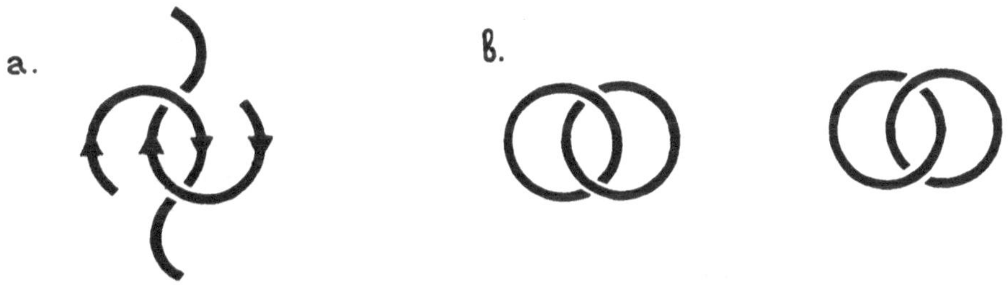

Fig. 5 a) The mutual arrangement of the B-loops.
b) The two enantiamorph forms of the "wedding rings" topology. The first one is presented in aspartyl proteases and in carboxypeptidase.

The illustrated topology can be called as "wedding rings" topology. If one takes into account the direction of the chain, then this topology can correspond to the two different structures and the two their enantiamorphs. One of these structures, present in aspartyl proteases, was called in our previous publications as "pepsin fold". It provides the formation of the antiparallel system of the two pairs of parallel strands. The structure with the same "wedding rings" topology can be observed in the region of the active site of carboxypeptidase (see, for example, Plate XIX of ref. (12)).

The B-layers of both domains form an important part of the active site in aspartyl proteases. At the turn of the first B-loops the active aspartic acid residues are located, while the hydrophobic surfaces of the layers in both domains are involved in the formation of the primary binding pockets for a polypeptide substrate.

There are few helices in pepsin; they are very short and the longest one involves only 10 amino acid residues. In addition to helical type regions designated as C-elements of the structural units there are also helical type turns in the irregular segments of a molecule (see Table 1).

Table 1

Helices or helical type turns in pepsin

The N-terminal domain	Helical type turn C_1	59-63
	Helical type turns at the top of the second B-loop	108-113
	Helix C_2	136-143
The C-terminal domain	Helix C_1	225-236
	Helical type turn C_2	304-308
	Left handed helical type turns	269-273

The differences between the structures of pepsin and mould aspartyl proteases

Comparing the three dimensional structures of pepsin and the aspartyl protease isolated from the primitive fungus Rhizopus chinensis (4) only small local differences are detected. The comparison was made by the method similar to that of Rossmann and Argos (13); root mean square deviation of C-α positions presented in both enzymes was 2.2 Å. Th biggest deviations were found for the first D-loop and the second A-hairpin of the C-terminal domain.

There are differences of the chain conformation at places of insertions and deletions. However, in most cases insertions or deletions involve only one residue. The exception is insertion 291-294 in pepsin. Because of this insertion the more prolonged turn of the second B-loop of the C-terminal domain forms a flap above the S_1' site. This flap is absent in mould enzymes.

There are additional residues at the N-terminus of mould aspartyl proteases as compared to pepsin, and here we have observed an important difference in their structure. The N_1-terminal segment of these enzymes is completely extended and forms a good β-strand of the interdomain sheet. In pepsin this segment has a turn at the end which submerges N-terminal isoleucine side chain into hydrophobic interior between Val-165 and Leu-167, near Phe-31 and Val-91, which are in the main hydrophobic nucleus of the domain (Fig.6). The charged NH_2-terminus of pepsin forms an ion pair with the carboxyl group of Glu-4 oriented inside the molecule. It is also close to peptide group 165-166 of the M_1-strand (Fig.6). This region of the M_1-strand and Glu-4 carboxyl are in the vicinity of the active-site-B-loop of the N-terminal domain. We do not think that the formation of this ion pair is a universal property of aspartyl proteases which have zymogens, as there is variation of the length of the

N-terminal segment and Glu-4 is not a conservative residue. One can see that the formation of the internal ion pair requires conformational changes during the activation process: both A-loops of the domain must move in order to submerge N-terminal isoleucine into hydrophobic interior. Description of these changes will be possible after the establishment of pepsinogen structure, which is under the study (14). Pepsin has two small loops closed by cystine bridges Cys-206-210 and Cys-45-50 which are absent in penicillopepsin. At position 206-210 penicillopepsin has a deletion. The segment 45-50 of this enzyme forms the same loop as pepsin but without the cystine bridge (15).
The internal ion pair Asp-11 - Arg-308 present in pepsin as well as in penicillopepsin (3).

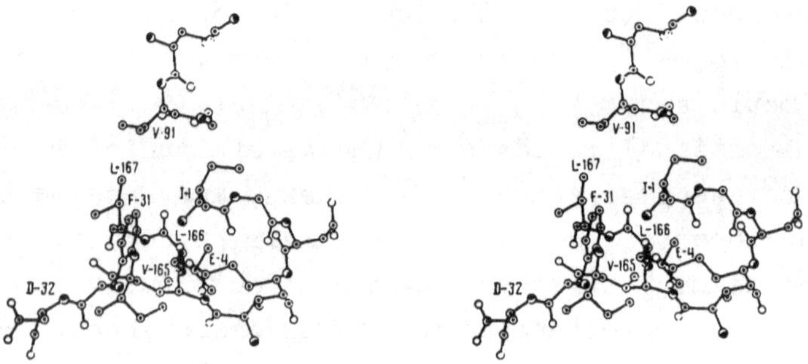

Fig.6 Arrangement of groups around the charged NH_2-terminus of pepsin

Structure of the active site

The two active-site aspartic acid residues of pepsin, Asp-32 and Asp-215, are located at the tops of the two homologous B-loops inside the cleft between the domains. The positions of these loops in the three dimensional structure of the enzyme can be recognized in Fig.1

As mentioned earlier, the actual object of our studies is ethanol-inhibited pepsin. The presence of ethanol molecules at the active site is the consequence of the crystallization procedure. Maxima, corresponding to ethanol molecules, are visible in the first 3 Å resolution maps, calculated with the use of isomorphous replacement method. However, only the 2 Å resolution work shows them convincingly. Several investigation indicated the inhibitory properties of ethanol to be the result of bond formation between the hydroxyl group of an ethanol molecule and one of the active carboxyls (16,17). One ethanol molecule is really observed close enouph to OD2 atom of Asp-215 to form a hydrogen bond; the second ethanol molecule forms a hydrogen bond with water molecule bound to Asp-32.

The arrangement of water molecules and the carboxyls of the active Asp-32 and Asp-215 is slightly different in monoclinic pepsin crystals than that in penicillopepsin (15). The distance between the active carboxyls is too large (3.5 Å) to form a hydrogen bond. However, the binding of ethanol could induce conformational changes. It means that the active carboxyls of aspartyl proteases have some freedom for the movement. Stereoview of the pepsin active site based on the atomic coordinates of the ethanol-inhibited form is presented in Fig.7.

The active carboxyl of Asp-215 forms hydrogen bonds with the oxygen of Thr-218 side chain and the main chain NH of Gly-217. The active carboxyl of Asp-32 forms a hydrogen bond with the main chain NH of Gly-34. The side chain oxygen of Ser-35 is too far from this carboxyl for hydrogen bond formation.

Several hydrophobic pockets are located around the active site. One of them in the C-terminal domain is outlined by residues Ile-213, Ile-301, and Leu-299 that are homologous in aspartyl proteases. This pocket described as the S_1' binding site in penicillopepsin has some specific features in pepsin: there is a flap, 292-298, above this site, which is absent in mould aspartyl proteases. This flap can restrict position of a substrate side chain. The other pocket in the C-terminal domain of pepsin is formed by residues Ile-192, Trp-190, Val-214, Val-184 and Val-321. This pocket is large enouph to accomodate the phenolic ring of a substrate after a small movement of segment 186-190.

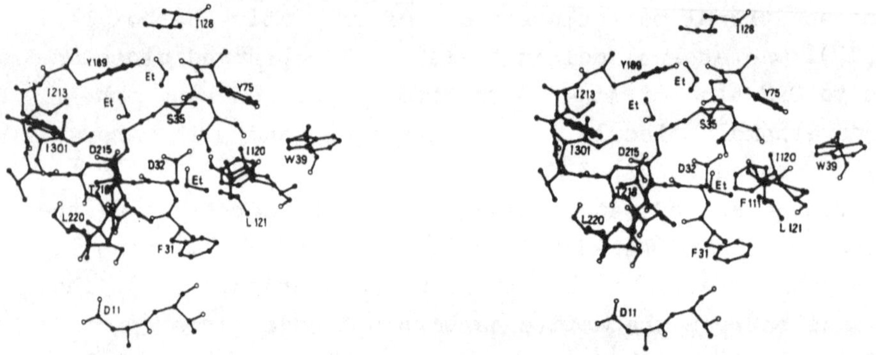

Fig.7. The active site of pepsin (stereoview).

The third pocket is located in the N-terminal domain and includes Ile-120, Trp-39, Tyr-75, and Ile-73. This pocket was described in penicillopepsin as the S_1 binding site. In pepsin it contains additional hydrophobic residues, Phe-117 and Ile-30, the latter being in close contact with Ile-120. There are two tyrosines near the pepsin active site: Tyr-75 and Tyr-189. The hydroxyl group of Tyr-75 is turned away from the active carboxyls.

The experiments on the binding of a substrate-like inhibitor, phenylalanyl-diiodotyrosine methyl ester, to pepsin in monoclinic crystals showed that the phenolic rings of the inhibitor occupy the two pockets in the C-terminal domain mentioned above (7) (See also poster presented by A.Gus-

tchina and N.Andreeva).
The inspection of the binding pockets in pepsin and their comparison with those in mould enzymes permits important conclusions. These pockets are rather large, especially in mould aspartyl proteases where 292-298 flap is absent. There exists a large variety of possibilities to accomodate hydrophobic side chains of a dipeptide substrate in such large pockets in different orientations and positions with the formation of equally good hydrophobic contacts. If one also takes into account the presence of several other hydrophobic pockets in these enzymes, then the low probability of productive binding of short dipeptide substrates, especially by the fungal proteases, becomes understandable.
In pepsin the position of a dipeptide substrate is partly restricted by 292-298 flap. Amino acid residues of this flap, for example Ser-296, may form bonds with some substrate side chains. It can explain the activity of pepsin towards some dipeptides.
The binding of polypeptide substrate side chains in the secondary sites of aspartyl proteases may induce conformational changes in the primary pockets. In other words restriction of a substrate position in the primary binding sites which undergo conformational changes could explain the rapid increase of hydrolysis rate by the increase of a a substrate length.
Data presented in this lecture were published in the cited papers of authors and in papers (18-20).

Aknowledgements: We are grateful to Drs. V.Jurkin and Yu. Lysov for the help in the preparation of stereoviews.

References

1. Andreeva, N.S., Gustchina, A.E., Fedorov,A.A., Volnova,

T.V., Shutzkever, N.E.: Adv. Exp. Med. Biol. 95, 23-31 (1977).

2. Andreeva, N.S., Fedorov, A.A., Gustchina, A.E., Risculov, R.R., Safro, M.G., Shutzkever, N.E.: Mol. Biol. (Moscow) 12, 704-717 (1978).

3. Hsu, I.-N., Delbaere, L.T.D., James, M.N.G., Hofmann, T.: Nature (London) 266, 140-145 (1977).

4. Subramanian, E., Swan, I.D.A., Liu, M., Davies, D.R., Jenkins, J.A., Tickle, I.J., Blundell, T.L.: Proc. Natl. Acad. Sci. U.S.A. 74, 556-559 (1977).

5. Zdanov, A.S., Andreeva, N.S.: Dokl. Akad. Nauk SSSR 272, 109-112 (1983).

6. Andreeva, N.S., Zdanov, A.S., Gustchina, A.E., Fedorov, A.A.: J. Biol. Chem. in press (1984).

7. Gustchina, A.E., Andreeva, N.S., Antonov, V.K.: Bioorg. Khim. 9, 1620-1624 (1983).

8. Tang, J., James, M.N.G., Hsu I.-N., Jenkins, J., Blundell, T.L.: Nature (London) 271, 618-621 (1978).

9. Tang, J., Sepulveda, P., Marciniszyn, J., Chen, K.C.S., Huang, W.-Y., Tao, N., Liu, D., Lanier, J.P.: Proc. Natl. Acad. Sci. U.S.A. 70, 3437-3439 (1973).

10. Blundell, T.L., Sewell, B.T., McLachlan, A.D.: Biochim. Biophys. Acta 580, 24-31 (1979).

11. Escher, M.C.: The graphic work, Ballantine Books, New York (1973).

12. Lipscomb, W.N., Reeke, G.N., Hartsuck, J.A., Quiocho, F. A., Bethge, P.H.: Phil. Trans. Roy. Soc. London Ser. B 257, 177-214 (1970).

13. Rossmann, M.G., Argos, P.: J. Mol. Biol. 105, 75-95 (1976).

14. Rao, S.N., Koszelak, S.N., Hartsuck, J.A.: J. Biol. Chem. 252, 8728-8730 (1977).

15. James, M.N.G., Sielecki, A.R.: J. Mol. Biol. 163, 299-361 (1983).

16. Antonov, V.K.: Adv. Exp. Med. Biol. 95, 179-198 (1977).

17. Nakatani, H., Hiromi, K., Satoi, S., Oda, K., Murao, S., Ichishima, S.: Biochim. Biophys. Acta 391, 415-421 (1975).

18. Andreeva, N.S., Gustchina, A.E.: Biochim. Biophys. Res. Commun. 87, 32-42 (1979).

19. Bakulina, V.M., Borisov, V.V., Melik-Adamjan, W.R., Shutzkever, N.E., Andreeva, N.S.: Kristallogr. 13, 44-48 (1968).

20. Andreeva, N.S., Zdanov, A.S., Fedorov, A.A., Gustchina, A.E., Shutzkever, N.E., Nikitina, L.V., Gorjunov, A.I.: Mol. Biol. (Moscow) 18, 313-322 (1984).

THE HIGH RESOLUTION STRUCTURE OF ENDOTHIAPEPSIN

Tom Blundell, John Jenkins*, Laurence Pearl and Trevor Sewell#

Laboratory of Molecular Biology, Department of Crystallography, Birkbeck College, University of London, Malet Street, London WC1E 7HX, UK

Vibeke Pedersen

Institut for Biokemisk Genetik, Københavns Universitat, Øster Farimagsgade 2A, 1353 København K, Denmark

*Present address :

Biozentrum der Universitat Basel, Abteilung Strukturbiologie, Klingelbergstrasse 70, CH-4056 Basel, Switzerland

#Present address :

Department of Biochemistry, University of Cape Town, Rondebosch, South Africa 7700

Introduction

Aspartic proteinases include not only the mammalian enzymes pepsin, chymosin, renin and cathepsin D, but also the homologous extracellular microbial enzymes penicillopepsin, rhizopuspepsin and endothiapepsin (from Endothia parasitica) (1). Medium resolution X-ray studies of the microbial enzymes (2,3) and pepsin (4) showed that the aspartic proteinases have similar three-dimensional structures in which the two aspartates lie at the centre of a deep and extended active site cleft (2-8). Although the earlier studies indicated that the aspartates (9) were in significantly different environments such that Asp 32 is ionised and Asp 215 protonated (10,11), high resolution studies (12,13) show that the two aspartyl groups are equivalent and related by the local two-fold axis previously identified as relating the two topologically

Aspartic Proteinases and their Inhibitors

equivalent domains of this bilobal enzyme (14,15). In this paper we describe the structure of endothiapepsin refined at 2.1Å resolution to an agreement value of 0.16 and based on the amino acid sequence (V. Pedersen, unpublished results). We describe the secondary structure and hydrogen bonding in terms of the topologically equivalent structural motifs and discuss the intricate hydrogen bonding network at the active site. A later paper (16) concerns the implications of this structure for the mechanism of aspartic proteinases.

Secondary and Tertiary Structure

Figure 1 shows the three-dimensional structure of endothiapepsin. Figure 2 shows a two-dimensional representation of

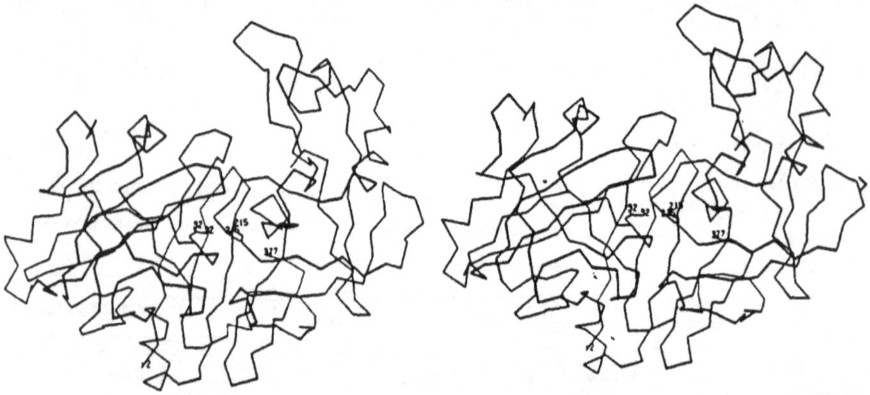

Figure 1 : A stereo view of endothiapepsin

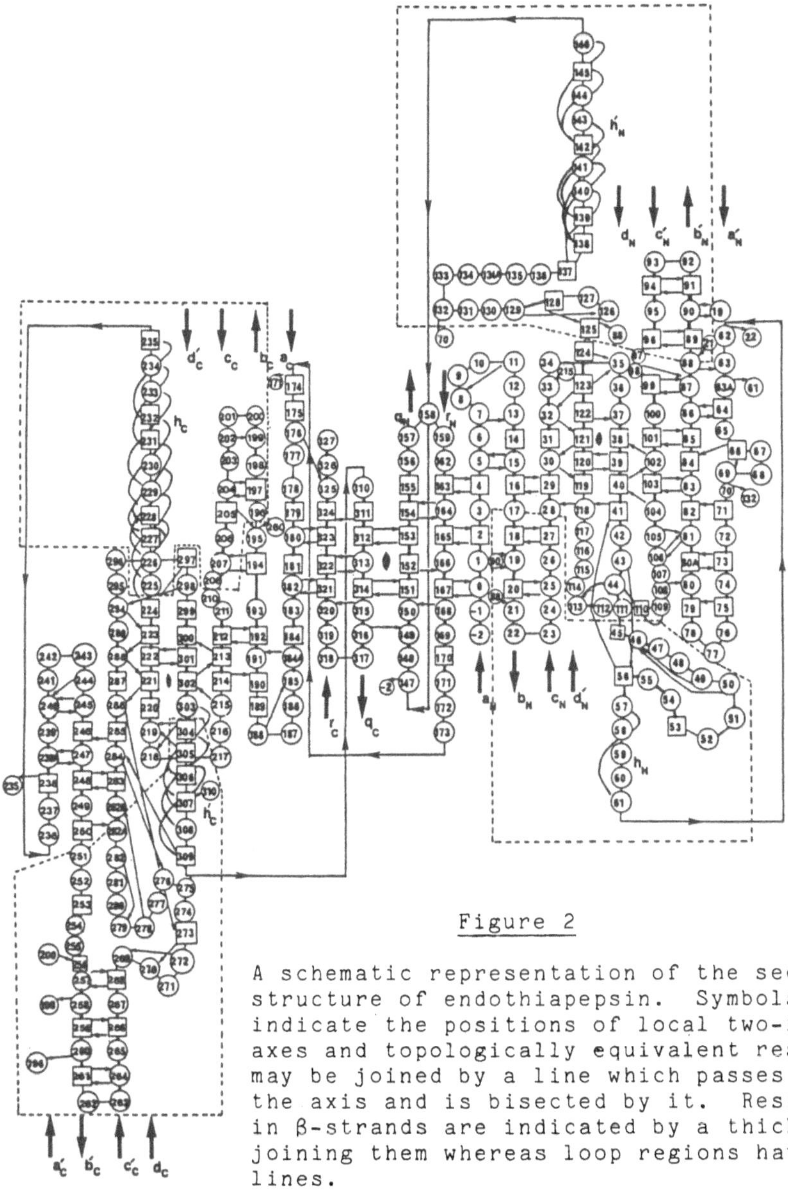

Figure 2

A schematic representation of the secondary structure of endothiapepsin. Symbols o indicate the positions of local two-fold axes and topologically equivalent residues may be joined by a line which passes through the axis and is bisected by it. Residues in β-strands are indicated by a thick line joining them whereas loop regions have thin lines.
Residues which have sidechains which contribute to the core are enclosed in square boxes.

the secondary structure arranged to demonstrate the topological equivalence of the two halves of the sequence each of which make up a domain as described earlier (9,15). Topologically equivalent strands are labelled with the same letter (a, b, c etc.) but with subscripts N or C indicating whether they are in the NH_2- or COOH- terminal domains. Within each domain there is a further topological equivalence between strands abcdh and a'b'c'd'h'. Local two-fold axes between the lobes and within the lobes are indicated (see ref. 15). Hydrogen bonds are indicated where the H---O distance is calculated to be less than 2.8Å (expected distances 1.8Å), if the N-H---O angle is greater than 110° and N---OC angle is greater than 100°.

Figure 2 shows that the secondary structure may be described as a 20-stranded, distorted β-sheet in which each strand has at least two (and usually many more) hydrogen bonds. Strands a, b and c are antiparallel and hydrogen bonded. Strand d is antiparallel to c but hydrogen bonded as a parallel β-sheet to c'; conversely strand d' (related by the local two-fold axis) is antiparallel to c' and hydrogen bonded to c. The strands q_N and r_N are antiparallel and the sheet is extended through hydrogen bonds to strand a_N. These three strands are in turn hydrogen bonded to the topologically equivalent q_C, r_C and a_C strands to form a six-stranded sheet.

Of the topologically equivalent helical regions h'_C, h'_N and h_C are well defined α-helices of between six and twelve residues while helix h_N comprises only one turn.

The tertiary structure is formed by the twisting of the β-sheets and the folding of extended strands b,c and the helices on to the main sheet. For instance, in the NH_2- terminal domain strands b_N and c_N are folded over so that b_{N2} and c_{N1} lie at approximately 90° across the β-sheet comprising b_{N1}, c_{N2}, d'_N and d_N. In a similar way strands b'_{N2} and c'_{N1} are folded on to the sheet b'_{N1}, c'_{N2}, d_N and d'_N. Hydrogen bonds are then formed between strand b_{N2} and the topologically equivalent antiparallel strand b'_{N2} so that strands c_{N1}, b_{N2}, b'_{N2}, c'_{N1} form an antiparallel sheet packed on to the sheet involving b_{N_1}, c_{N_2}, d'_N, d_N, c'_{N_2}, b'_{N_1}. The helices h'_N and h_N are also folded on to this sheet to complete the second layer of the sandwich. A similar arrangement occurs in the COOH- terminal domain. In Figure 2 the folds in the sheets are indicated by dotted lines.

Although the relative positions of the two layers of the sandwich for each of the four motifs are rather strongly conserved, the connecting strands vary tremendously. For example, the bend at $a_C b_C$ is two residues longer than the simpler bend at $a'_C b'_C$. Furthermore $a_N b_N$ is more open and five residues longer than $a'_C b'_C$ while $a'_N b'_N$ is considerably longer giving rise to a "flap" of 19 extra residues which extends over the active site cleft. There is also considerable variation between the four motifs where strands b and c fold over as described above. This is achieved most economically in b_N and c_N while the insertion of one residue in b'_N and c'_N leads to a β-bulge at the fold. There are further insertions in strand b_C and c_C while strands b'_C and c'_C have extensive insertions involving 28 extra residues mostly in strand c'_C. These insertions result in a distortion of the strands a'_C and b'_C relative to c'_C when compared to the topologically equivalent strands in the other three motifs. Note that strands c are generally longer than b as c is on the inside of the molecule at the bend.

The bend between strands c and d involves crossing an intervening strand in the sheet. This is achieved economically and identically between c_N and d_N, and c_C and d_C. These strands are part of the active site deep in the cleft (see next section) and have homologous sequences. However, the residues between strands c'_N and d'_N, and c'_C and d'_C lie on the outside of the enzyme and the polypeptides are considerably extended and of different lengths.

Finally there is great variation in the structural organisation between strands, d, and helices, h. Whereas helices h_C and h'_C derive directly from the β-strands d_C and d'_C the corresponding helices h_N and h'_N have large insertions involving extended loops.

In summary each domain comprises a two-layer sandwich with a pseudo dyad. The four topologically equivalent motifs form the basic structural elements but there are additional residues, sometimes forming loops which lie on the surface of the enzyme. The two domains are brought together by the sheet structure involving strands $a_N r_N q_N q_C r_C a_C$. However, this is achieved through a twist in the strands a_N (between residues 4 and 5) and a_C (between 184 and 185). This results in the positioning of residues 32 and 215 in the cleft between the domains in close proximity to each other and to the dyad relating the two domains. It also gives the appearance of a three-layered sandwich when viewed from certain directions, although the strands and the sheets of arq and $c'_1 b'_2$ (the outer layers of the three-layered sandwich) are approximately at right angles.

The Active Site

Endothiapepsin has an intricate hydrogen bond network involving the two catalytic aspartic sidechains (32 and 215 in pepsin numbering) which comprise the most symmetrical part of the structure (Figure 3a). The aspartyl carboxyl groups lie in a plane with rms deviation <0.1Å. There is a further peak of electron density lying equidistant to the carboxyls and <0.1Å out of the plane (Figure 3b). This may be a water molecule or, in view of the negative charge shared by the two carboxylates, an ammonium (NH_4^+) or oxonium (H_3O^+) ion.

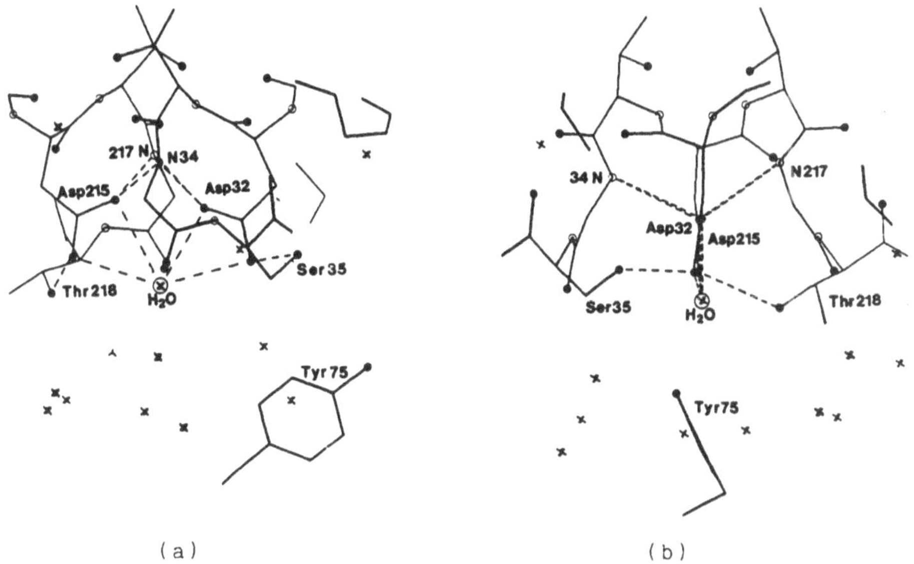

Figure 3

(a) Hydrogen bonding network in the active site of endothiapepsin involving the similar sequences Asp 32 - Thr 33 - Gly 34 - Ser 35 and Asp 215 - Thr 216 - Gly 217 - Thr 218. Dashed lines indicate possible hydrogen bond contacts. The view is normal to the plane formed by the carboxyl groups of Asp 32 and Asp 215.

(b) As (a) viewed parallel to carboxyl plane. There is a two-fold axis relating the 29 corresponding atoms in residues 31 to 35 to those in residues 214-218, calculated using the program EZIFIT of Andrew McLachlan. The axis rotation is 178.8° giving an rms fit between the two sides of 0.38Å. The bound water lies <0.1Å off the axis.

The close proximity of the two carboxyls (Oδ2 of Asp 32 is 2.9Å from Oδ2 of Asp 215) indicates that a proton must lie between them, but the position of the proton in this hydrogen bond cannot be defined by X-ray analysis. The bound water molecule or ion is within hydrogen bonding distance of all four carboxyl oxygens. Both the 'inner' carboxyl oxygens have possible contacts with the peptide nitrogens of the two-fold axis related glycine residues: Gly 34 and Gly 217, although the orientations and distances of these contacts are not favourable for strong hydrogen bonding. Although these interacting glycine peptide nitrogens make no other direct contact, both peptide oxygens, with which they are involved, make hydrogen bonding contacts: O33 directly to a water molecule, and O216 to a sidechain carboxyl oxygen of Asp 304, the other oxygen of which is hydrogen bonded to a water molecule. The 'outer' carboxyl oxygens make hydrogen bond contacts with sidechain alcohol oxygens: Asp 32 to Ser 35 and Asp 215 to Thr 218. The symmetrical arrangement of the two aspartates, the surrounding conserved amino acid residues and the bound molecule indicate that the structure may be the average of two structures so that the hydrogen atom occupies positions close to the Oδ2 atoms of Asp 32 and Asp 215 for equal times. There is nothing in the endothiapepsin structure to differentiate between the two aspartates. In a similar way the molecule bound to the carboxylates, while occupying the same positions, may occupy two different hydrogen bonded orientations.

Close by the two aspartyl groups, the sidechains of the dyad related threonine groups, Thr 33 and Thr 216, are directed towards the hydrophobic core and are involved in a curious interaction we have called a 'fireman's grip' (13). The sidechain Oγs (hands) make possible hydrogen bond contacts with the mainchain bearing the corresponding threonines (arms) as shown in Figure 4. In the case of Thr 33 the contacts are to the peptide oxygen of 214 and possibly the peptide nitrogen of 216, while the dyad related Thr 216 contacts O31 and

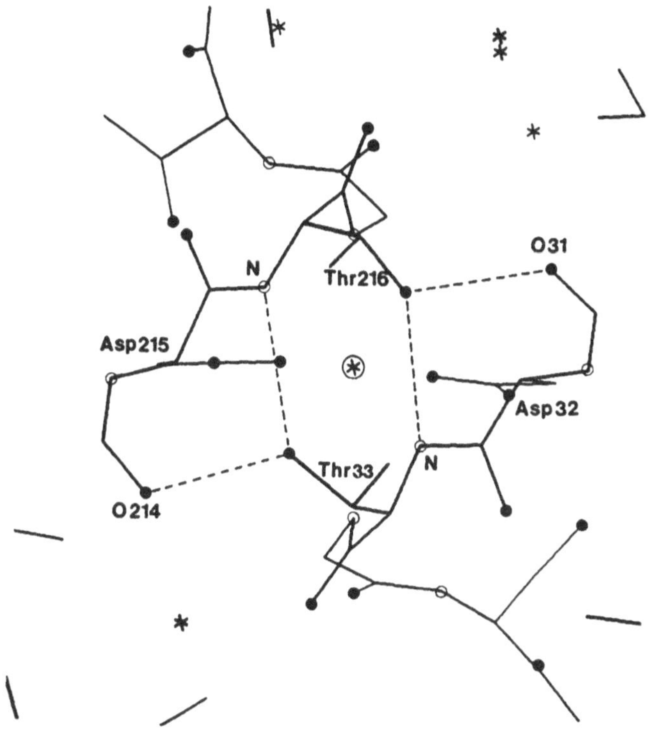

Figure 4

The 'fireman's grip' viewed along the dyad.
The dotted lines indicate probable hydrogen bonds.

possibly N33. The mainchains themselves are out of step and too far apart to make a direct secondary structural interaction of the kind found between strands of β-sheet. In view of the relatively low thermal parameters of the aspartyl groups and their bound solvent, it seems likely that the 'fireman's grip' serves to provide added structural stability for the catalytic centre.

The structures of both endothiapepsin and penicillopepsin show conclusively that with a distance of ∿2.9Å between the carboxylate oxygens there must be a hydrogen bond between them. Loss of the proton in this hydrogen bond would result in a close proximity of two negatively charged carboxylates which would be restrained from moving apart by the extensively hydrogen bonded arrangement. This would increase the free energy of the deprotonated state, make the loss of the proton more difficult and so increase the pK of this proton. With one proton shared in a hydrogen bond between the two carboxylates, the addition of a second proton would be less easy unless the hydrogen bond was broken and the carboxylates displaced in a similar way. This situation parallels that of maleic acid in which $pK_1 = 1.91$ and $pK_2 = 6.33$ with an hydrogen-bonded intermediate species and electrostatic repulsion in the dicarboxylate. Therefore we suggest that the pH profile of aspartic proteinases results from abnormal pKas of the dicarboxylic acid system as a whole.

References

1. Tang, J.: Acid Proteinases: Structure, Function and Biology, Plenum Press, New York 1977.

2. Subramanian, E., Liu, M., Swan, I.D.A., Davies, D.R., Jenkins, J.A., Tickle, I.J., Blundell, T.L.: Proc. Natl. Acad. Sci. (USA) 74, 556-559 (1977).

3. Hsu, I-N., Delbaere, L.T.J., James, M.N.G.: Nature 266, 140-145 (1977).

4. Andreeva, N.S., Gustchina, A.E., Federov, A.A., Shutzkever, N.E., Volnova, T.V.: Acid Proteinases: Structure, Function and Biology (J. Tang, ed.), pp. 23-31, Plenum Press, New York 1977.

5. Blundell, T.L., Jenkins, J.A., Khan, G., Roychoudhury, P., Sewell, B.T., Tickle, I.J., Wood, E.A.: Fed. Eur. Biochem. Soc. 52, 81-94 (1979).

6. Jenkins, J.A., Tickle, I.J., Sewell, B.T., Ungaretti, L., Wollmer, A., Blundell, T.L.: Acid Proteinases: Structure, Function and Biology (J. Tang, ed.), pp. 43-60, Plenum Press, New York 1977.

7. James, M.N.G.: Can. J. Biochem. 58, 251-271 (1980).

8. James, M.N.G., Hsu, I-N., Hofmann, T., Sielecki, A.: Structural Studies on Molecules of Biological Interest (G. Dodson, J.P. Glusker, D. Sayer, eds.) pp. 350-389, Clarendon Press, Oxford 1981.

9. Tang, J., Sepulveda, P., Marciniszyn, J., Chen, K.C.S., Huang, W-Y., Tao, N., Liu, D., Lanier, J.P.: Proc. Natl. Acad. Sci. (USA) 70, 3437-3439 (1973).

10. James, M.N.G., Hsu, I-N., Delbaere, L.T.J.: Nature 267, 808-813 (1977).

11. Blundell, T.L., Jones, H.B., Khan, G., Taylor, G.L., Sewell, B.T., Pearl, L.H., Wood, S.P.: Enzyme Regulation and Mechanism of Action (P. Mildner, J. Ries, eds.) pp. 281-288, Pergamon Press, Oxford 1980.

12. James, M.N.G., Sielecki, A.: J. Mol. Biol. 163, 299-361 (1983).

13. Pearl, L.H., Blundell, T.L.: FEBS Letters 174, 96-101 (1984).

14. Tang, J., James, M.N.G., Hsu, I-N., Jenkins, J.A., Blundell, T.L.: Nature 271, 618-621 (1978).

15. Blundell, T.L., Sewell, B.T., McLachlan, A.D.: Biochim. Biophys. Acta 580, 24-31 (1979).

16. Pearl, L.H.: this volume, pp. 189-195 (1984).

X-RAY DIFFRACTION STUDIES ON PENICILLOPEPSIN AND ITS COMPLEXES: THE HYDROLYTIC MECHANISM

Michael N.G. James, Anita R. Sielecki

Medical Research Council of Canada
Group in Protein Structure and Function
Department of Biochemistry, University of Alberta
Edmonton, Alberta, Canada, T6G 2H7

Theo Hofmann

Department of Biochemistry
University of Toronto
Toronto, Ontario, Canada

Introduction

X-ray crystallography provides a static, three-dimensional picture of enzyme molecules. If data at sufficiently high resolution (> 2.0 Å) are available, exhaustive least-squares refinement of the molecular parameters, can produce an exquisitely detailed view of the enzyme, its active site and the bound solvent molecules (1). However, it is unlikely that the structure of the native enzyme alone would be sufficient for deducing the productive binding modes of substrates. An ultimate goal would be to predict not only the enzyme's structure, but also the structure of the productive enzyme-substrate complex, in order to design molecules that would disrupt the normal catalytic activity. This goal is far away. The most powerful probe of protein structure, that of X-ray crystallography, at present can only provide the structural details of inhibitors and substrate analogue complexes that enable us to explore possible catalytic pathways.

Penicillopepsin is a fungal aspartic proteinase of molecular weight ∿ 34,500. There are 323 amino acid residues in the molecule. The three dimensional structure of penicillopepsin has been refined at 1.8 Å resolution using restrained parameter least-squares (2) to a crystallographic

Aspartic Proteinases and their Inhibitors

R-factor of 0.136 (3). The binding sites on penicillopepsin have been identified by the crystallographic analysis of the complex with the pepstatin analogue Iva-Val-Val-StaOEt (Iva = isovaleryl; StaOEt = the ethyl ester of statine [(4S, 3S)-4-amino-3-hydroxyl-6-methylheptanoic acid]) (4,5). This observed binding mode has been used to explain the trypsinogen activation data of microbial aspartic proteinases and to propose a productive binding mode for a good substrate (6). The complex structure has also been refined at 1.8 Å resolution to an R of 0.131.

In order to verify the proposed binding of a P_1 lysyl side chain to penicillopepsin and to provide additional data for the analysis of the catalytic mechanism, we have also determined the structure of the complex of penicillopepsin with Iva-Val-Val-LystaOEt. This additional pepstatin analogue was synthesized by F. Salituro and Dr. D. Rich at Wisconsin (7). The complex was prepared by Dr. T. Hofmann at Toronto. It has a lysyl side chain at the P_1 position, and LystaOEt is the ethylester of 4S,3S-4,8-diamino-3-hydroxyloctanoic acid.

The observed mode of binding of these tetrahedral intermediate analogue inhibitors to penicillopepsin has led to a proposal for productive substrate binding (6). From this it has been possible to explore possible mechanistic pathways for peptide bond hydrolysis by the aspartyl proteinases. The active site of penicillopepsin is formed in such a way that it is unlikely that the catalytic pathway involves covalently bound intermediates (9), supporting earlier proposals by Fruton (10). In the present discussion, the electrophilic and nucleophilic components of the enzymes active site are those suggested previously (9). The proton shared by the carboxyl groups of Asp33(32)* and Asp213(215) is the electrophile, and the nucleophile is a weakly bound water molecule, O284 of the native penicillopepsin structure; it is hydrogen bonded to $O^{\delta 1}$ of Asp33(32). Thus, we favor a general acid-general base mechanism of hydrolysis. The proton donor that makes the nitrogen of the scissile bond a better leaving group is not easily identified and we conclude that the proton could come from the bulk solvent at acid pH values. The stereochemistry of the hydrolytic cleavage of peptide bonds by aspartyl proteinases conforms to the principle of least nuclear motion (11).

* The numbering scheme is the sequential residue numbers of penicillopepsin. The equivalent pepsin numbers are in parentheses.

Experimental

The present discussion is based on the crystal structures of native penicillopepsin and two of its inhibitor complexes. The experimental details for the solution and refinement for two of these structures have been reported (3,4,5). For the third, a single crystal of the complex of Iva-Val-Val-LystaOEt with penicillopepsin was grown from $(NH_4)_2SO_4$. The space group is the same as that of native penicillopepsin and the isomorphous crystals of the complex with Iva-Val-Val-StaOEt. However, the unit cell dimensions changed substantially, with the largest changes in <u>c</u> and in the monoclinic cell angle β (Table 1). Significant differences in the

Table 1

Crystal Data for Penicillopepsin and its Complexes

Parameters	Native penicillopepsin	Penicillopepsin + Iva-V-V-StaOEt	Penicillopepsin + Iva-V-V-LystaOEt
Space Group:	C2	C2	C2
Unit cell:			
a (Å)	97.38	97.24	97.57
b (Å)	46.61	46.50	46.47
c (Å)	65.42	65.65	66.39
β (°)	115.39	115.34	116.15
Resolution Range:	8.0-1.8 Å	8.0-1.8 Å	6.0-1.8 Å
Reflections Used:	21,962	21,197	22,108
Total No. Atoms*	2363	2401	2402
No. Solvent Atoms†	319	322	170
No. cycles L.S.	86	21	12
R-factor§	0.136	0.131	0.144

* The number of atoms represents the non-hydrogen atoms of penicillopepsin plus those of the pepstatin analogues.

† There are many fewer solvent atoms in the case of the Lysta inhibitor because not all solvents have been selected and the refinement is not complete.

§ Defined as $R = \Sigma||F_o|-|F_c||/\Sigma|F_o|$.

structure factor amplitudes have resulted from these cell dimension changes. Intensity data for the crystal of the complex have been collected to 1.8 Å resolution on an Enraf-Nonius CAD4 diffractometer. Experimental details are similar to those of reference 3. Comparisons of the structure factor amplitudes from the crystal of the Iva-Val-Val-LystaOEt complex with those of native penicillopepsin and those of the complex with Iva-Val-Val-StaOEt show changes of 35.5% and 32.6%, respectively. The difference electron density map using the refined native phases showed positive density that was easily interpretable for most of the inhibitor. Twelve cycles of restrained parameter least-square refinement have been carried out. The initial model for this refinement consisted of all of the atoms of penicillopepsin in the conformation observed for the refined Iva-Val-Val-StaOEt complex (5), with the coordinates of the pepstatin analogue including only up to the C^{β} carbon of the statine residue. The present R factor is 0.144 for the data with $I \geqslant 2\sigma_I$ in the 6.0 to 1.8 Å resolution range (Table 1). The solvent structure has not been fully determined. Only 170 water molecules are included in the present model. However, the essential details of the molecular complex of Iva-Val-Val-LystaOEt can be deduced from the $2|F_o|-|F_c|$, α_c map computed after the initial 10 cycles of least squares. Further refinement is in progress.

Results and Discussion

The binding cleft of aspartyl proteinases is characterized by a major groove in the molecule that extends 28-32 Å across the junction of the N- and C-terminal domains (Figure 1). An extended hairpin β-loop (residues 72-82) projects across this cleft forming a flexible "flap" that effectively encloses inhibitor molecules when bound at the active site (4,5). To the left of the active site aspartyl residues Asp33(32) and Asp213(215) there is an extensive hydrophobic patch that includes residues Leu284(284), Tyr274(275), Leu218(220), Leu220(222), Ile293(297), Phe295(299), Ile297(301), Ile211(213), Phe190(189) and Ile129(128). These residues make contact to residues of substrates or inhibitors in the P_4, P_2, P_1' and P_3' positions. Residues on the "flap" that make interactions with P_1, P_2 and P_3 residues of a substrate are Tyr75(75), Gly76(76) and Asp77(77).

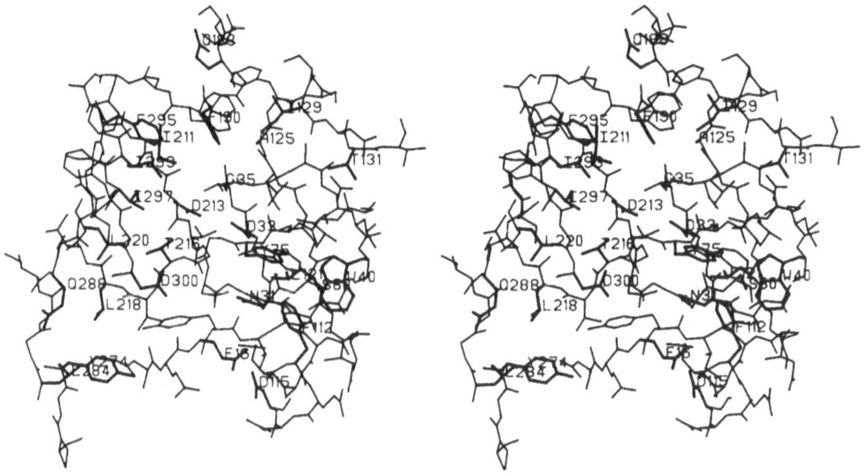

Figure 1. A stereo-optic view of the residues constituting the active site and substrate binding cleft of penicillopepsin. The residue numbering in this and following figures is the sequential residue numbering of penicillopepsin. Residues that figure prominently in the mechanism, or that are important for binding, have been highlighted with a thicker line.

Asn31(30), Phe112(111) and Leu121(120) also provide binding interactions for the P_1 residues of a substrate. These interactions have already been discussed in our analysis of the Iva-Val-Val-StaOEt complex structure (4,5). There is a complicated solvent structure associated with the substrate binding cleft of penicillopepsin (Figure 2). The solvent molecules are integrated into an extensive tight hydrogen bonding scheme that surrounds the carboxyl groups of Asp33(32) and Asp213(215). This hydrogen bonding scheme has been discussed (3) and is reflected in the relative rigidity of the residues comprising the active site region. Hydrogen atom positions have been assigned in order to aid in the interpretation of a hydrogen bonding pattern for residues of the active site (Figure 3). From the pK data available for pepsin (10,12) and the structural results for penicillopepsin, we suggest that the carboxyl group with a net negative charge is that of Asp33(32), at pH 2-4. The hydrogen bonding pattern depicted in Figure 3 is intended to show this situation. Also, it shows the positions of 3 solvent sites in the native enzyme O39, O61 and O284. In the

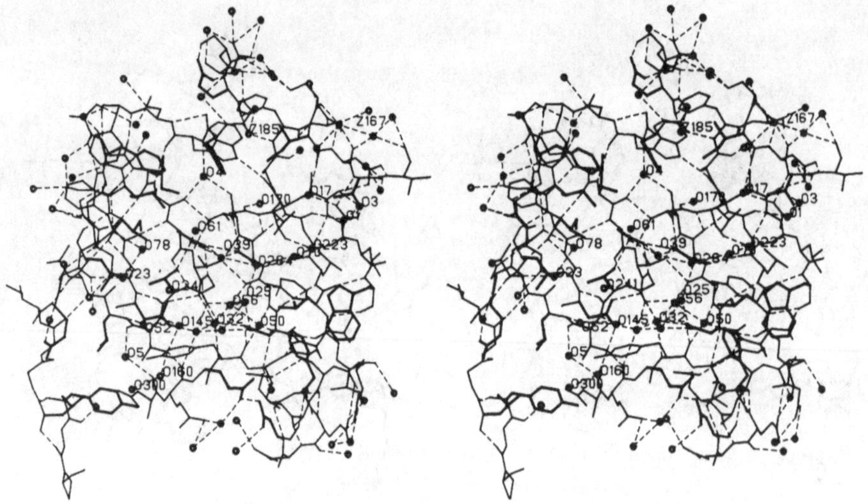

Figure 2. A stereoscopic drawing of the active site shown in Figure 1 with the refined solvent positions and hydrogen bonding interactions included. Highlighted residues are those of Figure 1 and the solvent sites (designated by O and Z with numbers) in the substrate binding cleft include both internal solvents and those tightly bound on the surface. Note the paucity of solvent sites near the residues that constitute the hydrophobic patch next to Asp213(215). Waters O4 and O56 are in almost identical environments and are related by the approximate molecular dyad axis (3). The residues of the flap have been removed from this view for clarity.

mechanistic proposals that follow, the nucleophile is an OH^- ion generated from O284; O39 and O61 are displaced by substrate when it binds.

Binding of Iva-Val-Val-StaOEt

The refined crystallographic structure at 1.8 Å resolution of Iva-Val-Val-StaOEt bound to penicillopepsin (4,5), has provided several very important results to the interpretation of the aspartyl proteinase mechanism. Similar experiments with pepstatin binding to Rhizopus pepsin (13) have been used to deduce a substrate binding mode and mechanistic pathway. This mechanism differs from the one that we have advanced. Both of these studies have confirmed that the tetrahedral 3-C atom of statine, with its

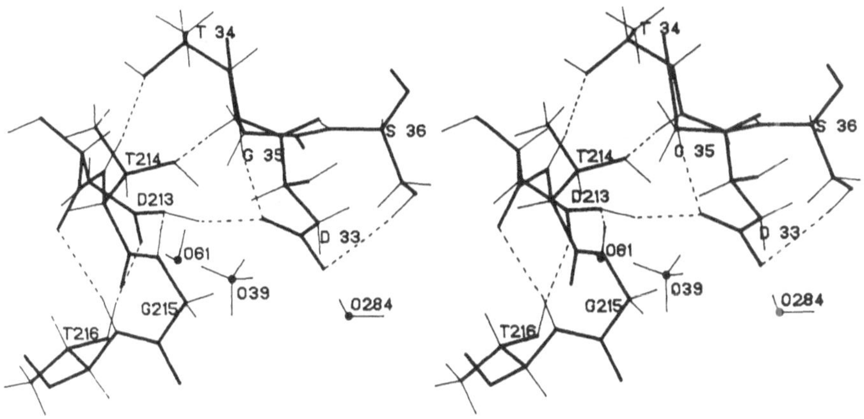

Figure 3. A stereoscopic view of the intricate hydrogen-bonding environment of the active site aspartic acid residues. The proton shared by Asp33(32) and Asp213(215) is shown covalently attached to Asp213(215) although the X-ray crystallographic data does not provide proton positions at this resolution. Hydrogen bonding is indicated by dashed lines drawn from the hydrogen to the acceptor atom. O39 is represented here as an ammonium ion, solvent O61 and O284 as water molecules.

hydroxyl functional group, is an excellent mimic of the tetrahedral intermediate in the reaction pathway of a good substrate. Secondly, the binding mode of Iva-Val-Val-StaOEt shows the detail of how residues in the P_4, P_3, P_2, P_1 and P_1' positions of a substrate would bind to aspartyl proteinases (Figure 4). The crucial interactions between the enzyme and residues in the P_1 and P_1' positions have been conserved faithfully in substrate modelling experiments (6,8). Finally, the mobility of the "flap" has been demonstrated experimentally. This external β-loop of penicillopepsin moves (maximally, ∿ 4 Å at Gly76) upon binding of Iva-Val-Val-StaOEt, thus confirming the importance of enzyme conformational flexibility for catalytic hydrolysis by allowing for optimal substrate binding interactions.

The details of the hydrogen bonding between the pepstatin analogue and penicillopepsin are shown in Figure 4. Thr219 provides two hydrogen bonding interactions to Val 1; its NH donates a proton to the C=O of P_3 Val 1 and the O^{γ} atom of the side chain is the recipient of a hydrogen bond from the NH of P_3 Val 1. Hydrogen bonds involving the P_2 Val residue are all from the residues of the "flap"; Asp77$O^{\delta 1}$ to the NH of Val2 and both

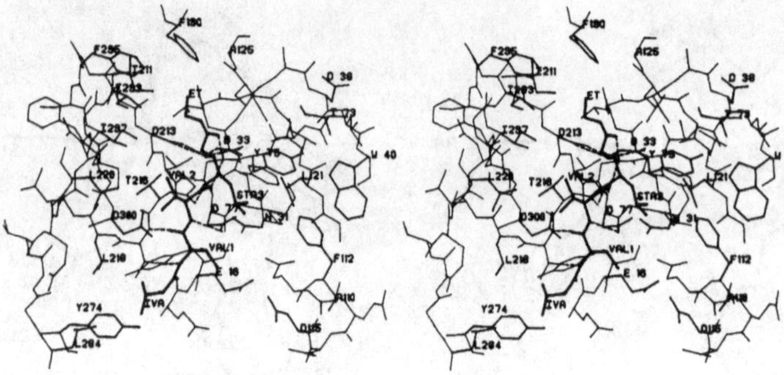

Figure 4. Details of the binding interactions of Iva-Val-Val-StaOEt to the active site residues of penicillopepsin. Hydrogen bonds are dashed, full thicker lines highlight residues of the enzyme important in substrate binding. The flap is in the closed position in the complex.

Gly76NH and Asp77NH provide hydrogen bonding to the C=O of this P_2 residue. The statine group makes the most extensive interactions with penicillopepsin. There are 33 non-bonded contacts that are $\leqslant$ 4.0 Å from atoms of P_1 Sta to the enzyme (5). The 3-OH group forms a complex interaction with the two aspartyl groups. The NH of P_1 Sta forms a hydrogen bond to the C=O of Gly215(217), in what resembles an analogous interaction well characterized for the case of the serine proteinase family. The carbonyl oxygen of the highly conserved Ser214 residue of the serine proteinases is the recipient of a hydrogen bond that strengthens (shortens) as the substrate approaches the tetrahedral intermediate. It is clear that an exactly analogous role in the aspartic proteinases is being provided by the C=O of Gly215(217). We also propose that the C=O of Gly35(34) fulfills a similar binding interaction to the NH of the P_2' residue of a substrate. This group would then be analogous to the C=O of residue 41 in the serine proteinases.

Binding of Iva-Val-Val-LystaOEt

The crystallographic structure of Iva-Val-Val-StaOEt bound to penicillopepsin has allowed us to explore the P_1 subsite specificity of the enzyme (6). The observation that trypsinogen is activated by microbial aspartic proteinases indicates that a carboxylate group(s) on the enzyme, in the region of the P_1 binding site, must provide lysyl specificity. We tried to model this interaction by replacing the leucyl side chain of P_1 Sta by a lysyl side chain and suggested that the carboxyl-carboxylate pair of Glu16 (13) and Asp115(114) would be the anionic site of binding (6). Based on this prediction, Rich et al. (7) have synthesized the inhibitor Iva-Val-Val-LystaOEt and have found that it binds more tightly to penicillopepsin ($K_I = 4 \times 10^{-10}$ M) than does Iva-Val-Val-StaOEt ($K_I = 4.7 \times 10^{-8}$ M). The crystal structure of the complex shows convincingly that the anionic site on penicillopepsin is the carboxylate of Asp77(77) rather than the Glu16(13)-Asp115(114) pair. The electron density of the inhibitor is shown in Figure 5. This map was computed with calculated phases from the partially refined structure at cycle 10, and with coefficients

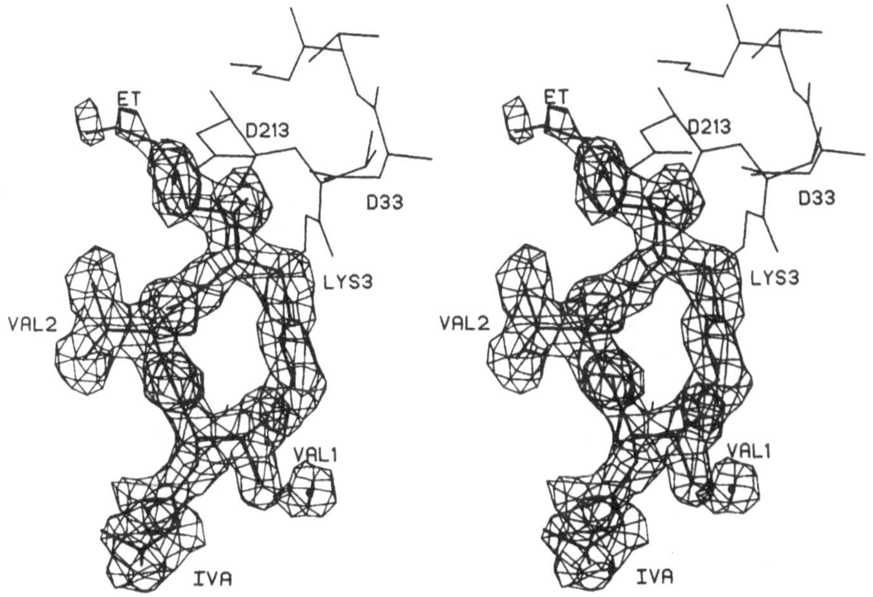

Figure 5. Electron density map in the region of the bound inhibitor Iva-Val-Val-LystaOEt. Except for the ethyl ester, the electron density contours are very clearly indicating the bound molecular conformation. This is a $2|F_o|-|F_c|$, α_c map computed after cycle 10 of the refinement. The contour level shown here is $0.4e\text{Å}^{-3}$. A bound water molecule, O87, is also shown with its density.

$2|F_o|-|F_c|$ (F_o, F_c, are the observed and calculated structure factors, respectively). The density for the lysyl side chain is unambiguous (the contributions of the atoms beyond C^{β} were not included in the structure factor calculation or in the preceding refinement cycles). The hydrogen-bonding interactions of the ε-NH_3^+ group are three-fold; to $O^{\delta 1}$ of Asp77, to O^{γ} of Ser79 and to a water molecule O87 (Figure 6). The hydrogen-bonding and inhibitor-enzyme interactions involving other residues of the pepstatin analogue are essentially the same as observed for Iva-Val-Val-StaOEt, and described in the previous section. A detailed comparison of these two structures will have to await the completion of the refinement of the present complex.

The additional ion-pair and hydrogen bonding from the ε-amino group of P_1 lysine to residues on the flap have reduced the atomic thermal vibration factors (B, Å^2) by a large amount. The B factors of residues 75-80 are approximately 1/2 of those in the crystal structure of the complex.

Two further points are worth noting. The conformation of the ethyl ester is not well defined at present (Figure 5, ref. 5) in either complex structure due to static or dynamic disorder for this group. Secondly, there is

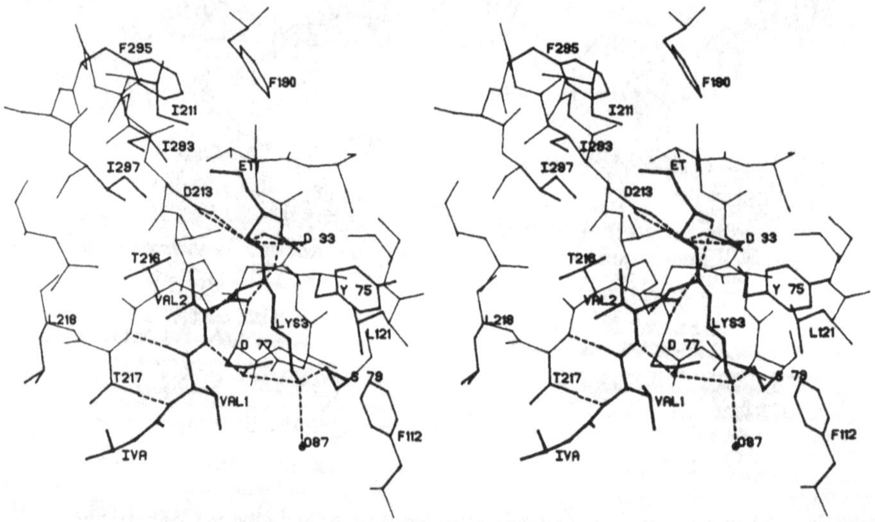

Figure 6. Hydrogen bonding details and the observed bound conformation of Iva-Val-Val-LystaOEt in the binding cleft of penicillopepsin shown in stereo. The ε-amino group of P_1 lysyl side chain forms hydrogen bonds to $O^{\delta 1}$ of Asp77, $O\gamma$ of Ser79 and to water O87. Van der Waals interactions are from this side chain to Tyr75, Leu121 and Phe112.

a further adjustment of the "flap" as a result of the P_1 lysyl side chain interactions (∿ 0.6 Å maximum displacement from the position in the Iva-Val-Val-StaOEt complex). It would appear that the presence of this additional electrostatic interaction has brought about relatively small but fairly extensive structural changes that have also resulted in quite markedly changed (up to 1.5% in c) unit cell parameters of the crystal (Table 1).

The Catalytic Pathway of Peptide Hydrolysis

Based on the inhibitor binding experiments described above, it has been possible to construct a model of a substrate in a productive binding mode (6,8). In this productive mode, we propose that the carbonyl-oxygen atom of the scissile bond points in towards the two carboxyl groups of Asp33(32) and Asp213(215) displacing solvent atoms O39 and O61 of the native structure (Figure 3). However, it is possible still to accommodate solvent O284, at a position ∿ 2.8-3.0 Å from the carbonyl-carbon atom, poised for incipient nucleophilic attack as an OH^- ion.

The catalytic mechanism of aspartyl proteinases has a dominant electrophilic character (10). We have suggested in earlier proposals (9) that the electrophile is the proton that is shared by Asp33(32) and Asp213(215). Thus, the initiation of the catalytic event would be the protonation of the carbonyl-oxygen atom by this proton. Possibly concerted with this protonation event are the electronic rearrangements of nucleophilic attack by OH^- (9), that lead to the formation of the tetrahedral intermediate.

We have now examined the stereochemistry of these initial events by molecular model building experiments (8). The model of the tetrahedral intermediate developed was based on the structure of the pepstatin analogue Iva-Val-Val-StaOEt, using the P_3, P_2, P_1 and P_1' binding site interactions we have described above.

Figure 7 shows the model of the tetrahedral intermediate. The substrate is Ac-Ala-Lys-Tyr-Ala-NH_2 where the scissile bond is that between Lys-Tyr. Both the carbonyl-carbon atom and the scissile bond nitrogen atom must be tetrahedral, as shown, as the nucleophilic attack on the carbonyl carbon atom proceeds. Water O284 is located in an ideal position relative to the

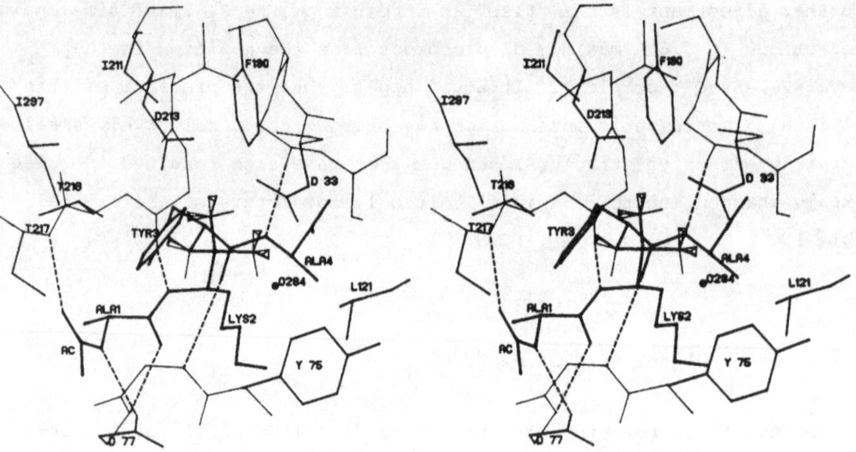

Figure 7. A close-up view of the stereochemistry of the tetrahedral intermediate resulting from protonation of the substrate carbonyl oxygen followed by the nucleophilic attack of OH^- (from O284) on the carbonyl carbon atom of the P_1 lysyl residue. Lone pair electrons on the two OH groups of the tetrahedral carbon and on the nitrogen of the scissile bond are represented by small pyramidal figures. The tetrahedral intermediate stereochemistry was deduced from the productive substrate binding mode and the geometry of the enzymes active site. Lone pair orbitals on the nitrogen and the carbonyl oxygen are oriented antiperiplanar to the direction of bond formation from O284.

carbonyl carbon for this role (14). The water molecule is assisted in the attack by the general base COO^- of Asp33(32). This group is then protonated.

Formation of the tetrahedral intermediate in this scheme, is sterically consistent with the principles of the stereoelectronic effect for amide hydrolysis (15). Thus, for the formation of the tetrahedral intermediate, as well as for its breakdown, there should be two lone pair electron orbitals on the two heteroatoms that are antiperiplanar to the bond undergoing formation or cleavage. Figure 7 shows that both the OH group (corresponding to the original C=O group) and the tetrahedral nitrogen atom, have lone pair orbitals in this required antiperiplanar conformation (that is antiperiplanar to the bond direction O284 to the C atom).

An alternative conformer is required for cleavage of the C-N bond of the tetrahedral intermediate. Additionally, protonation of the nitrogen atom prior to cleavage is necessary to make it a good leaving group. The alternatives for protonation of the nitrogen are two-fold: (a) from bulk solvent with or without inversion of configuration of the tetrahedral nitrogen, and (b) proton transfer via Asp213 from the OH group. It is not simple to distinguish between these two alternatives. Figure 8 depicts the first of these possibilities, in which protonation is from bulk solvent and the OH rotamers of the tetrahedral carbon have lone pair orbitals antiperiplanar to the C-N bond.

Extending these speculations on the hydrolytic mechanism of aspartyl proteinases of microbial origin to the more interesting and physiologically important enzymes of the same family, such as renin, is difficult. There are sufficient amino-acid changes in the regions of the active site residues to expect dramatically altered three dimensional structures. Certainly, structural differences must be responsible for the very different pH optima of the mammalian enzymes. However, the very extensive amino acid sequence homology for the whole molecules would suggest homologous structures so

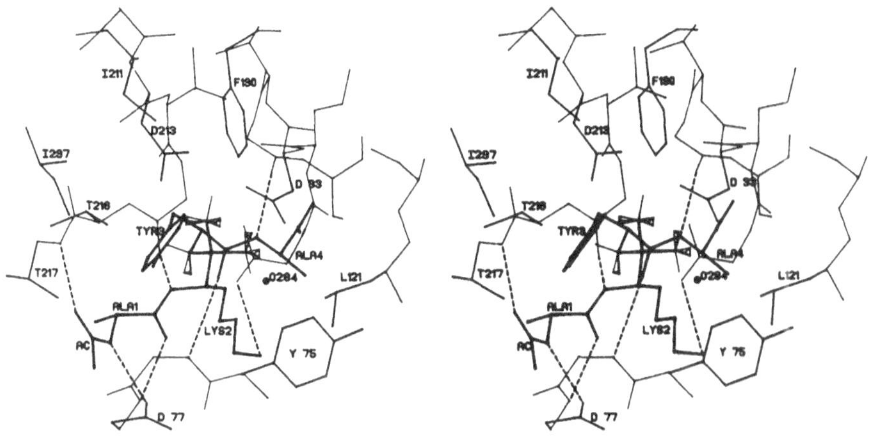

Figure 8. A stereo-optic view of a possible second tetrahedral intermediate in the reaction pathway of aspartyl proteinases. This is a similar view to that in Figure 7 but there are two lone pair orbitals on the two oxygen atoms of the tetrahedral carbon oriented antiperiplanar to the C-N bond according to the proposals of the stereoelectronic effect (15). Additionally the configuration of the nitrogen atom is proposed in order that the lone pair orbital be more accessible to bulk solution for ease of protonation thus making the nitrogen a good leaving group. Direct hydrogen transfer from the carbonyl oxygen via the carboxyl group of Asp213(215) to the leaving group nitrogen is also possible but with the nitrogen in the configuration shown in Figure 7.

that some of our inhibitor binding data can be directly applicable to those enzymes. Subtle structural differences can lead to profound functional differences. It will be exciting to discover those responsible for the differences between mammalian and microbial aspartic proteinases.

Acknowledgements

It is a pleasure to thank Mr. Frank Salituro and Dr. Dan Rich of Wisconsin for the synthesis of Iva-Val-Val-LystaOEt. Koto Hayakawa prepared the crystals that we used. We thank Mae Wylie for the preparation of the manuscript. This research is generously supported by the Medical Research Council of Canada by grants to the Group in Protein Structure and Function at the University of Alberta.

References

1. James, M.N.G., Sielecki, A., Brayer, G.D., Delbaere, L.T.J., Bauer, C.-A.: J. Mol. Biol. 144, 43-88 (1980).
2. Hendrickson, W.A., Konnert, J.H.: In "Computing in Crystallography", (Diamond, R., Ramaseshan, S., and Venkatesan, K., Eds.), pp. 13.01-13.23, Indian Academy of Sciences, Inter. Union of Crystallography. Bangalore, India (1980).
3. James, M.N.G., Sielecki, A.R.: J. Mol. Biol. 163, 299-361 (1983).
4. James, M.N.G., Sielecki, A., Salituro, F., Rich, D.H., Hofmann, T.: Proc. Natl. Acad. Sci. USA 79, 6137-6141 (1982).
5. James, M.N.G., Sielecki, A.R., Moult, J.: In "Peptides, Structure and Function", Proceedings of the 8th American Peptide Symposium, (Hruby, V., and Rich, D., Eds.), Pierce Chemical Company, Rockford, Illinois, pp. 521-530 (1983).
6. Hofmann, T., Hodges, R.S., James, M.N.G.: Biochemistry 23, 635-643 (1984).
7. Rich, D.H., Salituro, F.G., Holladay, M.W.: In "Peptides, Structure and Function", Proceedings of the 8th American Peptide Symposium, (Hruby, V., and Rich, D., Eds.), Pierce Chemical Company, Rockford, Illinois, pp. 511-520 (1983).
8. James, M.N.G., Sielecki, A.R.: Biochemistry (Submitted, 1984).
9. James, M.N.G., Hsu, I.-N., Delbaere, L.T.J.: Nature (London) 267, 808-813 (1977).

10. Fruton, J.S.: Adv. Enzym. 44, 1-36 (1976).

11. Hine, J.: Adv. Phys. Org. Chem. 15, 1-61 (1977).

12. Knowles, J.R.: Phil. Trans. Roy. Soc. B257, 135-146 (1970).

13. Bott, R., Subramanian, E., Davies, D.R.: Biochemistry 21, 6956-6962 (1982).

14. Bürgi, H.B., Dunitz, J.D., Shefter, E.: J. Am. Chem. Soc. 95, 5065-5067 (1973).

15. Deslongchamps, P.: Tetrahedron 31, 2463-2490 (1975).

STRUCTURE OF THE ACTIVE SITE OF PEPSIN AND ITS COMPLEXES WITH INHIBITORS

Alla Gustchina, Natalia Andreeva
Institute of Molecular Biology, the USSR Academy of Sciences, Vavilova 32, Moscow 117984

During X-ray analysis of the three-dimensional structure of porcine pepsin the enzyme crystals growing from 20% ethanol solution were used. At the region of the active site of the enzyme the additional electron density maxima have been observed in the proximity of active groups (Fig. 1). The interpretation of this additional maxima was done by the use of the difference Fourier synthesis calculated with coefficients (2 Fobs - Fcalc) at the final steps of the refinement procedure of pepsin structure at 2Å resolution (1). It was found that ethanol molecules - competative inhibitor of aspartyl proteinases - are located at the active site of pepsin (2,3). The hydroxyls of two ethanol molecules are hydrogen bonded to oxygen atoms of active Asp 215 and Asp 32 (via the water molecule in last case). As the consequence the probable in the

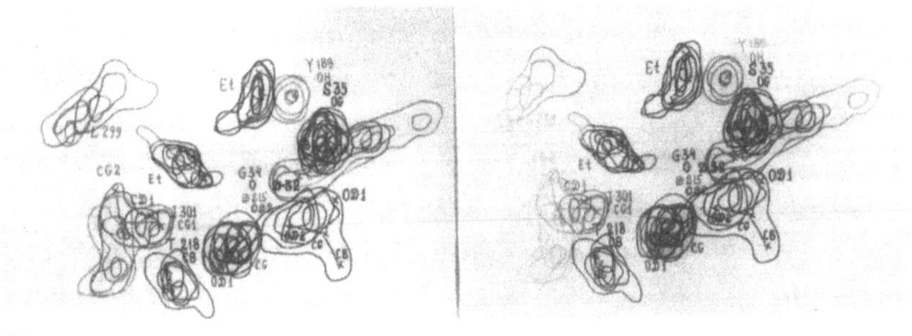

Fig. 1 Section of the difference Fourier synthesis showing the presence of ethanol molecules.

Aspartic Proteinases and their Inhibitors

native enzyme hydrogen bond between oxygen atoms of active aspartic acid residues **is** distorted and their distance increases up to observed value 3.5Å. The hydrophobic moieties of ethanol molecules contact residues Leu 299, Ile 301 and Ile 120 forming primary binding sites S_1 and S_1'.

On the base of this data the scheme of interaction of the ethanol molecules with active groups of the enzyme is proposed which explains the inhibiting properties of this compound. This scheme takes into account the known data on the different pK values of the two active aspartic residues and the properties of alcohol molecule to be in water solution as well donor as acceptor (4).

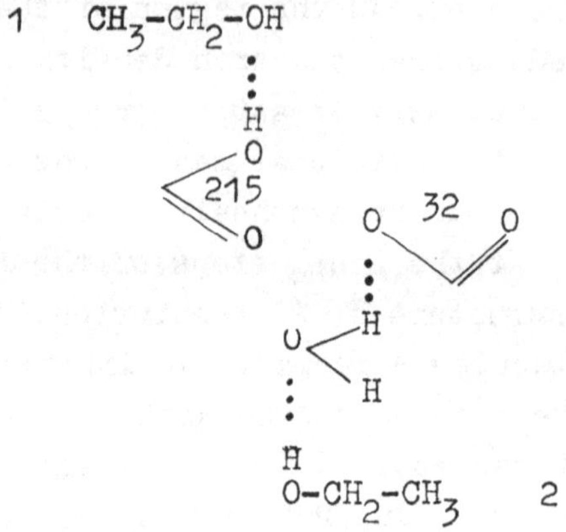

The formation of the hydrogen bond between the first ethanol molecule and OD_2 atom of Asp 215 decreases probability of hydrogen bonding of this atom to OD_2 atom of Asp 32. Therefore the distance between closed oxygen atoms of catalytic active aspartic acid residues increases. The binding of the second ethanol molecule via water molecule with OD2 atom of Asp 32 also decreases the probability of hydrogen bond forming between OD2 atoms of residues Asp 32 and Asp 215 and as a result the negative charge on OD2 atom Asp 32 is partially neutralized and nucleophilic properties of this residue must be weaker.

This scheme is in a good agreement with some biochemical

data (5-8). The presence of additional structural elements at the hydrophobic pockets S_1' and S_1 of pepsin - loop 292-298 above the S_1 site and residue Phe 117 at the S_1 site, restricting the arrangment for the substrate side chains, explains the occurence of the hydrolysis by pepsin of such short substrates as dipeptides.

Upon complexformation of pepsin with inhibitors 1,2 - epoxy -(3-p-iodophenoxy)propane covalently **binds** to the enzyme. The aromatic groups of these compounds occupy the primary binding sites S_1 and S_1', thereby including small conformational movements of the (292-298) loop in the S_1' region and a significant movement in the (68-83) loop in the S_1 region (9). In the structure of the pepsin complex with dipeptide Phe - $TyrI_2$-OME the iodinated tyrosine ring occupies the S_1' binding site while the phenylalanine ring is located in the pocket formed by Val 214, Ile 192, Trp 190 and Val 321 (10). The arrangement (see Fig. 2) of phenylalanine ring in this pocket is predetermined by the necessity of significant conformational movements upon binding of the aromatic groups at the S_1 site of pepsin, which is difficult, owing to a particular packing of molecules in pepsin crystals.

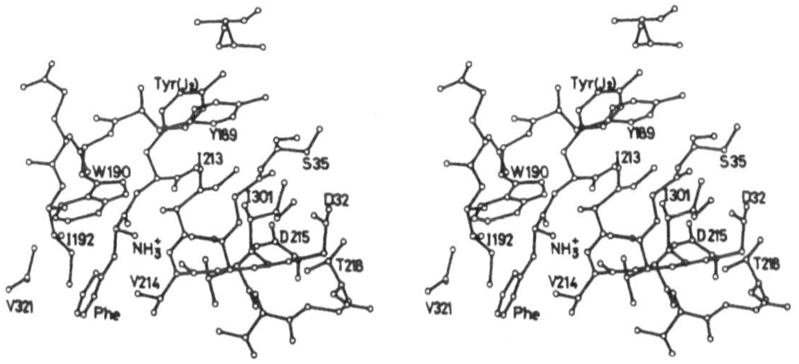

Fig. 2 The scheme of the arrangment of Phe-$TyrI_2$-OMe molecule bound to pepsin

References

1. Zdanov, A.S., Andreeva, N.S.: Dokl. Akad. Nauk SSSR 272, 109-112 (1983).
2. Tang, J.: J. Biol. Chem. 240, 3810-3815 (1965).
3. Gustchina, A.E., Andreeva, N.S., Antonov, V.K.: Bioorg. Khim. 9, 1620-1624 (1983).
4. Jorgensen, W.L., Madura, J.D.: J. Am. Chem. Soc. 105, 1407-1413 (1983).
5. Antonov, V.K.: Adv. Exp. Med. Biol. 95, 179-198 (1977).
6. Nakatani, H., Kitagishi, K., Hiromi, K.: J. Biochem. (Tokyo) 87, 563-571 (1980).
7. Nakatani, H., Hiromi, K., Satoi, S., Oda, K., Murao, S., Ichishima, E.: Biochim. Biophys. Acta 391, 415-421 (1975).
8. Kitagishi, K., Nakatani, H., Hiromi, K.: J. Biochem. (Tokyo) 87, 573-579 (1980).
9. Andreeva, N.S., Gustchina, A.E., Fedorov, A.A., Volnova, T.V., Shutzkever, N.E.: Adv. Exp. Med. Biol. 95, 23-31 (1977).
10. Andreeva, N.S., Gustchina, A.E., Fedorov, A.A.: Dokl. Akad. Nauk SSSR 259, 1261-1264 (1981).

THE DETERMINATION OF THE THREE - DIMENSIONAL STRUCTURE OF CHYMOSIN

Mark Safro, Nataliya Andreeva, Alexander Zdanov
Institute of Molecular Biology, the USSR Academy of Sciences, Vavilova 32, Moscow 117984

Chymosin - is the important member of the aspartyl protease family. The practical significance of this enzyme in food industry concernes its activity in milk clotting process, i.e. rapid enzymic cleavage of specific bond in κ- casein, stimulated the investigations of its three - dimensional structure. The using of a multiple isomorphous method had been met with difficulties because of the ambiguous interpretation of the difference Patterson map (1). At the same time amino acid sequence homology of chymosin and pepsin is approximately 80% (2,3). This fact and the knowledge of the pepsin structure at 2Å resolution (4) became the basis of studies of chymosin structure by molecular replacement method. Kindly sent to us experimental data on chymosin by T.Blundell and C.Bunn were used at this stage.

At the first step during low resolution studies, we have used the Fourier inversion of electron density map for calculations of structure factor of a model molecule. The molecule was isolated from electron density map by a polygons method. The maximum of the rotation function pepsin versus chymosin had a rather specific shape with two submaxima. The greater one was used for determination of mutual orientation of molecules (5).

The coordinates of pepsin molecule, placed in $2\vec{a}_p, 2\vec{b}_p, 2\vec{c}_p$ unit cell were used for structure factor calculations in the work with a 3Å resolution data. Besides that, structural data

Aspartic Proteinases and their Inhibitors

of Rhizopus chinensis pepsin (6) were used for refining of orientational parameters. Three types of rotation function maps were calculated, using Crowther Fast rotation algorithm :

$$[C]_{1-2} \backsim \{\text{Chymosin exper. data(1) - Pepsin model data (2)}\}$$
$$[C]_{1-3} \backsim \{\text{Chymosin exper. data(1) - Rhizopus model data(3)}\}$$
$$[C]_{2-3} \backsim \{\text{Pepsin model data (2) - Rhizopus model data(3)}\}$$

where $[C]$ represents a corresponding rotation matrix. Solution of equation :

$$[C]_{1-3} = [C]_{1-2} \cdot [C]_{2-3} ,$$

proved the correctness of the maximum(first maximum in the Table 1) at 5.5Å resolution.

Table 1

Summary of rotational and translational parameters for chymosin, pepsin and Rhizopus chinensis pepsin.

Type of data	Rotational parameters		Translational parameters
	Rossmann-Blow procedure $\theta_1^\circ, \theta_2^\circ, \theta_3^\circ$	Crowther procedure $\alpha^\circ, \beta^\circ, \gamma^\circ$	Crowther-Blow procedure $\Delta x, \Delta y, \Delta z$
1 - 2	312,77,28(127)* 305,80,65(122)	105,50,140(140) 100,55,110(146)	-0.33,-0.25,-0.275 No solution
1 - 3	248,55,143(90) 249,58,170(90)	155,50,50 (56) 155,55,85 (90)	
2 - 3	168,142,80(80) 0, 30,270(50)	-5,85,142(85) 105,85,162(75)	Unambiguous sol.

*) The arbitrary scale was used in each case. The numbers in parentheses are the two biggest maxima in a RF map.

The components of translation vector $\vec{t}$ were determined by Crowther - Blow method (7). Should be noted, that the translation function was unable to choose one out of two possible space groups I222 and $I2_12_12_1$ in this case. On the sections of the translation function (TF) calculated at x=0,y=0,z=0 was found only one molecular position with large peaks in all three sections which corresponds to group I222. However, only one was the biggest within its own section (with the changing of the resolution limits, peak on the other section became the biggest). In this connection we had analysed the solution of TF, taking account of the requirement of packing. The number of possible intermolecular contacts between α - carbons disposed at a distance < 5Å, was equal to six. The solution of TF corresponded to four intermolecular contacts. The components of TF vector (see Table 1) were refined by minimization of R - factor and it value made up 53.7%.

The refinement of the chymosin structure was carried out by simultaneous minimization of expression $\sum(|F_o| - |F_c|)^2$ and conformational energy according to the Jack-Levitt method(8). The pepsin molecule with the replacement of side chains by Ala in the regions, where differences in amino acid sequences occurred, was used as a starting model.

After four cycles of the refinement a Fourier map with coefficients $(2|F_o| - |F_c|)$ was calculated. It provided the evidence of reasonable quality of the current phases and was able to trace a polypeptide chain of molecule (Fig. 1). One of the differences which is revealed in the chymosin and pepsin structures was found near the N-terminus. Residues Gly and Glu in -1 and -2 positions were identified on the map. The difference in the structure of the loop 289-297, where chymosin has a deletion (four residues long), was also revealed. Replacement of Leu 299 (pepsin) by Lys and Leu 220 by Lys are interesting because their importance for enzyme activity. Now position of Lys 299 is not quite clear, but hydrophobic part of Lys 220 is directed to Ile 301 and terminus

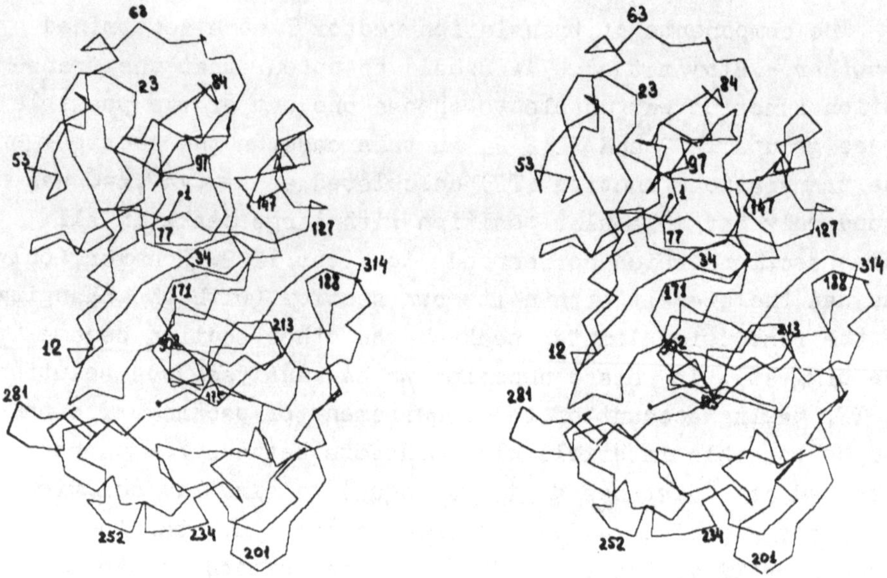

Fig. 1 Stereo-drawing of the conformation of chymosin chain

of Lys 220 is close to the active site of molecule. Some side chains which have distinctly different structure in pepsin and chymosin were detected. For instance, side chains of Tyr 19, His 74, Glu 160, Met 164 taken as Ala were visible. Current R-factor value is 40% at 3Å resolution.

References

1. Bunn, C.W., Camerman, N., Tsai, L.T., Moews, P.C., Baumber, M.E.: Phil. Tran. Roy. Soc. London Ser B 257, 153-158 (1970).
2. Foltmann, B., Pedersen, V.B., Kaufmann, D., Wybrandt, G.: J. Biol. Chem. 254, 8447-8456 (1979).
3. Sepulveda, P., Marciniszyn Jr., J., Liu, D., Tang, J.: J. Biol. Chem. 250, 5082-5088 (1975).
4. Zdanov, A.S., Andreeva, N.S.: Dokl. Akad. Nauk SSSR 272, 109-112 (1983).

5. Safro, M.G., Andreeva, N.S.: Dokl. Akad. Nauk SSSR 247, 758-761 (1979).

6. Subramanian, E., Swan, I.D.A., Liu, M., Davies, D.R., Jenkins, J.A., Tickle, I.J., Blundell, T.L.: Proc. Natl. Acad. Sci. U.S.A. 74, 556-559 (1977).

7. Crowther, R.A., Blow, D.M.: Acta Crystallogr. 23, 544-548 (1967).

8. Jack, A., Levitt, M.: Acta Crystallogr., Sect. A 34, 931-935 (1978).

THE EXTENDED BINDING CLEFT OF ASPARTIC PROTEINASES AND ITS ROLE IN PEPTIDE HYDROLYSIS.

Laurence Pearl

Laboratory of Molecular Biology,
Department of Crystallography,
Birkbeck College,
University of London,
Malet Street, London WC1E 7HX

The rate of peptide hydrolysis by an aspartic proteinase shows a marked dependance on the amino acid composition and length of the peptide substrate being hydrolysed. Within a series of related substrates, the rate of hydrolysis (Kcat), can vary by several orders of magnitude, in contrast with the binding constant (equivalent to Km) which is comparatively insensitive (1,2). Between different enzymes, the catalytic selectivity shown to a given substrate can also vary considerably.

If we make the reasonable assumption that the mechanism of catalysis is the same for all substrates, then we may suggest that the rate of hydrolysis is in some way a function of the nature of the interactions in the extended binding cleft, and appears to be dependent on the number of specificity subsites occupied and the properties of the substrate side chains that bind at the subsites and the enzyme residues that form them.

TRANSFER OF BINDING ENERGY

The binding of a substrate or inhibitor to an enzyme, is

Aspartic Proteinases and their Inhibitors

accompanied by an energy change, sufficiently exothermic as to considerably favour the bound state. The exothermicity results from the interactions between the groups that constitute the ligand and those that form the binding site. For an enzyme such as an aspartic proteinase, which is capable of interacting with eight or more residues of a peptide substrate, it is positively disadvantageous to express all the available energy of interaction as stabilisation of the bound complex, which would serve to increase the activation barrier to achievement of the transition state, and make the productive reaction less likely. It is however highly advantageous if the enzyme can transfer the interaction energy, extra to that required for favourable binding, in to performing work that serves to lower the activation barrier to the reaction. Thus the large differences in rates of hydrolysis between different substrates may be explicable as the differences in extra binding energy for these subtrates, and thus the degree to which they may promote such work.

ENZYME CONFORMATIONAL CHANGES

For the surplus binding energy to promote catalysis, it must exert an influence on the groups that catalyse the reaction or on the bond to be hydrolysed, or both. An induced fit model suggests that the conformation of the enzyme is changed to that required for activity, by the binding of the substrate (3). As the interactions that provide the energy to do this in aspartic proteinases, are generally distant from both the scissile peptide bond and the catalytic aspartates, there must exist a pathway for the transfer of that energy. Furthermore, this pathway should account for the additive effect of binding in different subsites, on the rate of hydrolysis. Generally, the rate enhancement for a larger more specific substrate over that observed for a smaller substrate is cumulative. For example, in porcine pepsin, the kcat observed for the hydrolysis of Z-Gly-Gly-Phe-Phe-OP4P is greater than that for Z-Gly-Phe-Phe-OP4P which is in turn greater than that for Z-Phe-Phe-OP4P.

The rate dependence phenomena in aspartic proteinases have been conventionaly explained by an induced fit model. However, most systems that are seen to display activity inducing conformational changes, such as the swinging Tyr 248 in carboxypeptidase A, do so as binary switches, being stabilised in one state or the other by the absence or presence of the substrate. Clearly such a single system could not satisfy the gradations of activity shown by aspartic proteinases. It is possible to propose a set of such switches corresponding to each subsite, which could cooperatively produce gradated effects, but it would still be neccesary to show how these communicated their effect to the active site.

The small conformational changes that have been observed in the available inhibitor complex structures (4,5,6) do not noticeably effect the conformation or environment of the catalytic groups in any way that would influence their ability to catalyse hydrolysis. It would be unwise to totally dismiss the possibility of protein conformational changes having a role in the rate dependence phenomena, but a far more satisfactory model results from consideration of the other component in the catalytic reaction.

SUBSTRATE CONFORMATIONAL CHANGES

The strain hypothesis or 'Circe effect' proposed by Jencks

(7) suggests that the binding forces between the enzyme and its substrate are used to cause bond distortions which facilitate the reaction. In the case of hydrolysis of a peptide, the only bond in the substrate whose distortion could facilitate the reaction, is the peptide bond between the P1 and P1' substrate residues. The strain hypothesis suggests that the observed binding energy for a ligand which is distorted on binding, is less than for a ligand that can bind in the same conformation without distortion. The difference being the energy required to cause the distortion. Comparison of Km for substrates with the Ki for homologous reduced peptide inhibitors (8), shows a difference in expressed binding energy of the order of 5 kcal/mole, suggesting that there is indeed a strain mechanism in operation (9).

Of all the ways in which the stable trans-planar peptide group can be distorted, rotation of the C-N bond is the least energetically demanding, and would have important effects upon the susceptibility of the carbonyl carbon to nucleophilic attack, the magnitude of these effects being a function of the degree of rotation away from planarity. In an inhibitor with a simple single bond replacing the C-N bond (such as a reduced peptide inhibitor), there would be no inherent barrier to rotation, and the equivalent conformation could be achieved without expenditure of energy.

A mechanism of binding induced substrate distortion avoids the problems of transfer of binding energy faced by an enzyme induced fit model, as the side chains occupying the specificity subsites and the scissile peptide itself are directly conformationally linked. Thus we propose that torsional distortion of the scissile peptide bond in the bound complex, decreases the delocalisation of the peptide nitrogen lone-pair, making the carbonyl carbon more susceptible to nucleophilic attack and the nitrogen a better proton acceptor; these factors will lower the barrier to the rate limiting step of formation of the transition state and thus promote catalysis. We further propose that this distortion is stabilised by the energy of interaction of the substrate with the subsites forming the extended binding cleft. The degree of torsional distortion and therefore rate of hydrolysis, being a function of the length and composition of the peptide substrate.

CATALYTIC MECHANISM

The active sites in the high resolution refined structures of penicillopepsin (10) and endothiapepsin (11), are sufficiently similar as to allow superposition of their atomic coordinates. The differences in mechanism proposed from consideration of these two structures, are due to different interpretations of the data rather than differences in the data themselves. It should be emphasised that the torsional distortion hypothesis described above, would apply to both mechanistic models.

The active site of endothiapepsin contains two aspartic acid residues, 32 and 215 (pepsin numbering), whose carboxyl side chains are coplanar and make a mutual oxygen-oxygen contact of 2.9 angstroms, suggesting that they are hydrogen bonded. A peak of electron density lies equidistant to the two carboxyl groups, and the geometry suggests it to be a water molecule tightly bound to the two aspartates. We have suggested, that the geometry of the water-carboxyl interactions is the time average of two hydrogen bond structures, in which the proton involved in the 32-215 hydrogen bond is equally likely to be close to either carboxyl group. We also suggest that such a carboxyl-water system will have the effect of polarising the water molecule and transfering some of the net negative charge of the two carboxyls to the water oxygen, therebye increasing its nucleophilicity. Thus, in contrast with James et al (12,13) we suggest that it is most likely to be this water molecule that acts as the primary nucleophile, and that the role of the aspartic acid pair is to provide a potentiating binding site for a water molecule, in close proximity to the peptide bond to be hydrolysed.

Nucleophilic attack by the oxygen of the bound water will be accompanied by transfer of one of its protons to the developing negative charge on the peptide nitrogen. It is possible that this transfer is mediated by one of the carboxyl groups. The resultant tetrahedral intermediate may be stabilised, in a manner analogous to the 'oxyanion hole' of serine proteinases, by interaction with the peptide nitrogens of residues 76 and 77 which are part of the 'flap' and are directed down towards the catalytic centre. Productive decay of this transition state will produce the free products, and the enzyme may be restored to activity by binding of a water to replace the original bound water which is incorporated into the free carboxyl product.

REFERENCES

1. Sachdev, G.P., Fruton, J.S.: Proc. Natl. Acad. Sci. USA 72, 3424-3427 (1975).

2. Medzihradszky, K., Voynick, I.M., Medzihradszky-Schweiger M., Fruton, J.S.: Biochemistry 9, 1154-1162 (1970).

3. Koshland, D.E., Neet, K.E.: Ann. Rev. Biochem. 37, 359-410 (1968).

4. James, M.N.G., Sielecki, A.R., Salituro, F., Rich, D.H., Hofmann, T.: Proc. Natl. Acad. Sci. USA 79, 6137-6141 (1982).

5. Bott, R., Subramanian, E., Davies, D.R.: Biochemistry 21, 6956-6962 (1982).

6. Hallett, A., Jones, D.M., Atrash, B., Szelke, M., Leckie, B.J., Beattie, S., Dunn, B.M., Valler, M.J., Rolph, C.E., Kay, J., Foundling, S.I., Wood, S.P., Pearl, L.H., Watson, F.E., Blundell, T.L.: this volume, pp 467-478 (1984).

7. Jencks, W.P.: Catalysis in Chemistry and Enzymology, 294-308, McGraw-Hill, New York 1969.

8. Szelke, M., Leckie, B., Hallett, A., Jones, D.M., Sueiras, J., Atrash, B., Lever, A.F.: Nature (London) 299, 555-557 (1982).

9. Kay, J., Foundling, S.E.: personal communication.

10. James, M.N.G., Sielecki, A.R.: J. Mol. Biol. 163, 299-361 (1983).

11. Pearl, L.H., Blundell, T.L.: FEBS Lett. 174, 96-101 (1984).

12. James, M.N.G., Hsu, I-N., Delbaere, L.T.J.: Nature (London) 267, 808-813 (1977).

13. James, M.N.G., Sielecki, A.R.: this volume, pp 163-177 (1984).

ZYMOGENS FOR ASPARTIC PROTEINASES

Structure predictions from amino acid sequences

Karen G. Welinder, Lars Mikkelsen, Bent Foltmann
Institute of Biochemical Genetics, University of Copenhagen
Ø. Farimagsgade 2A, DK-1353 København K, Danmark

Introduction

All known aspartic proteinases from vertebrates are synthesized as proenzymes, whereas proenzymes not have been found for aspartic proteinases from fungi. The proenzymes are stabilized in an inactive conformation by a N-terminal segment consisting of 40 to 47 amino acid residues. Irreversible activation takes place through a limited proteolysis that removes the N-terminal segment, also called the activation segment or the propart.

X-Ray crystallography has shown that the tertiary structures of aspartic proteinases from mammals and fungi are highly homologous, but no structure of a corresponding proenzyme has been solved by X-ray crystallography. The aim of this communication is to predict the secondary structure of the proparts from their homologous amino acid sequences. We have used the primary structures of 8 proenzymes and made predictions with 3 different methods. The results suggest that half of the propart is folded in 2 α-helices, each with a hydrophilic and a hydrophobic side.

Furthermore, we have investigated the pattern of hydrophilic and hydrophobic groups along the amino acid sequences in 5 different families of aspartic proteinases. The patterns are very similar in both proparts and enzymes, and support that the proparts are folded similarly. Comparison of penicillopepsin and the vertebrate enzymes shows a few hydrophobic surface sequences which occur in the latter only. These hydrophobic surface patches might participate in contacts with the propart.

Aspartic Proteinases and their Inhibitors

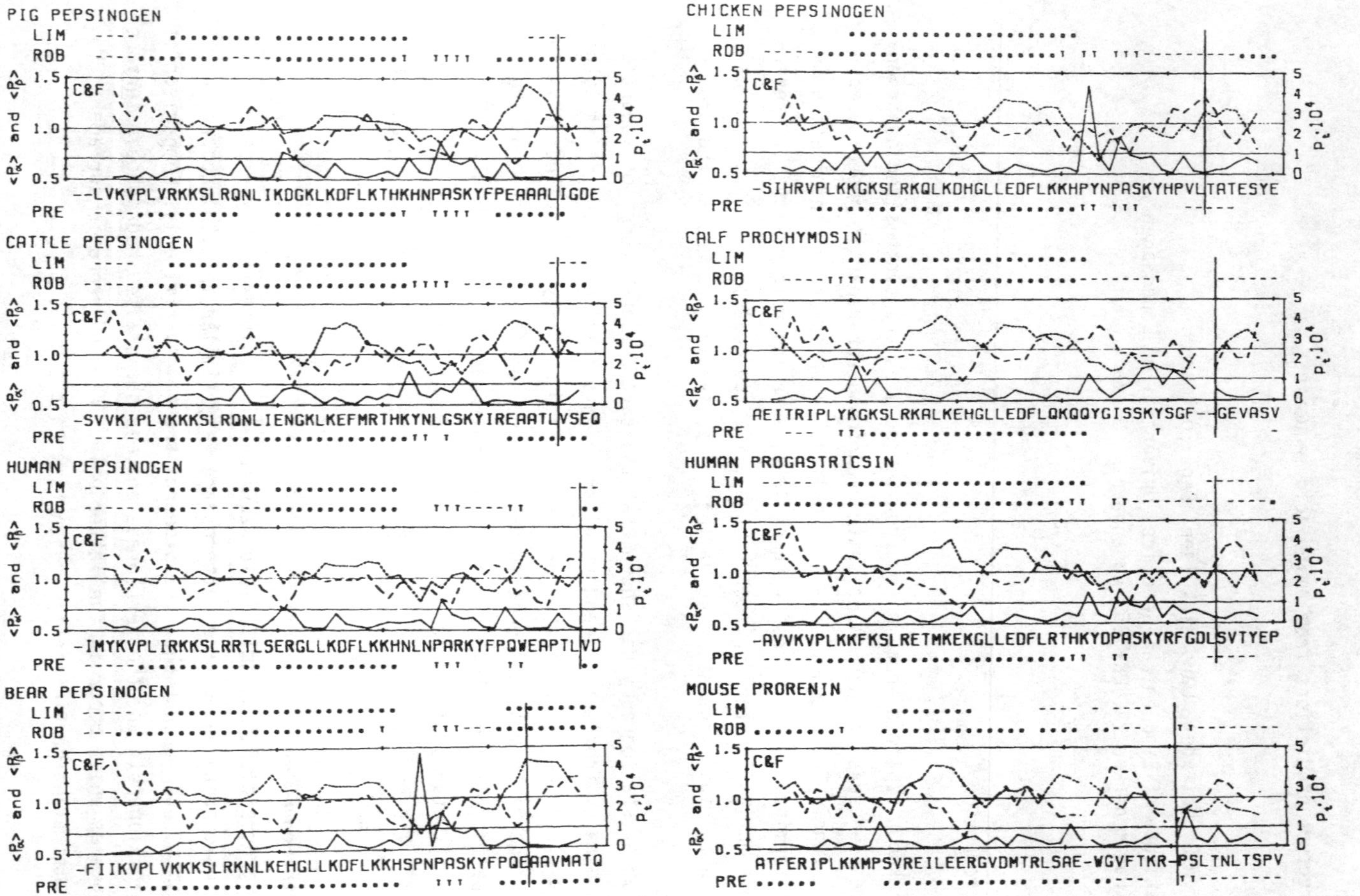
PIG PEPSINOGEN
--LVKVPLVRKKSLRQNLIKDGKLKDFLKTHKHNPASKYFPEAAALIGDE
CATTLE PEPSINOGEN
-SVVKIPLVKKKSLRQNLIENGKLKEFMRTHKYNLGSKYIREAATLVSEQ
HUMAN PEPSINOGEN
-IMYKVPLIRKKSLRRTLSERGLLKDFLKKHNLNPARKYFPQWEAPTLVD
BEAR PEPSINOGEN
-FIIKVPLVKKKSLRKNLKEHGLLKDFLKKHSPNPASKYFPQEAAVMATQ
CHICKEN PEPSINOGEN
-SIHRVPLKKGKSLRKQLKDHGLLEDFLKKHPYNPASKYHPVLTATESYE
CALF PROCHYMOSIN
AEITRIPLYKGKSLRKALKEHGLLEDFLQKQQYGISSKYSGF--GEVASV
HUMAN PROGASTRICSIN
-AVVKVPLKKFKSLRETMKEKGLLEDFLRTHKYDPASKYRFGDLSVTYEP
MOUSE PRORENIN
ATFERIPLKKMPSVREILEERGVDMTRLSAE-WGVFTKR-PSLTNLTSPV
LIM
ROB
PRE
C&F
<P_α> and <P_β>
$p_t \cdot 10^4$

Methods

Prediction of the secondary structure of proteins from their amino acid sequences is carried out with 3 different methods of relatively good records (1). The algorithm of Lim (2) is used in a computer-assisted form written by Dr. Lenstra, University of Leiden, and predicts α-helix and β-strand only. The methods of Garnier, Osguthorpe and Robson (3) and of Chou and Fasman (4) predict α-helix, β-strand, turn and coil. The "ROB" program is obtained from Dr. Garnier and is run with the unbiased decision constants $DC_H = -75$ and $DC_S = -87.5$.

Hydropathy plots are drawn using the hydropathic indices of amino acid residues derived by Kyte and Doolittle (5). Indices of a span of 7 consecutive residues are averaged and plotted over the central residue.

Results and Discussion

Fig. 1 shows predictions of the secondary structure of the 50 N-terminal amino acid residues in 8 zymogens of aspartic proteinases. Each protein is predicted by 3 different methods and the composite result is summarized in line PRE. The sequences and the predictions are aligned. Taken together these 8x3 predictions suggest that the proparts in all zymogens from the N-terminus are folded in a β-strand (4-5 residues), a long helical segment

---

Fig. 1. Predictions of the secondary structure of the propart segments of 8 aspartic proteinases. The zymogen source is shown in the first line and is followed by LIM and ROB predictions. β-Strand or extended structure is indicated by (----), helix by (oooo), and turn by (T). The Chou and Fasman prediction is shown as a plot of helix (····), strand (----) and turn (——) tendencies (4). Predictions of helix and strand require potentials above 1.0 (axis to the left) and turn probability above 10^{-4} (axis to the right). A composite prediction, PRE, requires that 2 of the 3 methods agree and is shown under the amino acid sequence (one-letter notation). Amino acid sequences are from ref. 6 (pig and cattle pepsinogens and calf prochymosin), and ref. 7-11 (listed in the same order as the sequences). Every tenth position is marked on the x-axes. The proparts start at p1 to p3 (unknown for prorenin), and the active enzymes after the vertical bar. The left part of the figure shows 4 examples of mammalian pepsinogens (EC 3.4.23.1), the right part an avian pepsinogen together with prochymosin (EC 3.4.23.4), progastricsin (EC 3.4.23.3), and prorenin (EC 3.4.23.15).

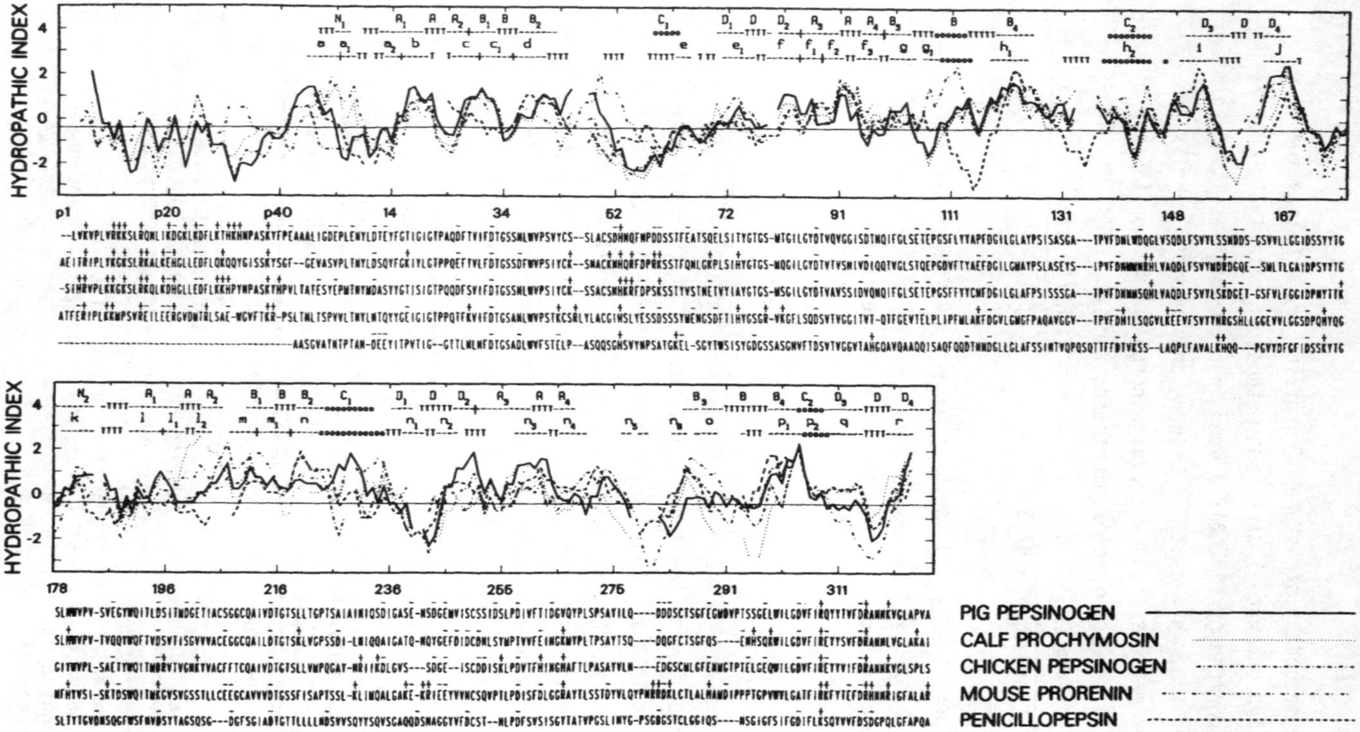

Fig. 2. Hydropathy plots and aligned sequences of 4 aspartic zymogens and penicillopepsin (6, 9,11,12). The notation is indicated in the lower right corner. The upper part of the figure includes the propart and the first domain of the active enzymes; the lower part shows the second domain of the enzymes. Residues are numbered at each twentieth position, p1, p20 and p40 within the propart, and the residue numbers of pig pepsin are indicated along the sequences of the enzymes. The secondary structures observed in the crystallographic models of pig pepsin (13) and penicillopepsin (12) are shown above the plots, (---) for strand, (ooo) for helix and (TTT) for turn residues, and labelled in accordance with these papers.

(20-26 residues) and a turn, maybe except for prorenin. The C-terminal ends of the proparts are largely predicted coil. Secondary structure predictions for the enzymes (not shown) gave very much β-strand and turn, but a few percent of helix only, consistent with observations in the crystal structures of pepsins.

The helical segment mentioned consists of 2 predicted helices. The Lim method consistently (except for prorenin) terminates the first helix after leucine p18, i.e. where pig and cattle pepsinogens are cleaved during activation. The C&F plot and ROB data (details, which are not shown) agree that tetrapeptide p19-22, which ends with an invariant Gly, is a turn candidate. All sequences of the first predicted helix show a conserved pattern of hydrophobic side chains 3 or 4 residues apart: p7-8, p11 (where lysine is present the hydrophobic character may be maintained, as the long side chain can bend the amino group away), p14 and p18. Within the second predicted helix a similar pattern consists of p23-24, and p27-28. In conclusion these predictions suggest that the proparts in aspartic zymogens fold very similarly and are dominated by two helices 15-20Å long. In prorenin proline p12 prevents prediction of a regular α-helix. Therefore, presumably this propart deviates in minor details only.

A plot of the hydropathic character along an amino acid sequence is a very good measure of peptide segment exposure or burial (5). A positive index indicates burial. In Fig. 2 the hydropathy profiles of pig pepsinogen A, calf prochymosin, chicken pepsinogen, mouse prorenin 1 and penicillopepsin are compared. The proparts are rather exposed, i.e. on the surface of the zymogens. These plots furthermore support that all proparts fold in the same way, possibly except p28-36 of prorenin. Within the enzyme part the profiles are also amazingly similar and clearly delineate exposed and buried segments.

Two segments, 112-117 and 129-133 , which are located on the surface of the enzymes, at each side of the Tyr-75 'flap' of the substrate binding site, are hydrophobic in the vertebrate enzymes but hydrophilic in penicillopepsin. Segment 129-133 is extented by 3 more hydrophilic residues in penicillopepsin. These differences could merely be the result of a different substrate specificity in penicillopepsin. But we suggest that these segments, in particular the first one, are essential for attachment of the hydrophobic patches of the propart, as strong binding of the propart to the substrate-binding site by hydrophobic and ionic forces under physiological storage would suppress enzyme activity efficiently.

References

1. Kabsch, W., Sander C.: FEBS Lett. 155, 179-182 (1983).
2. Lim, V.I.: J. Mol. Biol. 88, 873-894 (1974).
3. Garnier, J., Osguthorpe, D.J, Robson, B.: J. Mol. Biol. 120, 97-120 (1978).
4. Chou, P.Y., Fasman, G.D.: Adv. Enzymol. 47, 45-148 (1978).
5. Kyte, J., Doolittle, R.F.: J. Mol. Biol. 157, 105-132 (1982).
6. Foltmann, B., Pedersen, V.B. in "Acid Proteases" (Tang, J. ed.) Plenum-Press, New York, pp. 3-22 (1977).
7. Sogawa, K., Fujii-Kuriyama, Y., Mizukami, Y., Ichihara, Y., Takahashi, K.: J. Biol. Chem. 258, 5306-5311 (1983).
8. Kageyama, T., Moriyama, A., Takahashi, K.: J. Biochem. (Tokyo) 94, 1557-1567 (1983).
9. Baudys, M., Kostka, V.: Eur. J. Biochem. 136, 89-99 (1983).
10. Foltmann, B., Jensen, A.L.: Eur. J. Biochem. 128, 63-70 (1982).
11. Holm, I., Ollo, R., Panthier, J.-J., Rougeon, F.: EMBO J. 3, 557-562 (1984).
12. James, M.N.G., Sielecki, A.: J. Mol. Biol. 163, 299-361 (1983).
13. Andreeva, N.S., Zdanov, A.S., Gustchina, A.E., Fedorov, A.A.: J. Biol. Chem. in press (1984).

CHEMICAL APPROACHES TO THE MECHANISM OF ASPARTIC PROTEINASES

Vladimir K.Antonov
Shemyakin Institute of Bioorganic Chemistry, USSR Academy of Sciences, 117988 Moscow

A great variety of chemical approaches are applied to study the mechanism of enzyme action. Not only chemical modification of an enzyme or a substrate is used, but also all the methods employed for the investigation of non-enzymic reactions as well as a series of specific techniques for the reactions catalyzed by the enzyme. Here the knowledge of the enzyme spatial structure provides a success, however, the lack of this information does not influence understanding of many important characteristics of the mechanism of action.

The data obtained in our laboratory will help me illustrate the ways for solution of some problems of aspartic proteinases functioning. I would like to bring your attention to the following questions:

1) Orientation of substrate in the pepsin active site,
2) Types of nucleophilic groups of aspartic proteinases and problem of the "acyl-and amino-enzyme" intermediates,
3) Mechanism of transpeptidation reaction catalyzed by aspartic proteinases, and
4) Formation of tetrahedral intermediate during catalysis.

In principle, the structure of an enzyme-substrate complex can be established by means of crystallographic analysis. Nevertheless, in case of pepsin and some other enzymes with long active site cleft and with some peculiarities of the arrangement of enzyme molecules in the crystal, it is difficult to obtain crystalline complexes and to interpret electron

Aspartic Proteinases and their Inhibitors

density maps (1). Therefore complementary chemical data are of much value.

The dimension of the pepsin active site and other its features can be elucidated on the basis of a statistic analysis of splitting a variety of peptide sequences by this enzyme. I will not discuss here the methodology of this approach - pertinent data have been published (2,3). It should only be mentioned that the analysis resulted in obtaining indexes of specificity ($S_{i,j}$), which became positive or negative depending on preferableness or undesirability of the particular residue in the given position of the substrate (Table 1).

Table 1. Some Indexes of Specificity ($S_{i,j}$) for Pepsin

P_3	P_2	P_1	P'_1	P'_2
Ser +1.53	Asn +1.70	Leu +2.46	Trp +1.52	Arg +1.45
Pro +1.45	Glu +1.63	Phe +2.28	Tyr +1.52	Thr +1.36
Gly +1.45	Ala +1.36	Trp +1.67	Ile +1.40	Ile +1.34
Thr +1.34	Val +1.24	Glu +1.49	Phe +1.39	Ala +1.33
Trp -1.70	Lys -1.32	Gly -1.52	Asn -1.25	Tyr -1.37
Phe -2.03	Pro -1.43	Val -1.95	Arg -1.37	Asp -1.69
Met -2.16	Leu -1.58	Ile -5.25	Gly -1.82	Pro -1.89
Lys -2.90	Tyr -3.04	Arg -6.23	Thr -2.31	Phe -5.15

An absolute value of the index characterizes the degree of such preferableness/undesirability, and an average absolute value ($\bar{S}_j$) for each position is the rigidity of the active site loci to the substrate structure. As shown on Fig.1 a substantial deviation of the average index of specificity from unity is observed in positions P_3-P'_2 of the substrate. Thus pepsin specifically binds 5-6 residues in the polypeptide chain of the substrate, while locus S_1 of the enzyme reveals the most rigid requirements to the structure of the side chain of bound residue.

When considering the locus, certain limitations for the formation of a productive enzyme-substrate complex take place. Proline is an absolutely undesirable residue of the substrate, isoleucine and valine have rather low indexes of specificity. Thus branching at β-C-atom of the substrate is unacceptable, and, apparently, steric hindrance for efficient binding of these substrates exist in the active site. Besides, arginine residue and those of other positively charged amino acids are undesirable at S_1 locus of pepsin. Penicillopepsin and some other aspartic proteinases activating trypsinogen have specificity for a lysine residue in the S_1 binding site. They differ from pepsin by having aspartic acid residue at position 114 instead of tyrosine at this position in pepsin (cf.4). Locus S'_1 is also hydrophobic, but less specific

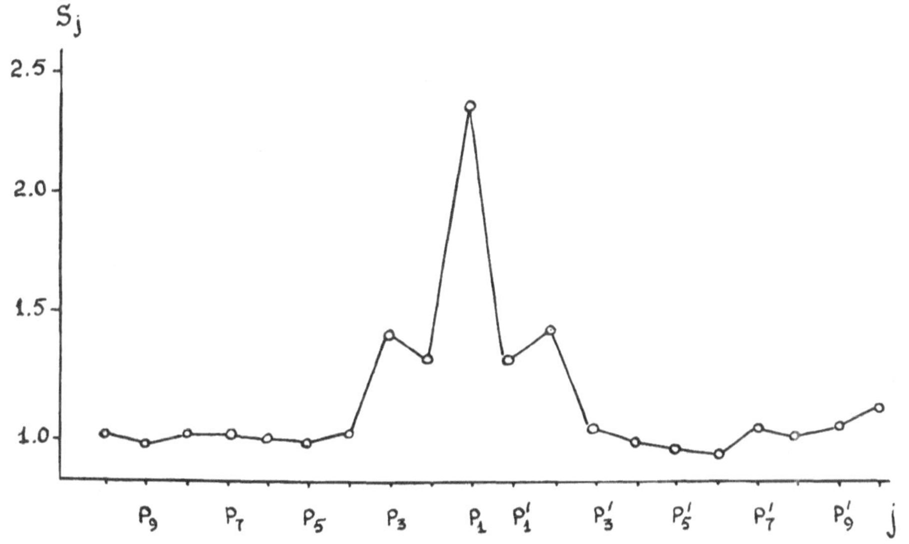

Fig.1 Dependence of the average specificity index S_j on the position of residue (P) in the polypeptide chain of the substrate.

than S_1 : indexes of specificity for S'_1 are correlated with the π-constant of the hydrophobicity.

Of much importance for the formation of the productive complex, at least of short substrates, is the possibility of hydrogen bond formation. Substitution of NH group in position P_1 of the substrate for an oxygen atom transforms the substrate into a competitive inhibitor. The same was observed when amide or ester group in position P'_1 has been replaced by an alcohol grouping (Table 2).

Table 2. Inhibition of pepsin by some substrate analogues

Substrate and Inhibitors	k_{cat} s^{-1}	$K_m \cdot 10^3$ M	$K_i \cdot 10^3$ M	Ref.
$CH_3CONHCH(CH_2Ph)CONHCH(CH_2Ph)COOMe$	0.1	1.9	-	5
$CH_3COOCH(CH_2Ph)CONHCH(CH_2Ph)COOMe$	-	-	1.8	5
$Z\text{-}HisNHCH(CH_2Ph)CONHCH(CH_2Ph)CH_2OH$			0.24	6

What is the orientation of a substrate cleaved group in the complex with the enzyme? To solve this problem we have used the ability of ethanol to bind to the pepsin crystals. Ethanol was shown to be a competitive inhibitor of hydrolysis $ZPhe(NO_2)AlaAmp$ and a noncompetitive one of hydrolysis $ZAlaPhe(NO_2)Apm$ (Fig. 2), i.e. in the latter case a triple complex, enzyme-substrate-ethanol can be formed (7). It is clear that in this case a short side chain is entrapped in the locus binding ethanol and the ethanol molecule is located in locus S_1 (Fig. 3). Consequently, in order to clear up the substrate orientation in the active site, the location of the region binding alcohol on the enzyme pertinent to, for example, a catalytically active residue of aspartic acid must be known. Only recently it became possible (8).

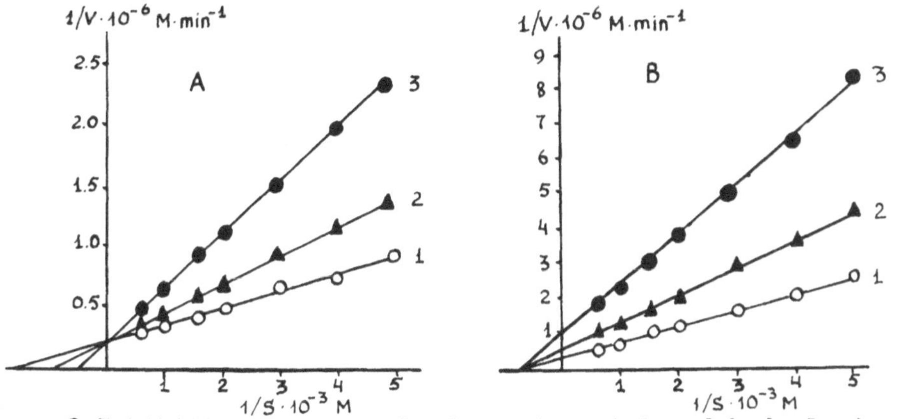

Fig.2 Inhibition by ethanol of pepsin catalyzed hydrolysis of: (A) ZPhe(NO_2)AlaApm and ZAlaPhe(NO_2)Apm (Apm - γ-morpholinopropylamide); 0.1 M acetate buffer, pH 4 [EtOH] =0 (1), 2.5 (2) and 5 (3) vol.%.

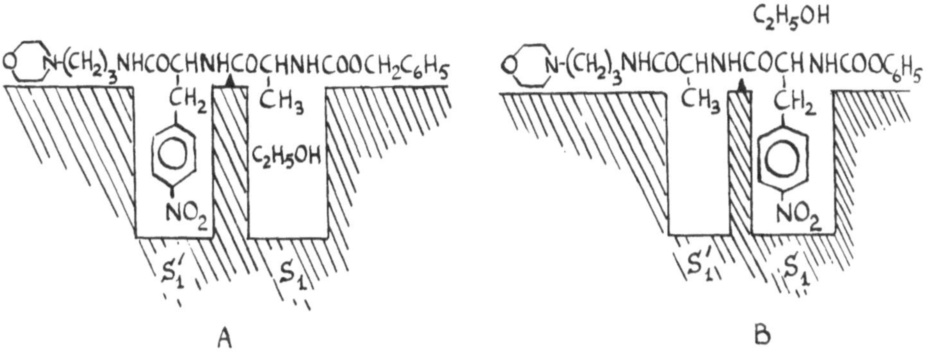

Fig. 3 Schematic presentation of comlexes formed by pepsin with EtOH and substrates: (A) ZAlaPhe(NO_2)Apm, (B) ZPhe(NO_2)AlaApm.

Undoubtedly, valuable information on the substrate location in the complex can be obtained by means of chemical modification of the enzyme. Modification of pepsin with p-nitrophenyldiazonium chloride leads to azo-coupling with Tyr-189 residue (9). Methyl phenylalaninate effectively protects the residue from modification, while acetylphenylalanine fails.

We may conclude therefore the tyrosine-189 residue enters locus S'_1 of the enzyme active site; that is in good agreement with the X-ray data (1).

All these data permit modeling the substrate binding to pepsin (Fig. 4).

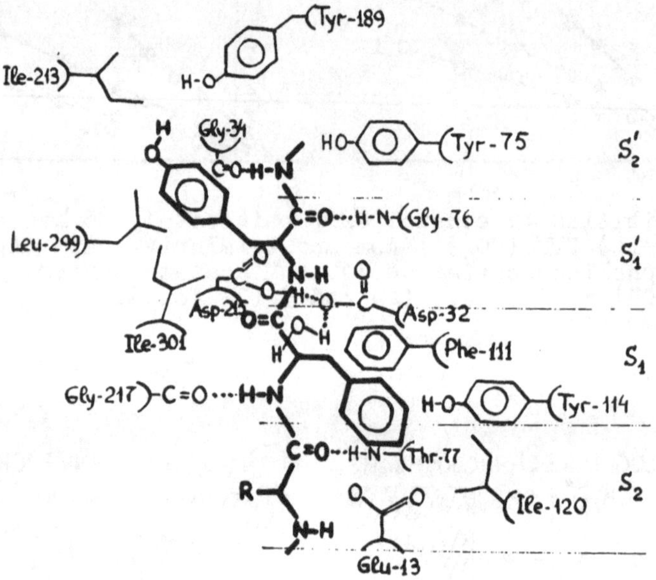

Fig.4 Model of the substrate binding to pepsin.

The next item concerns a nucleophilic group attacking the substrate carbonyl carbon. As known pepsin catalyzes transpeptidation reactions of the acyl and aminotransfer type (Scheme 1).

Scheme 1

$$E\text{-}COOH + RCONHR^1 \rightleftharpoons [E\text{-}CO\text{-}O\text{-}COR] \xrightarrow{R^2NH_2} E\text{-}COOH + RCONHR^2$$

$$E\text{-}COOH + RCONHR^1 \rightleftharpoons [E\text{-}CONHR^1] \xrightarrow{R^2COOH} E\text{-}COOH + R^2CONHR^1$$

For a long time the catalytically active carboxyl groups of

Asp-32 and Asp-215 residues were assumed to function via the covalent type of catalysis,i.e.via pepsin forms of acyl-enzyme,containing an anhydride bond, and amino enzyme.The problem of distinguishing between the covalent and general base type of catalysis of hydrolysis of the amide bond remained unsolved for a long time.The lines of approach to this problem taken in our Laboratory recently are described below (10-12).

The covalent or nucleophilic mechanism differs from the general base one in the step at which water is involved in the reaction (Scheme 2). In first case (A), water participates in hydrolysis of intermediate acyl-enzyme, in the second (B)-at the initial stade of attacking carbonyl carbon of the substrate. The scheme shows that in covalent catalysis isotope-labeled water ($H_2{}^{18}O$)under certain conditions will be incorporated neither into substrate nor the product of the acyl transfer, while in general base catalysis such insertion should occur. Besides, pepsin and a number of proteases catalyze the oxygen exchange in the amino acid carboxyl group.

Scheme 2

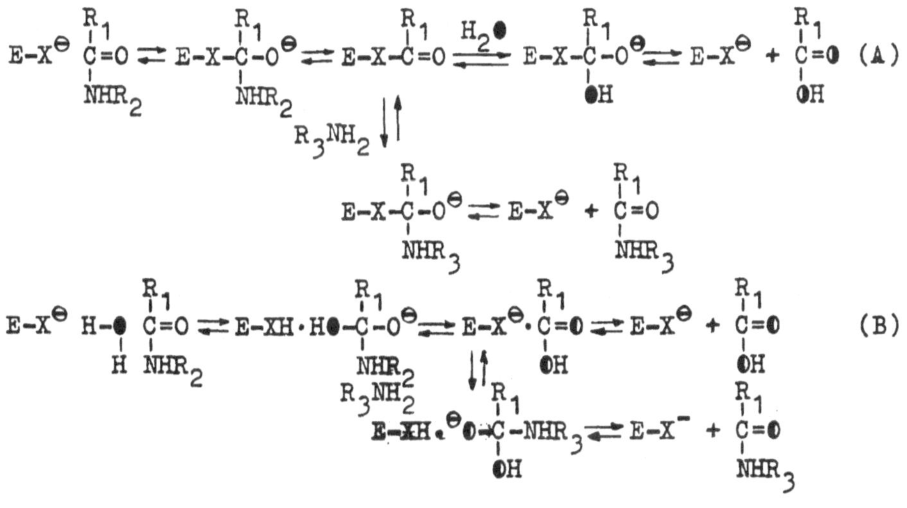

If amino acid or its derivative with an unsubstituted amino group is incorporated into the system, the synthesis of the peptide bond and its subsequent hydrolysis will also take place. In covalent catalysis the process does not change the rate of oxygen incorporation into the carboxyl group, but in case of the general base mechanism, when the rate for the "genuine" exchange and synthesis-hydrolysis are comparable, the rate of the lable insertion will increase (Scheme 3).

Scheme 3

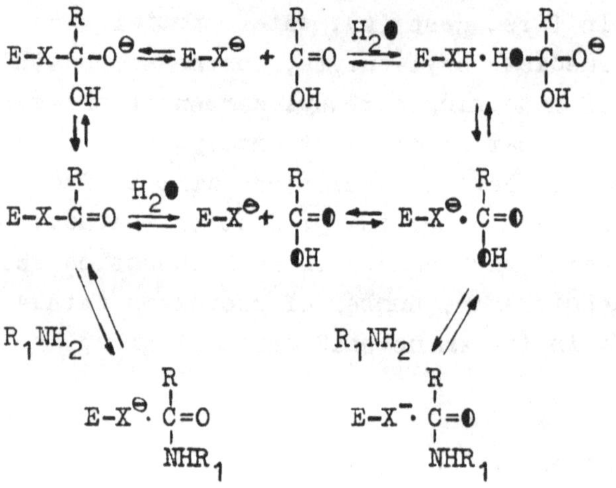

Tables 3 and 4 illustrate the reactions catalyzed by various proteases and the results of incorporation of a heavy oxygen label. Clearly serine and thiol enzymes function by the covalent type of catalysis, aspartic and metal-dependent - by general base catalysis.

It should be mentioned that when the nucleophilic group is represented by a carboxylate ion of the enzyme, the cleavage of the "acyl" or "amino" enzyme at protein carbonyl moiety will result in oxygen incorporation into the enzyme. To test this assumption we have elaborated a special procedure (13). After completion of the reaction in the heavy oxygen-labelled

Table 3. ^{18}O Content of the transpeptidation products and non-hydrolyzed substrates

Enzyme	Reaction studied	^{18}O Content in atom %	
		substrate	transpeptid. product
Chymotrypsin	$AcPheGlyNH_2 \rightleftharpoons AcPheOH + GlyNH_2$	0	
	$2LeuLeuNH_2 \rightleftharpoons LeuLeuLeu + LeuNH_2 + NH_3$	0	0
Papain	$ZGlyNH_2 + LeuGly \rightleftharpoons ZGlyLeuGly + NH_3$		0
	$ZGlyOMe + LeuGly \rightleftharpoons ZGlyLeuGly + MeOH$		0
Pepsin	$2LeuTyrNH_2 \rightleftharpoons LeuLeu + 2TyrNH_2$	13 ± 4	65 ± 12
Leucine aminopeptidase	$2LeuNH_2 \rightleftharpoons LeuLeuNH_2 + NH_3$	46 ± 5	48 ± 3
	$2\ [^{18}O]\text{-}LeuNH_2 \rightleftharpoons [^{18}O]\text{-}LeuLeuNH_2 + NH_3$		55 ± 5
Thermolysin	$2LeuLeuNH_2 \rightleftharpoons LeuLeu + 2LeuNH_2$	52 ± 5	47 ± 5

water the enzyme is subjected to the action of specific inhibitors which form ester bonds with the active center carboxyls (Scheme 4). The next step is mild alkaline hydrolysis and separation of the hydroxy-compounds where the heavy oxygen content is determined. In experiments with pepsin no label was detected in the enzyme. This is an additional proof for general base catalysis by pepsin.

If no covalent intermediate is formed in pepsin catalysis, a question arises - how to explain the capacity of the enzyme for catalyzing reactions of acyl- and amino-transfer. One must discriminate between the thermodynamic and kinetic aspects of the problem of transpeptidation. When the thermodynamics of the peptide bond formation from acyl amino acid

Table 4. Rates of oxygen exchange

Enzyme	Substrate and additions	^{18}O content in atom. %	Maximal rate of exchange $M^{-1}\cdot min^{-1}$	$\frac{V_s}{V_{s+a}}$
Chymotrypsin	AcPheOH	35±3	$1.8\cdot10^{-5}$	
	AcPheOH+AlaNH$_2$	22±3	$0.8\cdot10^{-5}$	0.45
	AcPheOH+GlyNH$_2$	16±2	$1.0\cdot10^{-5}$	0.63
Pepsin	AcPheOH	13±2	$2.5\cdot10^{-5}$	
	AcPheOH+PheAlaAlaOMe	29±3	$5.6\cdot10^{-5}$	2.2
Leucine aminopeptidase	LeuOH	21±1	$1.6\cdot10^{-3}$	
	LeuOH+NH$_4$Cl	19±1	$1.5\cdot10^{-3}$	0.9
	LeuOH+LeuNH(CH$_2$)$_2$OH	40±2	$3.5\cdot10^{-3}$	1.95
Carboxypeptidase A	AcPheOH	18±2	$7.9\cdot10^{-6}$	
	AcPheOH+PheOH	40±3	$1.8\cdot10^{-5}$	2.3

Scheme 4

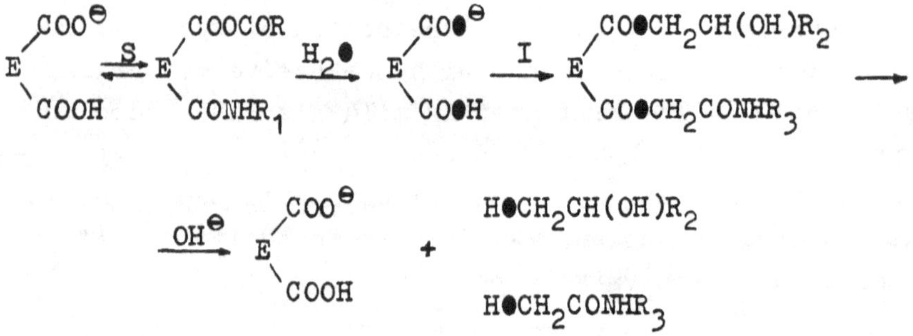

and from amino acid ester (or amide) at various concentrations of reagents is considered, the yield is very low:

$$\underset{(A)}{R\text{-}COO^{\ominus}} + \underset{(B)}{{}^{\oplus}H_3N\text{-}R^1} \overset{K}{\rightleftharpoons} \underset{(AB)}{RCONHR^1}$$

at $[A]_o=[B]_o=1mM$ and $K=1M$ $\quad [AB]_{eq}\leqslant 1\cdot10^{-6}M$

Actually, however, transpeptidation reactions often run with a 50% yield. But transpeptidation is an interaction of a peptide-donor with an acceptor, when a new peptide and amino acid derivative are formed :

$$\underset{(AB)}{RCONHR^1} + \underset{(C)}{R^2COO^{\ominus}} \underset{}{\overset{K=1}{\rightleftharpoons}} \underset{(BC)}{R^2CONHR^1} + \underset{(A)}{RCOO^{\ominus}}$$

$$\text{at } [AB]_o = [C]_o = 1\text{mM} \qquad [BC]_{eq} = 0.5\text{mM}$$

Thus, from the viewpointof the thermodynamics there is nothing surprising in transpeptidation.

The kinetic aspect of the problem is more complicated. The non-covalent enzyme complex with a free product must exist for a sufficiently long time for the acceptor to be bound and for a new amide bond to be formed. The pepsin-catalyzed reaction of transpeptidation of the amino transfer type may be taken as an example; data are available on the initial rates of the amide bond formation for this reaction (14). Using these data, one may calculate that the observed rate of transpeptidation can be reached if the rate constant for the dissociation of an enzyme-product complex will be about 10 s^{-1} for the system :

$$AcPheTyr + AcPhe(NO_2) \longrightarrow AcPhe(NO_2)\text{-}Tyr + AcPhe$$

This is at least by 3-4 orders of magnitude lower than the usually observed rates for dissociation of such complexes.

In order to solve this contradiction, one should suppose that at least two enzyme-product complexes exist on the route from an enzyme-substrate complex to products, and the equilibrium in the system is shifted toward the "external" (E_oP) complex :

$$\cdots \longrightarrow E_pP \underset{k_{-4}}{\overset{k_4}{\rightleftharpoons}} E_oP \overset{K_i}{\rightleftharpoons} E + P$$

$$\text{where } \frac{k_4}{k_{-4}} = K_q \gg 1 \qquad \text{and } K_{i\,(app)} = \frac{K_i K_q}{1 + K_q} \simeq K_i$$

In this case, in the experiments with inhibitions by the product the observed constant for dissociation of the enzyme -product complex ($K_{i(app)}$), would reflect only the formation of this "external" complex (K_i). Its conversion into the "internal" (E_pP) complex must be very slow, in above mentioned case slower than with $k \simeq 10\ s^{-1}$. Consequently, the "internal" complex which is responsible for transpeptidation can be formed from the substrate, but only to a very low extent from free reaction products and the enzyme.

Thus the transpeptidation reaction may proceed when no covalent intermediate compounds are formed. The fact that transfer reactions are observed in the system enzyme-substrate-acceptor gives no evidence of the covalent nature of intermediates.

Let's consider the formation of an tetrahedral intermediate in pepsin catalysis. Upon incubation of pepsin and some other proteases with substrates in heavy oxygen water an isotope of oxygen is observed to be incorporated into an amide group of the non-hydrolyzed substrate (Table 3). That seems to testify to the formation of a tetrahedral intermediate, its rate of decomposition on the route to products is lower than to the substrate. However, a detailed analysis of the kinetics of this process (Fig. 5) corroborated our expectations (15). Indeed, it was found that in case of leucine aminopeptidase and pepsin the heavy oxygen incorporation occurs simultaneously and in a comparable degree with incorporation of labelled amine reaction product into the substrate. Presumably, the oxygen exchange is due to the acyl transfer to the amino product formed during hydrolysis. These data suggest that heavy oxygen exchange with the substrate cannot prove the formation of tetrahedral intermediate in case of enzymatic hydrolysis.

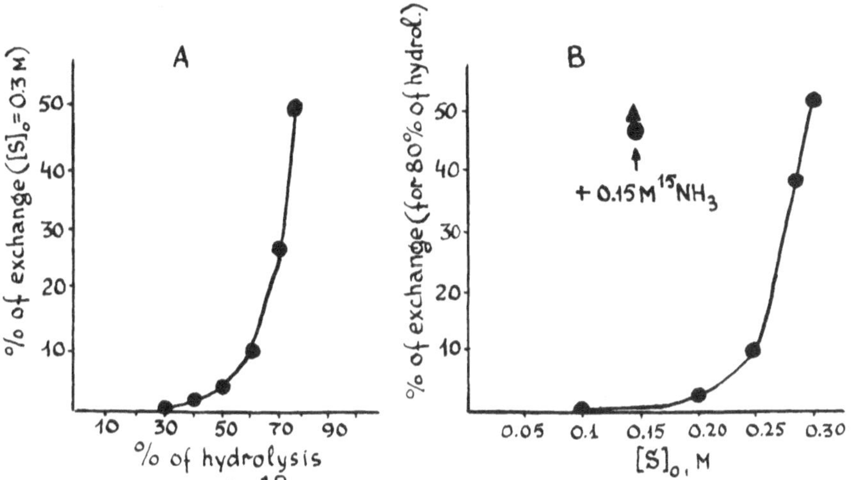

Fig. 5 Kinetics of ^{18}O incorporation into leucine amide catalyzed by leucine aminopeptidase. A: Dependence of the exchange on the degree of the substrate hydrolysis [S]:0.3 M. B: Dependence of the exchange on the substrate concentration at 80 % of hydrolysis. Arrow indicates the exchange in the presence of 0.15 M of $^{15}NH_3$ (Circle - oxygen; triangle - nitrogen).

Using quantum-chemical approaches we calculated the pK_a-values for the leaving amino group at different stages of the planar amide conversion into a tetrahedron. It appeared that this value becomes comparable to that of the proton donor long before the completion of the tetrahedron formation (Fig. 6). So protonation and cleavage of the leaving group can take place in an incompletely formed tetrahedron.

The degree of tetrahedron formation at any stage of nucleophilic attack (degree of amide bond pyramidalization) depends on the distance between a nucleophile and the carbonyl carbon and on nature of nucleophile(Fig.7). For the model system water-formate ion as a nucleophile and methylacetamide as a substrate at the distance equal to the sum of van-der-Waals radiuses the degree of pyramidalization is ca 0.02 Å . Thermal fluctuations noticeably bring together the reacting groups, simultaneously increasing electrophilicity of the carbonyl C-atom due to the increase of pyramidalization and the resonance destabilisation.

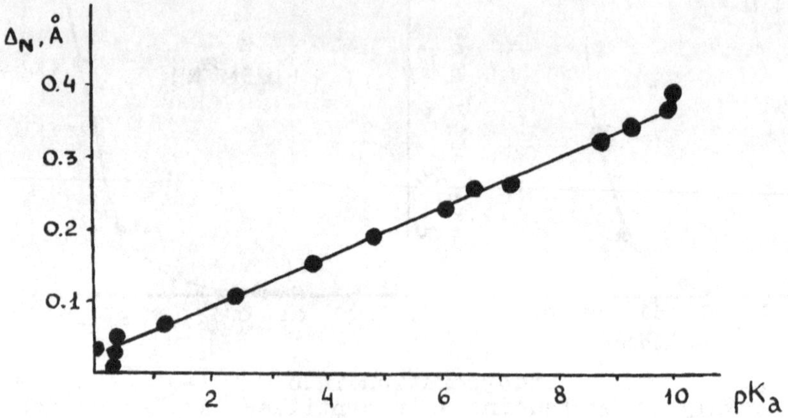

Fig.6. Relation between the degree of pyramidalization of the amide bond and pK_a of the amino-leaving group. Calculated by S. Alexandrov (CNDO-2 method).

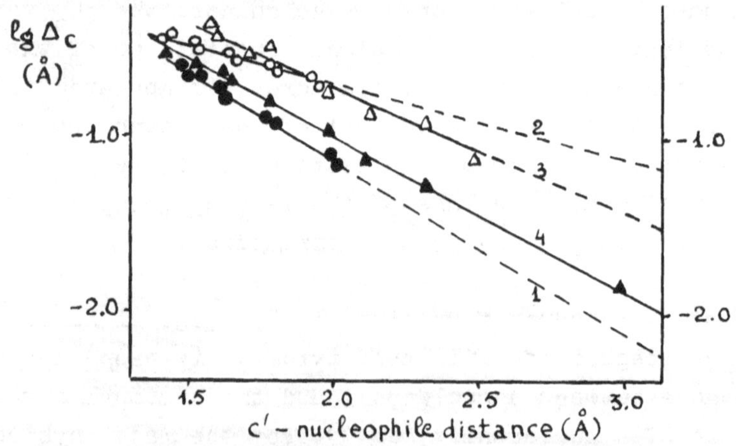

Fig.7. Dependence of the degree of pyramidalization of the amide bond on the distance between the carbonyl carbon of the substrate and the nucleophile for CH_3OH (1), CH_3O^- (2), CH_3S^- (3) and H_2O-$HCOO^-$ system (4). Calculated by CNDO-2 method.

As shown in Table 5 substantial pyramidalization exists even in certain enzyme-quasisubstrate complexes. Consequently, efficiency of the catalysis by proteases, in particular, by aspartic proteinases, is proposed to be caused by the loss of resonance stabilization of the cleaved amide group in the complex with the enzyme (22).

Table 5 Pyramidalization of the amide bond

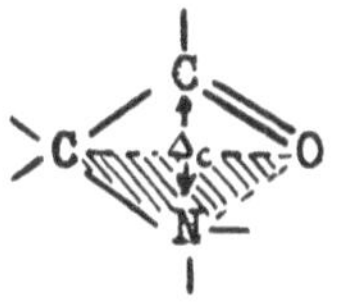

Enzyme	Ligand	Bond	Δ_C, Å*	Source of atomic coordinates
Chymotrypsin	BPTI**	Lys^{15}-Ala^{16}	0.38	16
Trypsin	BPTI	Lys^{15}-Ala^{16}	0.28	17
	BPTI(free)	Lys^{15}-Ala^{16}	0.008	18
Elastase	$CF_3COLysAlaNHC_6H_4CF_3$-p	CF_3CO-NH	0.042	19
Carboxypeptidase A	Gly-Tyr	CO-NH	0.012	20
Thermolysin	$C_6H_5(CH_2)_2COPheOH$	CO-NH	0.06	21

* Calculated from the atomic coordinates of the complexes.
**BPTI- Bovine pancreatic trypsin inhibitor

What stipulates the enzyme specificity from this viewpoint? If the formation of the productive enzyme-substrate complex is stepwise (as is the dissociation of the product, see above), whereupon the reactive capacity of the cleaved group becomes significantly higher in the complex than in solution, it is obvious that kinetic parameters experimentally determined will depend on the equilibrium between the complexes (Scheme 5).

Scheme 5

$$E+S \underset{k_{-1}}{\overset{k_1}{\rightleftharpoons}} E_oS \underset{k_{-2}}{\overset{k_2}{\rightleftharpoons}} E_pS \xrightarrow{k_3} E+P \qquad K_s = \frac{k_{-1}}{k_1}, \quad K_p = \frac{k_2}{k_{-2}+k_3}$$

$$k_{cat} = \frac{k_3 K_p}{1+K_p}, \quad K_{m(app)} = \frac{K_s}{1+K_p}$$

K_p	k_3	k_{cat}	$K_{m(app)}$
$\ll 1$	$\ll k_{-2}$	k_2k_3/k_{-2}	K_s
$\ll 1$	$\gg k_{-2}$	k_2	K_s
$\gg 1$	$\ll k_{-2}$	k_3	K_sk_{-2}/k_2
$\gg 1$	$\gg k_{-2}$	k_3	K_sk_3/k_2

When the equilibrium constant between E_oS and E_pS is less than unity (viz. the equilibrium is shifted toward E_oS) any change of the constant when changing the substrate will influence only the k_{cat} value, but $K_{m(app)}$ will be the same. When K_p becames higher than unity, k_{cat} reaches its maximal value and $K_{m(app)}$ decreases. The analysis of kinetic parameters of many substrates of pepsin is in agreement with this suggestion (Fig.8).

In conclusion, I would like to stress that chemical approaches, using wide variety of experimental and theoretical

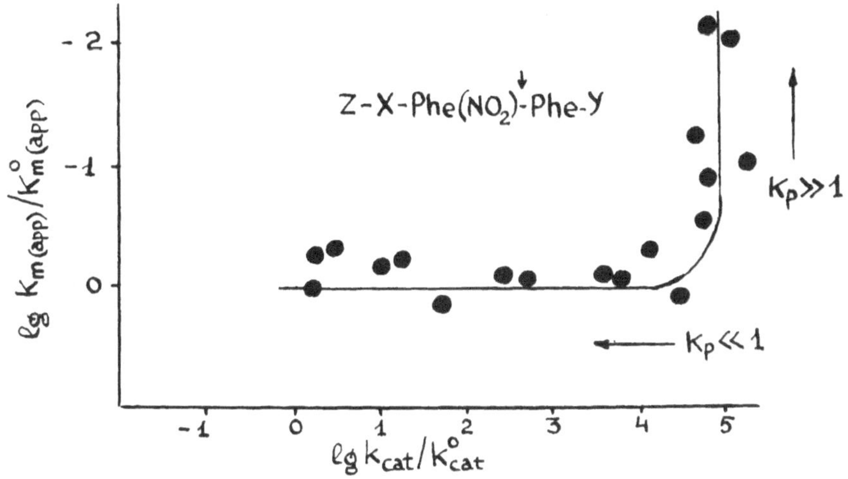

Fig.8 Plot of lg $K_{m(app)}$ vs. lg k_{cat} for series of substrates of pepsin.

methods, provide valuable information on the mechanism of enzymatic catalysis and together with the knowledge of the structure contribute to a better understanding of its mechanism

References

1. Andreeva,N.S., Gustchina,A.E., Fedorov,A.A.: Dokl. Akad. Nauk SSSR 259, 1261-1264 (1981).
2. Zinchenko,A.A., Rumsh,L.D., Antonov,V.K.: Bioorgan. khimia 2, 803-810 (1976).
3. Zinchenko,A.A., Rumsh,L.D., Antonov,V.K.: Bioorgan. khimia 3, 1663-1670 (1977).
4. Hofmann,Th., Hodges,R.S., James,M.N.G.: Biochemistry 23, 635-643 (1984).
5. Rumsh,L.D., Tichodeeva,A.G., Antonov,V.K.: Biokhimia 39, 899-902 (1974).
6. Inouye,K., Fruton,J.S.: Biochemistry 7, 1611-1615 (1968).
7. Antonov,V.K.: Adv. Exptl. Med. Biol. 95,179-198 (1977).
8. Gustchina,A.E., Andreeva,N.S., Antonov,V.K.: Bioorgan.

khimia 9, 1620-1624 (1983).

9. Zavada,L.L., Koslov,L.V., Volkova,L.I., Antonov,V.K.: Bioorgan. khimia 3, 1671-1681 (1977).

10. Lyakisheva,A.G., Ginodman,L.M., Antonov,V.K.: Molekul. biologia 7, 810-816 (1973).

11. Antonov,V.K., Ginodman,L.M., Rumsh,L.D., Kapitannikov,Yu. V., Barshevskaya,T.N., Yavashev,L.P., Gurova,A.G., Volkova,L.I.: Eur. J. Biochem. 117, 195-200 (1981).

12. Antonov,V.K., Ginodman,L.M., Kapitannikov,Yu.V., Barshevskaya,T.N., Gurova,A.G., Rumsh,L.D.: FEBS Lett. 88, 87-90 (1978).

13. Antonov,V.K., Ginodman,L.M., Rumsh,L.D., Kapitannikov,Yu. V., Barshevskaya,T.N., Yavashev,L.P., Gurova,A.G., Volkova,L.I.: Bioorgan. khimia 6, 436-446 (1980).

14. Antonov,V.K., Rumsh,L.D., Tikhodeeva,A.G.: FEBS Lett. 46, 29-33 (1974).

15. Kapitannikov,Yu.V., Yavashev,L.P., Ginodman,L.M., Antonov, V.K.: Bioorgan. khimia 9, 228-231 (1983).

16. Blow,D.: personal communication

17. Huber,R., Kukla,D., Bode,W., Schwager,P., Bartels,K., Deisenhofer,J., Steigemann,W.: J. Mol. Biol. 89, 73-101 (1974).

18. Huber,R.: personal communication

19. Hughes,D.L., Sieker,L.C., Bieth,J., Dimicoli,J.L.: J. Mol. Biol. 162, 645-658 (1982).

20. Quicho,F.A., Lipscomb,W.N.: Adv. Prot. Chem. 25, 1-78 (1971).

21. Kester,W.R., Matthews,B.W.: Biochemistry 16, 2506-2516 (1977).

22. Karpeiskii,M.Ya., Yakovlev,G.I., Antonov,V.K.: Bioorgan. khimia 6, 645-654 (1980).

INTERACTION OF ASPARTIC PROTEINASES WITH A NEW SERIES OF SYNTHETIC SUBSTRATES AND WITH INHIBITORS BASED ON THE PROPART OF PORCINE PEPSINOGEN

Ben M. Dunn, Benne Parten, Melba Jimenez

Department of Biochemistry and Molecular Biology, J. Hillis Miller Health Center, University of Florida College of Medicine, Gainesville, FL 32610 U.S.A.

Carole E. Rolph, Martin Valler and John Kay

Department of Biochemistry, University College, P.O. Box 78, Cardiff, CF1 1XL, Wales, U.K.

Introduction

It has been known for some time that much of the catalytic specificity and efficiency of Aspartic Proteinases derives from binding interactions that occur throughout the extended active site cleft. Although the primary specificity for pig pepsin is for aromatic side chains at the P1 and P1' positions, a substantial amount of data has been obtained to illustrate that a very large increase in catalytic activity results from elongation of a substrate from a di- or tripeptide to a hexapeptide (1-3). Fruton has termed this effect "Secondary Specificity" and it must arise from additional binding interactions that are present in subsites remote from the P1 - P1' locus. In order to probe this effect in more detail, and to obtain concrete data on the nature of the various subsites, substrates have been prepared with a length sufficient to fill the active site crevice. In this way it was hoped to avoid three problems that are present in all previous mechanistic work on these enzymes. Firstly, small substrates, i.e., di- or tri-peptides, are capable of binding in multiple modes or in multiple copies to the large

Aspartic Proteinases and their Inhibitors

active site cleft of Aspartic Proteinases. 2), small substrates are frequently subject to transpeptidation reactions (4,5), wherein multiple substrate binding certainly occurs, and where product appearance is a multi-step event that complicates kinetic analysis. 3), the comparison of kinetic data obtained with long substrates with that from short substrates suffers from differences in degrees of conformational freedom and, hence, entropy. Attempts to circumvent these difficulties had its basis in data presented by both Powers and Antonov at the first Aspartic Proteinase Conference in Oklahoma (6,7). Independent statistical analysis of cleavages obtained using pig pepsin in protein sequencing studies indicated those amino acid residues that are favored in seven subsites of the active site cleft, S4—>S3'. Although the data sets used in these analyses suffer from the broad spectrum of conditions used in various sequencing studies, nevertheless the predictions are useful as a general guide to subsite preferences. Therefore, a sequence of favored amino acids from the statistical analysis was chosen and a peptide was prepared (8) with the sequence Pro - Thr - Glu - Phe - Nph - Arg - Leu, where Nph stands for p-nitro-L-Phenylalanine. This peptide contains several hydrophilic residues to enhance aqueous solubility and thus circumvent another major problem intrinsic to most previous substrates. Since there is a requirement for hydrophobic groups at the P1 - P1' sites, many previous mechanistic studies were done with sparingly soluble substrates. This sequence also contains the Nph residue at the P1' position since Hofmann and Hodges (9) have shown that this provides a shift in absorbance upon hydrolysis of the Phe - Nph bond. This bond was cleaved by pig pepsin to give a change in absorbance of approximately 1500 M^{-1} at 300 nm. The change in absorbance as a function of time (initial velocity) may be conveniently measured as a function of substrate concentration to generate the kinetic parameters,

K_m and k_{cat}. Values of 130 μM and 90 sec^{-1} were obtained for Km and k_{cat} with pig pepsin at 37° and pH 3.1. A series of other Aspartic Proteinases were also studied with K_m's varying from a high of 2000 μM for human gastricsin to a low of 3 micromolar for Endothia parasitica proteinase. Likewise, k_{cat} values varied from a high of 173 sec^{-1} for human cathepsin D to a low of 1.4 sec^{-1} for Rhizopus chinensis Aspartic Proteinase. Values for other enzymes are reported later in this volume (p467). To expand this approach to find evidence for the nature of interaction at the subsites described above, a series of substrates with variation in one locus, the P3 position, have been prepared. A seventh peptide was also prepared with a Val in the P2 positions as a control for the effects of the P2 Glu in peptides I - VI.

Lys - Pro - X - Glu - Phe - Nph - Arg - Leu

X = Val	I
Pro	II
His	III
Lys	IV
Ala	V
Ile	VI

Lys - Pro - Thr - Val - Phe - Nph - Arg - Leu VII

Pro - Thr - Glu - Phe - Nph - Arg - Leu VIII

This series will provide evidence for the nature of interactions in the S3 and S2 subsites of Aspartic Proteinases. Since earlier kinetic work (1-3) indicated large effects arising from substitution away from the scissile bond and since data is now available on the crystal structures of several homologous Aspartic Proteinases (See Chapters by Andreeva, et al, Blundell, et al, and James and Sielecki in this volume), this approach seemed likely to generate valuable insights.

Materials and Methods

All peptides described were synthesized by standard methods of solid phase peptide synthesis (10). The products of synthesis have been purified by a combination of ion exchange chromatography on carboxymethyl-Sepharose with gradient elution (Ammonium Acetate, 0.01 to 1.0 M, pH 5.0) followed by reverse phase High Performance Liquid Chromatography (HPLC) using a Waters Instrument on C_{18} - microbondapak in a Z Module Radial Compression System. Methanol-water buffers were employed with ammonium acetate for buffering.

Each new substrate was tested for the position of cleavage by porcine pepsin by incubating approximately 0.5 mg of the peptide with 10 micrograms of enzyme for one hour at 37°. The sample was then injected on a C_{18} cartridge and the resulting peaks collected. In all cases, the initial peak due to the substrate (14.5 - 18 min, depending on sequence) was eliminated and a major new peak eluting at 12 min was observed. This peak was collected and shown to contain Nph, Arg, and Leu by amino acid analysis. The presence of Nph was also indicated by the large absorbance observed. In each case a second peak of much smaller absorbance was obtained at an earlier elution time (3 - 9 min, depending on sequence). These peaks were collected and shown by analysis to be the pentapeptide products, Lys - Pro - X - Glu - Phe. These observations establish the point of cleavage to be between the Phe and Nph in each case. In a variation of this experiment, aliquots were withdrawn from an incubation mixture at 15 min intervals after enzyme addition and run on an isocratic system (usually 40% Buffer B). In this fashion the progressive decrease in the peak due to substrate has been followed and the increase in the two peaks due to the pentapeptide and the tripeptide products was observed. Semilog plots of $A_{\infty} - A_t$ for decrease in Absorbance at 300 nm and for increasing peak height show that the conversion of substrate to the two products is a straight-forward reaction with no transpeptidation occurring.

Kinetic data for the enzyme catalyzed hydrolysis of these substrates was obtained by following the decrease in absorbance at 300 nm at 37° and 0.1 M ionic strength. Buffers of sodium formate were used for pH's between 2.5 and 4.1 and buffers of sodium acetate were used for pH's between 4.0 and 6.5. Between 6 and 12 assays were carried out at each pH at different concentrations of substrate to derive values of V_{max} and K_m. Initial slopes of the first 10% of the progress curves were obtained and expressed as Δ A/min. The data was plotted as (S) /ΔA /min versus (S). The values of k_{cat} were calculated from the resulting V_{max} values by substitution of enzyme concentration as determined from active site titrations using a known concentration of isovaleryl-pepstatin. Log k_{cat}, pK_m, and log k_{cat}/K_m were then analyzed by plotting versus pH as suggested by Dixon (11).

Studies of inhibitor peptides were carried out by preincubating enzyme and inhibitor peptide (37° or 25°) in buffer for four minutes before addition of substrate to initiate the reaction. For pepsin, substrate V was used for these assays. Assays were done at two or three different substrate concentrations and 5 - 8 different inhibitor concentrations and then analyzed by the Dixon Inhibitor plotting method (12). In all cases, the pattern of inhibition was competitive and the K_I value was obtained from the point of intersection of the resulting lines.

Results and Discussion

1) Kinetics of substrate hydrolysis

Kinetic results for hydrolysis of the substrates I - VI by pig pepsin at pH 3.5 are presented in Table Ia and for Endothia Parasitica Proteinase are presented in Table Ib. It may be seen that Val, Pro, Ala, or Ile are good substituents for pig pepsin with Ala giving a ratio k_{cat}/K_m nearly three fold higher than the original peptide. All the substrates are hydrolyzed readily by Endothia Parasitica Proteinase with very low values of Km observed. Although the k_{cat} values are modest, the low Km values result in very high specificity constants. This may be an example of tighter substrate binding leading to less efficient catalysis. By comparing the constants for pig pepsin with those for Endothia Paracitica Proteinase, it may be seen that some of the intrinsic binding energy in the pepsin case may be used to reach the reaction transition state. The substrates with positively charged residues, His and Lys, in the P3 position are less susceptible substrates for pepsin at this pH with lower k_{cat} and poor K_m values. This is reflected in the ratios, k_{cat} / K_m for these two substrates, III and IV, which are 15 to 110 fold lower than those of the other six substrates. Part of this difference derives from the ionic nature of the substrate and enzyme at the pH where the data in Table I were collected. To clarify this issue, the complete pH dependence of these parameters for several enzymes with selected substrates have been studied.

Table Ia

Kinetic Parameters for Pig Pepsin Catalyzed Hydrolysis of Synthetic Peptide Substrates[a]

Substrate	P3 Residue	k_{cat}, s^{-1}	K_m, mM	$k_{cat}/K_m, s^{-1}mM^{-1}$
I	Val	19	0.06	320
II	Pro	218	0.19	1170
III	His	10	0.42	22
IV	Lys	15	0.62	25
V	Ala	911	0.04	2770
VI	Ile	106	0.09	2180

a. Assays were performed at 37° as described in Materials and Methods in sodium formate buffers at pH 3.5.

Table Ib

Kinetic Parameters for Endothia Parasitica Aspartic Proteinase Catalyzed Hydrolysis of Synthetic Peptide Substrates[b].

Substrates	P3 Residue	k_{cat}, sec^{-1}	K_m, mM	$k_{cat}/K_m, s^{-1}mM^{-1}$
I	Val	20	0.003	6670
II	Pro	7	0.004	1750
III	His	--	-.---	----
IV	Lys	21	0.005	4250
V	Ala	16	0.002	8000
VI	Ile	16	0.003	5330

b. Individual values of k_{cat} and K_m are not reported for substrate III since binding was so tight that they could not be separated.

Examples of this type of analysis are shown in Figures 1 and 2. The profiles contain from zero to three breaks over the range investigated and detailed interpretation of these breaks will form the basis of the following discussion. The observed pK_a values observed for this interaction of a number of substrates and enzymes (taken from the break points of plots such as Figures 1 and 2) are collected in Table II for the systems studied. Our discussion of these results will follow the order given in Table II.

Substrate V, with an Ala in the P3 position and a Glu in the P2 position, gives a straightforward pH profile (Figure 1) with a single kinetically significant dissociation occurring in the plot of log k_{cat} versus pH. This can be interpreted to be an acid group in the enzyme substrate complex losing a proton as the pH is raised. This group must be present in the protonated form for effective catalysis. It is logical to assign this pK_a to one of the catalytic Aspartic acids in the active site of the enzyme, Asp 32 or 215. Based on the chemical reactivity of the two groups, the acid of higher pK_a has usually been considered to be Asp 215 (13). Based on this convention the observed pK_a of 5.1 is tentatively assigned to be derived from Asp 215. Since the most recent x-ray crystallographic results of these enzymes discussed earlier in this volume indicate that the two Aspartic Acids are in Hydrogen bonding distance and are coplanar, it has been suggested that they share a proton between them. In addition, there appears to be a tightly bound water molecule intimately associated with them. Thus, it would be more correct to assign the observed dissociation to the whole system of Asp 32 - Asp 215 - bound water since there is no means of distinguishing the source of the proton. Therefore, for the sake of simplicity, Asp 215 will be used throughout this chapter as representative of the entire catalytic apparatus. Returning to Figure 1, we also see

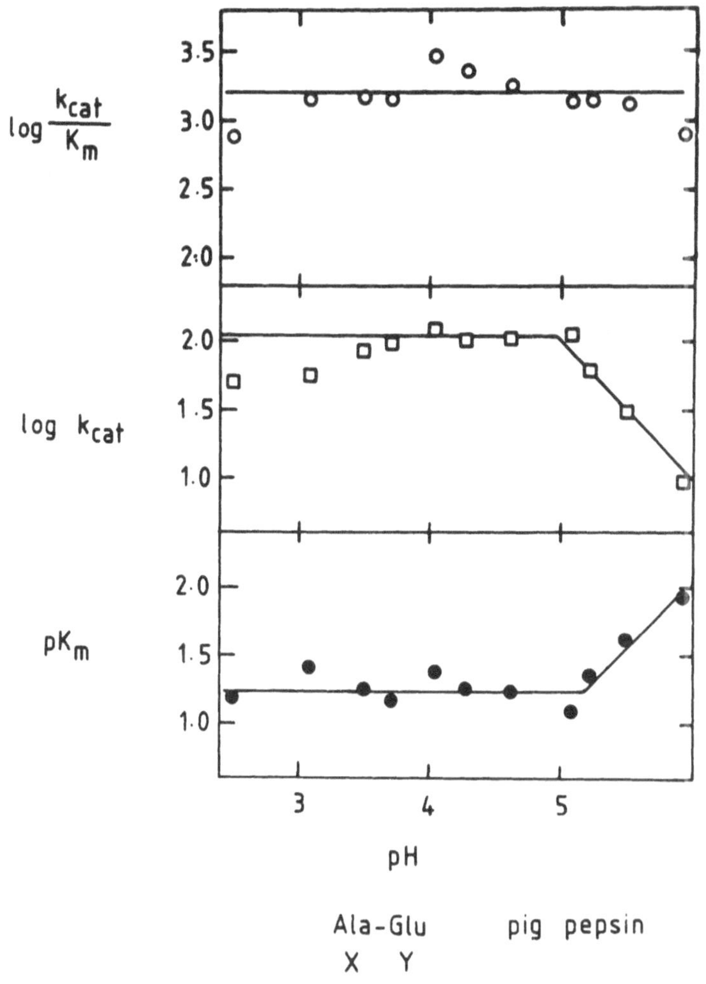

Figure 1. Plots of log k_{cat}/K_m, log k_{cat}, and pK_m versus pH for the pig pepsin catalyzed hydrolysis of substrate V.

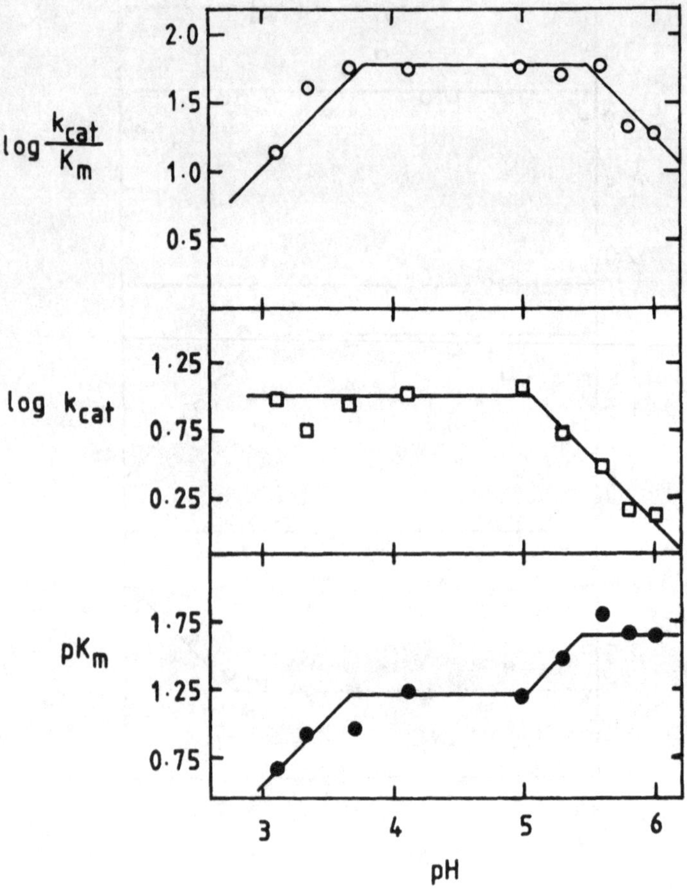

pig pepsin

His-Glu
X Y

Figure 2. Plots of $\log k_{cat}/K_m$, $\log k_{cat}$, and pK_m versus pH for the pig pepsin catalyzed hydrolysis of substrate III.

Table II

Kinetically Significant Dissociation Constants Derived from Inspection of Plots of Kinetic Parameters versus pH.

Enzyme	Substrates	$pK_a^{E.S}$	$pK_{a1}^{E \text{ or } S}$	$pK_{a2}^{E \text{ or } S}$
Pig Pepsin	V	5.1	-	-
	VII	5.4	3.7	-
	III	5.0	3.7	5.5
Chicken pepsin	V	5.1	-	-
	VII	4.9	-	-
Chymosin	VII	5.8	-	-
	V	5.2	3.3	-
	III	5.4	4.3	-
Endothia P.	VII	5.2	-	-
Proteinase	V	-	5.5	

that the break observed in the log k_{cat} plot is mirrored in the plot of pK_m versus pH. Since this occurs as an upward break in the pK_m profile, it can also be ascribed to a group in the E.S complex. It is noteworthy that the group that is observed in the bottom two panels of Figure 1 does not show up in the plot of log k_{cat}/K_m. This section shows kinetically significant dissociations occurring in the free enzyme or free substrate. Since no break occurs here, we must conclude that there is a substantial change in pK_a upon binding substrate. This could be of catalytic significance. A final noteworthy feature of the profiles in Figure 1 is the relatively large plateau between pH's 2.5 and 5.0. This illustrates the value of these substrates for general study of enzyme properties in this pH range.

In contrast to the observation on substrate V, hydrolysis of substrate VII with a Thr in the P3 position and a Val in the P2 position gives a second kinetically significant dissociation in the pH region studied (Table II). A break is observed in the plot of log k_{cat}/K_m versus pH at pH 3.7, arising from dissociation of a group in the free enzyme or free substrate. The log k_{cat} versus pH profile yields a break at pH 5.4 similar to that observed for substrate V. Finally, the pK_m versus pH plot, being the combination of the other two plots, contains both breaks. The interpretation of the new break, at pH 3.7, has been derived from inspection of models of the structures of the Aspartic Proteinases derived from the crystallography. In pig pepsin, Glutamic acid 13 is positioned in the S3 subsite of the active site cleft. It is apparent that this residue, in the carboxylate form, can form a hydrogen bond to the Thr side chain hydroxyl group strengthening the interaction between enzyme and substrate. The profile for the substrate with Ala in the P3 position does not contain this break since such an interaction is not possible.

The validity of this interpretation may be tested by consideration of the profiles for a third substrate with pig pepsin, substrate III, with His in the P3 position and Glu in the P2 position. As shown in Figure 2, three kinetically significant dissociations are now seen. First, there is a dissociation at pH 3.7, showing up in the log k_{cat}/K_m profile and in the pK_m profile. Again this may be attributed to Glu 13 forming an interaction with the side chain at P3 of the substrate. Once more, a catalytically significant break in the log k_{cat} profile with pK_a of 5.0 is seen which is mirrored in the pK_m profile. This can again be ascribed to the catalytic system of Asp 215. In addition, a second break in the log k_{cat}/K_m profile is observed which is due to a group in the free enzyme or free substrate. Since the enzyme is the same in all cases, this new break can be assigned to the substrate His residue. Dissociation of this group to give a neutral side chain diminishes the interaction with Glu 13 COO^- and gives the downward break observed. Again, the breaks observed in both upper panels are combined into the pattern seen in the pK_m profile. It is noteworthy here that the comparison is between the behavior of two substrates which differ only in the presence of a hydrogen ion, which represents a most subtle chemical modification.

Since the breaks observed at pH 3.7 have been attributed to the Glutamic acid at position 13 of the pepsin sequence, a natural question would be the kinetic behavior of an enzyme that did not have Glu at position 13. Since recent work has provided the sequence of several Aspartyl Proteases, such a test is readily possible. Chicken pepsin has a Serine in position 13 in its sequence (14). In Table II, two results are given for the Chicken pepsin catalyzed hydrolysis of substrates V and VIII. For V a pattern much like that observed for pig pepsin has been obtained with a single kinetically significant dissociation observable in the plot of log k_{cat} versus pH. Once again this may be attributed to the catalytic Asp 215. For Chicken pepsin substrate VII containing a Thr in the P3 position does not give a kinetically significant break at pH 3.7 in any of the profiles. Apparently, replacement of Glutamic acid at position 13 in pig pepsin by Serine in chicken pepsin eliminates the possibility of the same type of interaction observed for pig pepsin with substrates VII and III.

Similar experiments have been used to test for the possibility of significant interactions of an ionic nature in the S2 subsite of these enzymes. Examination the available primary sequences of several Aspartyl Proteinases in conjunction with a knowledge of the three dimensional structure permit detailed analysis of the possibility of such interaction. From examination of a computer graphics representation of the Chymosin molecule based on the known structure of the Aspartic Proteinase from Endothia Parasitica (by courtesy of L. Sibanda and T. Blundell, Birkbeck College, London) it has been observed that the S2 subsite of the enzyme could include the side chain of position 220. Residue 220 of chymosin is a Lysine (15), whereas pig pepsin and chicken pepsin, in common with most Aspartic Proteinases, have a Leucine in position 220. Therefore, substrates III, V, and VII were studied with Chymosin as the attacking enzyme.

Substrate VII serves as the control in this case, since it contains a Thr in the P3 position and a Val in the P2 position. Chymosin has a Gln in the S3 subsite at position 13 in the sequence and so again cannot display the behavior seen with pig pepsin. The profiles for Chymosin catalyzed attack on substrate VII were straightforward with only the catalytic break observed in the plot of log k_{cat} versus pH. This gave a value of 5.8 for the catalytic system. Next, however, it may be seen that substrate V which has an Ala in P3 and a Glu in P2 gave a second break in the profiles, defining a kinetically significant group of pK_a 3.3 (Table II). Since this shows up as a group in the free enzyme or free substrate, this pK_a has been attributed to the Glu at P2 in the substrate. This apparently can form an interaction with Lysine 220 of the enzyme S2 subsite. Curiously, this interaction leads to more efficient catalysis since it shows up in the log k_{cat} versus pH profile as well as the log k_{cat}/K_m profile. As further evidence for this interpretation, substrate III was studied with Chymosin. The results in Table II show again a kinetically significant interaction involving the basic form of a group with pK_a 4.3. Again, this may be attributed to the Glu of the substrate. Since this substrate has a different structure than substrate V, the difference in pK_a observed for the free substrate may be expected. It would be more difficult to rationalize this difference if it came from an enzyme group. However, the apparent catalytic group for the three different substrates with Chymosin only varies from 5.2 to 5.8.

As a last example of this analysis, the pH dependence of the kinetic parameters for the hydrolysis of two of the substrates catalyzed by Endothia Parasitica Proteinase has been studied. Substrate VII, with Thr in P3 and Val in P2, gives a routine profile with only the catalytically significant Asp 215 dissociation apparent. In particular, the plot of pK_m versus pH is flat. However, in the kinetics of substrate V with Endothia Proteinase, a downward deflection at higher pH in the plot of pK_m versus pH has been observed. This implies that there is a repulsion occurring between two groups leading to less effective binding. A reasonable interpretation of this phenomenon again derives from inspection of the Endothia Proteinase structure. An Aspartic Acid at position 77 (16) of the enzyme sequence is located on the so-called "flap" of structure that lies over the active site cleft. This residue could be in contact with the side chain of the amino acid in the P2 position of the substrate bound at the active site. If the Asp at position 77 and the Glu of the substrate were both in the COO^- form it is possible that a significant repulsion would occur. At intermediate pH, it is possible that a favorable interaction could take place due to a shared proton.

To summarize the results of these studies, several significant interactions involving ionic groups in subsites P3 and

P2 of these Aspartyl Proteases have been observed. The summation of such interactions in the seven or more subsites along the active site cleft could provide a rationale for the strength of interaction of various substrates and inhibitors with these enzymes. Furthermore, these interactions in the S3 and S2 subsites begin to offer a molecular explanation for some of the observations about "Secondary Specificity" that have been observed (1-3). If such effects are mirrored along the entire length of the active site crevice, then the exquisite specificity achieved by enzymes such as renin can be understood.

2) Inhibition of Pig Pepsin by Analogs of the Propart of Pig Pepsinogen

In addition to the insights derived from the studies of substrates described above, similar information can be obtained by examination of the nature of interactions of various peptide <u>inhibitors</u> of these enzymes. In particular, analogs based on the propart of pig pepsinogen have allowed a detailed study of inhibitor binding. Herriott initially reported that some low molecular weight component of the activation mixture resulting from acidification of pepsinogen was capable of inhibiting the milk clotting activity of pig pepsin (17). More recent studies have shown that the peptide consisting of the first 16 residues of pig pepsinogen (Table III) is responsible for the majority of this inhibition (18). The analogs that will be discussed here are listed on Table III. As seen there, a synthetic peptide of identical sequence could be prepared that reproduced the inhibition observed with the native fragment, thus making analog synthesis feasible. Due to the large number of basic residues (Lysines at position 3, 9, and 10 and Arginines at position 8 and 13) and the obvious possibility that the basic propart could bind to the very acidic enzyme surface by electrostatic interaction, a series of derivatives have been prepared (19) to test this hypothesis. Initially, the Arg residues were each replaced singly: [Thr^{8}]-(1-16) and [Thr^{13}]-(1-16) or in combination: [$Thr^{8,13}$]-(1-16). As seen in Table III, these replacements were without effect on the equilibrium position of binding to pig pepsin. In addition, chemical modification of the three lysines by carbamylation to produce homocittruline residues, [$HC^{3,9,10}$]-(1-16) has a minimal effect on binding. However, conversion of all basic residues to remove the positive charge has some effect since the peptide [$HC^{3,9,10}$ $Thr^{8,13}$]-(1-16) has a K_I value approximately 20 fold higher than the other derivatives.

Due to the relatively small effects seen from alteration of the basic residues, the next series of peptides tested the hypothesis that the hydrophobic residues in positions 4-7 of the inhibitor (-Val-Pro-Leu-Val-) are directly involved in interaction with the enzyme. In this series, only the 1-12 sequence was synthesized (10) since it was observed that the carboxy terminal residues are not essential. It may be seen in Table III that replacement of all four of the hydrophobic residues by Glycine, gives a derivative with binding reduced by about 300 fold. It may also be seen that the derivatives with two of the four hydrophobic residues replaced with Glycine, [$Gly^{4,5}$]-(1-12) and [$Gly^{6,7}$]-(1-12) give intermediate values of K_I. This data indicates strongly that hydrophobic interactions are directly involved in binding these inhibitors to pig pepsin. The structures of the homologous Aspartic Proteinases reported in earlier chapters in this volume have revealed extensive hydrophobic regions in the active site cleft. These regions should interact strongly with the hydrophobic sequence -Val-Pro-Leu-Val- of the (1-16) or (1-12) inhibitors.

Table III

Apparent K_I values for inhibition of pig pepsin by pepsinogen propart peptide analogs

Leu-Val-Lys-Val-Pro-Leu-Val-Arg-Lys-Lys-Ser-Leu-Arg-Gln-Asn-Leu

pig pepsinogen residues (1-16)=(1-16)

Peptide	K_I, μM
(1-16) (native)	0.25
(1-16)	0.20
[Thr^{8}]-(1-16)	0.35
[Thr^{13}]-(1-16)	0.26
[$Thr^{8,13}$]-(1-16)	0.40
[$HC^{3,9,10}$]-(1-16)	0.50
[$HC^{3,9,10}$ $Thr^{8,13}$]-(1-16)	5.3
(1-12)	0.1
[$Gly^{4,5}$]-(1-12)	4.0
[$Gly^{6,7}$]-(1-12)	1.0
[Gly^{4-7}]-(1-12)	30
(1-8)	18
[Val^{3}]-(1-8)	19
[Ser^{3}]-(1-8)	22
[Glu^{3}]-(1-8)	350

To seek more evidence for the mechanism of inhibition and the rate of interaction between peptide and protein, selectively α-dansylated derivatives of some of these peptides have been synthesized (20). The K_I values for the dansylated derivatives are equal to their free peptide counterparts. This is probably due to the separation between the amino terminus and the sequence -4-7- which binds in the active site. When these derivatives are mixed with pig pepsin, an energy transfer spectrum is generated rapidly. This has the characteristics of a spectral interaction between tryptophan residues of the enzyme and the dansyl group of the peptide. This spectral change occurs in two parts (21). The initial, rapid interaction leading to an energy transfer band is followed by a slower phase where the energy transfer band decreases. This second phase, which requires approximately one hour at 25° is exactly coincident with the slow appearance of inhibition observed with these peptide inhibitors. Apparently, the initial rapid event leading to a large degree of energy transfer is a simple collisional process giving a complex with the peptide loosely bound to the surface of the protein. It was also observed that the magnitude of the initial event, as measured by the magnitude of the fluorescence change, is pH dependent, increasing upon dissociation of a group with a pKa of about 4.7. Also, this process is reduced by increasing the ionic strength. These observations indicate that this collisional event is electrostatic in nature. The initial complex is not inhibited, presumably because the peptide is not in the active site at this point. Only in the second, slower phase does the peptide interact with the active site to give inhibition. The K_I values quoted in Table III were obtained at equilibrium by incubation of peptide and enzyme at an appropriate combination of temperature and time to complete the second phase of binding.

As described above, the equilibrium binding of these peptide inhibitors seems to be dominated by hydrophobic forces. However, there is an electrostatic component involved due to the following observation. Firstly, the equilibrium position of inhibition could also be evaluated from the magnitude of the <u>final</u> fluorescence intensity with the dansylated peptides (21). Again, stronger binding was observed at higher pH, with an apparent pKa of about 4.8. This indicates that a side chain carboxylate group of the enzyme is required for optimal binding. This agrees with Herriott's observation forty years ago, that inhibition by the activation fragment was better at pH 5.5 than at lower pH values (17).

To provide more evidence for this interpretation, a series of peptides based on the (1-8) sequence of the propart of pig pepsinogen has been synthesized. In Table III it can be seen that analogs with Lys, Ser, or Val in the third position are equivalent in binding to pig pepsin at pH 5.5. However, when a Glutamic acid occupies that position, the binding is reduced by about 20 fold. A straightforward interpretation of this observation is that the negatively charged Glutamic acid residue of peptide [Glu^3]-(1-8) is repelled by the negatively charged residue of the enzyme that is responsible for better binding at higher pH of inhibitors with a Lys in position 3. Indeed, the binding of peptide (1-8), with Lys in position 3, improves with increasing pH with an apparent pKa of 4.7. This agrees with the pH dependence seen by measuring fluorescence intensity at equilibrium in the dansylated peptide. Of further interest is the observation that the shortened peptides based on the (1-8) sequence give very rapid inhibition of pig pepsin, in contrast to the slow inhibition seen for (1-12) or (1-16) peptides. Apparently, the removal of Lys residues 9 and 10 reduces the <u>surface</u> binding that we observed spectrally for α-dansyl-(1-12). The smaller (1-8) peptides are thus able to bind to the active site more readily since they are not diverted to the surface first.

To summarize the observation made with peptide inhibitors of pig pepsin, there can be a rapid surface binding that is mainly electrostatic in nature. This is followed by a slower second phase in which the peptide finds the active site. The interaction at this stage is dominated by hydrophobic binding from the 4-7 sequence but it still contains an electrostatic element. These various observations can be correlated by the assumption that the peptide inhibitors bind in the active site cleft with an alignment to place Pro^{5} in the S1 subsite and the Lys^{3} in the S3 subsite. This would permit interaction with the Glutamic acid at position 13 of the pepsin sequence. This conclusion derives from the observation that the propart peptide only binds to those enzymes that have a Glu at position 13, the competitive nature of the inhibition, the importance of the -4-7- sequence, the pH effects observed for binding, and the effect of replacing Lys^{3} by Glu in the peptide.

The proparts of a variety of animal pepsinogens have now been sequenced (22). In most cases, the sequence contains a block of four hydrophobic residues such as the -Val-Pro-Leu-Val- sequence seen in pig pepsinogen. The Pro residue is always present while the Val and Leu residues are sometimes replaced by Ile. In three known zymogens, this hydrophobic block is only three residues long, prorenin, prochymosin, and monkey progastricsin. Interestingly, in all cases, this hydrophobic block is flanked by a Lys or Arg residue on each side. These residues must play a role in ensuring that the hydrophobic block can bind in the correct register to be properly aligned in its binding site.

In summary, a detailed study of the binding of substrates and inhibitors to Aspartic Proteinase has provided new evidence for the importance of interaction at subsites within the extended active site cleft. These interactions can be invoked to rationalize observation of "Secondary Specificity" such as those of Fruton (1-3) and also to begin to understand the often exquisite specificity of some of the Aspartic Proteinases.

References

1. Fruton, J.S.: Adv. in Enzymol. 33, 401-443 (1970).

2. Fruton, J.S.: in "The Enzyme" (Boyer, P.D., Ed.) 3rd Edition, Vol. 3, Academic Press, New York, pp. 119-164 (1971).

3. Fruton, J.S.: Adv. in Enzymol. 44, 1-36 (1976).

4. Takahashi, M., Hofmann, T.: Biochem. J. 147, 549-563 (1975).

5. Neumann, H., Levin, Y., Berger, A., Katchalski, E.: Biochem. J. 73, 33-41 (1959).

6. Powers, J.C., Harley, A.D., Myers, D.V.: in "Acid Proteases: Structure, Function, and Biology" (Tang, J., Ed.) Plenum Press, New York, 141-157 (1977).

7. Antonov, V.K.: in "Acid Proteases: Structure, Function, and Biology" (Tang, J., Ed.) Plenum Press, New York, 179-198 (1977).

8. Dunn, B.M., Kammermann, B., McCurry, K.R.: Anal. Biochem. 138, 68-73 (1984).

9. Hofmann, T., Hodges, R.S.: Biochem. J. 203, 603-610.

10. Dunn, B.M., Lewitt, M., Pham, C.: Biochem. J. 209, 355-362 (1983).

11. Dixon, M. and Webb, E.C.: "Enzymes", Second Edition, Academic Press, New York, 116-166 (1964).

12. Dixon, M.: Biochem. J. 55, 170 (1953).

13. Clement, G.E.: Prog. Bioorg. Chem. 2, 177-238 (1973).

14. Kostka, V., Keilova, H., Baudys, M.: in "Proteinases and Their Inhibitors: Structure, Function, and Applied Aspects" Pergamon Press, Oxford, 125-140 (1981).

15. Foltmann, B. Pedersen, V.B., Jacobsen, H., Kauffman, D., Wybrandt, G.: Proc. Natl. Acad. Sci. U.S.A. 74, 2321-2324 (1977).

16. Personal Communication from B. Foltmann and V.B. Pedersen.

17. Herriott, R.M.: J. Gen. Physiol. 21, 65-78 (1938).

18. Dunn, B.M., Deyrup, C., Moesching, W.G., Gilbert, W.A., Nolan, R.J., Trach, M.L.: J. Biol. Chem. 253, 7269-7275 (1978).

19. Dunn, B.M., Deyrup, C.: in "Peptides: Structure and Biological Function" (Gross, E., Meienhofer, J., Eds.) Pierce Chemical Co., pp. 161-164 (1979).

20. Dunn, B.M., Pham, C., Raney, L., Abayasekara, D., Gillespie, W., Hsu, A.: Biochemistry, 20, 7206-7211 (1981).

21. Dunn, B.M.: Arch. Biochem. Biophys. 214, 763-771 (1982).

22. Kageyama, T., Moriyama, A., Takahashi, K.: J. Biochem., 94, 1557-1567 (1963) and chapters by Takahashi, K. and by Welinder, K., Mikkelsen, L. and Foltmann, B. in this volume (pp 265-282 and 197-202, resp.).

KINETIC AND FLUORESCENCE STUDIES ON CHICKEN PEPSIN. THE USE OF CYS-115 AS AN ACTIVE SITE PROBE

Jan Pohl, Petr Štrop, Iva Pichová, Ivo Bláha,
Vladimír Kostka

Department of Biochemistry, Institute of Organic Chemistry and Biochemistry, Czechoslovak Academy of Sciences
166 10 Prague 6, Czechoslovakia

Introduction

A unique feature of catalysis by aspartic proteinases is the strong dependence of their hydrolytic efficiency (k_{cat}) on the secondary enzyme-substrate interactions (1), i.e. on the structure of the A and B moiety of substrates A-P1-P1´-B with P1-P1´ as the scissile bond. Hence, this specificity is achieved during the transition state of the reaction (1,2), and the key role of the conformational flexibility of the extended binding cleft of aspartic proteinases in this process has been hypothesized (3). In fact, the conformational flexibility of the "flap" region of the penicillopepsin molecule as a result of pepstatin analogs binding has been demonstrated in recent X-ray studies on crystals of penicillopepsin-inhibitor complexes (4).

Sequence analysis of chicken pepsin (CP) has demonstrated its high degree of homology with mammalian pepsins (5). The enzyme, however, is rather exceptional in its catalytic features: unlike mammalian pepsins it does not cleave their standard substrates (1), i.e. di- and tripeptides (6,7). Since, however, its general proteinase activity is the same as that of mammalian pepsins the key role of secondary interactions in the catalysis by chicken pepsin must be assumed. This enzyme is

Aspartic Proteinases and their Inhibitors

also unique in containing a free SH-group (6), that of Cys-115 (8). The labeling of the latter with bulky reagents makes tripeptide Z-His-Phe-Phe-OEt (ref.1), originally entirely resistant to chicken pepsin, a modest substrate of the modified enzyme (6,7).

This study will provide kinetic evidence of the strong dependence of the catalysis by chicken pepsin on its secondary interactions with synthetic substrates of the <Glu-His-Phe-Phe-X-Y-OMe type with Phe-Phe as the scissile bond. The Cys-115 residue is shown to be a part of the binding site of the enzyme. The binding of substrates and inhibitors to chicken pepsin with SH-group labeled with an extrinsic fluorescence probe (Acrylodan, ref.9) will be demonstrated to result in a conformational change in the neighborhood of the cysteine residue. Data will also be presented on the kinetics of binding of slow-binding chicken pepsin inhibitors, pepstatin and propart peptides.

Material and Methods

The peptides were synthesized by the solid phase method of peptide synthesis (10,11) and purified by free-flow electrophoresis and reversed-phase HPLC. All peptide substrates were fully convertible into the corresponding products by the pepsins. The isolation and purification of chicken and hog pepsin was effected by methods developed earlier (10,11). The native complex (1:1) of chicken pepsin with the chicken pepsinogen activation peptide p1-p42, chicken pepsinogen and its activation peptides p1-p26, p1-p42 and p27-p42 were isolated by Pichová and coworkers (12). Chicken pepsin contained 0.9-0.97 mol of SH-group titratable with 2,2'-dithiopyridine per mol of enzyme. The labeling of the enzyme with Acrylodan (a gift from Dr F.G. Prendergast, Mayo Foundation, Rochester, N.Y., ref.9) was achieved by the reaction of a 4-fold molar excess of the label over the enzyme (1mg/ml) in 0.1M sodium phosphate, pH 7.4, 1 mM in EDTA. Chicken pepsin modified by Ellman´s reagent (CP-EL) and 2,2´-dithiodipyridine (CP-GM) were prepared in a similar manner.

The kinetic parameters of the hydrolysis of Phe-Phe, Nph-Phe and Phe-Nph were determined spectrophotometrically (10,13). The K_i-values and types of inhibition were examined by methods recommended by Todhunter (14). The values of $k_{cat}=V/E_o$ are based on E_o determined by active site titration with iso-valeryl-pepstatin.

Results and Discussion

Chicken pepsin catalyzed hydrolysis of synthetic substrates

The k_{cat} and K_m values for chicken pepsin catalyzed hydrolysis of synthetic substrates were obtained with cationic histidine-containing (1) substrates cleaved at Phe-Phe (I-III) and at Nph-Phe (IV,V), and with an analog of the pepsin substrates designed recently by Dunn (13,15), peptide VI with Phe-Nph as the scissile bond. The values obtained with hog pepsin as a representative of mammalian pepsins are also listed.

The pH-profiles of k_{cat} and K_m (Fig. 1) are different for chicken and hog pepsin. The ability of chicken pepsin to cleave substrates I-V decreases with the decreasing pH, as evidenced by a decrease in k_{cat} and a simultaneous increase of K_m. For these reasons the kinetic constants were measured at pH 4.9 which is optimal for both enzymes (Table 1).

According to earlier observations (6) Z-His-Phe-Phe-OMe was entirely resistant to hydrolysis by chicken pepsin. The peptide behaves as a competitive inhibitor of the enzyme, the K_i being very close to the K_m-values for the substrates. After extension of the B-moiety of the substrate from -OMe to -Gly-Gly-OMe and replacement of Z by a pyroglutamyl group,

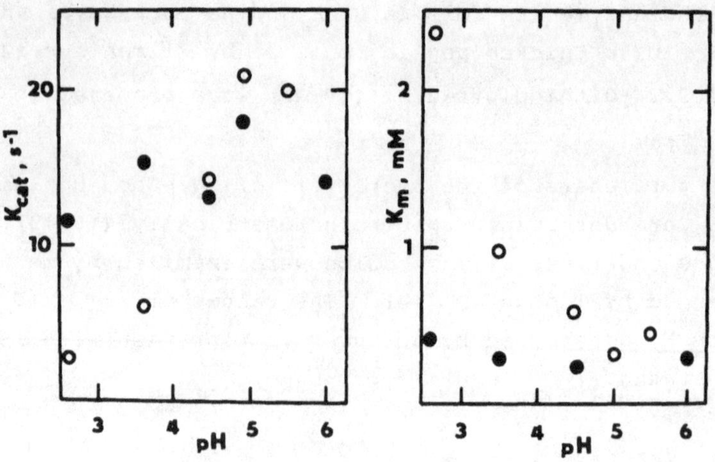

Figure 1 pH-profile of k_{cat} and K_m for hydrolysis of<Glu-His--Nph-Phe-Ala-Leu.NH_2 by hog and chicken pepsin. Measured at 30 °C in 0.1M sodim formate (pH 2.5-3.5) and in 0.1M sodium acetate (pH 3.5-5.5). (○) chicken pepsin, (●) hog pepsin.

the Phe-Phe bond of I was cleaved by chicken pepsin, however, the k_{cat} was approximately 40-times lower than with hog pepsin. The replacement of Gly-Gly by Ala-Ala in II and by Ala-Leu in III had a dramatic effect on the k_{cat} values measured with both enzymes. The increase of k_{cat} for hog pepsin catalyzed cleavage of substrates with aliphatic residues in their B-moiety was observed long ago (1). In the case of chicken pepsin its activity towards modified substrates III-VI became the same as that of hog pepsin in terms of k_{cat}. This shows that the energy from secondary interactions with the B-moiety of the substrate plays a dominant role in hydrolysis by chicken pepsin and is effectively utilized in the catalytic step.

Cleavage of synthetic substrates by chicken pepsin with modified Cys-115

The effect of labeling of Cys-115 in chicken pepsin with Ellman´s reagent and 2,2´-dithiodipyridine on k_{cat} and k_{cat}/K_m for substrates I-V, whose k_{cat}-values vary over 4 orders, is shown in Table 2. The modification has an activating effect on the cleavage of Phe-Phe in I-III, especially on k_{cat}. This activating effect is more marked with substrates showing a low k_{cat}, i.e. with Z-His-Phe-Phe-OEt (6), I, and II. In contrast, an inhibitory effect was observed on the hydrolysis of Nph-Phe in IV and V, as a result of the replacement of Phe in P1 by Nph, as follows from a comparison of the data measured with III and IV. The latter were to the same degree hydrolyzed by chicken pepsin and a small difference in k_{cat} was compensated by changes in the K_m, a phenomenon indicating nonproductive binding of a fraction of nitrated substrate IV (Table 1).

In contrast, modified pepsins hydrolyzed substrates III and IV with greatly different k_{cat}-values (122 s^{-1} and 7.7 s^{-1}, resp. for CP-GM), and with little effect on K_m, Table 2.

The modification markedly altered also the binding characteristics of the modified pepsins, however, only as regards inhibitors occupying also the secondary binding sites of the enzyme (Table 3). In contrast, the K_i´s for Phe-Phe-OMe were not affected by the modification. This shows that the modifying group does not interfere with the primary binding of the Phe-Phe unit of the substrate. Another piece of evidence in favor of this assumption emerges from the comparison of the reactivity of the SH-group toward 2,2´-dithiodipyridine in chicken pepsin itself and in complexes of the enzyme with ligands (Table 4). The reactivity is not influenced either by Phe-Phe-OMe and Phe-Phe-Ala-Ala-OMe or by substrates IV and V (not shown). Pepstatin has a mild promoting effect only.

The modification of complexes of chicken pepsin with Val-D-Leu-Pro-Phe-Phe-Val-D-Leu and with the activation peptide

Table 1

Kinetic parameters for chicken and hog pepsin catalyzed hydrolysis of synthetic substrates[a]

Substrate		Chicken pepsin			Hog pepsin			Hog pepsin
		k_{cat}	K_m	k_{cat}/K_m	k_{cat}	K_m	k_{cat}/K_m	k_{cat}/K_m
		s^{-1}	mM	s^{-1}/mM^{-1}	s^{-1}	mM	s^{-1}/mM^{-1}	
Z-His-Phe-Phe-OMe		-	0.3[b]	<0.0001	0.23	0.5	0.46	>5000
<Glu-His-Phe-Phe-Gly-Gly-OMe	I	0.01	2.0	0.005	0.43	2.8	0.19	38
<Glu-His-Phe-Phe-Ala-Ala-OMe	II	3.9	2.9	1.38	21	1.2	17.5	13
<Glu-His-Phe-Phe-Ala-Leu-OMe	III	31	0.55	56	54	0.55	98	1.8
<Glu-His-Nph-Phe-Ala-Leu-NH_2	IV	21	0.36	59	17	0.33	52	0.88
Gly-Gly-His-Nph-Phe-Ala-Leu-NH_2	V	34	0.23	148	18	0.14	130	0.88
Lys-Pro-Ala-Glu-Phe-Nph-Ala-Leu-OH	VI	74	0.18	415	92	0.03	3070	7.4

[a] 0.1M sodium acetate pH 4.9, 1 mM in EDTA, 30 °C. [b] K_i value, competitive inhibitor versus substrate IV.

Table 2

Effect of modification of Cys-115 in chicken pepsin on kinetic parameters of hydrolysis of synthetic substrates[a]

Substrate		Enzyme	Modified CP/Unmodified CP expressed as	
			k_{cat}-ratio	k_{cat}/K_m-ratio
Z-His-Phe-Phe-OEt		CP-NPS[b]	>400	>400
<Glu-His-Phe-Phe-Gly-Gly-OMe	(I)	CP-EL	4.4	8.4
		CP-GM	22	8.4
<Glu-His-Phe-Phe-Ala-Ala-OMe	(II)	CP-EL	8.0	8.8
		CP-GM	14	14
<Glu-His-Phe-Phe-Ala-Leu-OMe	(III)	CP-EL	2.0	1.6
		CP-GM	4.0	2.5
<Glu-His-Nph-Phe-Ala-Leu-NH_2	(IV)	CP-EL	0.50	0.18
		CP-GM	0.28	0.36
Gly-Gly-His-Nph-Phe-Ala-Leu-NH_2	(V)	CP-EL	0.17	0.29
		CP-GM	0.33	0.13

[a]The conditions of the experiment are described in Table 1. [b]p-nitrophenylsulfenyl-CP (ref.6)

Table 3
Effect of modification of Cys-115 in chicken pepsin on K_i-values[a]

Inhibitor	K_i, mM			
	CP	CP-EL	CP-GM	CP-PR
Phe-Phe-OMe	0.45	0.36	0.48	0.40
Phe-Phe-Ala-Ala-OMe	0.95	0.20	0.19	0.63
Val-D-Leu-Pro-Phe-Phe-Val-D-Leu	$5x10^{-5}$	$36x10^{-5}$	$91x10^{-5}$	$42x10^{-5}$

[a]The conditions of the experiment are described in Table 1. All the peptides are competitive inhibitors.

Table 4

Effect of inhibitors of chicken pepsin on reactivity of free SH-group of Cys-115 toward dithio-dipyridine

Ligand	L_o/K_i	K_i	10^3x $k_{obs.}$ 1st [b)] s^{-1}
none			8.6 $\pm$ 0.2
Phe-Phe-OMe	50	0.45mM	8.4
Phe-Phe-Ala-Ala-OMe	8	0.95mM	8.3
pepsinogen p1-p42 propart peptide	a)	<0.01 µM	0.009
Val-D-Leu-Pro-Phe-Phe-Val-D-Leu	a)	50 nM	0.22
Isovaleryl-pepstatin	a)	0.1 nM	8.8 $\pm$ 0.1

[a)] pepsin:inhibitor = 1:1. [b)] at 25 °C, pH 4.9; enzyme 20 uM dithio-dipyridine 1.7 mM p1-p42 proceeded at a substantially reduced rate. This seems to indicate a direct shielding of the SH-group by these inhibitors, a conclusion of which is also supported by a strong protection of the Acrylodan probe of the CP-PR complex with Val-D-Leu-Pro-Phe-Phe-Val-D-Leu against α,α -dipyridine formate quenching (see below) and by a 20-fold increase in K_i for Val-D-Leu-Pro-Phe-Phe-Val-D-Leu inhibition of CP-GM (Table 3). Although the effect of the modification of Cys-115 on the kinetics manifests itself by a change in primary specificity, this phenomenon is obviously somehow mediated by secondary interactions. The possibility that a conformational change in the neighborhood of Cys-115 is a result of the binding of ligands was checked with a chicken pepsin derivative bearing a fluorescent label on the SH-group.

Fluorescence measurements

The quantum yield of Acrylodan (9), a thiol-selective, polarity sensitive fluorescent probe was significantly increased after its reaction with Cys-115. The increase was paralleled by a 25 nm blue shift of the emission maximum with respect to the model compound. This shift indicates polarity changes around the probe in the enzyme. The binding of substrates and inhibitors brings about another increase of the quantum yield (Table 5) resulting from an additional polarity decrease in the neighborhood of Cys-115. The data on the binding of ligands of the A-Phe-Phe-B type show that the minimum pre-requisite of fluorescence increase is the presence of a bulky A-group. It should be noted in this connection that the tryptophan fluorescence of the enzyme was not influenced by the binding of pepstatin and some other ligands.

The accessibility of the probe in CP-PR and in its complexes with pepstatin and Val-D-Leu-Pro-Phe-Phe-Val-D-Leu was studied with the cationic quenchers, α,α-dipyridine formate and pyridine formate and with the anionic quencher, I^-.

As shown in Figure 2 the effectiveness of I^-, a small quen-

Table 5

Spectral properties of chicken pepsin with Cys-115 labeled with Acrylodan [a]

Compound	pH	λ_{max} nm	F/F_o at λ_{max} [d]	K_d
2-mercaptoethanol adduct of Acrylodan	4.9	532	1.00	
CP-PR	4.9	510	1.00	
CP-PR complexes with:				
Phe-Phe-OMe	4.9	510	1.03	
Phe-Phe-Ala-Ala-OMe	4.9	510	1.05	
<Glu-His-Phe-Phe-Ala-Ala-OMe	4.9	510	1.95	
<Glu-His-Phe-Phe-Gly-Gly-OMe	4.9	510	1.90	2.9 mM
Val-D-Leu-Pro-Phe-Phe-Val-D-Leu	4.9	508	1.71	1.2 uM
Asp-Ala-His-Phe-D-Leu-Ala-Val	4.9	510	1.74	10 mM
Isovaleryl-pepstatin	4.9	510	1.67[b]	0.1 nM
propart-peptide p1-p42	5.7	467	1.56[b]	0.37 uM
CP-PR	7.5	493	1.00	
CP-PR complexes with:				
propart peptide p1-p42	7.5	466	2.42[b]	0.1 nM
propart peptide p1-p26	7.5	480	1.69	0.44 uM
propart peptide p27-p42	7.5	485	1.63	1.67 uM

[a] at 25 °C. [b] after incubation for a suitable time period. [c] measured under conditions where less than 10 % of substrate was hydrolyzed. [d] value corresponding to saturation with ligand.

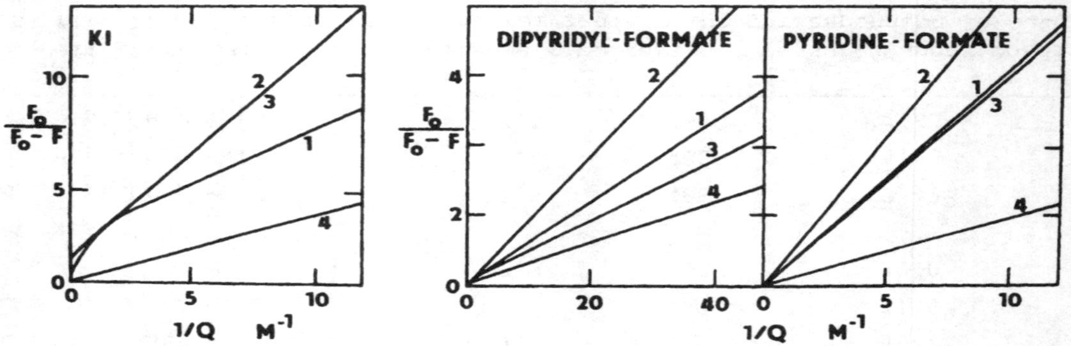

Figure 2. Lehrer plot of fluorescence quenching of modified chicken pepsin by KI and pyridinium ions. A - KI quenching, B - α,α-dipyridyl -formate quenching, C-pyridyl-formate quenching. 1 - chicken pepsin, 2 - chicken pepsin + Val-D-Leu-Pro-Phe-Phe-Val-D-Leu (0.02 M), 3 - chicken pepsin + pepstatin (0.02 mM), 4 - low molecular weight analog, 2-mercaptoethanol-Acrylodan. A - in 0.1 M sodium acetate pH 4.9, 1 mM EDTA, B - and C - in 0.1 M sodium formate pH 3.8, 1 mM EDTA. Enzyme concentration 740 nM, t=25 °C. Excitation at 386 nm, 1 nm band pass, emission at 500 nm, 7 nm band pass.

cher, is equally decreased in complexes of the enzyme with the two inhibitors. In contrast, the two cationic quenchers were strongly shielded in the complex of CP-PR with Val-D-Leu-Pro-Phe-Phe-Val-D-Leu. The probe is quenched more efficiently in the complex of chicken pepsin with pepstatin than in the enzyme itself. A parallel can therefore be drawn between the ability of the ligand to expose or shield the probe against cationic quenchers and the increase or decrease, respectively, of the reactivity of the SH-group to the probe. The possibility of a conformational change in the neighborhood of Cys-115 is suggested by the exposition of the probe due to pepstatin binding. Even evidence in favor of this possibility is the change in the near UV-CD spectrum in the range of the acrylodan absorption band of CP-PR during pepstatin binding (Figure 3).

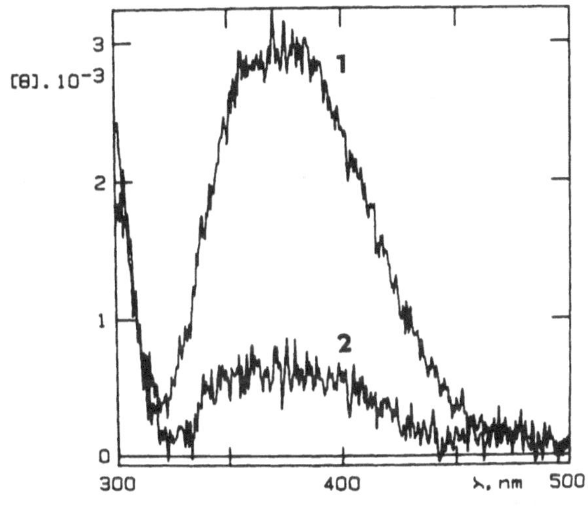

Figure 3. Changes in CD spectrum of labeled chicken pepsin during pepstatin binding. 1 - CD spectrum of chicken pepsin with Cys-115 labeled with Acrylodan, 2 - change in spectrum after addition of pepstatin (0.01 mM). Enzyme concentration 58 uM, in 0.1 M sodim acetate, pH 4.9, 1 mM EDTA, t=22 °C.

The possibility of a conformational change resulting from ligand binding brought us to the problem of localization of Cys-115 in the three-dimensional structure of the enzyme. X-ray diffraction studies on chicken pepsin have not been done so far. An inspection of the three-dimensional structures of other aspartic proteinases (16-19) has shown that, by analogy, residue 115 of chicken pepsin could be localized either in or very close to the specificity pocket of subsite S1 or S3. Insight in this problem could most probably be gained from computer graphics modeling (20) of this part of the molecule.

Fluorescence studies on ligand binding by chicken pepsin

Pepstatin binding. Pepstatin is a slow, tight binding inhibitor of hog pepsin (21,22). The rapidly formed initial complex EI is slowly converted into rigid complex EI´ according to scheme I

$$E + I \underset{k_2}{\overset{k_1}{\rightleftharpoons}} EI \underset{k_4}{\overset{k_3}{\rightleftharpoons}} EI' \; (k_4 \simeq 0) \qquad (I)$$

The reaction of pepstatin with chicken pepsin proceeds according to the same scheme (unpublished observations). The formation of the initial complex of CP-PR with pepstatin does not involve a change in quantum yield (Fig. 4) which is characteristic of the slow phase of the reaction only, i.e. the izomerization of the complex during which a time dependent increase of inhibition also takes place.

Binding of activation peptides

The interaction of chicken pepsin with propart peptides p1-p42, p1-p26 and p27-p42 is more complicated. The binding of these peptides at pH 5-7.5 is paralleled by a significant increase of the equilibrium fluorescence value (Table 5). Peptide p1-p42 behaves at pH 7.5 as a strong inhibitor of CP-PR form-

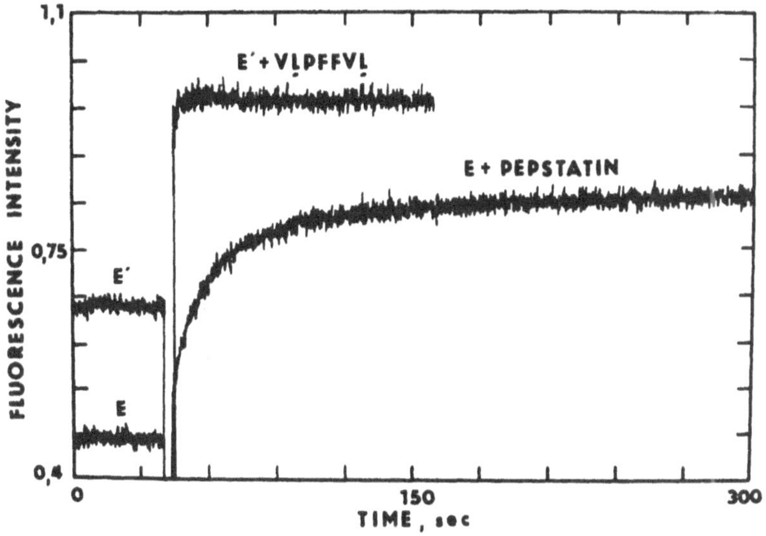

Figure 4. Time profile of pepstatin and Val-D-Leu-Pro-Phe-Phe-Val-D-Leu binding to chicken pepsin. Changes in intensity of fluorescence probe during binding. Enzyme 245 nM, pepstatin 615 nM, Val-D-Leu-Pro-Phe-Phe-Val-D-Leu 1.8 uM, t=15 °C. The remaining conditions same as in Fig. 2.

ing a 1:1 complex (Fig. 5). In contrast, weaker complexes result from the interaction of CP-PR with peptides p1-p26 and p27-p42. The simultaneous presence of the latter did not demonstrate a cooperativity in binding to CP-PR in terms of both fluorescence intensity and K_d. Hence, the strong inhibition exerted by p1-p42 is associated with the presence of intact Phe26-Leu27 bond in the inhibitor.

The formation of the 1:1 complex of p1-p42 with CP-PR is a slow process at pH 6-7.5 which complies with scheme I. After the two components are mixed together a slow fluorescence increase can be observed which is kinetically consistent with unimolec. isomerization. The latter is paralleled by a simultaneous inhibition increase (Fig. 6). The rate constant of this unimolecular process, k_3, is strongly dependent on ionic

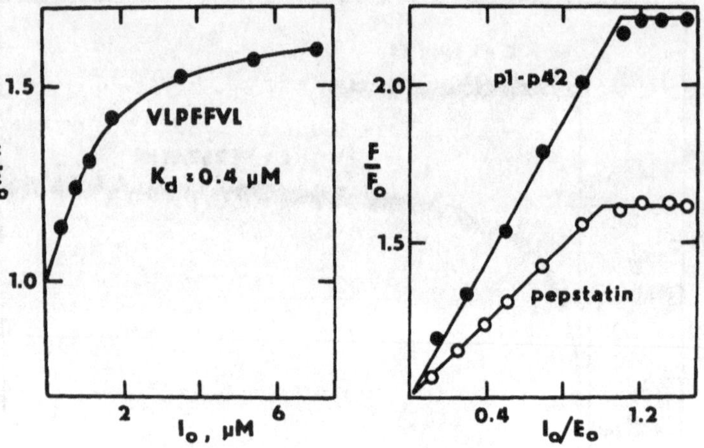

Figure 5. Titration of chicken pepsin by Val-D-Leu-Pro-Phe-Phe-Val-D-Leu (A) and with pepstatin and activation peptide p1-p42 (B). Changes in intensity of probe fluorescence during binding. The binding of the activation peptide p1-p42 was examined in 0.1 M sodium phosphate, pH 7.54, 0.1 mM in EDTA, at 25 °C. Excitation at 386 nm, 1 nm band pass, emission at 466 nm, 7 nm bandpass. Enzyme 0.3 uM (A) and 15 nM (B). The remaining conditions the same as in Fig. 2A.

strength and decreases with its increase (Fig. 6, inset). This points to the importance of attractive ionic interactions between basic residues of the inhibitor and pepsin carboxylates essential for productive positioning of the peptide in the final complex EI´.

The existence of initial complex EI has also been demonstrated in studies on pH and ionic strength profiles of initial fluorescence. The course of the latter is at I > 0.2 M and pH 6.1-7.5 compatible with unimolecular isomerization of EI. At I-values lower than 0.2 M, however, a rapid fluorescence increase (in 1s) is observed which is followed again by a slower phase c orresponding to the conversion of EI after the same equilibrium value has been obtained. Hence, the character of the initial complex as examined by fluorescence measurement, is manifold and dependent on ionic attractive forces.

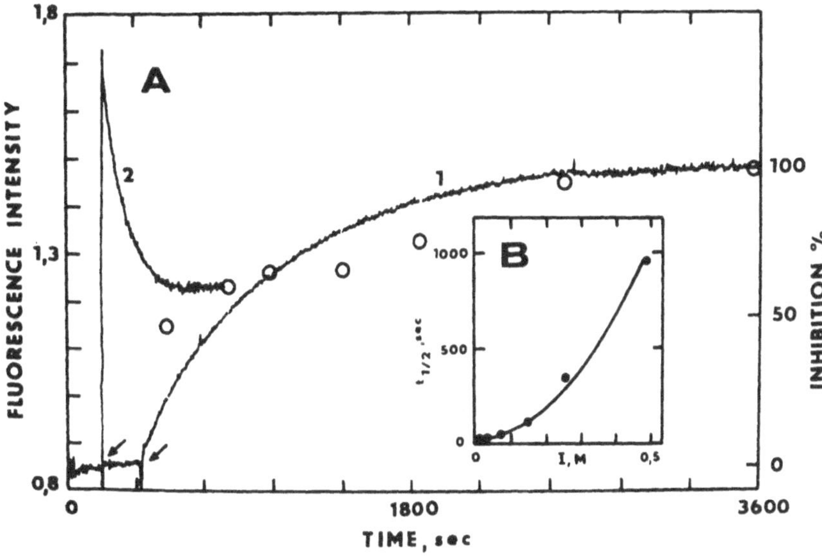

Figure 6. Time profile of binding of activation peptide pl-p42 by chicken pepsin assayed in terms of changes in probe fluorescence and enzymatic activity (A). A - enzyme 300 nM, peptide 750 nM, t=25 °C, 1 - 0.01 M sodium phosphate, pH 7.54 containing 0.1 mM EDTA and 0.48 M KCl. The residual activity (o) was determined in aliquots of reaction mixture taken at different time intervals and diluted 10 times with 0.1 M sodium acetate, pH 5.7, containing 70 uM substrate VI, 2 - 0.1 M sodium acetate, pH 5.7. The remaining conditions the same as in A. Inset B: 0.01 M sodium phosphate, pH 7.54, 0.1 mM in EDTA. The ionic strength was adjusted by addition of KCl as shown in the graph. The remaining conditions same as in A.

The fluorescence of the system undergoes a dramatic change at pH 5.9-6. Below pH 5.9 the emission maximum at 500 nm rapidly increases (F/Fo = 2.05) during the mixing. This increase is followed by a slow unimolecular isomerization process characterized by an equilibrium value of F/Fo = 1.4 (Figure 6). The rate of the slow phase is independent of ionic strength in the range 0.01-0.1 M.

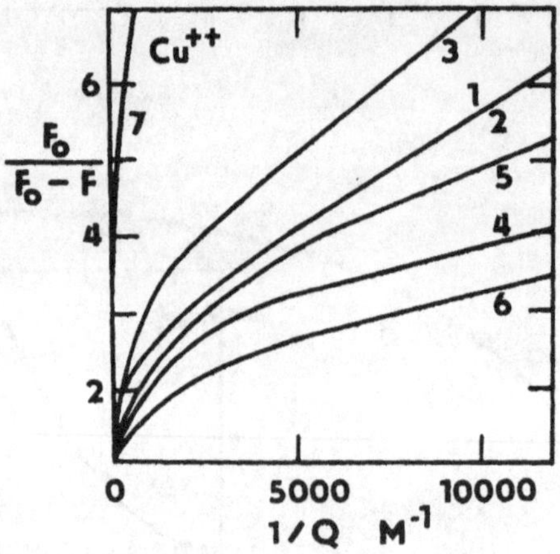

Figure 7. Lehrer plot of quenching of probe fluorescence, by Cu^{2+}-ions, in chicken pepsin, chicken pepsinogen and native chicken pepsin-activation peptide pl-p42 complex. 1 - chicken pepsin, 2 - chicken pepsin + Val-D-Leu-Pro-Phe-Phe-Val-D-Leu (0-019 mM), 3 - chicken pepsin + pepstatin (0.012 mM), 4 - chicken pepsin with β-carboxyl of Asp215 esterified with diazoacetylglycine ethyl ester, 5 - chicken pepsinogen, 6 - chicken pepsin-activation peptide pl-p42 complex (1:1 native complex), 7 - low molecular weight analog, 2-mercaptoethanol-Acrylodan. Measured in 0.1 M sodium acetate, pH 5.8. The remaining conditions the same as in Fig. 2.

Binding of heavy metals

It has been shown b y James (23) that the molecule of penicillopepsin has a binding site for heavy metals localized between Asp-33 and Asp-213. The very efficient fluorescence quenching by Cu^{2+}-ions of the probe bound to Cys-115 in chicken pepsin and pepsinogen and in the native chicken pepsin-activation peptide (pl-p42) complex (Fig. 7) demonstrates that these proteins also bind Cu^{2+}-ions to a site localized in the neighborhood of the probe. The dissociation constant is of the order of about 0.1 mM. The quenching is even more efficient after Asp 215 has been labeled by diazoacetylglycine ethyl ester. Hg^{2+}-ions are an even stronger fluorescence quencher of labeled chicken pepsin than Cu^{2+}-ions. We observed that Cu^{2+}-ions do not decrease the activity of chicken pepsin toward synthetic substrate III whereas Hg^{2+}-ions inhibit penicillopepsin (M.N.G. James, personal communication). These results indicate that the heavy metal binding directly to the active site of aspartic proteinases may be one of the general properties of these enzymes offering a possibility of investigating by fluorescence and NMR spectra measurements time-dependent changes of the active site both during the catalytic process and also during zymogen activation.

Acknowledgement. The authors thank Dr F.G. Prendergast of Mayo Foundation, Rochester, N.Y. for a sample of Acrylodan and Dr B.M. Dunn, University of Florida, Gainesville, for samples of peptides Lys-Pro-Ala-Glu-Phe-Nph-Arg-Leu and Lys-Pro-Val-Glu-Phe-Nph-Arg-Leu.

References

1. Fruton, J.S.: Adv. Enzymol. 44, 1-36 (1976).
2. Jencks, W.P.: Adv. Enzymol. 43, 219-410 (1975).
3. Fruton, J.S.: Mol. Cell Biochem. 32, 105-114 (1980).
4. James, M.N.G., Sielecki, A., Salituro, F., Rich, D.H., Hofmann, T.: Proc. Natl. Acad. Sci. USA 79, 6137-6141 (1982).
5. Baudyš, M., Kostka, V.: Eur. J: Biochem. 136, 89-99 (1983).
6. Becker, R., Schechter, Y., Bohak, Z.: FEBS Lett. 36, 49-52 (1973).
7. Schechter, Y., Rubinstein, M., Becker, R., Bohak, Z.: Eur. J. Biochem. 58, 123-131 (1975).
8. Pohl, J., Baudyš, M., Kostka, V.: Collect. Czech. Chem. Commun. 47, 3470-3474 (1982).
9. Prendergast, F.G., Meyer, M., Carlson, G.L., Iida, S., Potter, J.D.: J.Biol. Chem. 258, 7541-7544 (1983).
10. Pohl, J., Baudyš, M., Kostka, V.: Anal. Biochem. 133, 104-109 (1983).

11. Pohl, J., Zaoral, M., Jindra, Jr., A., Kostka, V.: Anal. Biochem. 139, 265-271 (1984).

12. Pichová, I., Pohl, J., Štrop, P., Kostka, V.: this volume pp.

13. Dunn, B.M., Kammermann, B., McCurry, K.R.: Anal. Biochem. 138, 68-73 (1984).

14. Todhunter, J.A.: Methods Enzymol. 63, PART A 383-411 (1979).

15. Dunn, B.M., Parten, B., Jimenez, M., Rolph, C.E., Valler, M., Kay, J.: this volume, pp.

16. Andreeva, N.S., Zdanov, A., Gustchina, A., Fedorov, A.: this volume, pp.

17. Bott, R., Subramanian, E., Davies, D.R.: Biochemistry 21, 6956-6962 (1982).

18. James, M.N.G., Sielecki, A.R.: this volume, pp.

19. Blundell, T., Jenkins, J., Pearl, L., Sewell, T., Pedersen, V.: this volume, pp.

20. Sibanda, B.L., Hemmings, A.M., Blundell, T.: this volume, pp.

21. Rich, D.H., Sun, E.T.O.: Biochem. Pharmacol. 29, 2205-2212 (1980).

22. Kitagishi, K., Nakatani, H., Hiromi, K.: J. Biochem. 87, 573-579 (1980).

23. James, M.N.G., Sielecki, A.: J. Mol. Biol. 163, 299-361 (1983).

MULTIPLICITY AND INTERMEDIATES OF THE ACTIVATION MECHANISM OF ZYMOGENS OF GASTRIC ASPARTIC PROTEINASES

Kenji Takahashi* and Takashi Kageyama
Department of Biochemistry, Primate Research Institute
Kyoto University, Inuyama, Aichi 484, Japan

Introduction

Pepsinogens are the zymogens of pepsins, the major gastric aspartic proteinases. They are converted to pepsins under acidic conditions. The reaction proceeds autocatalytically, releasing the so-called activation peptides from the NH_2-terminal part of the pepsinogen molecule (1). In porcine pepsinogen, these activation peptides are derived from the NH_2-terminal 44-residue segment. This activation was shown to follow predominantly an intramolecular mechanism below pH 3 (2-6). Essentially two reaction pathways are possible for the activation;i.e., the direct conversion pathway and the sequential conversion pathway. In porcine pepsinogen, evidence supporting the latter pathway has been presented by several investigators. Dykes and Kay reported that in the presence of pepstatin, a potent inhibitor of pepsin, the NH_2-terminal 16-residue peptide (residues 1-16) was released first, suggesting the sequential release of the activation segment (7). They obtained similar results using bovine, chicken, and canine pepsinogens and bovine prochymosin (8). We also isolated an intermediate form between pepsinogen and pepsin upon activation of human pepsinogen in the presence of pepstatin (9). Further, Christensen *et al.* reported that the initial cleavage

*Present address: Department of Biophysics and Biochemistry Faculty of Science, The University of Tokyo, Hongo, Tokyo 113, Japan

Aspartic Proteinases and their Inhibitors

of porcine pepsinogen occurred at the peptide bond Leu-Ile (16-17) on activation without pepstatin, forming the intermediate form (pseudopepsin) (10). These results indicate that porcine pepsinogen is activated to pepsin through intermediate form(s) by sequential release of the activation peptides. On the other hand, the direct conversion pathway to pepsin had not been substantiated since the intact activation segment had never been isolated.

In the present studies, we have investigated the course of activation of various pepsinogens by chemically identifying the activation products in order to obtain some information about the mechanism of activation of gastric aspartic proteinases. First, we have examined the process of activation of Japanese monkey pepsinogen and direct evidence has been obtained, showing that the activation in monkey pepsinogen occurs essentially by a one-step process (11). Secondly, we have investigated the process of activation of porcine pepsinogen, and isolated and characterized the released peptides and the intermediate protein species. The results demonstrate that one-step activation occurs in porcine pepsinogen along with the sequential process (12, 13). Thirdly, the course of activation of other pepsinogens, including bear pepsinogen, rabbit pepsinogen isozymes, and Japanese monkey progastricsin (pepsinogen C), has also been investigated.

Materials and Methods

Materials — Japanese monkey pepsinogen (major component, III-3) and progastricsin were purified as described previously (14). Porcine pepsinogen was purified chromatographically from a commercial sample (Sigma). Bear pepsinogen was the major component (II-1) of Asian black bear pepsinogens (15). Rabbit pepsinogen isozymes were obtained as described elsewhere (16).
Activation of pepsinogen — Pepsinogen was activated at pH 2.0

and 14°C as described previously (17). Unless otherwise specified, the concentration of pepsinogen in the activation mixture was 0.16 mg/ml. At desired intervals, aliquots were withdrawn to examine the extent of activation by sodium dodecyl sulfate (SDS)-polyacrylamide disc gel electrophoresis (18). The activation was stopped by the addition of 1 M NH_4OH to a final concentration of 0.2 M followed by lyophilization, or by raising the pH to near 5.5 by adding 5 M sodium acetate buffer, pH 5.5, containing an about 3-fold molar excess of pepstatin.

Purification of activation peptides — The lyophilized activation mixture was dissolved in 0.1 M sodium acetate buffer, pH 5.5, containing 8 M urea, and subjected to gel filtration on a column (1.6 x 150 cm) of Sephadex G-50 in the same buffer. Fractions containing a peptide mixture were purified by chromatography on a column (1.6 x 40 cm) of sulfopropyl(SP)-Sephadex in the same buffer containing 8 M urea. Peptides were eluted with a linear gradient of NaCl (0-0.75 M) using two 300-ml chambers. Salt- and urea-free peptides were prepared as described previously (17).

Isolation of pepsinogen, pepsin, and the intermediate forms — The activation mixture adjusted to pH 5.5 was applied to a column (1.15 x 25 cm) of DEAE-Toyopearl previously equilibrated with 0.1 M sodium acetate buffer, pH 5.5, containing 7 µM pepstatin. The adsorbed protein was eluted with a linear gradient of NaCl (0-0.5 M).

Results

Activation of monkey pepsinogen — When incubated at pH 2.0 and 14°C, pepsinogen was immediately converted to pepsin as described previously (17). Figure 1 shows the time course of the activation as analyzed by SDS-disc gel electrophoresis. Almost all pepsinogen appeared to be converted directly to pepsin within 7 min without formation of any intermediate

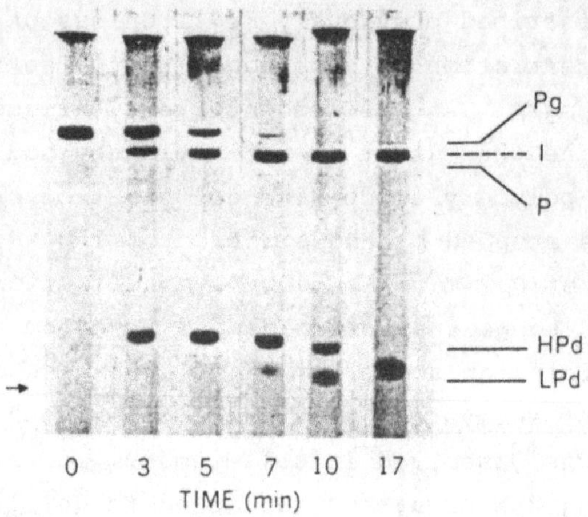

Fig.1. Time course of activation of monkey pepsinogen analyzed by SDS-disc gel electrophoresis. Aliquots were withdrawn at the indicated times, mixed with a one-fifth volume of 0.2 M Na_2HPO_4 containing 5% SDS and 5% 2-mercaptoethanol followed by heating at 90°C for 15 min, and subjected to electrophoresis. The concentration of acrylamide was 10%. The gel was stained with Coomassie brilliant blue. The arrow indicates the position of bromphenol blue. Pg, pepsinogen; I, intermediate form; P, pepsin; HPd, high molecular weight peptide; LPd, low molecular weight peptide.

species between pepsinogen and pepsin. The released peptides moved to near the dye front upon electrophoresis. A relatively high molecular weight peptide band appeared in the early period of activation and this band was gradually converted to a low molecular weight peptide band on further incubation. These results suggest that pepsinogen was converted directly to pepsin by releasing a large molecular weight peptide and that this peptide was further cleaved by pepsin formed to give a mixture of low molecular weight peptides.

Figure 2 shows gel filtration patterns on Sephadex G-50 of 3-, 5-, and 7-min activation mixtures. The protein was eluted at the void volume. On the other hand, peptides were separated into 3 fractions, designated as Fractions I, II, and III. With the progress of activation, the amount of peptide in Fraction I decreased while the amounts of peptide in Fractions II and

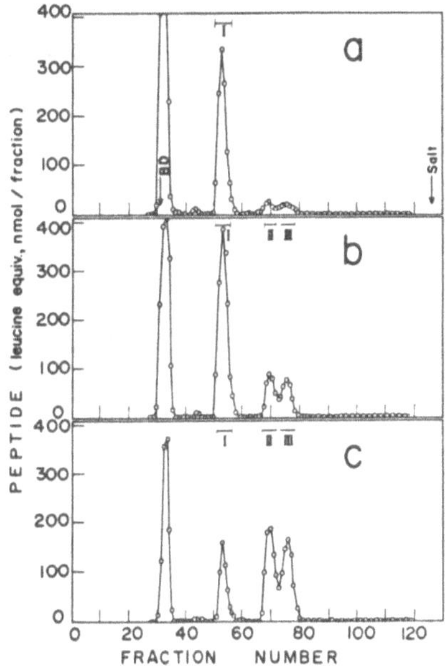

Fig.2. Sephadex G-50 gel filtration of activation mixtures. Fraction size, 3 ml. The fractions under the bars were pooled. a), b), and c) represent gel filtration patterns of activation mixtures after 3-, 5-, and 7-min incubations, respectively. The elution positions of blue dextran (BD) and inorganic salts are indicated by arrows.

III gradually increased. The changes in gel filtration patterns of peptides coincided well with those in SDS-disc gel electrophoresis shown in Fig.1. The high molecular weight peptide band corresponded to Fraction I, and the low molecular weight peptide band to Fractions II and III. Fraction I was further separated into two components (Fractions I-1 and I-2) by SP-Sephadex chromatography (data not shown). Fractions II and III were also purified by SP-Sephadex chromatography, and the latter was separated into two components (III-1 and III-2) (data not shown).

The amino acid compositions of acid hydrolysates and the NH_2-terminal amino acids of these purified fractions are shown in Table I. Fractions I-1 and I-2 were each composed of 47 residues and had the same amino acid composition except that

Table I. Amino acid compositions of the activation peptides.

Amino acid	Number of residues per molecule of peptide							
	I-1	I-2	II	III-1	III-2	Residues 1-47	Residues 1-25	Residues 26-47
Lys	6.8 (7)	7.7 (8)	3.5 (4)	2.7 (3)	3.6 (4)	7 (or 8)	4	3 (or 4)
His	1.8 (2)	1.7 (2)	0.9 (1)	0.9 (1)	0.9 (1)	2	1	1
Arg	2.9 (3)	2.7 (3)	2.6 (3)			3	3	0
Asp	4.0 (4)	4.0 (4)	2.0 (2)	2.0 (2)	2.0 (2)	4	2	2
Thr	1.2 (1)	0.8 (1)		0.9 (1)	0.9 (1)	1	0	1
Ser	2.8 (3)	2.7 (3)	1.8 (2)	0.9 (1)	0.9 (1)	3	2	1
Glu	2.9 (3)	2.0 (2)	1.3 (1)	2.1 (2)	1.0 (1)	3 (or 2)	1	2 (or 1)
Pro	4.3 (4)	3.5 (4)	0.6 (1)	2.7 (3)	2.6 (3)	4	1	3
Gly	1.2 (1)	1.1 (1)	1.1 (1)			1	1	0
Ala	3.1 (3)	2.8 (3)		3.4 (3)	3.2 (3)	3	0	3
Val	2.1 (2)	1.9 (2)	1.9 (2)			2	2	0
Ile	1.6 (2)	1.5 (2)	1.2 (2)			2	2	0
Leu	8.3 (8)	7.7 (8)	5.0 (5)	2.7 (3)	2.8 (3)	8	5	3
Tyr	0.5 (2)	0.5 (2)	0.5 (1)	0.7 (1)	0.6 (1)	2	1	1
Phe	2.2 (2)	2.1 (2)		1.6 (2)	1.9 (2)	2	0	2
Total	47	47	25	22	22	47	25	22
N-terminus	Ile	Ile	Ile	Phe	Phe	Ile	Ile	Phe
Yield (%) 3 min	46 (I-1 + I-2)		6	7 (III-1 + III-2)		Expected numbers of residues from the amino acid sequence (17)		
Yield (%) 5 min	58 (I-1 + I-2)		24	30 (III-1 + III-2)				
Yield (%) 7 min	20 (I-1 + I-2)		54	64 (III-1 + III-2)				

The values in parentheses are nearest integers (or assumed values).

Fraction I-1 contained one more glutamic acid and one less lysine than Fraction I-2. The amino acid compsitions were also the same as that of the activation segment as determined previously (17). Fraction II contained one major peptide composed of 25 residues, and its composition was the same as that of the NH_2-terminal half (residues 1-25) of the activation segment. Fraction III contained two major peptides (III-1 and III-2) each composed of 22 residues, and their compositions were the same as that of the COOH-terminal half (residues 26-47) of the activation segment. NH_2-terminal analyses of peptides supported these assignments. These results confirmed that the activation segment was released as a single intact polypeptide of 47 residues and then cleaved into smaller peptides.

<u>Activation of porcine pepsinogen</u> — Pepsinogen was activated at various concentrations at pH 2.0 and 14°C, and the activation process was analyzed by SDS-disc gel electrophoresis (Fig. 3). In all cases, pepsinogen disappeared rapidly after acidification. The resulting protein species were detected as two bands; one of them had the same molecular weight as the authentic pepsin and the other had a molecular weight intermediate between those of pepsinogen and pepsin. The intermediate form was relatively stable as compared with pepsinogen, but was gradually converted to pepsin during a long period of incubation. The formation of the intermediate form became predominant when the initial pepsinogen concentration decreased.

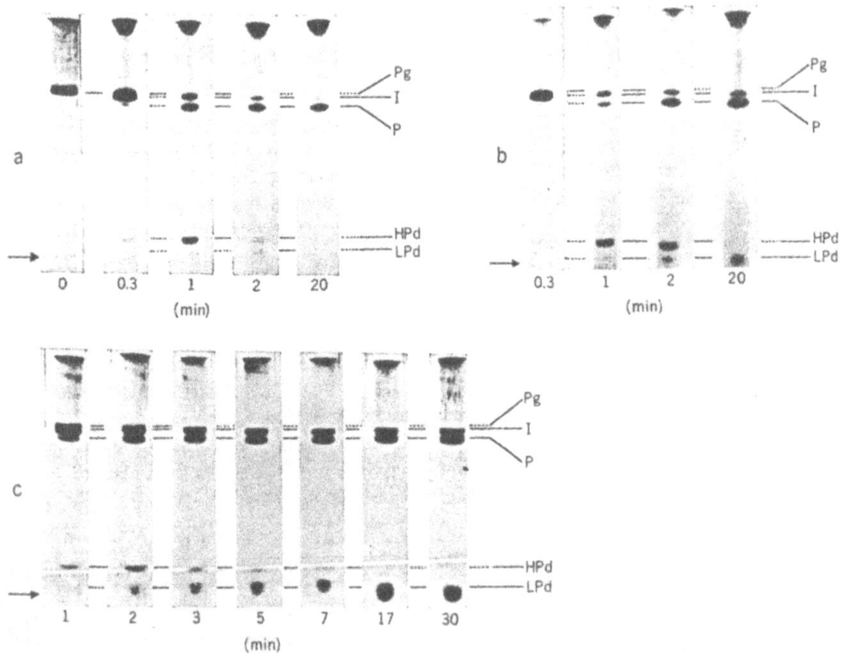

Fig.3. Time course of activation of porcine pepsinogen analyzed by SDS-disc gel electrophoresis. Initial pepsinogen concentration: (a) 1.6 mg/ml, (b) 0.8 mg/ml, and (c) 0.16 mg/ml. The arrow indicates the position of bromphenol blue. The symbols used are the same as in Fig.1.

Released peptides were also detected as two bands. The amount of peptide in the high molecular weight peptide band appeared to be maximum at 1 or 2 min and to decrease rather rapidly, while that in the low molecular weight peptide band appeared to increase gradually with the progress of incubation time. However, when the initial pepsinogen concentration was 1.6 mg/ml, both peptide bands were scarcely detectable after 2 min. This may be due to further cleavage of these peptides to smaller peptides, which were not retained in the gel.

Peptides released after 2-min and 30-min activation at a pepsinogen concentration of 0.16 mg/ml were isolated and characterized. The lyophilized reaction products were fractionated by Sephadex G-50 gel filtration (Fig.4). The protein mixture was eluted near the void volume, separated from peptides. Peptides in the 2-min activation mixture were separated into

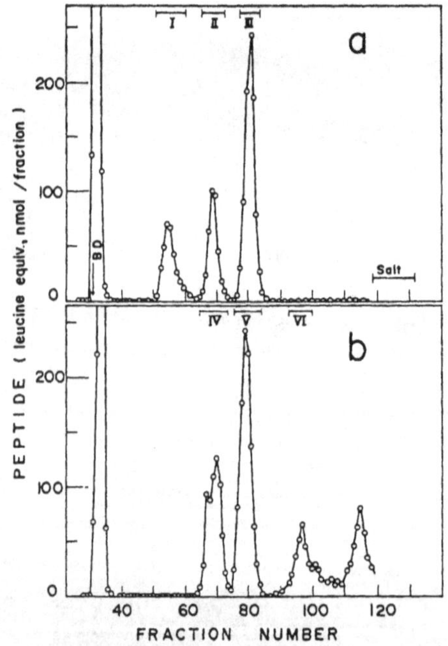

Fig.4. Sephadex G-50 gel filtration of activation mixtures. Activation time: (a) 2 min; (b) 30 min. Other conditions are the same as in Fig.2.

3 peaks (Fractions I, II, and III). Each fraction showed a single NH_2-terminal amino acid. The amino acid compositions of Fractions I, II, and III corresponded to those of residues 1-44, 17-44, and 1-16 of the activation segment, respectively (data not shown). Fraction I was thus deduced to be the intact activation segment. Upon further incubation until 30 min, Fraction I disappeared and peaks of smaller peptides appeared (Fig.4b). These peptides were isolated and identified (see Fig.7).

In order to isolate and characterize the protein species, the activation mixtures were analyzed by chromatography on DEAE-Toyopearl in the presence of pepstatin. After activation for 2 min, several peaks (Fractions A through F) appeared (Fig. 5b). Fraction B was eluted at the same position as that of authentic pepsinogen. Upon further incubation until 30 min, Fractions A and B disappeared and the relative contents of Fractions E and F increased as shown in Fig.5c as Fractions J and K. The amino acid compositions of some of these fractions were determined (data not shown). Fractions A and B had practically the same composition as pepsinogen, and Fractions E, F, J, and K the same composition as pepsin. The compositions of

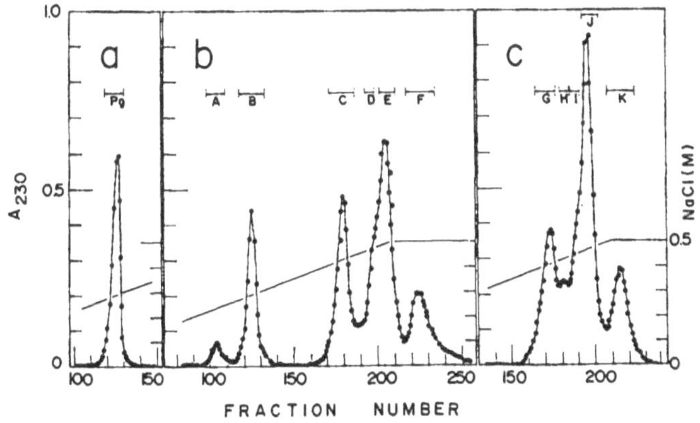

Fig.5. Chromatography of activation mixtures of pepsinogen on DEAE-Toyopearl. Fraction size, 3 ml. The fractions under the bars were pooled. (a) Authentic pepsinogen; (b) 2 min; (c) 30 min.

Fractions C, D, G, H, and I were intermediate between those of pepsinogen and pepsin, and those of Fractions C, D, G, and I were nearly identical with one another.

The NH_2-terminal sequences of these fractions were determined by manual Edman degradation. Analysis of the NH_2-terminal 5-residue sequence of Fraction B indicated that this fraction contained the NH_2-terminal sequences of both pepsinogen (Leu-Val-Lys-Val-Pro-) and pepsin (Ile-Gly-Asp-Glu-Pro-). Moreover, two protein bands corresponding to pepsinogen and pepsin, and one peptide band corresponding to the high molecular weight peptide were detected in Fraction B by SDS-disc gel electrophoresis (Fig.6). The peptide was isolated from Fraction B by adsorption with SP-Sephadex in the presence of 8 M urea. Amino acid analysis and NH_2-terminal analysis showed that the peptide was identical with the 44-residue intact activation segment. From these results Fraction B was judged to be a mixture of pepsinogen and a 1:1 complex of pepsin and the activation segment. On quantitative determination of the NH_2-terminal residues in Fraction B, the complex was estimated to occupy about 50% of Fraction B. These results indicate that the activation segment formed a tight complex with pepsin at pH 5.5 and that the complex was cochromatographed with pepsinogen at the same pH. When the activation was terminated by

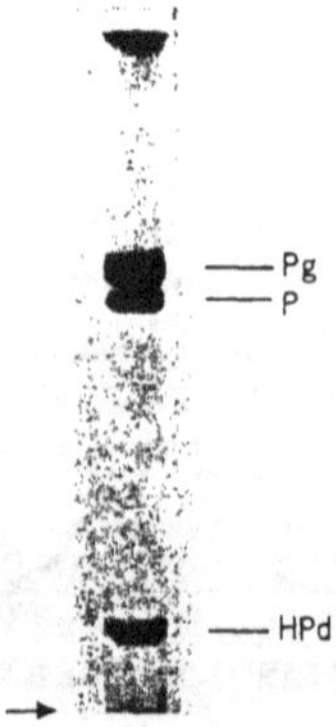

Fig.6. SDS-disc gel electrophoresis of Fraction B after DEAE-Toyopearl chromatography. The conditions and symbols used are the same as in Fig.1.

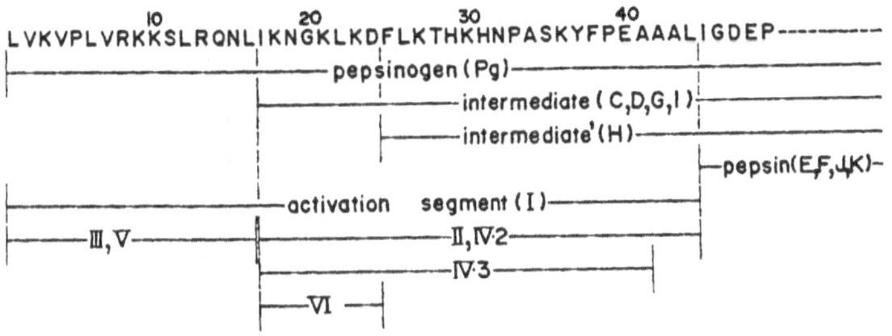

Fig.7. Assignment of protein species and peptides obtained on activation of porcine pepsinogen for 2 min and 30 min.

raising the pH to 5.5 and the activation mixture was chromatographed in the absence of pepstatin, almost all the intermediate forms (Fractions C, D, G, and I) were converted to corresponding pepsins during chromatography, whereas no notable change was observed in the pepsin-activation segment complex.

The NH_2-terminal 5-residue sequences of Fractions C, D, G, and I were identical with the sequence (Ile-Lys-Asn-Gly-Lys) starting from Ile-17 of pepsinogen, indicating that these fractions were formed by removal of the NH_2-terminal 16 residues from pepsinogen. Fractions E, F, J, and K had an NH_2-terminal 5-residue sequence identical with that of pepsin (Ile-Gly-Asp-Glu-Pro). These results suggest that the pepsinogen sample contained at least 2 isozymes. The major component in Fraction H was shown to have the NH_2-terminal sequence: Phe-Leu-Lys-. This component is therefore thought to have been generated by removal of the NH_2-terminal 24 residues from pepsinogen. Figure 7. shows the assignment of the protein species and the released peptides isolated and identified.

Activation of bear pepsinogen — Upon activation of bear pepsinogen at pH 2.0, almost all pepsinogen was converted to pepsin within 30 min. No intermediate form was observed while three activation peptides were produced (15). Peptide I was the intact activation segment of 42 or 43 residues, peptide II was the NH_2-terminal half (residues 1-25) of peptide I, and peptide III was the COOH-terminal half (residues 26-42 or 43) of peptide I. Peptides II and III are thought to be formed by further cleavage of peptide I. Thus the cleavage sites were restricted to mainly three peptide bonds, i.e., Asp-Phe(25-26), Glu-Ala(42-43), and Ala-Ala(43-44), and no appreciable cleavage occurred at Val-Met(45-46), which had been expected to be a site of cleavage from the sequence homology of the NH_2-terminal parts of various pepsins. Since no intermediate form could be detected during activation, bear pepsinogen appears to be activated mainly by a one step process (see Fig.11).

Activation of rabbit pepsinogens — Rabbit pepsinogen isozymes (I, II-1, II-2, II-3, II-4, and III) were also activated under the same conditions as above and the activation products were isolated and identified (16). The amino acid sequences of the NH_2-terminal regions of these pepsinogens are shown in Fig.8. The activation processes were significantly different among the pepsinogen groups. Conversion of pepsinogen I occurred within a few min, and the resulting protein species were detected as one band (Fig.9). In the earlier period of activation, a protein species (intermediate I-a) formed by releasing the NH_2-terminal 16 residues was identified as the major product. After 1-h activation a stable active intermediate species (intermediate I-b) was shown to be formed which lacked the NH_2-terminal 25 or 26 residues of the pepsinogen. No further conversion of this form occurred,however, upon further incubation for several hours. Judging from these results, pepsinogen I is thought to be converted sequentially to the stable active intermediate form under the conditions employed.

The activation process was essentially the same among pepsinogens II (Fig.9). Two intermediate forms were obtained in the case of pepsinogen II-1, the NH_2-termini of which were Leu-27 and Ala-35. On the other hand, one intermediate form starting from Tyr-26 was obtained in the case of pepsinogens II-2, II-3, and II-4. Pepsins derived from pepsinogens II had the same NH_2-terminal Ala, which corresponded to the residue-45 of pepsinogens. The released peptides were detected as two bands. The high molecular weight peptide band was shown to contain the intact activation segment. Thus two different pathways

	1	10	20	30	40	50	60
pepsinogen I	V/HKVPLVRKKSLRKNLIEKGLLQDYLKTHSPNPATKYFPNAAYA———————						
pepsinogens II-1	/VHKVPLVRKKSLRKNLIEKGLLQDYLKTHTPNLATKYFPKETFASVSTESLENYLDAEY—						
pepsinogens II-2	V/HKVPLVRKKSLRKNLIEKGLLQDYLKTHTPNPATKYFPKETFATVSTESLENYLDAEY—						
pepsinogens II-3	V/HKVPLVRKKSLRKNLIEKGLLQDYLKTHTPNPATKYFPKETFATVSTESLENYLDAEY—						
pepsinogens II-4	/VHKVPLVRKKSLRKNLIEKGLLQDYLKTHTPNPATKYFPKETFATVSTESLENYLDAEY—						
pepsinogen III	//HKVPLVRKKSLRKNLIEKGLLKDYLKTHTPNLATKYLPKAAFDSVPTETLENYLDTEY—						

Fig.8. The amino acid sequences of the NH_2-terminal regions of rabbit pepsinogens. The activation segments are enclosed. The amino acid residues at positions where substitution occurred are shown by italic letters.

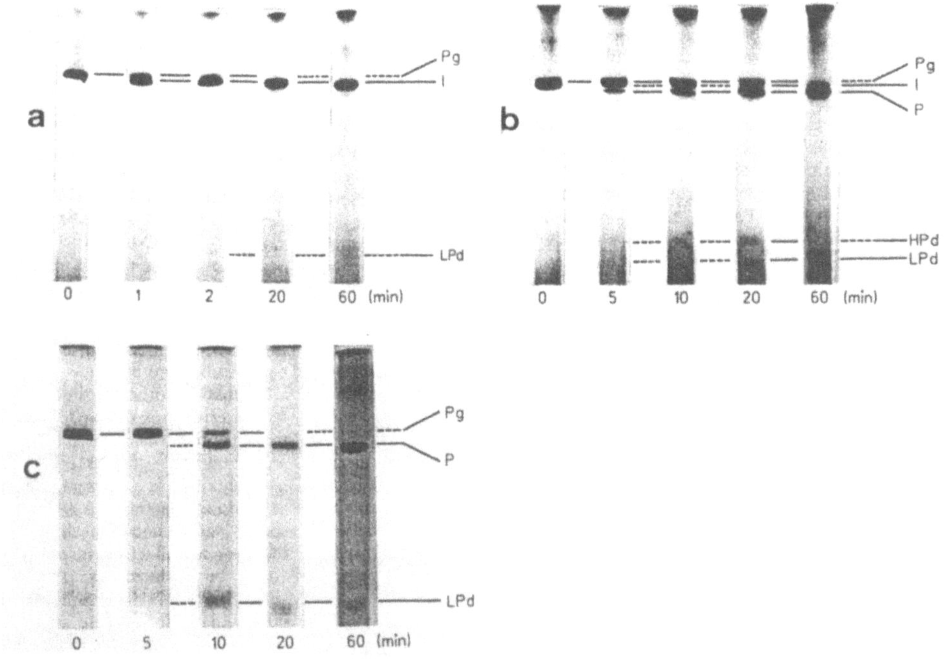

Fig.9. Time course of activation of rabbit pepsinogens analyzed by SDS-disc gel electrophoresis. The conditions and symbols used are the same as in Fig.1. (a) Pepsinogen I. (b) Pepsinogen II-2. The results with pepsinogens II-1, II-3, and II-4 were essentially the same. (c) Pepsinogen III.

occur in pepsinogen II group.

Unlike other pepsinogen isozymes, pepsinogen III appeared to be converted directly to pepsin since no intermediate form was detected (Fig.9). However, the high molecular weight peptide band corresponding to the intact activation segment was not detected during activation. The cleavage of the activation segment may have occurred simultaneously at a few peptide bonds including the bond connecting the activation segment and pepsin, thus generating pepsin and a mixture of shorter peptide fragments.

Activation of monkey progastricsin — In contrast with monkey pepsinogen, monkey progastricsin was shown to be activated exclusively by a two-step process (Fig.10). Conversion of progastricsin to the intermediate form was very fast and was complete within 1 min. This intermediate form was formed by releasing the NH_2-terminal 26 residues. The released peptides appeared exclusively as a low molecular weight peptide band. Two forms of gastricsin were obtained; the major one was formed by cleavage of Leu-Ser(43-44) and the minor one by cleavage of Phe-Gly(40-41). The latter was rather stable and was not converted to the former during the incubation. A similar two-step activation has been reported recently for human progastricsin (19).

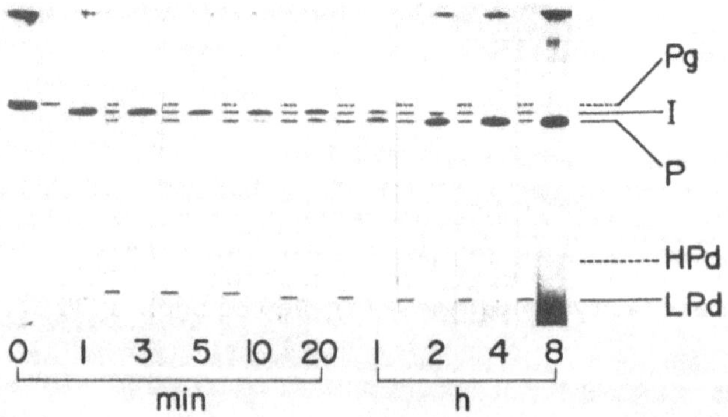

Fig.10. Time course of activation of monkey progastricsin analyzed by SDS-disc gel electrophoresis. The conditions and symbols used are the same as in Fig.1.

Discussion

Figure 11 shows the initial cleavage sites and resulting protein species isolated and identified in the present studies in the activation of various pepsinogens. The one-step activation pathway was found for the first time in the activation of monkey pepsinogen. Then it was also found to occur in some other pepsinogens either exclusively or simultaneously with the sequential process. In contrast with monkey pepsinogen, on the other hand, monkey progastricsin was found to be activated exclusively by the sequential process. Since all these zymogens are very homologous proteins, it is suggested that they may be activated to pepsin through either or both of the two activation pathways depending on the slight differences in the structure, especially, of the NH_2-terminal activation segment region and in the substrate specificity of the exposed active site.

The activation of porcine pepsinogen at pH below 3 was shown to proceed predominantly by an intramolecular mechanism by kinetic experiments (3). The possibility may exist that the one-step activation of pepsinogens also proceeds intramolecularly. This mechanism, however, must be reconciled with the result of X-ray crystallographic studies (20) showing that the NH_2-terminal of porcine pepsin is not located near the active site. It may be possible that the one-step activation occurs by an intermolecular reaction between two pepsinogen molecules. Indeed, Pedersen et al. reported that bovine prochymosin was activated at pH 2 predominantly by an intermolecular reaction between two zymogen molecules, although the first cleavage site in this case is the bond Phe-Leu(27-28) (21). Thus further elaborate studies are required to clarify whether each activation pathway occurs intramolecularly or intermolecularly.

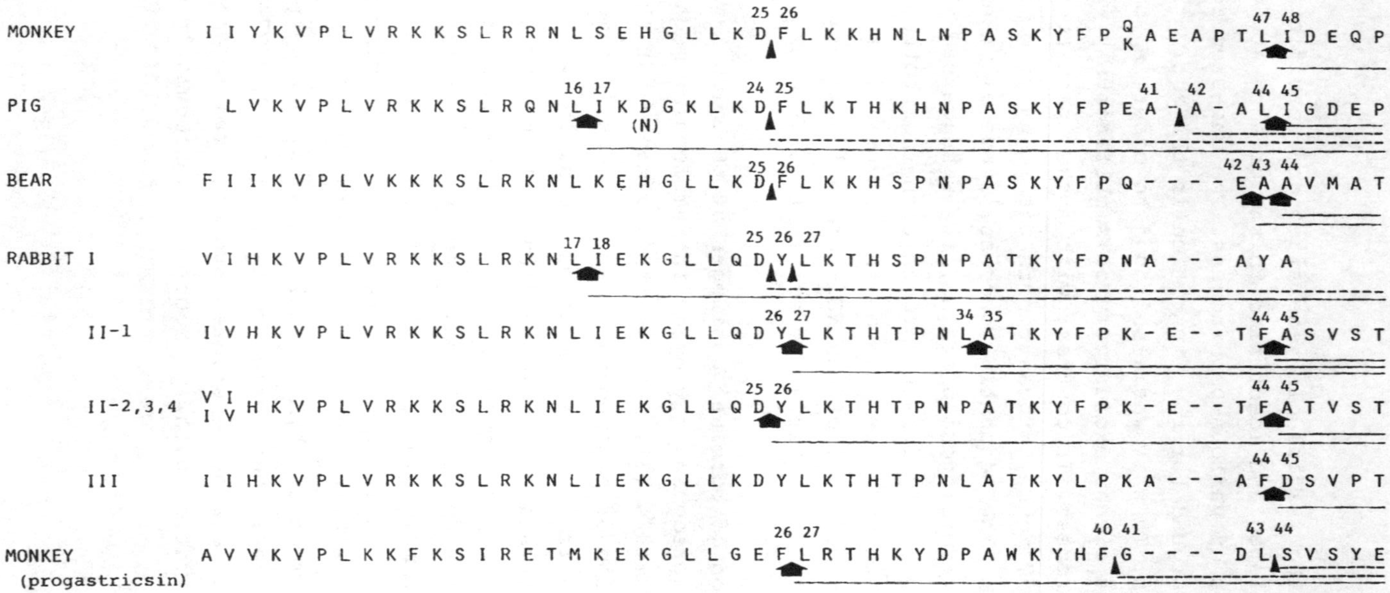

Fig.11. Initial cleavage sites and resulting protein species in the activation of various pepsinogens at pH 2.0 and 14°C. [arrow], major cleavage site; ▲, minor or secondary cleavage site. Underlines show the protein species isolated and identified.

Summary

The courses of activation of various pepsinogens were investigated in order to obtain some information about the mechanism of activation of gastric aspartic proteinases. When Japanese monkey pepsinogen was activated at pH 2.0, the activation segment of the NH_2-terminal 47 residues was released as a single intact polypeptide. This clearly showed that the pepsinogen was activated to pepsin directly. On the other hand, activation of porcine pepsinogen under the same conditions was found to proceed simultaneously by two different pathways. One is the direct conversion process of pepsinogen to pepsin, releasing the intact activation segment of 44 residues. The other is a sequential pathway, in which pepsinogen is activated through intermediate forms, releasing activation peptides stepwisely. Bear pepsinogen appeared to be activated by a one step pathway while rabbit pepsinogens were shown to be activated by either or both of the two pathways. Monkey progastricsin was shown to be activated exclusively by a two-step process. These results indicated that in pepsinogens either or both of these two activation pathways may occur depending on slight differences in the structure of the NH_2-terminal region and in the substrate specificity of the exposed active site.

References

1. Herriott,R.M.: J. Gen. Physiol. 45, 57-76 (1962).
2. Bustin, M., Conway-Jacobs, A.: J. Biol. Chem. 246, 615-620 (1971).
3. Al-Janabi, J., Hartsuck, J.A., Tang, J.: J. Biol. Chem. 247, 4628-4632 (1973).
4. McPhie, P.: Biochem. Biophys. Res. Commun. 56, 789-792 (1974).
5. Sanny, C.G., Hartsuck, J.A., Tang, J.: J. Biol. Chem. 250, 2635-2639 (1975).

6. Marciniszyn, J.,Jr., Huang, J.S., Hartsuck, J.A., Tang, J. : J. Biol. Chem. 254, 7095-7102 (1976).

7. Dykes, C.W., Kay, J.: Biochem. J. 153, 141-144 (1976).

8. Kay, J., Dykes, C.W.: in Acid Proteases: Structure, Function, and Biology (Tang, J. ed.) pp. 103-130, Plenum Press, New York 1977.

9. Kageyama, T., Takahashi, K.: J. Biochem.(Tokyo) 88, 571-582 (1980).

10. Christensn, K.A., Pedersen, V.B., Foltmann, B.: FEBS Lett. 76, 214-218 (1977).

11. Kageyama, T., Takahashi, K.: J. Biochem.(Tokyo) 92, 1179-1188 (1982).

12. Kageyama, T., Takahashi, K.: Biochem. Biophys. Res. Commun. 107, 1117-1122 (1982).

13. Kageyama, T., Takahashi, K.: J. Biochem.(Tokyo) 93, 743-754 (1983).

14. Kageyama, T., Takahashi, K.: J. Biochem.(Tokyo) 79, 455-468 (1976).

15. Kageyama, T., Moriyama, A., Takahashi, K.: J. Biochem. (Tokyo) 94, 1557-1567 (1983).

16. Kageyama, T., Takahashi, K.: Eur. J. Biochem. 141, 261-269.

17. Kageyama, T., Takahashi, K.: J. Biochem.(Tokyo) 88, 9-16 (1980).

18. Weber, K., Osborn, M.: J. Biol. Chem. 244, 4406-4412 (1969).

19. Foltmann, B., Jensen, A.L.: Eur. J. Biochem. 128, 63-70 (1982).

20. Andreeva, N.S., Gustchina, A.E.: Biochem. Biophys. Res. Commun. 87, 32-42 (1979).

21. Pedersen, V.B., Christensen, K.A., Foltmann, B.: Eur. J. Biochem. 94, 573-580 (1979).

CATHEPSINS D AND E : MOLECULAR CHARACTERISTICS AND MECHANISM OF ACTIVATION

Vito Turk, Tamara Lah, Vida Puizdar, Joža Babnik, Matjaž Kotnik, Igor Kregar

Department of Biochemistry, J.Stefan Institute, E.Kardelj University , Jamova 39, 61000 Ljubljana, Yugoslavia

Roger H.Pain

Department of Biochemistry, University of Newcastle upon Tyne, Newcastle upon Tyne, NE17RU, U.K.

Introduction

The superfamily of aspartic proteinases (EC 3.4.23) comprises a widely distributed group of extracellular and intracellular homologous enzymes of animal and microbial origin. They all exhibit their maximal enzyme activity at acid pH (the exception is renin) and have two Asp in the active site which are involved in the mechanism of catalysis. Whereas some of the extracellular enzymes, such as various pepsins, renins and microbial aspartic proteinases are well characterized including their amino acid sequence and crystal structure (summarized in 1 and in this book), it is only recently that the first amino acid sequence of an intracellular enzyme, porcine spleen cathepsin D has been determined (2, this book).

Cathepsin D (EC 3.4.23.5) is a lysosomal aspartic proteinase widely distri-

This paper is dedicated to Milan Osredkar, Professor of Physics, E.Kardelj University, Ljubljana, Yugoslavia, on the occasion of his 65th birthday.

Aspartic Proteinases and their Inhibitors

buted in all mammalian cells. It was originally known as Anson enzyme (3) and from 1960 as cathepsin D (4). Its physiological function is to degrade proteins under normal and pathological conditions (5). Cathepsin D may also have an important function in the degradation of structural proteins in CNS and may participate in the generation of biologically active peptides from precursors (5,6).

Another mammalian proteinase is cathepsin E (EC 3.4.23.-), an enzyme with only a limited distribution in some animals and organs. It was first isolated from rabbit spleen and bone marrow (7). The localization, characterization and physiological role of this enzyme have yet to be determined.

Here, we will describe the properties of bovine spleen cathepsin D and its autolysis, the structure and conformational changes arising from the binding of pepstatin, its denaturation and partial renaturation, as well as its activation from a precursor. In addition, some of the properties of cathepsin E from rat spleen are presented.

Cathepsin D

Purification and properties

Cathepsin D has been purified from various species and tissues using standard methods (reviewed in 8). A significant advance in the purification of this enzyme was achieved by the introduction of affinity chromatography using the substrate haemoglobin (9) or the inhibitor pepstatin (10, 11) as ligands covalently bound to Sepharose.

For the isolation of cathepsin D from bovine spleen we are currently using a slightly modified method based on that published previously (9). A homogenate at neutral pH is centrifuged and the resulting supernatant concentrated by ultrafiltration using UM-10 membranes (Amicon). Ammonium sulphate precipitation is omitted due to poor recovery at this step and on subsequent dialysis. The concentrated supernatant is applied to a haemoglobin-Sepharose column and the enzyme purified further as previously published (9).

Ion exchange chromatography on DEAE-cellulose separates three multiple forms (two major and one minor) of cathepsin D, all having M_r of about 44 000 determined by gel chromatography. pI values for these forms are about 5.6, 5.9 and 6.4. The first major form is the intact single polypeptide chain of cathepsin D (D_1) and the second is a non-covalently associated complex of two chains (D_2) of M_r 28 000 - 30 000 and M_r 14 000 - 15 000 (12) as determined by SDS-PAGE. These are termed the heavy and light chains respectively and, as will be shown, probably arise from specific autolytic cleavage of the single 44 K polypeptide form. They are joined solely by non-covalent interactions and co-migrate with the 44 K form on gel permeation chromatography.

It is known, that cathepsin D, isolated from various species and tissues varies in the number of multiple forms obtained (reviewed in 8). These multiple forms, in general, have similar M_r, immunological cross reactivity, amino acid composition, whereas they can differ in their proteolytic activity. The basis of this variety of multiple forms is not known. However, their appearance must be viewed in light of recent studies connected with early events of cathepsin D biosynthesis, post-translational modification and packaging into lysosomes (13-18). It has been established, that the recognition marker is mannose-6-phosphate, which is essential for targeting cathepsin D into lysosomes. A receptor specific for phosphorylated high-mannose oligosaccharides binds the modified precursors and transports them into a pre-lysosomal compartment where they dissociate from the receptor. Cathepsin D is then converted into the mature and catalytically active cathepsin D.

Our results show that mature cathepsin D, isolated from bovine spleen may be a single chain or a two chain enzyme. Both forms are glycosylated. Recently, the oligosaccharide structures have been determined for cathepsin D from porcine spleen (19,20). Both, cathepsin D light and heavy chain are glycosylated and the two glycosylation sites of cathepsin D are located at asparagine residues in the light chain and in the heavy chain (2). Differences in oligosaccharide content may lead to a variety of multiple forms. In addition, structural microheterogeneity at two amino acid positions in the porcine spleen cathepsin D molecule was observed (2). Therefore, the obtained results suggest that different carbohydrate content and minor structural differences in the cathepsin D molecule cause the appearance of multiple forms.

The single-chain and two-chain cathepsin D molecule have the same content of secondary structure. The near UV CD spectra, however, show a decrease in ellipticity for the two-chain enzyme (D_2). This indicates a small change in environment of one or more aromatic residues. No significant differences in fluorescence spectra were observed between the two forms. Also denaturation studies with GdmCl show that the stability of the two chain enzyme is closely similar to that of the single-chain enzyme. At 2.5 M GdmCl and above, the heavy and light chain are separated. There is, however, a difference in enzyme activity, the specific activity of the two chain enzyme being about half that of the single chain enzyme (21).

Autolysis

There have been contradictory reports concerning autolysis of cathepsin D, some of which suggested that autolytic cleavage of cathepsin D does not occur in vitro. For this reason we have investigated this process in more detail.

Many proteolytic enzymes, including pepsin (22) and papain (23) undergo autodigestion when susceptible bonds are cleaved. Recently we reported that autolytic cleavage of the single polypeptide chain cathepsin D occurs under in vitro conditions (24). The intact molecule was incubated at pH 3.5 and

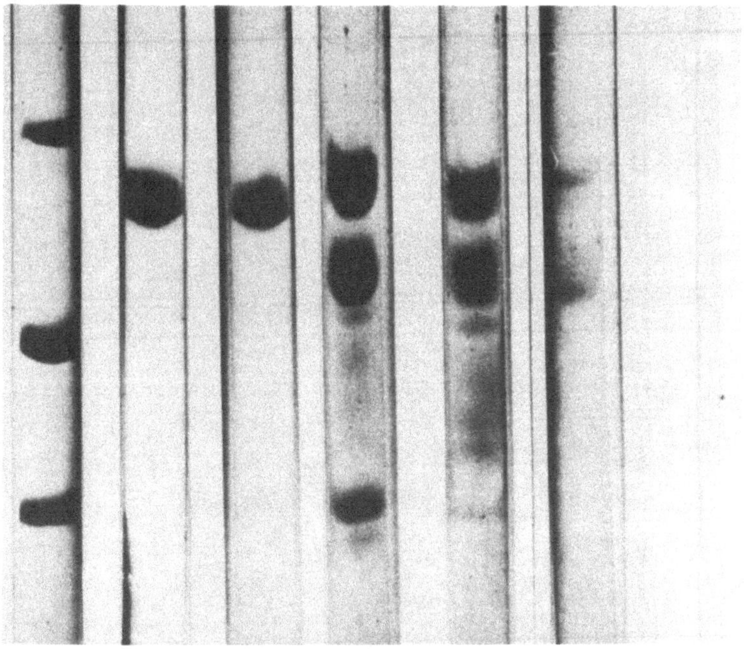

Fig. 1: Autolysis of cathepsin D followed by dodecyl-sulphate polyacrylamide gel electrophoresis. From left to right : standards (bovine serum albumin, chymotrypsinogen A and lysozyme); cathepsin D incubated at pH 6.9 for 72 h; cathepsin D - pepstatin complex incubated at pH 3.5 for 72 h; cathepsin D immediately after acidification to pH 3.5; cathepsin D incubated at 37° C for 4 h at pH 3.5; cathepsin D incubated for 10 days at 37° C at pH 3.5.

37° C, and the resulted fragmentation of cathepsin D to smaller fragments was followed by dodecyl-sulphate polyacrylamide gel electrophoresis as is seen in Fig. 1. The partial decomposition of the single chain into the main bands of M_r 30 000 and 15 000 appeared first, followed by further fragmentation with prolonged incubation. The addition of an equimolar amount of pepstatin to active cathepsin D, prior to incubation, prevented the splitting of the single chain enzyme. These results confirmed the occurence of fragmentation of single-chain enzyme in an acid environment, which is prevented by the addition of pepstatin, the specific inhibitor of aspartic proteinases.

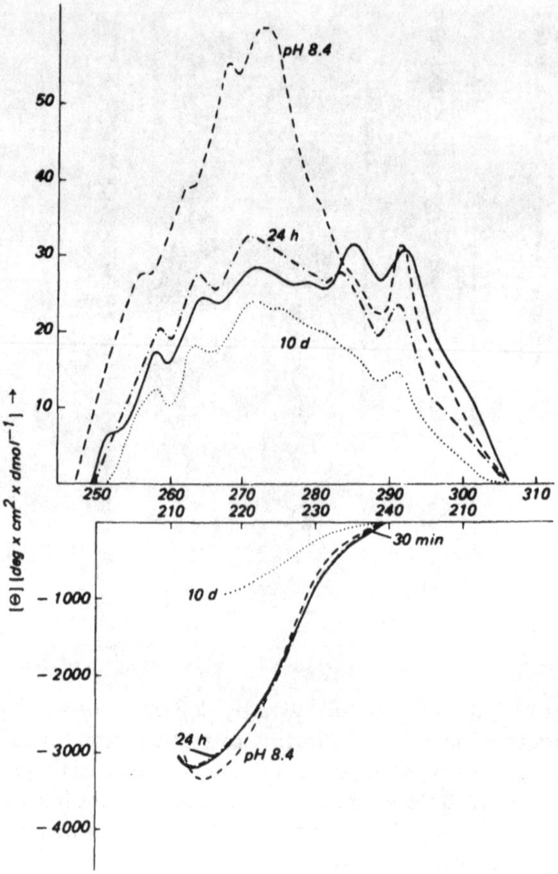

Fig. 2: Circular dichroism spectra of cathepsin D in the near UV region (above) and far UV region (below). The spectra were measured : —— immediately after acidification to pH 3.5 in the near UV region and 30 min after acidification in the far UV region; -•-• after 24 h incubation at pH 3.5; ••••• after 10 days incubation at pH 3.5; ---- spectrum of cathepsin D at pH 8.4 in Tris buffer.

Immediately after the acidification and fragmentation no significant changes in proteolytic activity were observed. Prolonged incubation showed loss of cathepsin D activity, while this was not observed when the enzyme was incubated at pH 6.9. The inactivation rate was dependent on protein concentration. Specific activity decreased faster at lower enzyme concentration (23).

We looked for possible conformational changes of cathepsin D during autolysis using near and far UV CD measurements being started immediately after acidification to pH 3.5 and continued then up to 10 days. The near UV CD spectrum changes immediately on lowering the pH. On prolonged incubation in acid, the far UV CD spectrum changes become to that of the denatured enzyme, while immediately after acidification no significant changes were observed in the secondary structure (Fig. 2).

Enzymatic autodigestion or autolysis is, in certain cases at least, related to the process of zymogen activation and it can occur in vivo or in vitro. Autolysis in vitro is strongly dependent on the reaction conditions, including pH, salt concentration, temperature and enzyme concentration. Our results show, that the splitting of the single chain to the two chain enzyme is a consequence of autolysis, at least in vitro. It is highly likely that this also occurs in vivo as can be deduced from the pulse chase experiments (13, 14, 18). The autolytic capability of cathepsin D is supported by the fact that pepstatin prevents fragmentation. Autolysis is a complex reaction process and we cannot draw any conclusions about the mechanism of limited proteolysis of cathepsin D so far. The presence of autolytic cleavage must be strictly monitored for any investigations involving specific activity and physical properties, particularly x-ray crystallography. For these reasons autolysis must be excluded from the purification procedures.

Binding of pepstatin

Pepstatin is a low molecular weight peptide, the most powerful reversible specific inhibitor of aspartic proteinases. It was isolated from the culture filtrates from various species of Actinomycetes (25). Due to its unusual potency, inhibiting porcine pepsin at about 1×10^{-10} M (26) and human cathepsin D at 5×10^{-10} M (27) and other remarkable properties, pepstatin has been widely used as a research tool in the studies of enzyme mechanism

(27, 28, this book), localization (29), biological function (reviewed in 8) and in affinity chromatography as a ligand (10,11,30, 31). It was reported that pepstatin is a good titrant for the determination of the molarity of solutions of cathepsin D (27). The binding of pepstatin is strongly pH-dependent as was shown by gel-chromatographic experiment (27) and CD spectroscopy (32). Both experiments show that cathepsin D binds pepstatin in the pH range from acid pH to pH 7.0. It was suggested that the binding site for substrate and pepstatin is abolished by a conformational change in the enzyme molecule (27).

Our earlier studies show that CD is useful method to follow structural studies and conformational changes of cathepsin D (32,33). We found that aromatic side chains are perturbed upon binding of pepstatin and diazoacetyl-D,L-norleucine methyl ester (DAN) to cathepsin D (32). Here, we present the interaction of bovine and chicken cathepsin D, hog and chicken pepsin, chymosin, Aspergillus niger aspartic proteinase and Claviceps purpurea aspartic proteinase with pepstatin. The interaction was carried out by dissolving the enzymes in the same buffer containing at least an equimolar amount of pepstatin at pH 3.5 (details are in preparation for publication).

The characteristic bands of cathepsin D in the aromatic CD spectra indicate contributions from Trp and Tyr and the fine structure between 250-270 nm may be assigned additionally to the Phe residues. Summarizing, it is evident that the investigated enzymes exhibit dichroic peaks in the near UV region at very similar wavelengths. On acidification the intensities of some bands change in both cathepsin D and Claviceps purpurea aspartic proteinase but not in pepsins and Aspergillus niger proteinase. These can be ascribed to changes in tertiary structure, or to changes in the environment of certain aromatic residues due to protonation of side chains. No significant changes were observed in the far UV CD spectra after acidification. All spectra in this region are similar and the dichroic minima range from 210-218 nm. The characteristically low content of α-helix and high content of β-structure

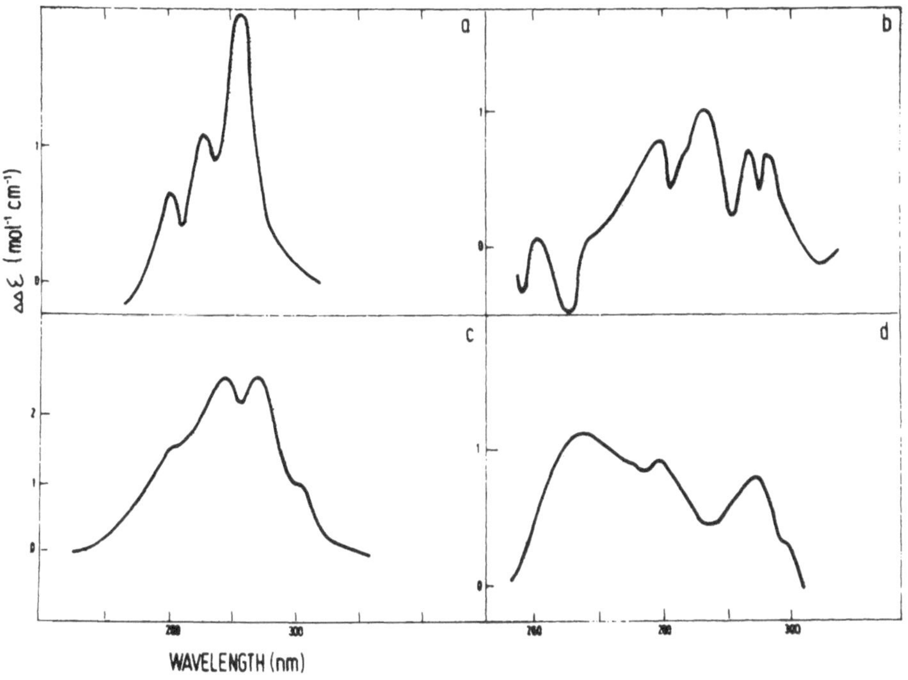

Fig. 3: Differential spectra in the near UV region obtained by subtraction of absorbances of free proteinase from proteinase-pepstatin complex which was formed by incubation at acid pH and 37 °C for 10 min : a) hog pepsin; b) chicken pepsin; c) bovine cathepsin D; d) chicken cathepsin D.

are in accordance with crystallographic data (34). These spectroscopic results show that the tertiary structure has been strongly conserved during evolution. This is in agreement with the high degree of homology in the primary structure in the superfamily of aspartic proteinases (this book).

We have followed the binding of pepstatin to different aspartic proteinases by circular dichroism in the near (Fig. 3) and far (Fig. 4) UV CD regions. On binding of pepstatin to bovine cathepsin D there is a pronounced increase in the intensities of Trp bands at 292, 285 and 280 nm indicating changes in the

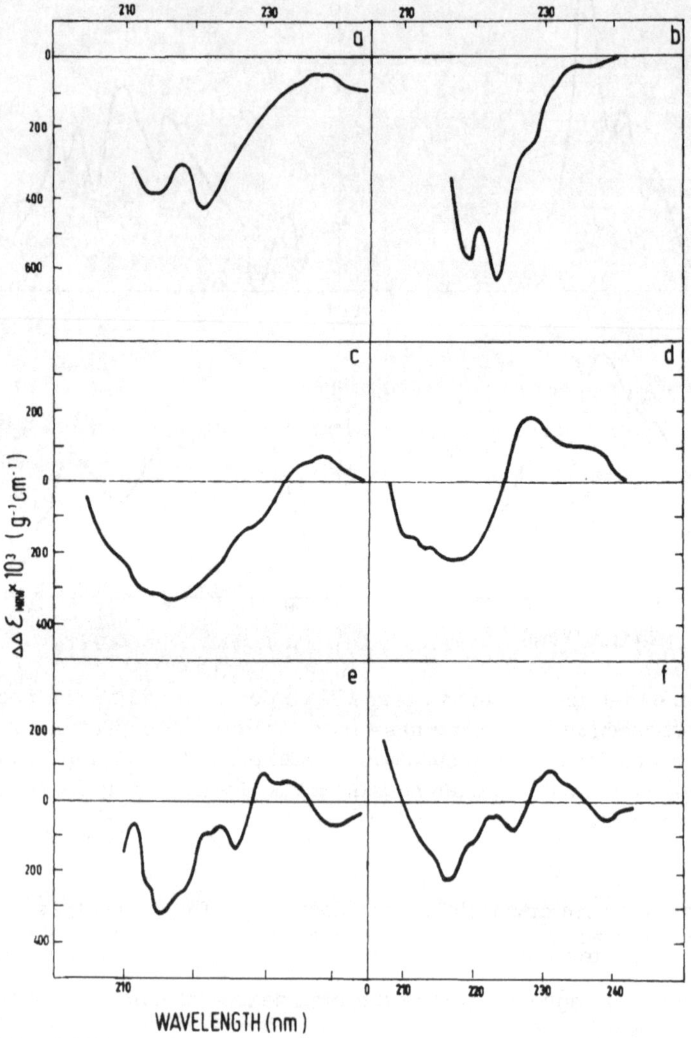

Fig. 4. Differential spectra in the far UV region, obtained similarly as described in Fig.3. : a) hog pepsin; b) chicken cathepsin D; c)bovine cathepsin D; d) calf chymosin; e) Aspergillus niger aspartic proteinase; f) Claviceps purpurea aspartic proteinase.

microenvironment of Trp in the active site and its vicinity. Comparing this result with other differential near UV CD spectra of different aspartic proteinases , as is seen in Fig. 3, we can conclude that the dichroism of aromatics is perturbed to different extent in different enzymes. Trp and Tyr dichroism is intense in the complex of all proteinases. Increased absorption bands of Tyr transition probably reflect movement of Tyr-75 residue and other Tyr into a more assymetric environment as in the active site of Penicillopepsin (35).

The differential far UV CD spectra show that the binding of pepstatin lowers the ellipticity of all aspartic proteinases (Fig. 4). Minima are in the region from 210-220 nm and are double in hog pepsin and chicken cathepsin D, very broad in bovine cathepsin D and chymosin and rather narrow in fungal proteinases. The general picture of reproducible minima in the different spectra, centered at about 217 nm suggests that there is some ordered structure induced in the enzyme-inhibitor complexes. This suggestion is supported by the crystallographically observed arrangement of the pepstatin fragment into the β-sheet structure (35). As an increase in negative ellipticity in the region from 210-218 nm was observed in all the proteinases tested we conclude that the ordered binding of pepstatin is common to all aspartic proteinases.

Unfolding and refolding of cathepsin D

During the course of our denaturation studies on cathepsin D, we have concentrated on attempts to demonstrate the reversibility of unfolding by guanidinium chloride (GdmCl) in order to see whether the enzyme exists as synthesized or whether post-translational modification has led to the loss of this property. It was shown that proteins which had been cleaved after biosynthesis and folding were no longer able to refold after denaturation (36). Such proteins included mainly those derived from a precursor. Pepsin cannot be refolded if once denatured (37) although pepsinogen can regain its native conformation spontaneously (38). Similarly cathepsin D lost its activity irreversibly after denaturation with 2.32 M GdmCl (12, 39).

The denaturation and renaturation of the single-chain cathepsin D has been studied by CD, fluorescence and enzyme activity measurement (for details see reference 21). From Fig. 5 it is evident, that cathepsin D exhibits at least two consecutive unfolding transitions in GdmCl. The first transition is steep and occurs between 0.5 M and 2 M GdmCl. The loss of enzyme activity is paralleled by loss of ellipticity at 291 nm. There is also decrease in the ellipticity at 220 nm, corresponding to 60% of the total change in ellipticity between 0 M and 6 M GdmCl. These results show a highly co-operative transition between 0 M and 2 M GdmCl in which cathepsin D activity, aromatic acid environment and backbone structure are affected in a concerted manner. The cathepsin D molecule is not completely unfolded under these conditions and further changes in ellipticity at 220 nm take place between 2 M and 6 M

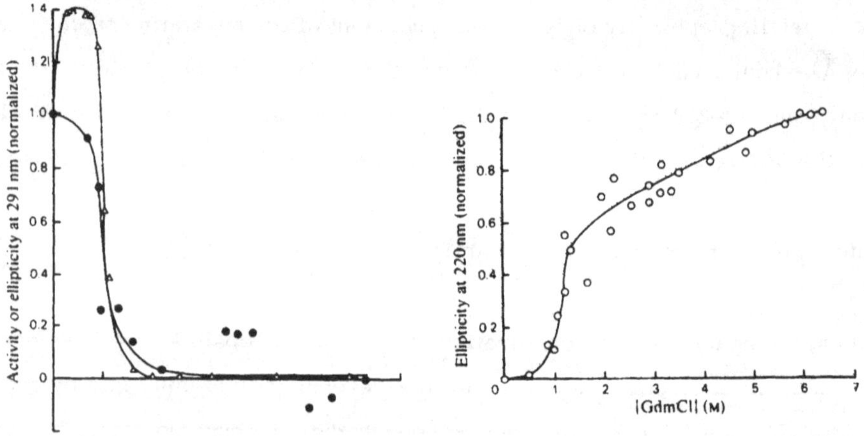

Fig. 5. Unfolding of cathepsin D_1 in GdmCl. Cathepsin D was diluted into GdmCl solutions at the stated concentrations and time-dependent reactions were allowed to go to completion. The values for activity (Δ), ellipticity in near UV (•) and far UV region (o) were normalized to that of the native enzyme, stored under the same conditions.

GdmCl. This second transition is broad and much less co-operative than is the first transition. These experiments show that cathepsin D unfolds in two main stages. Kinetics of these transitions indicate the existence of at least two intermediate forms between the native and the fully unfolded states.

From Fig. 5 it is also evident an increase of specific enzyme activity at 0.5 M GdmCl for about 40% of that of native enzyme. Control experiments showed that this increase in specific activity is not due to increased susceptibility of haemoglobin, as was suggested for the apparent activation after urea treatment (40). Our results suggest that the activation of cathepsin D is associated with a separate conformational transition, since the main transition involves loss of activity (Fig. 5). Our further experiments were performed in order to restore cathepsin D activity after denaturation. Cathepsin D denatured in GdmCl concentrations up to 6 M were adjusted back to zero concentrations of denaturant by dilution, by gel chromatography and by dialysis, but no recovery of activity was observed. It can be concluded, that the activity of cathepsin D is lost irreversibly on unfolding, but the loss of backbone ellipticity and of folded aromatic environment is 75% reversible (for details see reference 21). The results are in keeping with the fact that cathepsin D originates from a precursor by proteolytic cleavage with the loss of N-terminal peptide (14,18, 41).

Cathepsin D precursor and its activation

Our early investigations indicated the existence of cathepsin D precursor, when inactive protein was bound to pepstatin-Sepharose resin at acid pH and was eluted from the affinity column together with active enzyme. Separation of active cathepsin D from the inactive protein was achieved by gel chromatography (9,11). When this inactive protein was exposed to pH 2.5 - 3.5 for a very short time 1-2 min at 3° C, it converted itself into an active enzyme, releasing several activation peptides. These peptides inhibited cathepsin D at pH 5.0. Separation of these peptides was achieved by affinity chromatography on immobilized cathepsin D - Sepharose resin. One of the peptides was further purified and its amino acid composition suggested that this peptide might correspond to the activation peptide (residues 1-17) released during cow pepsinogen activation (41). These results supported our suggestion that cathepsin D precursor (cathepsinogen D or procathepsin D) can be activated by the mechanism, similar to the activation of pepsinogen. More details were published (42).

Our results were confirmed by the observed cleavage of cathepsin D precursor, isolated from human fibroblast in vitro, resembling the autocatalytic activation of pepsinogen (43). In addition, procathepsin D-II was purified from Japanese monkey lung and was converted to the active form upon treatment at acid pH (44).

The activation peptide, released from procathepsin D strongly inhibits cathepsin D and E, Aspergillus niger aspartic proteinase and does not inhibit chicken pepsin and pig pepsin. Similarly, the mixture of peptides released from pig pepsinogen does not inhibit cathepsin D and E. These preliminary results suggest some structural differences in the active site of aspartic proteinases of different origin.

Cathepsin E

Contrary to cathepsin D, cathepsin E is a much less investigated and characterized enzyme. After its discovery by Lapresle (7) further progress was made by the discovery that this enzyme is present in relatively high amount in rat spleen (45). Cathepsin E was isolated by a new purification procedure and characterized (45). Later we developed a method for the simultaneous isolation of cathepsin D and E (46). Almost an equal amount of both enzymes was found in rat spleen, whereas other rat organs contained cathepsin E only in traces.

Rat spleen cathepsin E is an acidic protein (pI 4.3) having a M_r of 90 000 which was determined by gel chromatography. By SDS-PAGE a M_r of about 45 000 was found which corresponds to cathepsin D. However, the amino acid composition of cathepsin E (45) is different from that of rat spleen cathepsin D (47). In addition, rat spleen cathepsin D and E show different immunological properties (48). On the basis of these differences we can conclude that cathepsin D and E are different enzymes. More investigations are needed to understand the biological role of this peculiar enzyme.

Acknowledgement

Chicken pepsin was a generous gift from prof.dr. V.Kostka, Prague, ČSSR, and chymosin was kindly given by prof.dr. B.Foltmann, Copenhagen,Denmark. This work was supported by a research grant from the Research Council of Slovenia, and in part by the NSF, USA.

References

1. Tang, J.: Acid Proteases: Structure, Function,and Biology, Plenum Press, New York - London 1977.

2. Shewale, J.G., Tang,J.: Proc.Natl.Acad.Sci.USA 81 , 3703-3707 (1984).

3. Anson, M.L.: J.Gen.Physiol. 20 , 565-574 (1937).

4. Press,E.M., Porter, R.R., Cebra,J.: Biochem.J. 74 , 501-514 (1960).

5. Barat,E., Patthy,A., Graf,L.: Proc.Natl.Acad.Sci.(Wash) 76 , 6120-6123 (1979).

6. Marks,N.: Subcellular Mechanisms in Reproductive Neuroendocrinology (Naftolin,F., Ryan,K.J., Davies,J. Eds), Elsevier, Amsterdam, pp.129-147, 1976.

7. Lapresle,C., Webb,T.: Biochem.J. 84 , 455-462 (1962).

8. Barrett,A.J.: Proteinases in Mammalian Cells and Tissues, North-Holland Publ.Co., Amsterdam, pp.209-248, 1977.

9. Smith,R., Turk,V.: Eur.J.Biochem. 48 , 245-254 (1974).

10. Kazakova,O.V., Orekhovich,V.N. : Biokhimiya 40, 969-972 (1975).

11. Kregar,I., Urh,I., Umezawa,H., Turk,V.: Croat.Chem.Acta 49 , 587-592 (1977).

12. Turk,V., Lah,T., Puizdar,V., Kregar,I., Pain,R.H. : Acta Med.Germ. 40 , 1439-1450 (1981).

13. Erickson,A.H., Blobel,G.: J.Biol.Chem. 254, 11771-11774 (1979).

14. Hasilik, A., Neufeld,E.F.: J.Biol.Chem. 255, 4937-4945 (1980).

15. Hasilik,A., Neufeld,E.F.: J.Biol.Chem. 255, 4946-4950 (1980).

16. Hasilik, A.: Trends Biochem.Sci. 5, 237-240 (1980).

17. Hasilik, A., von Figura, K.: Eur. J.Biochem. 121, 125-129 (1981).

18. Erickson, A.H., Conner,G.E., Blobel,G.: J.Biol.Chem. 256, 11224-11231 (1981).

19. Takahashi, T., Schmidt,P.G., Tang,J.: J.Biol.Chem. 258 , 2819-2827 (1983).

20. Nakao,Y., Kozutsumi,Y., Kawasaki,T., Yamashina,I., van Halbeck,H., Vliegenthart,J.F.G. : Arch.Biochem.Biophys. 229, 43-54 (1984).

21. Lah,T., Drobnič-Košorok,M., Turk,V., Pain,R.H.: Biochem.J. 218, 601-608 (1984).

22. Rajagopalan, T.G., Moore,S., Stein,W.H.: J.Biol.Chem. 241, 4940-4950 (1966).

23. Sluyterman, L.A.E., Wijdens,J.: Eur.J.Biochem. 113, 189-193 (1980).

24. Lah,T., Turk,V.: Hoppe-Seyler's Z.Physiol.Chem. 363, 247-254 (1982).

25. Umezawa,H., Aoyagi,T., Morishima,H., Matsuzaki,H., Hamada, M., Takeuchi,T.: J.Antibiot. 23, 259-262 (1970).

26. Kunimoto,S., Aoyagi,T., Nishizawa,R., Komai,T., Takeuchi,T., Umezawa,H.: J.Antibiot. 27, 413-418 (1974).

27. Knight,C.G., Barrett, A.J.: Biochem.J. 155, 117-125 (1976).

28. Marciniszyn, J., Hartsuck,J.A., Tang,J.: J.Biol.Chem. 251, 7088-7094 (1976).

29. Knight,C.G., Hornebeck,W., Matthews,I.T.W., Hembry,R.M., Dingle, J.T.: Biochem.J. 191, 835-843 (1980).

30. Murakami,K., Inagami,T., Michelakis,A.M., Cohen,S.: Biochem. Biophys.Res.Comm. 54, 482-487 (1973).

31. Huang,J.S., Huang,S.S., Tang,J.: J.Biol.Chem. 254, 11405-11417 (1979).

32. Turk,V., Kregar,I., Gubenšek,F., Ločnikar,P. : Protein Turnover and Lysosomal Function (Segal,H.L., Doyle,D.J., Eds.), Academic Press, New York, pp. 353-361 , 1978.

33. Turk,V., Lah,T., Puizdar,V., Kregar,I., Pain,R.H.: Acta Biol.Med. Germ. 40, 1439-1450 (1981).

34. Tang,J., James,M.N.G., Hsu,I.N., Jenkins,A.J., Blundell,T.L.: Nature 271, 618-621 (1978).

35. James,M.N.G., Sielecki,A.R.: J.Mol.Biol. 163, 299-361 (1983).

36. Anfinsen,C.B., Sheraga,H.A.: Adv.Prot.Chem. 29, 205-299 (1975).

37. Blumenfeld,O.O., Leonis,J., Perlman,G.E.: J.Biol.Chem. 235, 379-382 (1960).

38. Ahmad,F., McPhie,P.: Biochemistry 17, 241-246 (1978).

39. Lah,T., Turk,V., Pain.R.H.: Period.Biol. 85, 95-99 (1983).

40. Wojtowicz,M.B., Odense,P.: Comp.Biochem.Physiol.B 68, 215-220 (1970).

41. Foltmann, B., Pedersen,V.B.: Acid Proteinases : Structure, Function and Biology, Plenum Press, New York-London, pp. 3-22, 1977.

42. Puizdar,V., Turk,V.: FEBS Letters 132, 299-304 (1981).

43. Hasilik,A., von Figura,K., Conzelmann,E., Nehrkorn,H., Sandhoff,K.: Eur.J.Biochem. 125, 317-321 (1982).

44. Moriyama,A., Kageyama,T., Takahashi,K.: Eur.J.Biochem. 132, 687-692 (1983).

45. Yamamoto,K., Katsuda,N., Kato,K.: Eur.J.Biochem. 92, 499-508 (1978).

46. Turk,V., Kregar,I., Gubenšek,F., Popović,T., Ločnikar,P., Lah,T.: Enzyme Regulation and Mechanism of Action (Mildner,P., Ries,B.,Eds.), Pergamon Press, Oxford, pp. 317-330 , 1980 .

47. Yamamoto,K., Katsuda,N., Kato,K.: Seikagaku 50 , 889-890 (1978).

48. Yamamoto,K., Kamata,O., Katsuda,N., Kato,K. : J.Biochem. 87, 511-516 (1980).

ACTIVATION OF CHICKEN PEPSINOGEN AND CHICKEN PEPSIN - PROPART PEPTIDE (p1 - p42) COMPLEX

Iva Pichová, Jan Pohl, Petr Štrop, Vladimír Kostka
Department of Biochemistry, Institute of Organic Chemistry and Biochemistry, Czechoslovak Academy of Sciences, 166 10 Prague 6, Czechoslovakia

Introduction

The activation of zymogens of aspartic proteinases proceeds under acidic conditions either via the one-step conversion pathway or via a sequential mechanism (1, 2). When chicken pepsinogen (CPG) was activated in the presence of pepstatin a peptide derived from the N-terminus of the zymogen molecule and corresponding to positions p1 - p26 in the propart was isolated. This finding was interpreted as evidence in favor of the sequential activation mechanism involving the cleavage of the bond Phe 26 - Leu 27 as the first step with simultaneous formation of intermediary pseudopepsin (3). The isolation of the native chicken pepsin-activation peptide (p1 - p42) complex (4) and the recently reported multiplicity of the activation mechanism of mammalian pepsinogens (5) has led us to examine the conversion of CPG into the active enzyme and to compare the behavior of the native complex in this respect.

Aspartic Proteinases and their Inhibitors

Results

Sequence studies on CPG activation

A solution of CPG (370nM) in water, pH 7.5, was activated for 15-45 sec at 10 or 14°C. The activation was discontinued by increasing the pH to 5.5 by addition of 5M sodium acetate containing a 3-fold molar excess of pepstatin. The activation peptides were separated from the protein (a) by chromatography of the activation mixture on SE-Sephadex C-25 in 0.1M sodium acetate, pH 5.5, and were subsequently purified by HPLC, and (b) by resolution of the activation mixture by gel filtration on Sephadex G-50 in a denaturating solution and were desalted by gel filtration in 10% acetic acid. As can be seen in Figure 1A the exposure of CPG to pH 2 (10°C, 15 sec) gives rise to one new peak only (1, fractions 94-101). This peak disappears with prolonged incubation period and two new peaks, 2 and 3, can be observed (Figure 1B, C). Amino acid analysis and N-terminal end group analysis of the material contained in peak 1, 2, and 3 and of the peptides purified by reversed phase HPLC show that peak 1 contains the peptide corresponding to the whole propart (p1 - p42) of CPG (6). Peaks 2 and 3 contain peptides p1 - p26 and p27 - - p42, respectively, from the propart of CPG. The finding of the peptide corresponding to the whole propart of CPG leads us to assume that, under the conditions given and in the absence of pepstatin, the first bond cleaved during CPG activation is Leu 42 - Thr 43 and that pepsin is formed directly without any intermediates. The cleavage of the bond Phe 26 - - Leu 27 giving rise to peptides p1 - p26 and p27 - p42 therefore occurs in the subsequent step.

Kinetics of activation of CPG and of native complex

The activation of CPG at pH 1.5-3.5, as measured in terms of appearance of proteolytic activity determined with synthetic

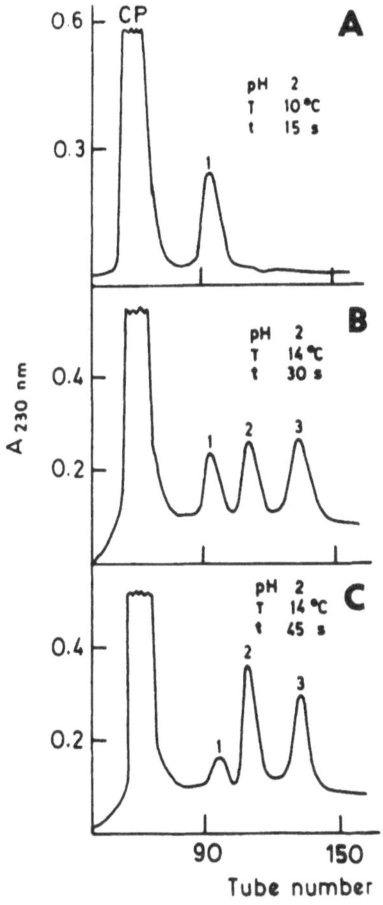

Figure 1. Gel chromatography of CPG activation mixtures on Sephadex G-50. Column 1.9x140 cm, eluted by 0.1M sodium acetate at pH 5.3, containing 8M urea; flow rate 12 ml/h, 2-ml fractions.

substrates (7, 8), is a unimolecular process. An entirely identical behavior during activation shows also the native complex (NC). The halflifes of appearance of activity of both compounds are shown in Figure 2. It can be seen that there is practically no difference in the rate of activation of CPG and NC at pH 1.5-2.0. The rate of NC activation at

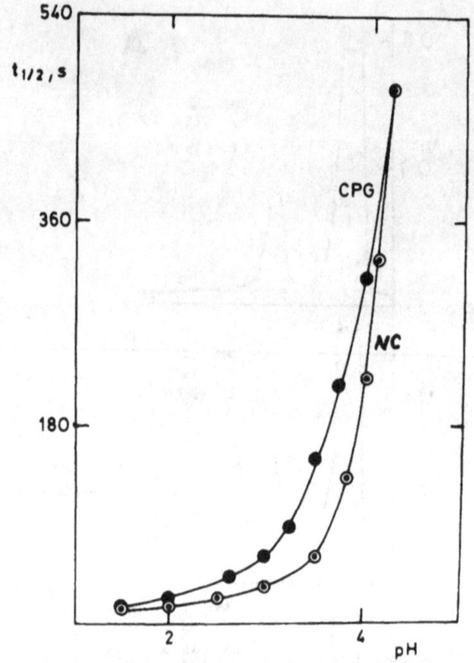

Figure 2. pH-profile of halflife of CPG and NC activation. CPG and NC (concentration 1.3nM in both cases) were activated at 14°C in the presence of 1mM substrate, <Glu-His-Phe-Phe--Ala-LeuOMe. The concentration of the product, Phe-Ala-LeuOMe, in aliquots withdrawn at time intervals was determined with fluorescamine (9).

pH 2.5 and higher is approximately 2 times lower and this indicates a difference between the two species. Identical results were obtained with derivatives of CPG and NC whose free SH-group of Cys 115 had been modified (8) by a fluorescent label, Acrylodan. As demonstrated elsewhere in this volume the probe is localized in the active site of chicken pepsin and significantly increases its quantum yield during binding of the activation peptide p1 - p42 (8). The dissociation of NC or the activation of labeled CPG can thus be detected by a time-dependent fluorescence decrease. Figure 3 shows that the time profile of fluorescence of the Acrylodan label during CPG and NC activation is very similar. It is therefore

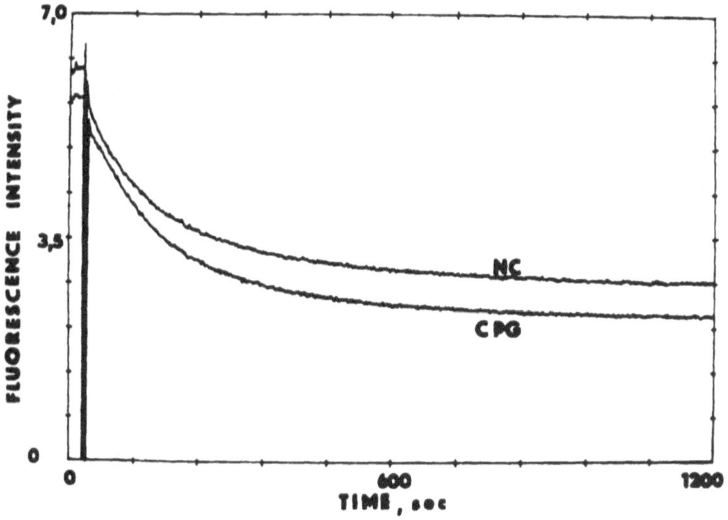

Figure 3. Time profile of fluorescence of Acrydolan-labeled CPG and NC during acidification of their solutions to pH 3.5. Temperature 20°C, excitation at 386 nm, emission at 500 nm.

evident from the above results that the rate-limiting step of activity appearance during CPG activation is not the cleavage of the bond Leu 42 - Thr 43 but the dissociation of the activation peptide or its simultaneous hydrolysis at Phe 26 - - Leu 27 or at other bonds. We are not able to make a final decision in this respect at present. Glick et al. (10) studying the activation of spin-labeled CPG arrived at the conclusion that the dissociation of the activation peptide is the rate-limiting step of the process. Hence, the CPG activation pathway proposed here complies with their scheme (10) except that the first bond cleaved in CPG is that between residues 42 and 43. This scheme also postulates a rapid conformational change occurring after acidification of hog pepsinogen. Evidence of a conformational change during CPG and NC activ-

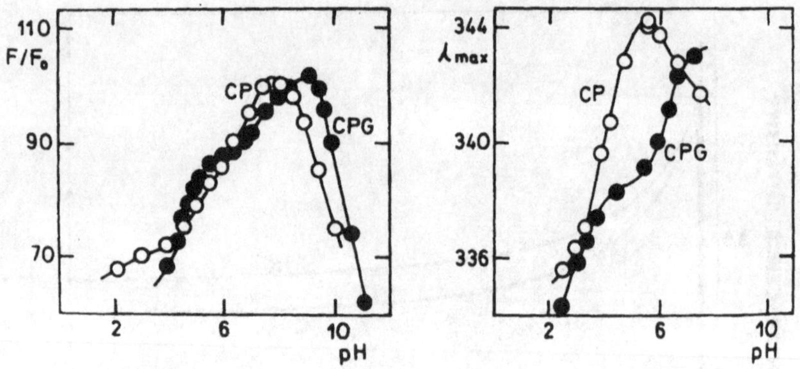

a b

Figure 4. Conformational changes of CPG and CP.
a - pH-profile of relative intensity of tryptophan fluorescence of CPG and CP; t 20°C, excitation at 280 nm, emission at 341 nm, b - pH-profile of change in emission maximum of CPG and CP fluorescence; conditions the same as under a.

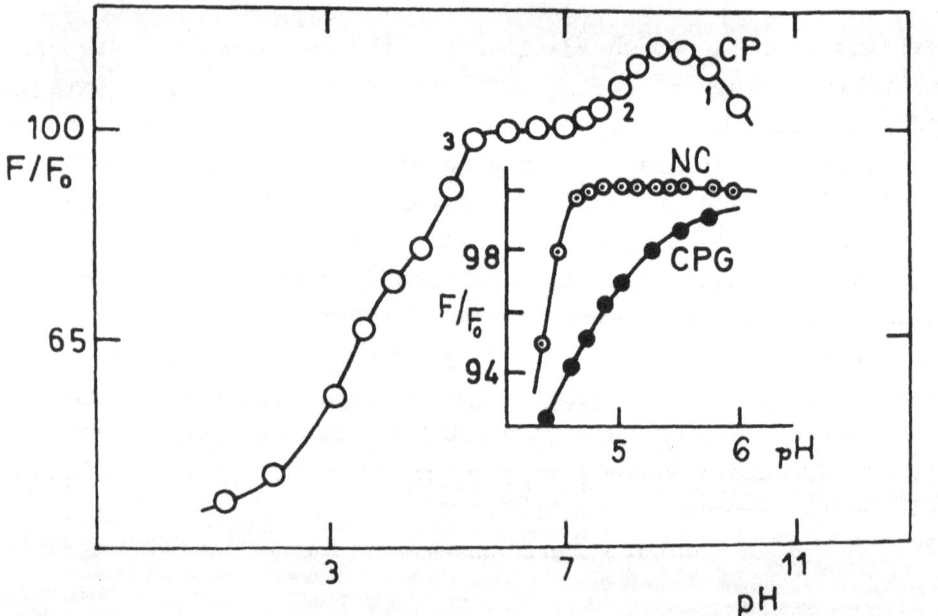

Figure 5. Conformational changes of CPG, CP, and NC labeled with Acrylodan at Cys 115. Spectrofluorometric measurement, excitation at 386 nm, emission at 500 nm; 1, 2, 3 - individual conformational changes. Inset: fluorescence of CPG and NC in the pH-range 4-6.

ation is indicated by pH-changes of intrinsic fluorescence or, alternatively, Acrylodan fluorescence of native or labeled, resp., CPG and NC (Figure 4 and 5).

The zymogens of aspartic proteinases differ in the sites at which their proparts undergo the first proteolytic attack. It is demonstrated in Figure 6 that this site depends, inter alia, on the distribution of basic amino acid residues along the propart:

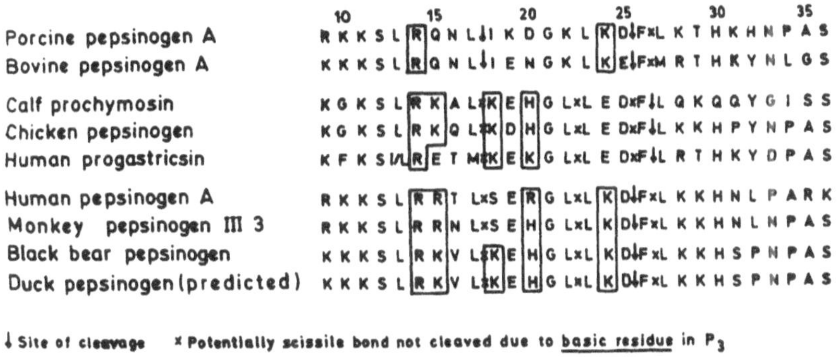

Figure 6. Sites of cleavage of proparts of zymogens of various aspartic proteinases and potentially scissile bonds which are not cleaved because of presence of basic residues in P_3 (x). For references see (5).

The presence of a basic residue in position P_3 in the vast majority of proparts excludes the possibility of cleavage of potentially very susceptible bonds, such as Leu 22 - Leu 23 or Phe 26 - Leu 27 in pepsinogens A, a feature which is typical of the subsite specificity of pepsin (11, 12).

References

1. Dykes, C.W., Kay, J.: Biochem. J. 153, 141-144 (1976).
2. Kageyama, T., Takahashi, K.: J. Biochem. (Tokyo) 92, 1179-1188 (1982).

3. Keilová, H., Kostka, V., Kay, J.: Biochem. J. 167, 855--858 (1977).
4. Kostka, V., Pichová, I., Baudyš, M.: this volume,pp 53-72
5. Takahashi, K., Kageyama, T.: this volume, pp. 265-282.
6. Baudyš, M., Kostka, V.: Eur. J. Biochem. 136, 89-99 (1983).
7. Pohl, J., Baudyš, M., Kostka, V.: Anal. Biochem. 133, 104-109 (1983).
8. Pohl, J., Štrop, P., Pichová, I., Bláha, I., Kostka, V.: this volume, pp. 245-264.
9. Udenfriend, S., Stein, S., Böhlen, P., Dairman, W.: Science 178, 871-872 (1972).
10. Glick, D.M., Valler, M.J., Rowlands, C.C., Evans, J.C., Kay, J.: Biochemistry 16, 3746-3750 (1982).
11. Antonov, V.K.: in Acid Proteases, Structure, Function, and Biology (Tang, J., ed.), pp. 179-198, Plenum Press, New York and London (1977).
12. Powers, J.C., Harley, D.A., Myers, D.V.: in Acid Proteases, Structure, Function, and Biology (Tang, J., ed.), pp. 141-157, Plenum Press, New York and London (1977).

CHICKEN PEPSIN - ACTIVATION PEPTIDE (p1 - p42) COMPLEX ISOLATED AND ARTIFICIALLY FORMED: A COMPARISON

Miroslav Baudyš, Iva Pichová, Jan Pohl, Vladimír Kostka
Department of Biochemistry, Institute of Organic Chemistry and Biochemistry, Czechoslovak Academy of Sciences, 166 10 Prague 6, Czechoslovakia

Introduction

The peptides liberated from the N-terminal parts of the pepsinogen molecules during their activation have been known to act as inhibitors of the newly formed pepsins (1-3) by forming complexes. Structural evidence of complex formation during the activation of hog pepsinogen and human progastricsin has been obtained recently (4, 5). A similar complex of chicken pepsin with the whole propart has been obtained during purification (6) of chicken pepsinogen which had not been exposed to conditions under which activation occurs. We became interested in a comparison of the properties of the "naturally occurring" complex with those of the "artificial" complex prepared from chicken pepsin and the activation peptide, also because the interaction of pepsin with the whole propart has not been studied as yet. Another problem interesting from the viewpoint of activation mechanism was the possible interconversion of both types of complexes.

Aspartic Proteinases and their Inhibitors

Results and Discussion

Chromatographic stability of complexes

The activation peptide (p1-p42) (AP) was prepared from the natural (native) complex (NC) by gel chromatography on Sephadex G-50 equilibrated with 8M urea, pH 5. Its homogeneity was verified by amino acid analysis and sequence analysis as compared with the known propart structure (7). Ion exchange chromatography of pepsinogen and NC on DEAE-Sephacel at pH 7.4 using a gradient of NaCl shows that the two species cannot be resolved. The artificial complex (AC) prepared by 1-h incubation at the same pH of an equimolar mixture of chicken pepsin with AP, was stable under the conditions of ion exchange chromatography and was coeluted with NC. The formation and stability of AC, prepared by the procedure described above at different pH-values, were also examined by gel permeation chromatography at pH 6.5, 7.0, and 7.5 on Sephadex G-50. Sequence analysis and SDS-PAGE showed the stability of AC at pH 7.0 and 7.5; the peak emerging in the hold--up volume at pH 6.5 obviously contained a small quantity of pepsin. NC, however, is stable under these conditions. For the lack of a sufficient quantity of AP the stability during gel permeation chromatography in alkaline solutions above pH 9.0 was assayed with NC only; the latter is stable up to pH 10.5. Incomplete resolution of AP from the hold-up peak was observed at pH 11; the dissociation is complete at pH 11.5.

Alkaline inactivation of CPG, NC, and AC

The alkaline stability of chicken pepsinogen (CPG), NC, and AC was compared also quantitatively at several pH-values (Table 1) by determination of residual activity toward hemoglobin at pH 2 (8).

Table 1

k_{inact} (min^{-1}, for $t \to 0$) of CPG, NC and AC

	CPG	NC	AC[x]
pH 10.0	stable	stable	stable
pH 10.5	10^{-4}	10^{-4}	0.5×10^{-3}
pH 11.0	5.1×10^{-3}	6.5×10^{-3}	4.1×10^{-1}
pH 11.5	1.6×10^{-1}	2.8×10^{0}	$\sim 10^{1}$

[x] preformed at pH 8.0 at a AP/pepsin ratio of 1.1 with one-hour preincubation

All three species are stable at pH 10.0, yet already at pH 10.5 a difference between NC and AC can be observed in the inactivation rate expressed by the first order rate constant for $t \to 0$ (see below). The inactivation of pepsinogen and of NC at pH 11.0 (Figure 1) follows the first order kinetics with very similar k_{inact}. No dependence on the concentration of AP was observed. In contrast AC shows a biphasic inactivation kinetics which is dependent on the concentration of AP. Assuming the validity of the kinetic scheme

$$P - AP \underset{}{\overset{k_{off}}{\rightleftharpoons}} P + AP \xrightarrow{k_{inact}^{pepsin}} P_{inact} + AP \qquad /1/$$

where P is pepsin, the total k_{inact} for the first inactivation phase ($t \to 0$)

$$k_{inact}^{-1} = \frac{AP}{k_{inact}^{pepsin} \times K_D} + k_{off}^{-1} \qquad /2/$$

is dependent on the concentration of AP, the K_D being the dissociation constant of the complex. This linear dependence has been verified with AC (Figure 1) at pH 11.0 over the range of molar AP/P ratios of 1.1 to 10 with k_{off} 4.5 x x 10^{-1} min^{-1}. The second slow inactivation phase proceeds most likely via a "nonspecific" mechanism

$$P - AP \longrightarrow P_{inact} + AP \qquad /3/$$

where k_{inact} for $t \rightarrow \infty$, $1.5 \times 10^{-2}\,min^{-1}$ approximates k_{inact} for NC; this indicates that the overall unfolding of all three species is similar. The situation is more complicated at pH 11.5. The semilogarithmic plot for CPG shows a slight deviation from linearity during the initial phase. The inactivation of both complexes follows the biphasic kinetics. All three species, however, are inactivated at different

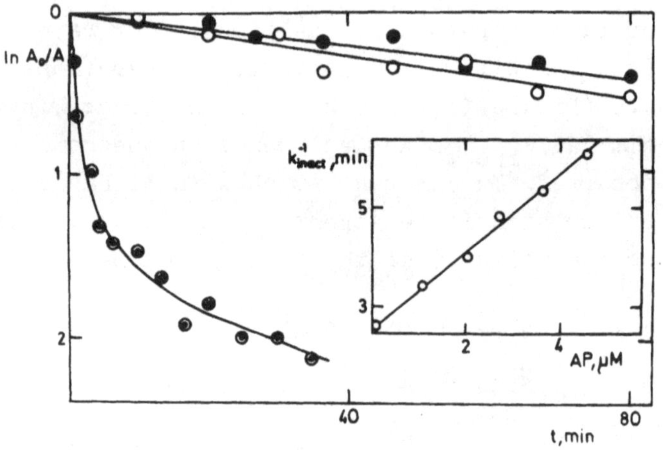

Figure 1. Inactivation of CPG (●), NC (○), and AC (◉) at pH 11.0 and 25°C. The initial concentration of all three species was 0.46 μM. A/A_0 is the relative residual activity toward hemoglobin at pH 2.0 and 37°C. AC was prepared by 30-min incubation of AP with pepsin (molar ratio 1.1) in 0.01M sodium phosphate buffer, pH 8.0. The samples of CPG and NC were treated in the same manner. After the incubation the pH was raised to 11.0 by the addition of alkali. The dependence of the reciprocal value of total inactivation constant at the same pH for $t \rightarrow 0$ of AC (0.46 μM) on the concentration of free AP is shown in the inset. AC was prepared at a varying excess of AP as described above.

rates (Table 1). The k_{inact} ($t \rightarrow \infty$) for NC is 1.4×10^{-1} min^{-1}, i.e. very close to the k_{inact} value for pepsinogen; this indicates that the second inactivation phase should proceed via mechanism /3/. Because of high inactivation rates the AP concentration dependence of the rate of the first inactivation phase could not be measured reliably; it seems, however, that this rate of NC inactivation is independent of the concentration of AP. If the inactivation of NC proceeded according to mechanism /1/ the inactivation characteristics of the two complexes should be the same at this pH, which is not the case. We have also examined the possibility of the conversion of AC into NC by incubating the former for up to 24 h at pH 7.0-8.5, yet without any change in the inactivation characteristics of AC. These experiments indicate that the conversion of AC into NC cannot take place in the alkaline pH-range.

Kinetics of interaction of AP with chicken pepsin

The kinetics of interaction of chicken pepsin and its AP was examined with a synthetic substrate, Lys-Pro-Ala-Glu-Phe-Nph--Ala-Leu (9). The activity was assayed by measurement of changes in absorbance at 300 nm. The degree of inhibition was time-dependent and the enzyme was therefore preincubated with AP at the given pH for 30 min; this is sufficient for maximal inhibition. The residual activity was determined at pH 5.9. The complex preformed at pH 6.5-7.5 in 0.01M phosphate buffer was not stable at this pH and dissociated, the traces, however, indicated a clear lag phase lasting for several minutes on the average. The dissociation was slow enough to enable us to determine the fraction of free pepsin in the preincubation mixture. AP is a titrant of chicken pepsin at pH 7.0 and 7.5 (Figure 2) and most likely also at pH 6.5, in accordance with the results of the gel permeation chromatography experiments. The inhibition is still time-dependent at pH 6.15; a 15-min preincubation period is satisfactory, weaker inhibition occurs even without the preincubation. We were also able to determine the inhibition type in these experiments after preincubation of the enzyme with AP. The latter is an apparently competitive inhibitor of $K_i = 1.0 \times 10^{-6}$M (10). The inhibition occurs even at pH 5.9 where the inhibition no longer is time-dependent, K_i determined from IC_{50} (11) is of the magnitude of 10^{-5}M. The K_i-value for pH 5.9, determined from steady state values after dissociation of the complex preformed at pH 7.5, is of the order of 10^{-6}M. It may thus be assumed that the initial complex formed at pH 5.9 is different from that resulting from the breakdown of the tight complex at the same pH, preformed at pH 7.5. The time dependence of the inhibition degree at pH 6.15 and higher demonstrates that after the initial complex has been formed, which at least at pH 6.15 is partly inhibited, a conversion into the tight complex takes place. This finding corresponds to the scheme designed for the interaction of hog pepsin with shorter peptides derived from the N-terminal part of the prosequence of hog pepsinogen (12), the important difference being, however, that even the initial complex is partly inhibited in the case of chicken pepsin. Moreover, the K_i for AP and chicken pepsin is higher by several orders than the K_i values for the interaction of hog pepsin with shorter peptides mentioned above; this suggests the essential role of the C-terminal part of AP played in its tight binding to the pepsin molecule.

Interest deserves the behaviour of NC and AC during activation (dissociation) at pH 6.0-5.2. Whereas all the activity possible is obtained from AC at pH 5.2, the halflife being 70 s, NC yields about 10% of activity (Figure 3) only under identical conditions. This demonstrates the higher stability of NC. Moreover, NC is stable at pH 5.9 under the conditions of the assay with immeasurable peptic activity.

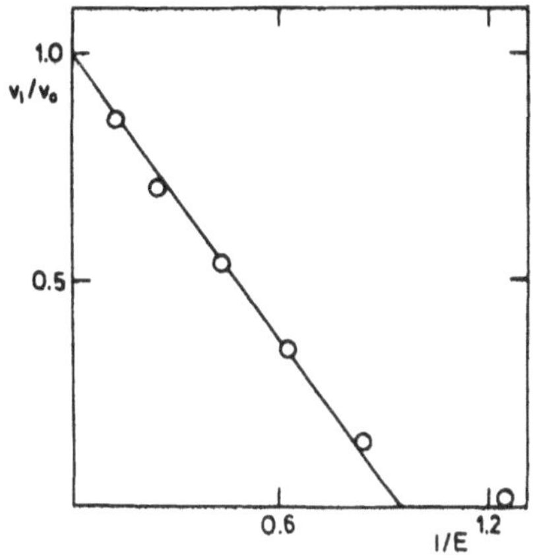

Figure 2. Titration of chicken pepsin by AP at pH 7.5. The enzyme (1 µM) was preincubated for 30 min with AP (0-1.23 µM) in 0.01M sodium phosphate buffer, pH 7.5 at 25°C. The fractional residual activity was determined with a synthetic substrate in McIlvain buffer at pH 5.9 and 30°C, making use of the fact that the complex breakdown at pH 5.9 is slower than the time necessary for the determination of the residual activity.

Inhibition of other aspartic proteinases by chicken AP

The inhibition of hog pepsin by chicken AP at pH 5.5 was investigated and time dependence of the inhibition was observed. The initial complex was inhibited (Figure 4). The K_i determined under steady state conditions (40-min preincubation) is 1.5×10^{-9}M and the inhibition is of the apparent noncompetitive type. Here, too, the C-terminal part of the activation peptide is essential for its tight binding to the molecule of hog pepsin.

The inhibition of human pepsin A was again time-dependent at pH 5.5 and the K_i calculated from IC_{50} (11) is 1.0×10^{-9}M, i.e. very similar to hog pepsin.

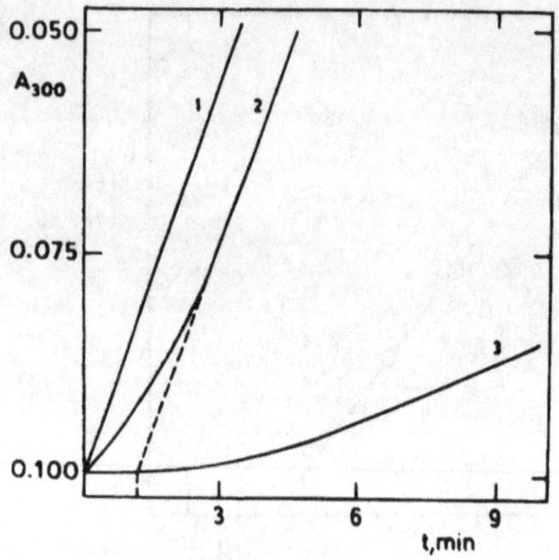

Figure 3. Time profile of appearance of peptic activity assayed by measurement of change in $A_{300\ nm}$ in McIlvain buffer at pH 5.2 and 30°C and a substrate concentration of 0.205mM. 1 - potential activity of AC (80nM) after activation at pH 2.2, determined at pH 5.2, 2 - AC (80nM) preformed at pH 7.5, 3 - NC (80nM). The intersection of the extrapolated linear part of the activation curve for AC with the time axis indicates - on condition of first order reaction - the halflife of "activation (dissociation)".

We have also tried to detect the complex of AP and bovine spleen cathepsin D (13). The enzyme was preincubated with AP over the pH-range 7.0-8.0 and the residual activity was determined at pH 5.3. No inhibition was detected. This finding, however, does not eliminate the possibility of interaction of cathepsin D with AP because the pH of the assay was probably too low (cf. activation of artificial complex of chicken pepsin and AP). The activity of cathepsin D toward the synthetic substrate was immeasurable, however, at a higher pH.

The inactivation experiments with NC and AC revealed an important difference between the two complexes in strong-

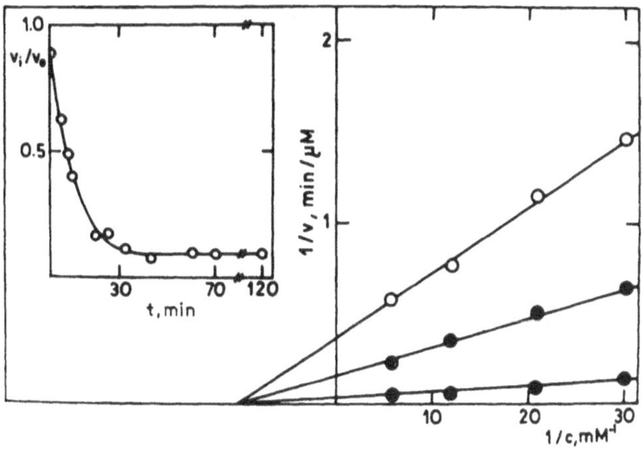

Figure 4. Lineweaver-Burk plot of the inhibition of hydrolysis by hog pepsin (active concentration 3.0nM) of synthetic substrate in 0.1M sodium acetate at pH 5.5 and 30°C by AP. Hog pepsin was preincubated with AP 40 min in the absence of the substrate, the concentration of AP being 0 (●), 10 nm (◉), and 20 nm (○). Inset: time-dependent inhibition of hog pepsin (3 nm) by AP (10 nm). The same substrate aliquot (concentration in the assay 0.17mM) was always added after the given preincubation period.

ly alkaline solutions. This difference was evidenced in kinetic experiments and in experiments with the activation of NC and AC at pH-values decreasing down to pH 5.0. We did not detect moreover a transition of AC to NC. We have also obtained evidence in our inactivation and chromatographic experiments that NC can mimick pepsinogen in many respects. NC is characterized best as a stable intermediate of the pepsinogen type in the chicken pepsinogen activation cascade. A novel finding is the tight binding of AP to chicken pepsin and to mammalian pepsins. Obviously, the C-terminal part of AP is necessary for the formation of tight complexes with vertebrate pepsins in general. Unlike with mammalian pepsins of the A type the association in the case of chicken pepsin and possibly also in the case of cathepsin D occurs at higher pH-values.

References

1. Herriott, R.M.: J. Gen. Physiol. 22, 65-78 (1939).
2. Bohak, Z.: Eur. J. Biochem. 32, 547-554 (1973).
3. Dunn, B.M., Deyrup, C., Moesching, W.G., Gilbert W.A., Nolan, R.J., Trach, M.L.: J. Biol. Chem. 253, 7269-7275 (1978).
4. Foltmann, B., Jensen, A.L.: Eur. J. Biochem. 128, 63-70 (1982).
5. Kageyama, T., Takahashi, K.: Biochem. Biophys. Res. Commun. 107, 1117-1122 (1982).
6. Kostka, V., Pichová, I., Baudyš, M.: this volume,pp 53-72
7. Baudyš, M., Kostka, V.: Eur. J. Biochem. 136, 89-99 (1983).
8. Ryle, A.P.: Methods Enzymol. 19, 316-336 (1970).
9. Pohl, J., Štrop, P., Pichová, I., Bláha, I., Kostka, V.: this volume, 245-264.
10. Dixon, M.: Biochem. J. 55, 170 (1953).
11. Cha, S., Agarwal, R.P., Parks, R.E.: Biochem. Pharmacol. 24, 2187-2197 (1975).
12. Dunn, B.M., Pham, C., Raney, L., Abayasekara, D., Gillespie, W., Hsu, A.: Biochemistry 20, 7206-7211 (1981).
13. Pohl, J., Zaoral, M., Jindra, A., Kostka, V.: Anal. Biochem. 139, 265-271 (1984).

RENIN AND GENERAL ASPARTYL PROTEASES: DIFFERENCES AND SIMILARITIES IN STRUCTURE AND FUNCTION

Tadashi Inagami, Kunio Misono, J.-J. Chang, Yukio Takii, Colin Dykes,
Department of Biochemistry, Vanderbilt University School of Medicine
Nashville, TN 37232, U.S.A.

Introduction

Little is known about putative specific proteases implicated in prohormone-to-active peptide conversion for many hormones. Renin catalyzes the formation of the decapeptide angiotensin I from angiotensinogen but shows no general protease activity. It performs this function in endocrine and vascular cells as well as in blood.

Studies on renin have been impeded by its exceedingly small quantity in tissues, lack of a rapid in vitro assay method, lack of pure preparations and confusion due to its instability and renin-like activity of cathepsin D. Successful application of a radioimmunoassay method to renin activity determination (1), and isolation of pure renin in stable forms opened the way for rigorous biochemical investigation of this enzyme which plays a key role in blood pressure regulation.

Purification

Pepstatin was found to strongly inhibit acid proteases but renin was inhibited with a moderate ID_{50} value (2). Exploiting this moderate affinity to renin, pepstatin-aminohexyl-Sepharose (3-8) was successfully applied to the first purification of renal renin from hog kidney (5,7). This method consisted of the affinity chromatography and four additional steps of conventional fractionation procedures by column and batch processes. The

Aspartic Proteinases and their Inhibitors

principle of the affinity techniques was applied to the purification of human (9-12), rat (13,14) and dog renal renin (15) and hog pituitary renin.

Table 1. Purification of Renin from Human Kidney[a]

Purifications step	Total protein mg	Specific Activity Gu/mg protein	Purification	Yield %
Haas, step 7	1930	0.3	150	100
Hemoglobin-Sepharose	140	3.6	1,800	88
Pepstatin-Sepharose	4.2	76	38,000	55
Sephadex G-100	0.50	280	140,000	24
Octapeptide-Sepharose	0.11	550	270,000	10
DEAE-cellulose	0.066	830	420,000	9.7

[a] Yokosawa _et al_., (9), courtesy of Academic Press.

Isolation of pure renin from human kidney, first accomplished in our laboratory, is illustrated in Table 1 (9). For tissues containing renin at high concentrations conventional chromatographic methods were used. Mouse submandibular gland (16,17) and human renin secreting tumors (18) served as the sources of such purification studies. Recent application of a new synthetic renin inhibiting peptide as affinity ligand appears to be an interesting alternative (19). An essential factor contributing to the successful purification was the protection of renin from proteolysis by the addition of a mixture of protease inhibitors to eliminate traces of proteases which destroy partially purified renin at a low concentration (5,7).

The purification laid the groundwork for clarification of much confusion which had persisted in basic and clinical research on renin. The characterization of the active site of renin ensued. The large scale purification method permitted determination of its amino acid sequence. Antibodies raised with pure renin greatly aided the identification, localization and purification of renin and its distinction from non-specific renin like activity.

Properties of Renin

Renin purified by the methods discussed above satisfied multiple criteria of purity that include polyacrylamide electrophoresis, isoelectric focusing, column chromatography, sedimentation analysis and immunodiffusion. Renin in a given species or organ consists of multiple isoenzymes with different pI's (over a range of 4.8 to 6.4 (7,8,14,16,17,20,21). The heterogeneity may be partly due to structural variation in the carbohydrate moiety. Difference in the extent of protein processing may also account for the heterogeneity. The molecular weight of renin determined by sedimentation equilibrium was approximately 36,000 (7,16,17) in agreement with values calculated from amino acid sequences (22-24). By gel filtration apparently higher values, approximately 40,000, have been observed (7,8,10-19). This may be due to asymmetry of renin molecule or carbohydrate in it.

Various renal renin preparations are retained by columns containing concanavalin A linked to Sepharose, providing evidence that the enzyme is a glycoprotein (7,10,13,15,18,21,25). This conclusion has been firmly proven for the pure hog preparation by direct demonstration of glucosamine by chromatography (7). In contrast to renal renins, those isolated from the mouse submandibular gland (17) and bovine pituitary gland (26) are apparently not glycoproteins since they do not bind to Sepharose gels coupled by various types of plant lectins.

The optimal pH at which purified renin catalyzes the cleavage of a heterologous angiotensinogen or synthetic peptide substrate can vary considerably depending on the species of animals from which the enzyme is isolated. A homologous substrate usually exhibits an optimal pH in the neutral or slightly acidic range. This situation seems to differ somewhat from that of renin on its endogenous substrate in intact plasma. It may be that plasma contains a component which interacts with renin or angiotensinogen and thereby modifies the characteristics of the reaction (8,13). Properties of renin from various sources are summarized in Table 2.

Table 2. Properteries of Pure Renin

	Hog[a] Kidney	Rat[b] Kidney	Dog[c] Kidney	Human[d] Kidney	Mouse[e] Submandib Gland	Hog[f] Pituitary
MW						
Sed. Equil	36,400			40,000	36,000	
Gel Filtration	42,000	36,000-40,000	42,000	42,000	38,000	36,000
SDS Gel		36,000-38,000	36,000	50,000 40,000		36,000
pI	5.2	5.0-5.2		5.7	5.4-5.9	5.25
Glycoprotein	Yes	Yes	Yes	Yes	No	No
pH optimum						
hog angiotensinogen	6.0-7.2	5.5-6.5				6.5-7.5
rat angiotensinogen		6.0-7.5				6.5-7.5
sheep angiotensinogen					8.3	
Octapeptide	6.5	3.5			5.4	
Tetradecapeptide					6.5	

[a] Refs. 2 & 7. [b] Refs. 13 & 14. [c] Ref. 15. [d] Refs. 9 & 10. [e] Refs 16 & 17. [f] Ref. 26.

Structure

Although the very small quantity of renin in the kidney and other tissues seemed to be an insurmountable barrier to the determination of their structure, the development of a large scale purification method of mouse submandibular gland renin (17) permitted the determination of its complete amino acid sequence by direct Edman degradation by Misono et al. in our laboratory (22,27). Mouse submandibular gland renin was found to consist of a light chain of 5,000 dalton and a heavy chain of 30,000 dalton linked by a disulfide bridge (27). The complete amino acid sequences (Fig. 1) of these peptide chains revealed an extensive sequence homology with porcine pepsin determined by Tang and his associates (28). The light chain with 48 amino acid residues were found to have a 46% sequence identity with the carboxyl terminal region of porcine pepsin and 46% of the amino acid sequence of 288 residue heavy chain were identical with the amino terminal side of porcine pepsin.

Rougeon et al. (29) cloned cDNA of the mouse submandibular gland renin, and Panthier et al. (30) deduced amino acid sequence of the preprorenin from the nucleotide sequence of the cDNA. Although this original work contained mistakes in nucleotide sequence determination and deduction of amino acid sequence, the structure obtained after correction (23) showed complete agreement with the result of the direct determination (22). The nucleotide sequence of renin cDNA yielded additional information which was not obtained by the direct method. An -Arg-Arg- sequence was found to exist between the heavy and light chains in the single chain precursor of the two-chain renin. The preprorenin was found to contain additional 63 amino acid residues extending from the amino terminal of active renin and the prepro sequence is connected to active renin by the paired basic amino acid residue sequence -Lys-Arg- (23,30) (Figure 2). It is interesting to note that the pro-sequence or the inhibitor sequence of the renin zymogen shows little similarity with the corresponding sequence of pepsin (31), suggesting the activation mechanism may be different. This is one of the important features differentiating renin from acid proteases.

Using human renin cDNA Imai et al. (24) deduced the amino acid sequence of human preprorenin. Greater than 70% sequence identity was found between the amino acid sequences of mouse and human preprorenin. Although the structure of active human renin was not determined directly, the amino acid sequence of the active renin was deduced by analogy with the mouse enzyme.

	1 · 10 · 20
MOUSE RENIN	SER SER LEU THR ASP LEU ILE SER PRO VAL VAL LEU THR ASN TYR LEU ASN SER GLN TYR TYR GLY GLU ILE GLY ILE
HUMAN KIDNEY RENIN	GLN PRO MET LYS ARG LEU THR LEU GLY ASN THR THR SER SER VAL ILE LEU THR ASN TYR MET ASP THR GLN TYR TYR GLY GLU ILE GLY ILE
PIG PEPSIN	ILE GLY ASP GLU PRO LEU GLU ASN TYR LEU ASP THR GLY TYR PHE GLY THR ILE GLY ILE

	21 · 30 · 40
MOUSE RENIN	GLY THR PRO PRO GLN THR PHE LYS VAL ILE PHE ASP THR GLY SER ALA ASN LEU TRP VAL
HUMAN KIDNEY RENIN	GLY THR PRO PRO GLN THR PHE LYS VAL VAL PHE ASP THR GLY SER SER ASN VAL TRP VAL
PIG PEPSIN	GLY THR PRO ALA GLN ASP PHE THR VAL ILE PHE ASP THR GLY SER SER ASN LEU TRP VAL

	41 · 46 46A 46B · 50 · 60
MOUSE RENIN	PRO SER THR LYS CYS SER ARG LEU TYR LEU ALA CYS GLY ILE HIS SER LEU TYR GLU SER SER ASP
HUMAN KIDNEY RENIN	PRO SER SER LYS CYS SER ARG LEU TYR THR ALA CYS VAL TYR HIS LYS LEU PHE ASP ALA SER ASP
PIG PEPSIN	PRO SER VAL TYR CYS SER --- --- SER LEU ALA CYS SER ASP HIS ASN GLN PHE ASN PRO ASP ASP

	61 · 70 · 80
MOUSE RENIN	SER SER SER TYR MET GLU ASN GLY ASP ASP PHE THR ILE HIS TYR GLY SER GLY ARG VAL
HUMAN KIDNEY RENIN	SER SER SER TYR LYS HIS ASN GLY THR GLU LEU THR LEU ARG TYR SER THR GLY THR VAL
PIG PEPSIN	SER SER THR PHE GLU ALA THR SER GLN GLU LEU SER ILE THR TYR GLY THR GLY SER MET

	81 · 90 · 100
MOUSE RENIN	LYS GLY PHE LEU SER GLN ASP SER VAL THR VAL GLY GLY ILE THR VAL THR --- GLN THR
HUMAN KIDNEY RENIN	SER GLY PHE LEU SER GLN ASP ILE ILE THR VAL GLY GLY ILE THR VAL THR --- GLN MET
PIG PEPSIN	THR GLY ILE LEU GLY THR ASP THR VAL GLN VAL GLY GLY ILE SER ASP THR ASN GLN ILE

	101 102 102A · 110 · 120
MOUSE RENIN	PHE GLY GLY VAL THR GLU LEU --- PRO LEU ILE PRO PHE MET LEU ALA GLN PHE ASP GLY VAL
HUMAN KIDNEY RENIN	PHE GLY GLY VAL THR GLU MET --- PRO ALA LEU PRO PHE MET LEU ALA GLU PHE ASP GLY VAL
PIG PEPSIN	PHE GLY --- LEU SER GLU THR GLU PRO GLY SER PHE LEU TYR TYR ALA PRO PHE ASP GLY ILE

	121 · 130 · 140
MOUSE RENIN	LEU GLY MET GLY PHE PRO ALA GLN ALA VAL GLY GLY VAL THR PRO VAL PHE ASP HIS ILE
HUMAN KIDNEY RENIN	VAL GLY MET GLY PHE ILE GLU GLN ALA ILE GLY ARG VAL THR PRO ILE PHE ASP ASN ILE
PIG PEPSIN	LEU GLY LEU ALA TYR PRO SER ILE SER ALA SER GLY ALA THR PRO VAL PHE ASP ASN LEU

	141 · 150 · 160 160A 160B
MOUSE RENIN	LEU SER GLN GLY VAL LEU LYS GLU LYS VAL PHE SER VAL TYR TYR ASN ARG GLY --- --- --- PRO HIS LEU LEU
HUMAN KIDNEY RENIN	ILE SER GLN GLY VAL LEU LYS GLU ASP VAL PHE SER PHE TYR TYR ASN ARG ASP SER GLU ASN SER GLN SER LEU
PIG PEPSIN	TRP ASP GLN GLY LEU VAL SER GLN ASP LEU PHE SER VAL TYR LEU SER SER ASN --- --- --- ASP ASP --- ---

	161 · 170 · 180
MOUSE RENIN	GLY GLY GLU VAL VAL --- LEU GLY GLY SER ASP PRO GLU HIS TYR GLN GLY ASP PHE HIS
HUMAN KIDNEY RENIN	GLY GLY GLN ILE VAL --- LEU GLY GLY SER ASP PRO GLN HIS TYR GLU GLY ASN PHE HIS
PIG PEPSIN	SER GLY SER VAL VAL LEU LEU GLY GLY ILE ASP SER SER TYR TYR THR GLY SER LEU ASN

	181 · 190 · 200
MOUSE RENIN	TYR VAL SER LEU SER LYS THR ASP SER TRP GLN ILE THR MET LYS GLY VAL SER VAL GLY
HUMAN KIDNEY RENIN	TYR ILE ASN LEU ILE LYS THR GLY VAL TRP GLN ILE GLN MET LYS GLY VAL SER VAL GLY
PIG PEPSIN	TRP VAL PRO VAL SER VAL GLU GLY TYR TRP GLN ILE THR LEU ASP SER ILE THR MET ASP

	201 · 210 · 220
MOUSE RENIN	SER SER THR LEU LEU CYS GLU GLU GLY CYS GLU VAL VAL VAL ASP THR GLY SER SER PHE
HUMAN KIDNEY RENIN	SER SER THR LEU LEU CYS GLU ASP GLY CYS LEU ALA LEU VAL ASP THR GLY ALA SER TYR
PIG PEPSIN	GLY GLU THR ILE ALA CYS SER GLY GLY CYS GLN ALA ILE VAL ASP THR GLY THR SER LEU

	221 · 230 · 240
MOUSE RENIN	ILE SER ALA PRO THR SER SER LEU --- LYS LEU --- ILE MET GLN ALA LEU GLY ALA LYS GLU
HUMAN KIDNEY RENIN	ILE SER GLY SER THR SER SER ILE GLU LYS LEU --- --- MET GLU ALA LEU GLY ALA LYS ---
PIG PEPSIN	LEU THR GLY PRO THR SER ALA ILE --- ALA ILE ASN ILE GLN SER ASP ILE GLY ALA SER GLU

	241 · 250 · 260
MOUSE RENIN	LYS ARG LEU HIS GLY TYR VAL VAL SER CYS SER GLN VAL PRO THR LEU PRO ASP ILE SER
HUMAN KIDNEY RENIN	LYS ARG LEU PHE ASP TYR VAL VAL LYS CYS ASN GLU GLY PRO THR LEU PRO ASP ILE SER
PIG PEPSIN	ASN SER ASP GLY/ASP GLU MET VAL ILE SER CYS SER SER ILE ASP SER/GLN LEU PRO ASP ILE VAL

	261 · 270 · 278 278A 278B 279 (H-chain)
MOUSE RENIN	PHE ASN LEU GLY GLY ARG ALA TYR THR LEU SER SER THR ASP TYR VAL LEU GLN TYR PRO ASN
HUMAN KIDNEY RENIN	PHE HIS LEU GLY GLY LYS GLU TYR THR LEU THR SER ALA ASP TYR VAL PHE GLN GLU SER TYR SER SER LYS LYS
PIG PEPSIN	PHE THR ILE ASN GLY VAL GLN TYR PRO LEU SER PRO SER ALA TYR ILE LEU GLN --- --- ASP --- --- --- ---

	(L-chain) 280 281 282 · 290 · 300
MOUSE RENIN	ASP LYS LEU CYS THR VAL ALA LEU HIS ALA MET ASP ILE PRO PRO PRO THR GLY PPO VAL TRP
HUMAN KIDNEY RENIN	--- --- LEU CYS THR LEU ALA ILE HIS ALA MET ASP ILE PRO PRO PRO THR GLY PRO THR TRP
PIG PEPSIN	ASP ASP SER CYS THR SER GLY PHE GLU GLY MET ASP VAL PRO THR SER SER GLY GLU LEU TRP

	301 · 310 · 320
MOUSE RENIN	VAL LEU GLY ALA THR PHE ILE ARG LYS PHE TYR THR GLU PHE ASP ARG HIS ASN ASN ARG
HUMAN KIDNEY RENIN	ALA LEU GLY ALA THR PHE ILE ARG LYS PHE TYR GLU GLU PHE ASP ARG ARG ASN ASN ARG
PIG PEPSIN	ILE LEU GLY ASP VAL PHE ILE ARG GLN TYR TYR THR VAL PHE ASP ARG ALA ASN ASN LYS

	321
MOUSE RENIN	ILE GLY PHE ALA LEU ALA ARG
HUMAN KIDNEY RENIN	ILE GLY PHE ALA LEU ALA ARG
PIG PEPSIN	VAL GLY LEU ALA PRO VAL ALA

Fig. 1. Amino acid sequence of mouse submandibular gland renin, human renal renin and porcine pepsin - The mouse renin amino acid sequence was determined by Misono et al. (22), and Misono and Inagami (27). Human renin structure was deduced from the nucleotide sequence of human renal renin cDNA by Imai et al. (24). The

porcine pepsin sequence was determined by Sepulveda *et al.* (28). The residue numbering is that of porcine pepsin. Amino acid residues of mouse and human renin are arranged in such a way to maximize homology with that of porcine pepsin.

	-65 -60
MOUSE PRE·PRO·RENIN	Met-Asp——Arg-Arg-Arg-Met-Pro-
HUMAN PRE·PRO·RENIN	Met-Asp-Gly-Trp-Arg-Arg-Met-Pro-
PIG PEPSINOGEN	
	-50
MOUSE PRE·PRO·RENIN	Leu-Trp-Ala —— Leu-Leu-Leu-Leu-Trp-Ser-Pro-Cys-Thr-Phe-Ser-Leu-
HUMAN PRE·PRO·RENIN	Arg-Trp-Gly-Leu-Leu-Leu-Leu-Leu-Trp-Glu-Ser-Cys-Thr-Phe-Gly-Leu-
PIG PEPSINOGEN	Leu-Val-Lys-
	-40 -30
MOUSE PRE·PRO·RENIN	Pro-Thr —— Gly-Thr-Thr-Phe-Glu-Arg-Ile-Pro-Leu-Lys-Lys-
HUMAN PRE·PRO·RENIN	Pro-Thr-Asp-Thr-Thr-Thr-Phe-Lys-Arg-Ile —— Phe-Lys-Arg-
PIG PEPSINOGEN	Val-Pro-Leu-Val-Arg-Lys-Lys-Ser-Leu-Arg-Gln-Asn-Leu-Ile-
	-25 -20 -15
MOUSE PRE·PRO·RENIN	Met-Pro-Ser-Val-Arg-Glu-Ile-Leu-Glu-Glu-Arg-Gly-Val-Asp-
HUMAN PRE·PRO·RENIN	Met-Pro-Ser-Ile-Arg-Glu-Ser-Leu-Lys-Glu-Arg-Gly-Val-Asp-
PIG PEPSINOGEN	Lys-Asp-Gly-Lys-Leu-Lys-Asp-Phe-Leu-Lys-Thr-His-Lys-His-
	-10 -5 -1
MOUSE PRE·PRO·RENIN	Met-Thr-Arg-Leu-Ser-Ala-Glu-Trp-Asp-Val-Phe-Thr-Lys-Arg-ACTIVE MOUSE RENIN
HUMAN PRE·PRO·RENIN	Met-Ala-Arg-Leu-Gly-Pro-Glu-Trp-Ser-Gln-Pro-Met-Lys-Arg-ACTIVE HUMAN RENIN
PIG PEPSINOGEN	Asn-Pro-Ala-Ser-Lys-Tyr-Phe-Pro-Glu-Glu-Ala-Ala-Ala-Leu-ACTIVE PIG PEPSIN

Fig. 2. Amino acid sequences of the prepro-sequences of mouse submandibular gland renin and human renin as deduced from nucleotide sequence of their respective cDNA's (23, 24, 30) are compared with the pro-sequence of porcine pepsinogen (85-87).

The Structure of Active Site

The highly specific action of renin in cleaving the single unique leucyl peptide bond in angiotensinogen aroused interest in the catalytic mechanism and type of protease to which renin belongs. Using type specific inhibitors of the acid protease developed in studies of pepsin, the nature of the active site of renin was explored. Diazoacetyl-D,L-norleucine methyl ester (DAN) in the presence of cupric ion (32) was found to inactivate pure mouse submandibular gland renin (33) with concomitant stoichiometric incorporation of the reagent (34). Hog (33) and human renal renin (35) were also shown to be inhibited by DAN-Cu^{++}. Another group of reagents, aliphatic epoxides (36) were also found to inhibit mouse renin (34). DAN and the epoxide 1,2-epoxy 3-nitrophenoxypropane were incorporated into renin molecule independently, presumably by esterifying two different carboxyl groups. These two reagents had been shown to inhibit porcine pepsin (32,36) in similar manners, DAN-Cu^{++} esterifying the β-carboxyl group of Asp-215 (37) and epoxide esterifying Asp-32 in an independent manner (38). Furthermore, chemical modification of tyrosine residues in renin with three different tyrosine-specific reagents, tetranitromethane, acetylimidazole and diazonium-1-H-tetrazole led to inactivation of renin (34). Pepsin had been shown to be sensitive to inhibition by the chemical modification of tyrosyl residues (39-41). These chemical modifications inhibitory to renin enzyme activity were suppressed by pepstatin, which presumably protected the active site from attack by these compounds (34). These observations common to aspartyl proteases and renin indicated that renin and aspartyl proteases share common features in the structure and function of their active site.

This conclusion received further support as the amino acid sequences of renins were elucidated. Extensive sequence identity was found in the vicinity of the two catalytically essential aspartyl residues Asp-32 and Asp-215 (see for review by Inagami *et al.*, (42)). The identity extends from acid proteases of fungal origin to gastric digestive enzymes.

On the other hand, important differences exist between the catalytic features of renin and aspartyl proteases. These are the exceedingly stringent substrate specificity and neutral pH optimum of renin while aspartyl proteases show very broad substrate specificity and acidic pH optima.

The close mechanistic relationship of renin and acid proteases is also reflected by the low level of renin-like activity of pepsin (43) and cathepsin D (14,44). Indeed, cathepsin D was shown to be identical to brain isorenin (45) and to pseudorenin (46). The specific activities of these enzymes for the generation of angiotensin I from angiotensinogen at neutral pH are orders of magnitude lower than that of renin. For example, at pH 7.0 rat kidney cathepsin D generates angiotensin I from rat angiotensinogen at a rate 1/1000 of that of rat renin (14). In contrast, the rate of formation of the decapeptide from the tetradecapeptide substrate catalyzed by these enzymes is comparable to that resulting from the action of renin (44). The optimal pH for the cathepsin D-dependent release of angiotensin I from either angiotensinogen or the tetradecapeptide is below 5.5, whereas the activity of purified renin with these substrates is maximal at pH 6-7 (14,45).

Substrate Specificity

Although renin is closely related to acid proteases, its uniqueness is in the highly specific regulatory role and its stringent substrate specificity. This specificity is illustrated by the observation that with radioactive hemoglobin as substrate pure renin from mouse submaxillary gland exhibited no proteolytic activity at a concentration 10,000 times higher than that required to demonstrate such activity for cathepsin D (17).

Systematic studies on the substrate specificity of renin were carried out by Skeggs et al. (47) with synthetic peptides whose structures were based on that of renin tetradecapeptide. They found that the octapeptide, His-Pro-Phe-His-Leu-Leu-Val-Tyr, was the minimum sequence which could function effectively as substrate. These results indicate that the substrate binding site of the enzyme is likely to include 7 to 8 subsites extending toward amino as well as carboxyl sides of the scissible bonds (Leu-Leu) of the substrate.

The action of renin on its substrate is dependent on the species from which these macromolecules are derived. Human renin is capable of cleaving human, porcine, canine, bovine and goat angiotensinogens, whereas animal renins, while generally active with heterologous animal substrates, are very inefficient in catalyzing hydrolysis of human angiotensinogen. Human angiotensinogen contains Val-Ile-His-Thr-Glu in positions 11-15 (48). This contrasts with the structure Leu-Val-Tyr-Ser previously identified in corresponding positions in horse angiotensinogen (49), and Leu-Tyr-Tyr-Ser in rat angiotensinogen (50). These differences may account for the inefficiency of animal renins to cleave human substrate.

Renin Zymogen - Prorenin

From the analogy with various proteases, an inactive form of renin exists in the forms of zymogen and/or an inactive complex with an inhibitor. It is also possible that regulatable, partially active renin exists. Acidification of kidney extract (51) was found to cause increase in renin activity. More reproducible evidence for the activation of renin was obtained upon successive dialyses of human amniotic fluid in acid pH then in neutral pH (52).

Renin activity in human plasma (53) and rabbit (54) and hog (55) kidney extract was also increased by the dialysis in acidic buffers. The treatment of amniotic fluid or plasma with various proteases (56,57) as well as prolonged cold storage (58,59) also caused augmentation of renin activity. The early phase of studies on prorenin has been reviewed in several articles (8,60-63).

Central issues related to activation of renin were whether the increase in renin activity in these fluids or extract was due to full expression of enzyme activity of partially active renin, activation of zymogen, or dissociation of a renin-inhibitor complex. Another important question concerned the significance of the activation in the pathophysiology of hypertension and implication in the assay of renin activity. The source of inactive renin was not clear.

Observation of an almost undetectable plasma renin level which can be brought to an assayable activity level by the acid dialysis in one case of patients with Wilms' tumor (57) suggested the presence of a completely inactive form. Such was explicitly demonstrated by separation of a completely inactive renin from the active enzyme in our laboratory. The former was activatable upon trypsin treatment while the latter showed no potential for further activation (65,66).

Evidences were obtained that cold activation, which occurred at a low temperature slightly above freezing, and acid activation were both due to inactivation of protease inhibitors which in turn permitted expression of endogenous proteases (67,68). Acidification of human plasma was shown to cause transient and reversible activation (69,70) suggesting conformational change of a zymogen-like molecule as a mechanism of acute and transient acid activation. This reversible activation did not seem to be caused by proteases since molecular weight was not reduced. The absence of reduction in molecular weight also eliminates the possible mechanism of activation involving dissociation of a renin-inhibitor complex (70).

Trypsin activatable inactive renin was separated from active renin from human plasma as well as human kidney extract. Immunoaffinity chromatography using anti-human renin IgG-Sepharose (20,71) pepstatin-Sepharose, Cibacron Blue-agarose (20,72,72) and octyl-Sepharose (71) permitted the separation of inactive renin from human kidneys and demonstrated the kidney as a source of inactive renin. We have shown that inactive renin isolated from hog kidney in a partially purified form was not activated upon treatment with dissociative reagents such as urea, detergents and high concentrations of salts (73). This was contrary to the earlier report of activatable renin in the kidney being an inhibitor-renin complex

(54,55,74). This was additional evidence that kidney contains an inactive form of renin which is a zymogen. This pursuit of the nature of prorenin culminated in the complete purification of inactive renin from hog kidney which consisted of multiple steps of affinity chromatography by Takii and Inagami (75). A more than 3 million-fold purification was obtained by a combination of affinity chromatographic steps which included octyl-Sepharose, Affigel-Blue (Cibacron Blue-agarose), concanavalin A-Sepharose for renin-zymogen and pepstatin-Sepharose used for the removal of active renin (Table 3). Takahashi et al. (76) also reported the purification of prorenin from hog kidney by a combination of similar affinity techniques by a 700,000-fold purification.

Table 3. Purification of Renin Zymogen from Hog Kidney[a]

Step	Total Protein (mg)	Specific[c] Activity	Purification	Yield (%)
Crude Extract[b]	1,600,000	0.0015[d]	1	100
DEAE Batch	125,000	0.016	10	82
Pepstatin-Sepharose	60,000	0.03	20	75
DEAE-Sephacel	18,800	0.08	53	63
Octyl-Sepharose	1,000	1.43	950	59
Sephadex G-100	208	5.29	3,530	46
Affi-Gel Blue	16.0	62.9	41,900	42
Con A-Sepharose	0.750	1,000	666,700	31
Sephadex G-100	0.170	2,400	1,600,000	17
DEAE-Sephacel	0.030	5,000	3,333,000	6.3

[a] Taken from Ref. 75, courtesy of Academic Press; [b] Prepared from 20 kg of fresh hog kidney; [c] Activity of inactive renin as defined by, inactive renin = total renin activity after activation - renin activity before activation; expressed as μg angiotensin I/mg of protein/hr; [d] Determined after removal of contaminating renin by using affinity chromatography

The presence of the proenzyme was also demonstrated without complete purification. Takii et al. (77) have partially purified inactive renin-zymogen from rat kidney, electrophoresed in a SDS gel and transferred it to a nitrocellulose membrane by an electrophoretic blotting and detected it by immunostaining by unconjugated peroxidase method. The presence of trypsin activatable renin in the partially purified fraction was confirmed. The molecular weight of the renin-zymogen was 48,000.

In order to identify renin precursors, products of in vitro translation of renin mRNA were identified. Translation products of mouse submandibular gland renin mRNA were shown to have a molecular weight of 45,000-50,000 by Poulsen et al. (78), Dykes et al. (79), Pratt et al. (80) and Morris et al. (81). Similarly rat kidney mRNA was translated in vitro and products with approximate molecular weight of 50,000 were identified (78,81). Since renin in the kidney and submandibular gland is secreted, these in vitro translation products must contain a signal sequence and, consequently is preprorenin. Their molecular weight of 48,000-50,000 seems to be somewhat greater than that deduced from the nucleotide sequence of cDNA, presumably due to a systematic error in the SDS polyacrylamide gel electrophoresis.

The preprorenin should be processed to active renin in the tissues of its synthesis. In mouse submandibular gland, a translate with a molecular weight of 44,500 was produced and was converted to a 40,000 dalton protein in a time-dependent manner (80). Using the same tissue, prorenin which can be detected in mouse submandibular gland was shown to have a molecular weight of 43,000 which was converted to active renin of 38,000 dalton by Catanzaro et al. (83). Pratt et al. (84) recorded a similar observation with the mouse submandibular gland. The first observable product of 44,000 dalton, presumably a prorenin, was rapidly converted to 38,000 dalton protein, presumably active renin. This protein underwent slow conversion to a two-chain form. This mechanism is compatible with the result of renin structure studies in which the molecular weight of preprorenin was deduced to be 45,000 and prorenin 43,500 (30). The two-chain active renin consisting of 31,000 and 5,500 dalton fragments (27,30) may have been derived from one-chain renin of 37,000 dalton by limited proteolysis.

The molecular weights of prorenin (renin zymogen) found in the kidneys of human (20,64,71,72), hog (73,75,76), and rat (77) are greater than 44,000. This seems to be due to the carbohydrates in the kidney renin molecule while the submandibular gland does not have carbohydrate.

Biosynthetic studies by gel-electrophoresis under denaturing conditions or deduction of an amino acid sequence from a nucleotide sequence of cDNA does not provide information on enzyme activity, whether dormant or active. However, the fact that a portion of the prosequence is an inhibitor of a moderate strength (23) provides evidence that the protein with the prosequence may be an inactive zymogen.

The mechanism and agent involved in the activation of renin-zymogen is not yet known. Most likely the activation takes place upon cleavage of the pro-sequence at the arginyl peptide bond of -Lys-Arg- adjacent to the amino terminal of the active enzyme. The protease involved in this activation is likely to be a neutral protease of a serine or cysteine protease type in contrast to the well-known auto-activation of the zymogens of aspartyl proteases.

Concluding Remarks

Studies on renin have made considerable progress in the last 15 years. During this period renins have been purfied from various sources, their properties have been characterized, features of its active site have been determined, their structures have been clarified and the inactive renins in the kidney and plasma have been identified as renin-zymogen and their structures have also been deduced.

Through these studies it has become clear that renin has many points of similarity with aspartyl protease and belongs to the family of aspartyl proteases with regard to the overall structure and the catalytically essential functional groups. However, many significant properties distinguish renin from general aspartyl proteases. These are: i) highly stringent substrate specificity limited exclusively to angiotensinogen and to its single unique leucyl peptide bond, ii) neutral rather than acidic pH optimum, iii) unique amino terminal sequence of renin zymogen, iv) a putative activation mechanism by a neutral trypsin-like enzyme rather than aspartyl protease, and v) a unique physiological role in the production of an important hormonal peptide and exquisite regulation of its concentration. This last point deserves further elaboration since the regulation of renin seems to depend on cellular process. This enzyme has been known to function in blood plasma, and its release from the kidney has been known to be tightly controlled by multiple regulatory mechanisms. Furthermore, it has become clear that renin also exists in various non-renal tissues. In organs like the brain, adrenal, testis and pituitary gland, renin functions within these neuroendocrine cells and produces angiotensins by an intracellular mechanism (63). In these tissues it is angiotensins rather than renin that is secreted. Locally formed angiotensin II seems to play a paracrine or autocrine function while renin remains within the endocrine cells. The intracellular mechanism of renin action is that of one of the specific peptide hormone producing enzymes of endocrine cells. Thus, the formation of certain peptide hormones seems to be mediated by an aspartyl protease rather than by a serine or cysteine protease.

Acknowledgment

We are indebted to Vicki Garrett and Susan Heaver for assistance in preparing the manuscript. This work was supported by Research Grants HL-14192, HL-22288 and Hl-24112 from National Institutes of Health.

References

1. Haber, E., Koerner, T., Page, L.B., Kliman, B, Purnode, A.: J. Clin. Endocrinol. 29, 1349-1355 (1969).
2. Aoyagi, T., Morishima, H., Nishizawa, R., Kunimoto, S., Takeuchi, T., Umezawa, H.: J. Antibiotics. 25, 689-94 (1972).

3. Murakami, K., Inagami, T., Michelakis, A.M., Cohen, S.: Biochem. Biophys. Res. Commun. 54, 482-487 (1973).

4. Corval, P., Devaux, C., Menard, J.: FEBS Lett. 34, 189-192 (1973).

5. Murakami, K., Inagami, T.: Biochem. Biophys. Res. Communs. 62, 757-763 (1975).

6. Devaux, C., Menard, J., Sicard, P., Corval, P.: Eur. J. Biochem. 64, 621-627 (1976).

7. Inagami, T., Murakami, K.: J. Biol. Chem. 252, 2978-2983 (1977).

8. Inagami, T.: In "Biochemical Regulation of Blood Pressure" (R.L. Soffer, ed) pp. 39-71 (1981).

9. Yokosawa, H., Inagami, T., Haas, E.: Biochem. Biophys. Res. Commun. 83, 3067-312 (1978).

10. Yokosawa, H., Holladay, L.A., Inagami, T., Haas, E., Murakami, K.: J. Biol. Chem. 255, 3498-3502 (1980).

11. Slater, E.E., Strout, H.V.: J. Biol. Chem. 256, 8164-8171 (1981).

12. Higaki, J., Hirose, S., Ogihara, T., Imai, N., Kisaragi, M., Murakami, K., Kumahara, Y.: Life Sci. 32, 1591-1598 (1983).

13. Matoba, T., Murakami, K., Inagami, T.: Biochim. Biophys. Acta 526, 560-571 (1978).

14. Figueiredo, A.F.S., Takii, Y., Tsuji, H., Kato, K., Inagami, T.: Biochemistry 22, 5476-5481 (1983).

15. Dzau, V.J., Slater, E.E., Haber, E.: Biochemistry 18, 5224-5228 (1979).

16. Cohen, S., Taylor, J.M., Murakami, K., Michelakis, A.M., Inagami, T.: Biochemistry 11, 4286-4293 (1972).

17. Misono, K.S., Holladay, L.A., Murakami, K. Kuromizu, K., Inagami, T.: Arch. Biochem. Biophys. 217, 574-581 (1982).

18. Galen, F.X., Devaux, C., Guyenne, T., Menard, J., Corvol, P.: J. Biol. Chem. 254, 4848-4855 (1979).

19. McIntire, G.D., Leckie, B., Hallett, A., Szelke, M.: Biochem. J. 211, 519-522 (1983).

20. Chang, J.J., Kisaragi, M., Okamoto, H., Inagami, T.: Hypertension 3, 509-515 (1981).

21. Lauritzen, M., Damsgaard, J.J., Rubin, I., Lauritzen, E.: Biochem. J. 155, 317-323 (1976).

22. Misono, K.S., Chang, J.J., Inagami, T.: Proc. Natl. Acad. Sci. USA. 79, 4858-4862 (1982).

23. Corvol, P., Panthier, J.J., Foote, S., Rougeon, F.: Hypertension 5, I-3-I-9 (1983).

24. Imai, T., Miyazaki, H., Hirose, S., Hori, H., Hayashi, T., Kageyama, R., Ohkubo, H., Nakanishi, S., Murakami, K.: Proc. Natl. Acad. Sci. USA. 80, 7405-7409 (1983).

25. Printz, M.P., Dwarshak, R.T.: Biochim. Biophys. Acta 494, 162-171 (1977).

26. Hirose, S., Ohsawa, T., Inagami, T., Murakami, K.: J. Biol. Chem. 257, 6316-6321 (1982).

27. Misono, K.S., Inagami, T.: J. Biol. Chem. 257, 7536-7540 (1982).

28. Sepulveda, P., Marciniszin, J., Jr., Liu, D., Tang, J.: J. Biol. Chem. 250, 5082-5088 (1975)

29. Rougeon, F., Chambraud, B., Foote, S., Panthier, J-J., Nageotte, R., Corvol, P.: Proc. Natl. Acad. Sci. USA. 78, 6367-6371 (1981).

30. Panthier, J-J., Foote, S., Chambraud, B., Strosberg, A.D., Corvol, P., Rougeon, F.: Nature 298, 90-92 (1982).

31. Harboe, M., Andersen, P.M., Foltman, B.: J. Biol. Chem. 249, 4487-4494 (1974).

32. Rajagopalan, T.G., Stein, W.H., Moore, S.: J. Biol. Chem. 241, 4295-4297 (1966).

33. Inagami, T., Misono, K., Michelakis, A.M.: Biochem. Biophys. Res. Commun. 56, 503-509 (1974).

34. Misono, K.S., Inagami, T.: Biochemistry 19, 2616-2622 (1980).

35. McKown, M., Gregerman, R.I.: Life Sci. 16, 71-79 (1975).

36. Tang, J.: J. Biol. Chem. 246, 4510-4517 (1971).

37. Lundblatt, R.L., Stein, W.H.: J. Biol. Chem., 244, 154-160 (1969).

38. Cheng, K.C.S., Tang, J.: J. Biol. Chem. 247, 2566-2574 (1972).

39. Kozlov, K.V., Kogan, G.A., Zavada, L.L.: Biokhimiya 34, 1257-1265 (1968)

40. Lokshina, L.A., Orekovich, V.N.: Biokhimiya 31, 143-150 (1968)

41. Hollands, T.R., Fruton, J.S.: Biochemistry 7, 2045-2053 (1968)

42. Inagami, T., Misono, K.S., Inagaki, T., Takii, Y.: In "Proteinase Inhibitors" (N. Katunuma, H. Holzer and H. Umezawa, eds). pp. 251-259, 1983, Springer Verlag, Berlin, Tokyo.

43. Paiva, A.C.M., Grandino, A.: Experientia 29, 154-155 (1973).

44. Hackenthal, E., Hackenthal, R., Hilgenfeldt, U.: Biochim. Biophys. Acta 522, 574-588 (1978).

45. Hackenthal, E., Hackenthal, R., Hilgenfeldt, U.: Biochim. Biophys. Acta 522, 561-573 (1978).

46. Dorer, F.E., Lentz, K.E., Kahn, J.R., Levine, M., Skeggs, L.T.: J. Biol. Chem. 253, 3140-3142 (1978).

47. Skeggs, L.T., Lentz, K.E., Kahn, J.R., Hochstrasser, H.: J. Exp. Med. 128, 13-34 (1968).

48. Tewksbury, D.A., Dart, R.A., Travis, J.: Biochem. Biophys. Res. Commun. 99, 1311-1315 (1980).

49. Skeggs, L.T., Lentz, R.E., Kahn, J.R., Shumway, M.P.: J. Exp. Med. 106, 439-453 (1958).

50. Bouhnik, J., Clauser, E., Strosberg, D., Frenoy, J.-P., Menard, J., Corvol, P.: Biochemistry 20, 7010-7015 (1981).

51. Haas, E., Lumfrom, H., Goldblatt, H.: Arch. Biochem. Biophys. 42, 368-381 (1953).

52. Lumbers, E.R.: Enzymologia 40, 329-336 (1971).

53. Day, R.P., Luetscher, J.A.: J. Clin. Endocrinol. Metab. 38, 923-926 (1974).

54. Leckie, B.J.: Clin. Sci. 44, 301-304 (1973).

55. Boyd, G.W.: Circ. Res. 35, 426-438 (1974).

56. Morris, B.J., Lumbers, E.R.: Biochim. Biophys. Acta 285, 385-391 (1972).

57. Day, R.P., Luetscher, J.A.: J. Clin. Endocrinol. Metab. 40, 1085-1093 (1975).

58. Osmond, D.H., Ross, L.J., Scaiff, K.D.: Can. J. Physiol. 51, 705-708 (1973).

59. Sealey, J.E., Laragh, J.H.: Circ. Res. 36/37 (Suppl. I) 10-16 (1975).

60. Overturf, M.L., Druilhet, R.E., Kirkendall, W.M.: Life Sci. 24, 1913-1924 (1979).

61. Inagami, T., Murakami, K.: Biomed. Res. 1, 456-475 (1980).

62. Sealey, J.E., Atlas, S.A., Laragh, J.H.: Endocrine Rev. 1, 365-391 (1980).

63. Inagami, T.: Neuroendocrinology 375, 475-482 (1982).

64. Leckie, B.J.: Clin. Sci. 60, 119-130 (1981).

65. Yokosawa, N., Takahashi, N., Inagami, T., Page, D.L.: Biochim. Biophys. Acta 569, 211-219 (1979).

66. Takii, Y., Takahashi, N., Inagami, T., Yokosawa, N.: Life Sci. 26, 347-353 (1980).

67. Atlas, S.A., Laragh, J.H., Sealey, J.E.: Clin. Sci. Mol. Med. 55, 135s-137s (1978).

68. Osmond, D.H., Loh, A.Y.: Lancet I, 102 (1978).

69. Leckie, B.J., McGhee, N.K.: Nature 288, 702-705 (1980).

70. Hseuh, W.A., Carlson, E.J., Israel-Hagman, M.: Hypertension 3 (Supple I), I-22-I-29 (1981).

71. Inagami, T., Okamoto, H., Ohtsuki, K., Shimamoto, K., Chao, J., Margolius, H.S.: J. Clin. Endocrinol. Metab. 55, 619-627 (1982).

72. Atlas, S.A., Sealey, J.E., Dharmgrongartama, B., Hesson, J.E., Laragh, J.H.: Hypertension 3 (Suppl. I): I-30-I-36 (1981).

73. Takii, Y., Inagami, I.: Biochem. Biophys. Res. Commun. 94, 182-188 (1980).

74. Leckie, B.J., McConnell, A.: Circ. Res. 36, 513-519 (1975).

75. Takii, Y., Inagami, T.: Biochem. Biophys. Res. Commun. 104, 133-140 (1982).

76. Takahashi, S., Miura, R., Miyake, Y.: J. Biochem. 92, 559-567 (1982).

77. Takii, Y., Figueiredo, A.F.S., Inagami, T.: Hypertension in press (1985).

78. Poulsen, K., Vuust, J., Lykkegard, S., Nilsen, A.H., Lung, T.: FEBS Lett. 98, 135-138 (1979).

79. Dykes, C.W., Chat, K., Taylor, J.M., Inagami, T.: Biomed. Res. 1, 565-568 (1980).

80. Pratt, R.E., Dzau, V.J., Ouellette, A.J.: Nucleic Acid Res. 9, 3433-3449 (1981).

81. Morris, B.J., Catanzaro, D.F.: Clin. Exp. Pharmacol. Physiol. 8, 441-445 (1981).

82. Poulsen, K., Vuust, J., Lung, T.: Clinic. Sci. 59, 297-299 (1980).

83. Catanzaro, D.F., Mullins, J.J., Morris, B.J.: J. Biol. Chem. 258, 7364-7368 (1983).

84. Pratt, R.E., Ouellette, A.J., Dzau, V.J.: Proc. Natl. Acad. Sci. USA. 80, 6809-6813 (1983).

85. Ong, E.B., Perlmann, G.E.: J. Biol. Chem. 243, 6104-6109 (1968).

86. Pedersen, V.B., Foltmann, B.: FEBS Lett. 35, 255-256 (1973).

87. Stepanov, V.M., Baratova, L.A., Pugacheva, I.B., Belyanova, L.P., Revina, L.P., Timokhina, E.A.: Biochem. Biophys. Res. Commun. 54, 1164-1170 (1973).

COMPUTER GRAPHICS MODELLING AND THE SUBSITE SPECIFICITIES OF HUMAN AND MOUSE RENINS

B.L. Sibanda, A.M. Hemmings and T.L. Blundell

Laboratory of Molecular Biology, Department of Crystallography, Birkbeck College, University of London, Malet Street, London WC1E 7HX

Abstract

We have constructed models of human and mouse renins,based on their published sequences and the three-dimensional structure of endothiapepsin defined to high resolution by X-ray analysis. They show that the renins can adopt a three-dimensional structure similar to that of other aspartic proteinases with a few insertions and deletions occurring at the surface and mainly at turn regions. The catalytically essential aspartates are conserved and located at the centre of an extended cleft. These models offer an explanation for the neutral pH optima and subsite specificities of the renins. Furthermore certain residues on the periphery of the active site cleft are uniquely different in the renins and may play a part in recognising and binding angiotensinogen.

Introduction

Renin is a very specific aspartic proteinase which reacts with its substrate angiotensinogen to release angiotensin I. This is the first step in the conversion of angiotensinogen to angiotensin II. The latter is a cause of elevated blood pressure. Hence inhibition of the renin-angiotensin system, at the level of the renin-angiotensinogen reaction, may have important therapeutic implications. This then indicates the

Aspartic Proteinases and their Inhibitors

need for a detailed structure of renin as a basis for designing specific inhibitors.

In this short communication we present a summary of the work carried out in model building the renins and suggest explanations for subsite specificities and neutral pH optimum. More detailed accounts of this work (1,2,3) have been published elsewhere.

Methods

The sequences of human renin (4) and mouse renins (5,6) were aligned with the sequences of other aspartic proteinases. The tertiary structures of those aspartic proteinases defined by X-ray analysis indicated that most of the hydrophobic residues are confined to the core of the molecule and hydrophilic residues are exposed on the surface of the molecule . The known structures included endothiapepsin (7), penicillopepsin (8), rhizopuspepsin (9) and porcine pepsin (10) . The last two structures were used only to a limited extent as only α-carbon atom coordinates are available for porcine pepsin and rhizopuspepsin has not yet been sequenced .

The model building was carried out using FRODO (11), which has been modified by Dr. I.J. Tickle, on an Evans and Sutherland Picture System II. The tertiary structure of endothiapepsin which has been refined to 2.1Å was used as a starting point. All main chain atoms and invariant side chains were taken to be identical to those of endothiapepsin. Other side chains were placed in positions close to those of endothiapepsin, in places where tertiary interactions were optimised. Contacts less than van der Waals distances were avoided. The few insertions and deletions, which mainly occur at loop regions on the surface of molecules, were carried out utilising the existing tertiary structure data of other aspartic proteinases as suggested by Greer (12) and by incorporating the information gained from a study of β-hairpins carried out by B.L. Sibanda and J.M. Thornton (manuscript in preparation). Modelling using endothiapepsin as a starting point offered many advantages; for example, certain insertions which are present when comparing the renin sequences with the porcine pepsin sequence already exist in the endothiapepsin sequence. However, those insertions in the renin sequences compared to the endothiapepsin sequence are unique to the renins, for example residues 46a, 46b, those residues around 159 and residues 279-282 (pepsin numbering). Fig. 1a shows the human renin model built as described above.

Given the complexity of protein architecture, the inherent subjectivity of the model-building process, even in highly experienced hands, can result in models which display unaccountably high conformational energies. Thus, for each of these models we have utilised the empirical force field developed by Weiner et al (13) in conjunction with the program EMP, written by I. Haneef of our Laboratory, to allow for the in vacuo, Cartesian coordinate optimization for all atomic degrees of freedom with respect to the total intramolecular potential energy.

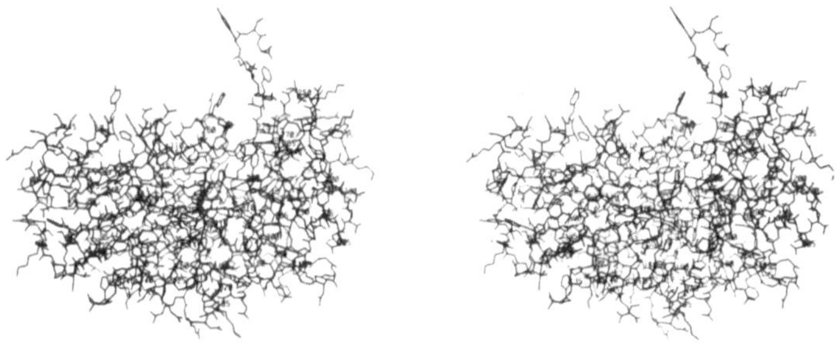

Figure 1a: Stereo view of the modelled tertiary structure of human renin viewed orthogonal to the active site cleft.

The analytical function representing the total intramolecular potential energy is based on discrete contributions from internal coordinate and pair-wise non-bonded interaction terms viz.

$$E(tot) = E(bond) + E(angle) + E(torsion) + E(improper\ torsion) + E(vdW) + E(charge) + E(h\text{-}bonded)$$

and involves use of extended atom potentials.

After generation of coordinates for those hydrogen atoms bonded to potential hydrogen-bond donor atoms in stereochemically feasible positions, the structures were subjected to conjugate-gradient energy minimization for 300 energy evaluations on the CRAY-1S at the University of London Computer Centre. This allowed for approximately 130 successful minimizing iterations for both models.

Discussion

The catalytic aspartates 32 and 215 of the aspartic proteinases are symmetrically arranged with their carboxyls coplanar (14,8). These aspartates are locked into position by hydrogen bonding to residues around them. This arrangement is thought to play a vital role in the catalytic mechanism of this family of enzymes (14). The residues involved in this arrangement are the same or conservatively varied in all aspartic proteinases sequenced so far. However residue 218 which is either a serine or a threonine is an alanine in human renin. As this is a serine in mouse renin, it clearly has no effect on the pH optimum of the renins. However, residue 304 which is mainly buried and close to the active site is an alanine in the renins; this may have the effect of raising the pH optimum of the renins (1).

Although the catalytic centre is conserved throughout this family, different specificities for different substrates result from specificity pockets (subsites) which lie along the length of the active site cleft. The cleft can accommodate up to eight residues (15,16). Subsites S4 to S1' (see table 1) were first characterised by model building and have now been confirmed by X-ray studies of inhibitors bound to penicillopepsin (16), rhizopuspepsin (18) and endothiapepsin (S. Foundling et al, this volume). The subsites on either side of the scissile bond are related by a psuedo dyad which also relates many residues between the two lobes of the aspartic proteinases (19,20,14). Therefore subsites S1 and S1' occupy topologically equivalent positions. The size of these two pockets is large even in the renins, which is rather surprising because the renins prefer to have smaller hydrophobic residues occupying these pockets. However, residues as big as phenylalanine have been shown to be easily accommodated in these two subsites (21). Subsites S4 to S2 are similar in human and mouse renins but different from those of other aspartic proteinases.

However, most differences between the renins and other aspartic proteinases occur in subsites S2' and S3'. Table 1 shows possible residues for these subsites. In the S3' subsite residue 189 is normally a phenylalanine or a tyrosine in other aspartic proteinases but a serine in mouse and a valine in human renin. Replacement with a small residue in this subsite would allow a large substrate residue to be accommodated; for example, in human renin a histidine occupies this subsite. In addition, in human renin, residue 127, which is in the vicinity of S3', is uniquely glutamate. The residue of the substrate (histidine) which occupies this binding pocket can easily be brought close to this acid group (3). Figure 1b shows the residues around the active site cleft.

Table 1

Possible specificity subsites in human and mouse renin		
	Residues close to substrate	Residues in proximity of substrate
S_4	Tyr 220 (Phe)	Thr 284, Met 10 (Leu) Asp 245 (Glu), Ser 219 Tyr 275
S_3	Thr 12 (Ser), Gln 13 Phe 111	Ser 219, Pro 110, Ala 114
S_2	Thr 77 (Ser), Ser 76 (Gly) Ala 2p8 (Ser)	His 288, Tyr 220 (Phe)
S_1	Tyr 75, Thr 77 (Ser) Val 120, Val 30 (Ile) Phe 111, Asp 32	Trp 39
S_1'	Leu 2p3 (Val) Asp 215	Ala 301, (Val), Ser 222, Thr 216 Thr 299 (Val), Val 189 (Ser)
S_2'	Ser 35, Leu 73 (Ile) Gln 128, Arg 74 (His)	
S_3'	Thr 187, Val 189 (Ser) Gln 191, Leu 213 (Val) Glu 127 (Ala)	

Residues in parenthesis indicate the mouse sequence where it differs from the human sequence.

The renins display a number of unique features on the surface of the molecule. For example, on the N-terminal lobe there is a hydrophobic patch which involves the following residues:

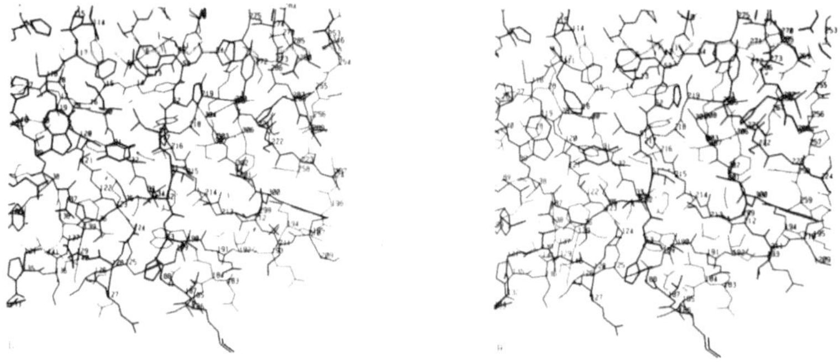

Figure 1b: Stereo view of the residues which constitute the inner surface of the active site cleft of human renin with angiotensinogen peptide bound

46b, 49, 80, 106, 107, 108, 109, 110, 112, 113 and 114 (in human renin these residues are: Leu, Ala, Val, Met, Pro, Ala, Leu, Pro, Met, Leu, Ala respectively). On the C-terminal lobe the surface is rather basic, in human renin the following residues constitute this basic region: Lys 240, Lys 241, Arg 242, Lys 249, Lys 282a and Lys 282b. These surfaces may play a role in interacting with angiotensinogen. The loop region 293 to 298 involves several prolines (P P P T G P), which make that part of the molecule fairly rigid. This loop region is at the edge of the active site cleft, and hence, may be involved in the substrate enzyme

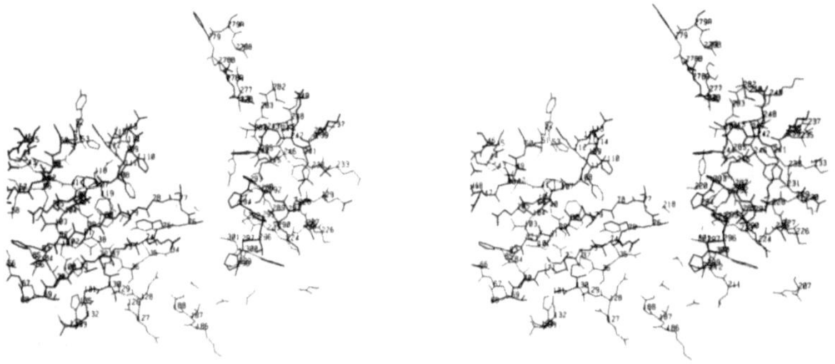

Figure 1c: Stereo view of the arrangement of residues at the periphery of the active site cleft.

interactions. Figure 1c shows the residues on the edge of the active site cleft.

Studies of this type enable unique differences among members of a family of proteins to be identified. In this case, exposed residues unique to the renins, which may be involved in interacting with the surface of angiotensinogen, have been identified. Furthermore, the study of the binding sites along the active cleft has revealed differences which may be utilised in designing specific inhibitors for human renin.

As can be seen from the results of the empirical force field calculations outlined in Table 2, the models for human and mouse renin attain relaxed conformations very similar in total intramolecular potential energy.

Our models show little change in gross structure; the small volume decrease based on calculated radii of gyration is most likely due to the neglect of solvation effects and is found to act isotropically in both cases. Due to the size of these systems, only structural waters have been included in the calculation. A future refinement of the procedure will be to include an estimation of solvent-based conformational dependence.

The carboxyl groups of the catalytic aspartates 32 and 215, found to lie about a plane with r.m.s. deviation of < 0.1 Angstrom (14) were modelled similarly for human and mouse renins. The orientation of this least-squares plane was found to be preserved in the relaxed models, notwithstanding an increase in r.m.s. deviation of ca. 40%.

Whilst the minimization procedures have yet to attain convergence, as evidenced by the root mean square (r.m.s.) first derivatives, most of the unfavourable atom-atom contacts have been removed. It has been demonstrated that the absence

Table 2

	HUMAN	MOUSE
Vtot/ Kcal/Mol		
ORIGINAL	6.94 10(5)	1.65 10(5)
FINAL	-4.84 10(3)	-4.97 10(3)
G(r.m.s.)/ Kcal/Mol/A		
ORIGINAL	2.45 10(5)	5.01 10(4)
FINAL	0.842	0.534
Volume Change/ %	-0.42	-0.43
D(r.m.s.)/ Angstrom	0.45	0.50

Statistics for empirical force field calculations

of unfavourable non-bonded contacts, though necessary is not sufficient to demonstrate the validity of a model-built structure (22). However, the all-atom r.m.s. displacements from original to relaxed structures are similar for both models, thus indicating a close similarity between these model-built structures and their relaxed forms.

Acknowledgements

We wish to thank Dr. I.J. Tickle for developing programs used on the Evans and Sutherland interactive computer graphics equipment, and Dr. L. Pearl for making available the endothiapepsin coordinates. We also thank the UK Science and Engineering Research Council for financial support.

References

1. Blundell, T.L., Sibanda, B.L. and Pearl, L.H., Nature, 304, 273-275 (1983).
2. Tickle, I.J., Sibanda, B.L., Pearl, L.H., Hemmings, A.M., and Blundell, T.L., X-ray Crystallography and Drug Action, eds. Horn, A.S. and De Ranter, C.J. (Clarendon Press) Oxford, pp 427-440 (1984).
3. Sibanda, B.L. and Blundell, T.L., FEBS Lett., 174, 102-111 (1984).
4. Hobart, P.M., Fogliano, M., O'Connor, B.A., Schaefer, I.M. and Chirgwin, J.M., Proc. Natl. Acad. Sci., USA (in press).
5. Panthier, J.-J., Foote, S., Chambrand, B., Strosberg, A.D., Corvol, P., Nature 298, 90-92 (1982).
6. Misono, K.S., Chang, J.-J. and Inagami, T., Proc. Natl. Acad. Sci., USA 79, 4858-4862 (1982).
7. Pearl, L.H., Sewell, T.A., Jenkins, J.A., Blundell, T.L. and Pederson, V. (1984), manuscript in preparation.
8. James, M.N.G. and Sielecki, A., J. Mol. Biol. 163, 299-361 (1983).
9. Subramanian, E.f Liu, M., Swan, I.D.A. and Davies, D.R., Acid Proteinases; Structure, Function and Biology, ed. Tang, J., pp 33-41, Plenum, New York (1977).
10. Andreeva, N.S., Gustchina, A.E., Federov, A.A., Shutzkever, N.E. and Volnova, T.V., Acid Proteinases; Structure, Function and Biology, ed. Tang, J., pp 33-41, Plenum, New York (1977).
11. Jones, T.A., J. Appl. Crystallogr. 11, 268-272 (1978).
12. Greer, J., J. Mol. Biol. 153, 1027-1042 (1981).
13. Weiner, S.J., Kollman, P.A., Case, D.A., Singh, U.C., Ghio, C., Alagona, G., Profeta, S. Jr. and Weiner, P., J. Am. Chem. Soc. 106, 765-784 (1984).
14. Pearl, L.H. and Blundell, T.L. FEBS Lett. 174 , 96-101 (1984).
15. Blundell, T.L., Jenkins, J.A., Khan, G., Roy Choudhury, P., Sewell, B.T., Tickle, I.J. and Wood, E.A., Fed. Eur. Biochem. Series 52, 81-94 (1979).
16. Blundell, T.L., Jones, H.B., Khan, G., Taylor, G., Sewell, B.T., Pearl, L.H. and Wood, S.P., Fed. Eur. Biochem. Series 60, 281-288 (1980).
17. James, M.N.G., Sielecki, A., Salituro, F., Rich, D.H. and Hofman, T., Proc. Natl. Acad. Sci. USA, 79, 6137-6141 (1982).

18. Bott, R., Subramanian, E. and Davies, D.R., Biochemistry 21, 6956-6962 (1982).

19. Tang, J., James, M.N.G., Hau, I.N., Jenkins, J.A. and Blundell, T.L., Nature 271, 618-621 (1978).

20. Blundell, T.L., Sewell, B.T. and McLachlan, A.D., Biochem. Biophys. Acta 580, 24-31 (1979).

21. Burton, J., Cody, R.J., Herd, J.A. and Haber, E., Proc. Natl. Acad. Sci. USA 77, 5476-5496 (1980).

22. Novotny, J., Broccoleri, R., Karplus, M., J. Mol. Biol. (in press).

CHANGES OF DIFFERENT FORMS OF ACTIVE AND INACTIVE RENIN UNDER STRESS IN RATS

Antonín Jindra, Jr.

2nd Department of Internal Medicine, Charles University
128 08 Prague, Czechoslovakia

Richard Kvetňanský

Institute of Experimental Endocrinology, Centre of Physiological Sciences, Slovak Academy of Sciences, 809 36 Bratislava, Czechoslovakia

Introduction

Biochemical characterization of plasma inactive renin and its in vitro activation was well documented in recent years /1, 2, 3, 4/. However, our knowledge of in vivo activation of plasma inactive renin and changes between different forms of renin is not so much advanced. We present a study in which we found out that immobilization stress in rats /5/ is a potent stimulus for the activation of plasma inactive renin. This fact was used for the investigation of interrelationship between various forms of renin in the plasma of Wistar rats.

Results

Renin activity was measured using 3 μM hog angiotensinogen /Miles/ by the method described previously /6/. Inactive renin concentration was calculated as total activity after acid activation /6/ minus activity before activation. Fig. 1 shows that active renin increased gradually during the initial 60 min of immobilization stress /IMO/. Total renin level did not change in the first 15 min of IMO. Consequently, in this interval inactive renin concentration promptly decreases, obviously due to its activation. The increase of total renin activity between the 15th and 60th min of IMO

Aspartic Proteinases and their Inhibitors

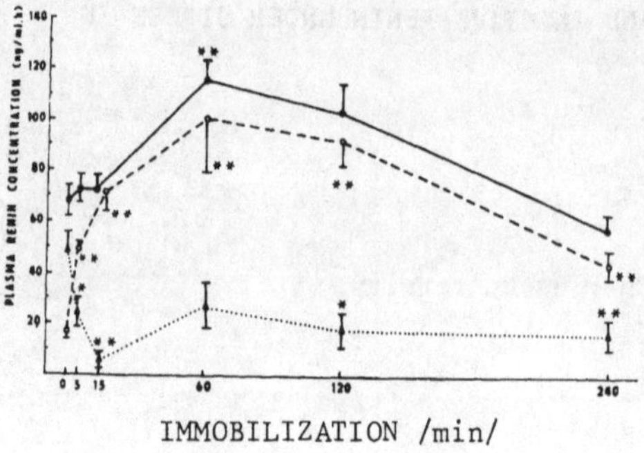

Fig. 1. Effect of acute immobilization stress on plasma concentration of total /●—●/, active /○- -○/ and inactive /▲···▲/ renin. Mean ± SEM for 8 animals. Statistical significance compared to controls:
*, $P < 0.05$; **, $P < 0.01$.

can be elucidated by the increase of renin secretion. Thus two mechanisms are responsible for the increase of active renin during acute immobilization stress - activation of inactive renin in the early phase, and renin secretion in the later phase.

Gel filtration profile of active and inactive renin in the pooled plasma from control rats is shown in Fig. 2A. Two higher peaks of inactive renin

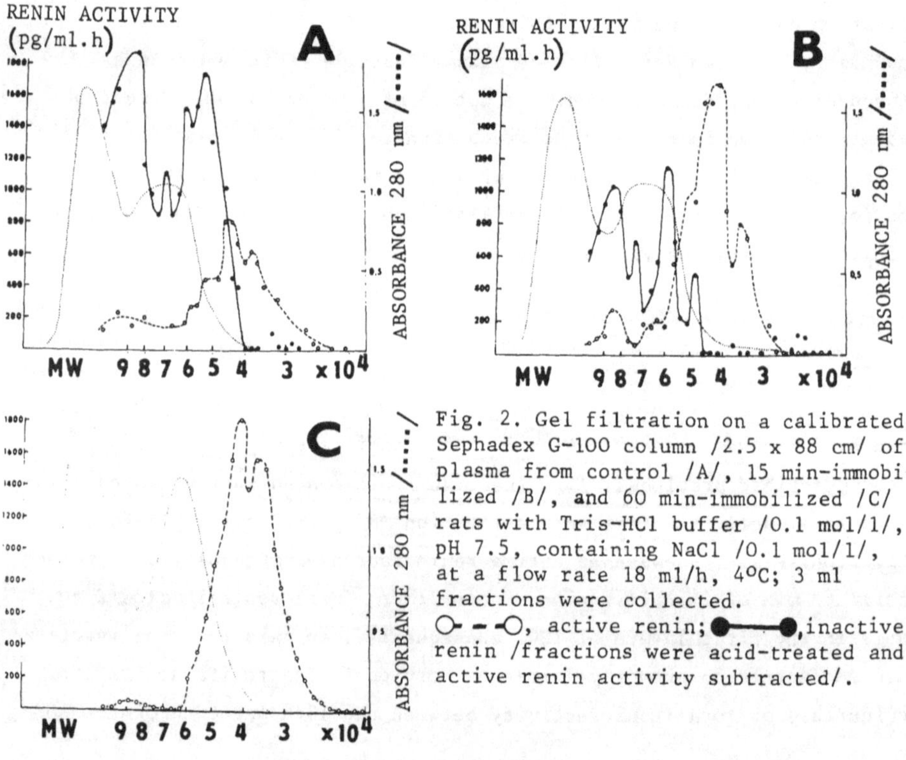

Fig. 2. Gel filtration on a calibrated Sephadex G-100 column /2.5 x 88 cm/ of plasma from control /A/, 15 min-immobilized /B/, and 60 min-immobilized /C/ rats with Tris-HCl buffer /0.1 mol/l/, pH 7.5, containing NaCl /0.1 mol/l/, at a flow rate 18 ml/h, 4°C; 3 ml fractions were collected.
○- -○, active renin; ●—●, inactive renin /fractions were acid-treated and active renin activity subtracted/.

were observed at M_r 85,000 and 50,000, and two lower peaks at 58,000 and 72,000. Low values of active renin activity were found at the same M_r - 85,000, 58,000 and 50,000 - but the highest peaks of active renin were at M_r 43,000 and 36,000.

In the pool of plasma from 15 min-immobilized rats /Fig. 2B/ the amount of inactive renin at M_r 85,000 and especially that at 50,000 was decreased compared to controls, and inversely, concentration of active renin form M_r 43,000 was much higher.

Fig. 2C shows that in the time of increased renin secretion in the 15th - 60th min of IMO mainly the level of active renin form of M_r 36,000 increased in plasma. These results suggest that in rats there are two forms of active renin in blood circulation. The form of M_r 43,000 originates obviously via activation of inactive renin. The lower molecular weight form M_r 36,000 is probably released directly to circulation in the time of increased renin secretion. There is an analogy in human plasma. In patients with enhanced renin secretion - renovascular hypertension /7/, or high-renin essential hypertension /8/ - lower molecular weight forms of active renin were found compared to normal or low renin subjects.

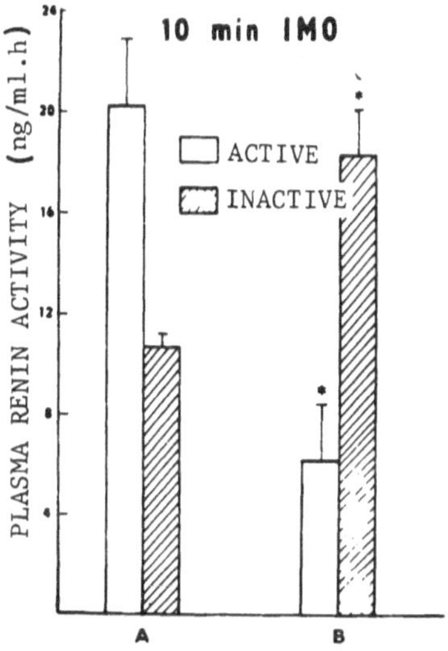

Fig. 3. Active and inactive renin activity in plasma from 10 min-immobilized rats /A/, and in the same plasma which was preincubated 10 min at 37°C /B/. Mean ± SEM of four experiments. ***, compared to A, $P < 0.05$.**

For the following experiment we used plasma from 10 min-immobilized rats with increased content of active and decreased content of inactive renin. Plasma was incubated 10 min at 37°C. Inactive renin was determined by trypsin activation /9/. This treatment caused a great decrease of active plasma renin activity, and conversely, increase of inactive renin activity /Fig.3/. The ratio of inactive /active renin in incubated samples was similar to that found in controls. The results suggest that stress-induced activation of inactive renin can be reversed by incubation of plasma at 37°C. The reversibility of renin activation was observed also in in vitro studies with human plasma /10, 11/ and amniotic fluid /12/

inactive renin. Leckie et. al. /10/ proposed that exposure of plasma to acid may result in a reversible dissociation of a renin-inhibitor complex. On the contrary, Atlas et. al. /11/ and Franks et. al. /12/ suggested that reversible acid-activation of inactive renin involves a conformational change in a sigle polypeptide chain, but these latter studies were performed on purified totally inactive renin. Our gel filtration results indicate that in the rat plasma most of inactive renin is not present as a totally inactive precursor of renin - there were small waves of renin activity under the peaks of inactive renin /Fig. 2A/. Furthermore, the stress-induced renin activation was obviously accompanied by a decrease of molecular weight. Therefore, it is higly probable that, at least in rats, most of plasma inactive renin is an enzyme-inhibitor complex with a capability of reversible dissociation during its in vivo activation.

References

1. Chang, J.J., Kisaragi, M., Okamoto, H., Inagami, T.: Hypertension 3, 509-515 /1981/.
2. Inagami, T., Okamoto, H., Ohtsuki, K., Shimamoto, K., Chao, J., Margolius, H.S.: J. Clin. Endocrinol. Metab. 55, 619-627 /1982/.
3. Sealey, J.E., Atlas, S.A., Laragh, J.H.: Clin. Sci. 63, 133-145 /1982/.
4. Derkx, F.H.M., Schalekamp, M.P.A., Bouma, B., Kluft, C., Schalekamp, M.A.D.H.: J. Clin. Endocrinol. Metab. 54, 343-348 /1982/.
5. Kvetňanský, R., Mikulaj, L.: Endocrinology 87, 738-743 /1970/.
6. Jindra, A., Jr., Kvetňanský, R.: J. Biol. Chem. 257, 5997-5999 /1982/.
7. Inagami, T., Murakami, K.: Biomed. Res. 1, 456-475 /1980/.
8. Ogawa, K., Matsunaga, M., Hamada, H., Oohashi, H., Pak, Ch.H., Morimoto, K., Hara, A., Kawai, Ch.: Jap. Circ. Journ. 47, 294-299 /1983/.
9. Glorioso, N., Madeddu, P., Dessi-Fulgheri, P., Fois, G. Meloni, F., Bandiera, F., Tonolo, G., Rappelli, A.: Clin. Sci. 64, 137-140 /1983/.
10. Leckie, B.J., Mc Ghee, N.K.: Nature 288, 702-705 /1980/.
11. Atlas, S.A., Hesson, T.E., Sealey, J.E., Laragh, J.H.: Clin. Sci. 63, 167-170 /1982/.
12. Franks, R.C., Bodola, F., Renthal, R.D., Hayashi, R.H.: Clin. Sci. 64, 481-486 /1982/.

MOUSE RENIN GENE STRUCTURE, EVOLUTION AND FUNCTION

Burt, D. W., Beecroft, L. J., Mullins, J. J., Pioli, D., George, H., Brooks, J., Walker, J. and Brammar, W. J.

ICI/Joint Laboratory
School of Biological Sciences
University of Leicester
Leicester LE1 7RH
ENGLAND, UK

Introduction.

"One often hears speakers discussing the renin-angiotensin-aldosterone system who pronounce the word *renin* as "wren-in". This is clearly an error. The correct pronunciation of *renin* is "ree-nin". The word *rennin*, which is correctly pronounced "wren-in", refers to an enzyme found in calves' stomachs that curdles milk and is used to make cheese. Rennin, which is also known as rennet or chymosin, has nothing whatever to do with hypertension. Renin is an enzyme produced in the kidney that is thought to play a role in the pathogenesis of at least some cases of hypertension. How important it is, has been and remains a subject of much speculation in the hypertension research community, but it would seem appropriate for all at least to know how to pronounce properly a word that is so widely used (1)."

The aspartyl protease renin cleaves the plasma protein angiotensinogen to initiate the production of vasoactive peptides and thus plays an important role in the maintenance of blood pressure (2). Other members of the aspartyl proteases include mammalian digestive enzymes such as pepsin (3), rennin (4) and cathepsin D (5) and those from microbial sources

Aspartic Proteinases and their Inhibitors

penicillopepsin (6), rhizopuspepsin (7) and endothiapepsin (7). These aspartyl proteases are characterised by two active site aspartates localised in two short amino acid stretches with homology to one another and separated by a distance equivalent to half of the polypeptide chain. These enzymes are closely homologous sharing a bilobal structure with an extended hydrophobic core, the two lobes being separated by a 2-fold axis of symmetry. These observations have led to a hypothesis that the aspartyl proteases represent an old gene family that has evolved by a process of gene duplication and divergence (8). Also, the ancestral gene may have evolved from duplication and fusion of an earlier gene, coding for a polypeptide having a fold similar to that of one lobe of pepsin (8). However, the lack of amino acid homology between lobes seems to support the view that the aspartate fold has evolved convergently in the two lobes. Alternatively, it may indicate that the tertiary structure is conserved in evolution to a greater extent than the primary structure of a protein. Comparison of the DNA sequences and structures of genes of the aspartyl proteases may provide evidence for or against these hypotheses.

The primary site of renin biosynthesis is the juxta-glomerular cells of the kidney, though the enzyme is also produced by other tissues (9). In particular, the sub-maxillary glands (SMG) of mice produce large amounts of renin from the convoluted granular tubules (9). Renin production from the SMG is under genetic, hormonal and developmental controls (10, 11 and 12) and is therefore an ideal system to examine the control of mammalian gene expression. Inbred strains of mice can be divided into two distinct groups on the basis of the levels of SMG renin activities (10). This difference can be mapped to a single genetic locus, *Rnr*, on chromosome 1 (13). Renin SMG cDNA clones have been used in hybridisation studies (14 and 15) and genomic cloning (16 and 17) to show that high renin-producers (eg. DBA/2) have two genes, *Ren*-1 and *Ren*-2, whereas low renin producing strains (eg. C57BL/10) contain a single gene. Nucleotide sequencing of genes and cDNAs suggest that the two genes of high renin-producer mice are differentially expressed (18), *Ren*-1 probably coding for a kidney renin and *Ren*-2 a SMG protein.

Isolation and initial characterisation of mouse renin genomic clones.

To study the structure of the mouse renin genes we constructed a genomic library of DBA/2 DNA in the vector λL47, screened it with a renin cDNA and purified possible clones (16). All of the renin genomic clones isolated from the DBA/2 mouse strain, a high renin-producer, fall into two distinct groups based on restriction mapping and heteroduplex analysis (Figure 1).

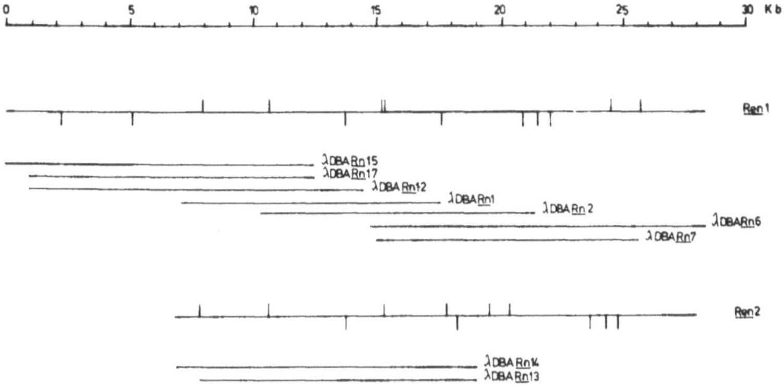

Figure 1. Restriction maps of overlapping renin genomic clones from the DBA/2 mouse. Only *Eco*RI (upward arrows) and *Hind*III (downward arrows) are shown for clarity.

Southern blotting of genomic DNA (14 and 15) and extensive genomic cloning indicate that, at least for the DBA/2 mouse strain, only two genes are present. The composite physical maps for the two regions derived from the maps of cloned genomic DNAs are shown in figure 2. Despite the differences seen in the restriction maps the *Ren*-1 and *Ren*-2 regions show a very high degree of homology, with an extra 3kb of DNA towards the centre of the *Ren*-2 map being the only detectable gross difference within the 13 kb region of homology. The location of the renin coding sequences within the genomic clones were determined by Southern transfer of restriction fragments, using SMG renin cDNAs and synthetic oligonucleotides as hybridisation probes. The cDNA pSMG199, containing mostly sequences from the 3' end of the renin mRNA (16), hybridised to

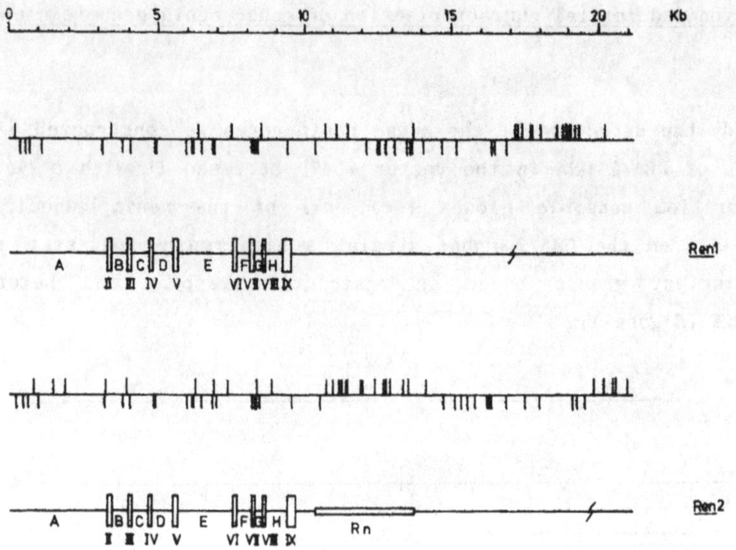

Figure 2. Restriction maps of the *Ren*-1 and *Ren*-2 genomic regions from the DBA/2 mouse. Exons II-IX are indicated as open boxes, introns A-H as lines between exons. Transcription of both genes is from left to right. Homologous restriction sites are marked below the DNA and non-homologous sites above. The 3kb interruption is shown as an open box on the 3' side of *Ren*-2. The right end of homology between the two regions is indicated with a slash (/). Symbols for restriction sites have been omitted for clarity.

Ren-1 at map coordinates 15-17 (Figure 1) and to the homologous *Ren*-2 region. A synthetic 19 base pair oligonucleotide, complementary to a region containing the translation start-site, hybridised to *Ren*-1 at map coordinates 6.5-7.5 (Figure 1), but failed to hybridised to any *Ren*-2 clones. These observations indicated that the beginning of the *Ren*-1 gene was present on the *Ren*-1 clones, but the homologous region on *Ren*-2 was absent from the present clones. The hybridisation data suggested that the *Ren*-1 gene and probably the *Ren*-2 gene occupy at least 10 kb of genomic DNA. Since the renin mRNA has been shown to be only 1600 nucleotides in length (19), we would expect the *Ren*-1 gene to contain at least 8.5 kb of non-coding sequences, as introns and 5' and 3' non-coding segments.

Exons	I	II	III	IV	V	Va	VI	VII	VIII	IX
mouse renin	245	151	124	119	197	-	120	145	99	343
human renin	?	151	124	119	197	9	120	142	99	355
human pepsin	106	163	118	119	200	-	117	145	99	314

Table 1. Comparison of the exon sizes for mouse renin, human renin (20) and human pepsin (21) genes. Sizes are in base pairs.

	mouse renin	ACG.GTAACT		CCTGGAGCAG.A	
I	human renin	ACG.GTAATT	IVS A	CCTGGGCCAG.G	II
	human pepsin	CAA.GTGAGT		CAAACCACAG.G	
	mouse renin	AAT.GTGAGT		CCCGCCACAG.A	
II	human renin	TGA.GTGCCT	IVS B	ACCCCCACAG.A	III
	human pepsin	GAT.GTGAGT		GCCTGGACAG.A	
	mouse renin	GTG.GTAAGA		CCTCTGCTAG.G	
III	human renin	GTG.GTGAGA	IVS C	CCCCTGCCAG.C	IV
	human pepsin	GCA.GTAAGT		GTCCTTGCAG.C	
	mouse renin	ACT.GTGAGT		TCTCTCACAG.G	
IV	human renin	ACC.GTAAGT	IVS D	CCTCCCACAG.G	V
	human pepsin	CAG.GTGGGC		CCCCACCCAG.G	
	mouse renin	CAG.GTGGGC		TTTCCTTTAG.G	
V	human renin	CAG.GTGGGG	IVS E	CCCCTGCCAG.G	VI
	human pepsin	CGC.GTAAGT		CTTTCCACAG.C	
	mouse renin	GGG.GTGGGT		GCCTCTGCAG.G	
VI	human renin	GGG.GTCAGA	IVS F	CTCCCCCAAG.G	VII
	human pepsin	CAG.GTGAGA		TTGCCCTCAG.C	
	mouse renin	GAA.GTAAGA		ATTCCCCCAG.T	
VII	human renin	GAT.GTAAGA	IVS G	CCCACCCCAG.T	VIII
	human pepsin	GAC.GTGAGA		CTCTTTCCAG.A	
	mouse renin	CAG.GTGAGG		TTCTTGCCAG.T	
VIII	human renin	CAG.GTGAGG	IVS H	TTCCTGCCAG.G	IX
	human pepsin	CAG.GTGAGG		CTTTTCTCAG.A	
	Consensus	CAG.GTAAGT		YYYYYYNYAG.G	

Table 2. Comparison of the exon/intron boundaries for mouse renin, human renin (20) and human pepsin (21) genes. Consensus sequences for 5' and 3' splice junctions are from Mount (22).

Comparison of the intron/exon structure of mouse and human renin genes with the gene for another aspartyl protease gene, human pepsin.

Initially the gross structure of the renin genes was determined by heteroduplex analysis of genomic and cDNA clones. The data indicated that both genes have a complex structure with at least eight introns and nine exons (data not shown). The precise position and sizes of the *Ren*-1 exons and introns were determined by DNA sequencing and restriction mapping.

If the aspartyl proteases represent a gene family evolved by a process of gene duplication then primary gene sequences and structures should be homologous. Comparison of exons sizes (Table 1) show remarkable conservation: most importantly any variation found consisted of deletion/substitution of codons, a result expected for homologous genes. The sequences at the intron/exon boundaries for mouse renin, human renin and human pepsin are summarised in table 2. No clear relationship other than conservation of exon/intron boundaries (22) is obvious. The gross structures of mouse renin and human pepsin genes are compared in figure 3. In general, the number of exons and introns have been conserved, with intron sizes being most variable, a feature found in most gene families (23). The structure of the human renin gene is of added interest because it seems to have an extra exon, only 9 bp in length. This could be a case of exon-creation during the past 80 million years since rodent/primate divergence (24).

Comparison of primary DNA sequences from different aspartyl proteases.

Comparison of primary amino acid sequences reveals homology between renin and other aspartyl proteases of 34-43 %. Here we compare DNA sequences. Two-dimensional plots representing all of the homologous segments between two sequences are referred to as "dot-plots" (25). The horizontal and

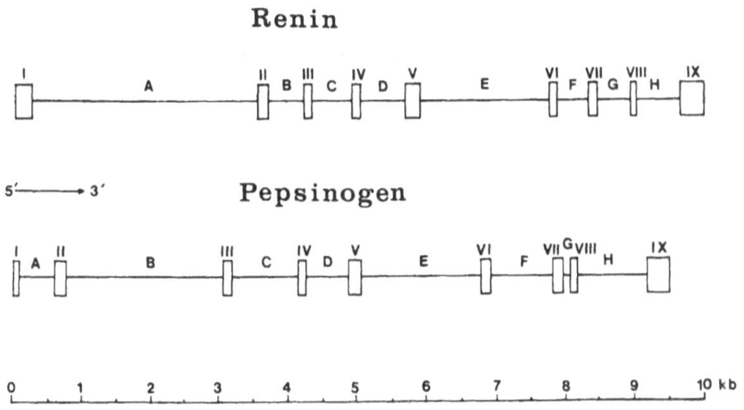

Figure 3. Comparison of the structures of mouse renin and human pepsin (21) genes. Exons I-IX are indicated as open boxes and introns A-H as lines between exons. Transcription of both genes is from left to right.

vertical axes of the plot correspond to positions within the respective sequences. Each point (x,y) within the plot is marked with a dot if the corresponding x-th position of one sequence matches the y-th position of the other. Extended homologies between the two sequences will appear as a diagonal string of dots. This strategy produces a considerable background of chance matches, however, this can be suppressed by requiring that individual dots correspond to a short "span" of homology. The size of the region compared for a single dot is referred to as a "window" and the minimum number of individual matches required within the window to produce a dot is known as the "stringency". The size of the window should be an odd number so that the dot is correctly positioned at the centre of the window. In dot-plots, deletions and insertions cause the diagonal string of dots to break and shift to a new diagonal. When such a gap falls in a search window, it may prevent detection of homology. Gaps cause less of a problem when large windows are used, though this can also increase the execution time considerably. An algorithm, the diagonal-traverse search (25), reduces the execution time required in the use of large windows (upto 100 base pairs). One drawback in using large windows is the appearance of regions of apparent homology, upto the length of the window. These arise because once a matched window has been found the probability of an adjacent match is high. However, choosing suitable stringencies reduces this problem and only statistically significant regions of homology are seen. A consequence of this solution is a decrease in the ability to detect weak homologies. The results of comparing mouse renin, human renin, human pepsin and bovine rennin are shown (Figure 4) and clearly demonstrate homologies between these genes.

These dot-plots were used to align gene sequences and because only exon and cDNA sequences were available in most cases, the alignments only show coding regions, starting from the ATG initiation codons (Figure 5). Immediately we can see that the exon/intron positions have been conserved and predictions are possible for the bovine rennin gene. Alignments were possible for exons III to IX, indicating at least for these exons a common ancestor. The fact that these exons code for the mature enzyme suggests that they have been subject to strong functional constraints during evolution. The overall DNA sequence homologies found between exons III and IX are summarised (Table 3).

	human renin	human pepsin	bovine rennin
mouse renin	78.9 %	54.2 %	51.8 %
human renin		53.0 %	52.0 %
human pepsin			64.8 %

Table 3. Overall sequence homology observed between pairs of aspartyl protease genes. References: human renin (20), human pepsin (21) and bovine rennin (26).

The sequences of exons I and II are more variable, suggesting that in contrast to the strong functional constraints seen in the regions coding for the mature enzyme there seems to be more freedom for variation in these exons. The extent of sequence divergence observed between pairs of aspartyl proteases should be related to the divergence times that separate these genes. It was therefore of interest to determine the rates of evolution between pairs of genes.

Figure 4. Comparison of aspartyl protease genes using dot-plots. Window size is 29 base pairs and stringency 55 %.

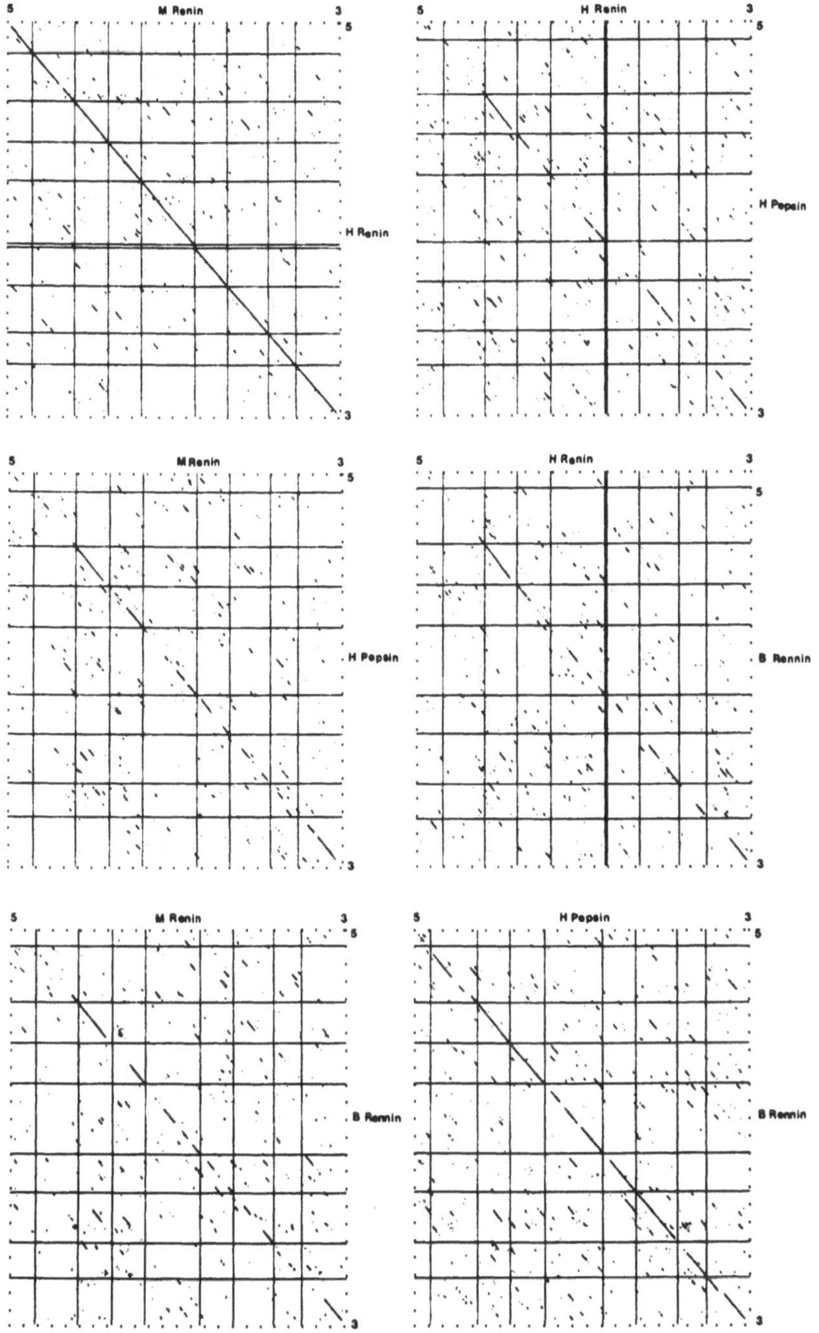
M Renin
H Renin
H Renin
H Pepsin
M Renin
H Pepsin
H Renin
B Rennin
M Renin
B Rennin
H Pepsin
B Rennin
5
3

```
         M D - R R R M P L W A L . . - . W S P C T F S L P T R T A T F E R I P L K
M Renin : ATGGAC---A GGAGGAGGAT GCCTCTCTGG GCACTCTTGT TG---CTCTG GAGTCCTTGC ACCTTCAGTC TCCCAACACG CACCGCTACC TTTGAACCAA TCCCACTCAA
H Renin : ATGGATGGAT GGAGAAGGAT GCCTCGCTGG GGACTGCTGC TGCTGCTCTG GGGCTCCTGT ACCTTTGGTC TCCCGACAGA CACCACCACC TTTAAACGGA TCTTCCTCAA
H Pepsin: ATGAAGTGG- --CTGCTGCT GCTGGGTCTG GTGGCGCTCT CTGAGTGCAT CATGTAC--- ---------- ---------- ---------- -----AAGG TCCCCCTCAT
B Rennin: ATGAGGTGTC TCGTGGTGCT ACTTGCTGTC TTCGCTCTCT CCCAGGGC-- ---------- ---------- ---------- ---------- -----GCTG AGATCACTAG
                 10         20         30         40         50         60         70         80         90        100        110

                                                                                                                            1
          K M P S V R E I L E E R G V D M T R L S A E R G V F T K R P S L I N L T S
M Renin : GAAAATGCCC TCTGTCCGGG AAATCCTGGA GGAGCGGGGA GTGGACATGA CCAGGCTCAG TGCTGAAAGG GGCGTATTCA CAAAGAGGCC TTCCTTGATC AATCTTACCT
H Renin : GAGAATGCCC TCAATCCGAG AAAGCCTGAA GGAACGAGGT GTGGACATGC CAAGGCTTGG TCCCGAGTGG AGCCAACCCA TGAAGAGGCT GACACTTGGC AACACCACCT
H Pepsin: CAGAAAGAAG TCCTTGAGGC GCACCCTGTC CGAGCGTGGC CTGCTGAAGG ACTTCCTGAA GAAGCACAAC CTCAACCCAG CCAGAAAGTA CTTCCCCCAG TGGGAGGCTC
B Rennin: GATCCCTCTG TACAAAGGCA AGTCTCTGAG GAAGGCGCTG AAGGAGCATG GGCTTCTGGA GGACTTGCTG CAGAAACAGC AGTATGGCAT CAGCAGCAAG TACTCCGGCT
                120        130        140        150        160        170        180        190        200        210        220

                                          10                   20                   30
          P V V L T - - - - N Y L N A Q Y Y G E I G I G T P P Q T F K V I F D T G
M Renin : CCCCCGTGGT CCTCACC--- ---------A ACTACCTGAA TGCCCAGTAC TACGGCGAGA TTGGCATCGG TACCCCACCC CAGACCTTCA AAGTCATCTT TGACACGGGT
H Renin : CCTCCGTGAT CCTCACC--- ---------A ACTACATGGA CACCCAGTAC TATGGCGAGA TTGGCATCGG CACCCCACCC CAGACCTTCA AAGTCGTCTT TGACACTGGT
H Pepsin: CCACCCTGGT AGATGAACAG CCCCTGGAGA ACTACCTGGA TATGGAGTAC TTCGGCACTA TCGGCATCGG AACTCCTGCC CAGGATTTCA CCGTCGTCTT TGACACCGGC
B Rennin: TCGGGGAGGT GGCCAGCGTG CCCCTGACCA ACTACTTGGA TAGTCAGTAC TTTGGGAAGA TCTACCTCGG AACCCCGCCT CAGGAGTTCA ACGTGCTGTT TGACACTGGC
                230        240        250        260        270        280        290        300        310        320        330

                 40               a b   50                       60
          S A N L W V P S T K C S R L Y L A C G I H S L Y E S S D S S S Y M E N G S
M Renin : TCGGCCAACC TCTGGGTGCC CTCCACCAAG TGCAGCCGCC TCTACCTTGC TTGTGGGATT CACAGCCTCT ATGAGTCCTC TGACTCCTCC AGCTACATGG AGAACGGGTC
H Renin : TCGTCCAATG TTTGGGTGCC CTCCTCCAAG TGCAGCCGTC TCTACACTGC CTGTGTGTAT CACAAGCTCT TCGATGCTTC GGATTCCTCC AGCTACAAGC ACAATGGAAC
H Pepsin: TCCTCCAACC TGTGGGTGCC CTCAGTCTAC TGCTCCAGT- -----CTTGC CTGCACCAAC CACAACCGCT TCAACCCTGA GGATTCTTCC ACCTACCAGT CCACCAGCGA
B Rennin: TCCTCTGACT TCTGGGTACC CTCTATCTAC TGCAAGAGC- -----AATGC CTGCAAAAAC CACCAGCGCT TCGACCCGAG AAAGTCGTCC ACCTTCCAGA ACTTGGGCAA
                340        350        360        370        380        390        400        410        420        430        440

          70                    80                    90                    100
          D F T I H Y G S G R V K G F L S Q D V V T V G G I T V T Q T F G E V T E L
M Renin : CGACTTCACC ATCCACTACG GATCAGGAAG AGTCAAAGGC TTCCTCAGCC AGGACGTGGT GACTGTGGGT GGAATCACTG TGACACAGAC CTTTGGAGAG GTCACCGAGC
H Renin : AGAACTCACC CTCCGCTATT CAACAGGGAC AGTCAGTGGC TTTCTCAGCC AGGACATCAT CACCGTGGGT GGAATCACGG TGACACAGAT GTTTGGAGAG GTCACGGAGA
H Pepsin: GACAGTCTCC ATCACCTACG GCACCGGCAG CATGACAGGC ATCCTCGGAT ACGACACTGT CCAGGTTGGA GGCATCTCTG ACACCAATCA GATCTTCGGC CTGAGCGAGA
B Rennin: GCCCCTGTCT ATCCACTACG GGACAGGCAG CATGCAGGGC ATCCTAGGCT ATGACACCGT CACTGTCTCC AACATTGTGG ACATCCAGCA GACAGTAGGC CTGAGCACCC
                450        460        470        480        490        500        510        520        530        540        550

               110  112  114               120                   130                   140
          P L I P F M - L A K F D G V L G M G F P A Q A V G G V T P V F D H I L S
M Renin : TGCCCCTGAT CCCTTTCATG ---CTGGCCA AGTTTGACGG TGTCCTAGGC ATGGGCTTTC CTGCTCAGGC CGTTGGCGGG GTTACCCCTG TCTTTGACCA CATTCTCTCC
H Renin : TGCCCGCCTT GCCCTTCATG ---CTGGCCG AGTTTGATGG GGTTGTGGGC ATGGGCTTCA TTGAACAGGC CATTGGCAGG GTCACCCCTA TCTTCGACAA CATCATCTCC
H Pepsin: CGGAACCTGG CTCCTTCCTG TATTATGCTC CCTTCGATGG CATCCTGGGG CTGGCCTACC CCAGCATTTC CTCCTCCGGG GCCACACCCG TCTTTGACAA CATCTGGAAC
B Rennin: AAGAGCCCGG GGACGTCTTC ACCTATGCCG AATTCGACGG GATCCTGGGG ATGGCCTACC CCTCGCTCGC CTCAGAGTAC TCGATACCCG TGTTTGACAA CATGATGAAC
                560        570        580        590        600        610        620        630        640        650        660

                      150                            a  160                   170
          Q G V L K E E V F S V Y Y N R - - - G S H L L G G E V V L G G S D P Q H Y
M Renin : CAGGGGGTGC TGAAGGAGGA AGTGTTCTCT GTCTACTACA ACAG------ ---GGGTTCC CACCTGCTGG GGGGCGAGGT GGTGCTAGGA GGCAGTGACC CGCAGCATTA
H Renin : CAAGGGGTGC TAAAAGAGGA CGTCTTCTCT TTCTACTACA ACAGGAATTC CGAGAATTCC CAATCGCTGG GAGGACAGAT TGTGCTGGGA GGTAGCGACC CCCAGCATTA
H Pepsin: CAGGGCCTGG TTTCTCAGGA CCTCTTCTCT GTCTACCTCA GCGG------ ---CGAT--- GACCAGAGTG GCAGCGTGGT GATCTTTGGT GGCATTGACT CTTCTTACTA
B Rennin: AGGCACCTGG TGGCCCAAGA CCTGTTCTCG GTTTACATGG ACAG------ ---GAAT--- GGCCAGGAGA GC---ATGCT GACGCTGGGG GCCATCGACC CTTCCTACTA
                670        680        690        700        710        720        730        740        750        760        770

                 180                   190                   200                   210
          Q G N F H Y V S I S K T D S W Q I T M K G V S V G S S T L L C E E G C A V
M Renin : CCAAGGCAAT TTTCACTATG TGAGCATCAG CAAGACTGAC TCCTGGCAGA TCACGATGAA GGGGGTGTCT GTGGGGTCTT CCACCCTGCT ATGTGAAGAA GGCTGTGCGG
H Renin : CGGAGGGAAT TTCCGGTATA TCAACCTCAT CAAGACTGGT GTCTGGCAGA TTCAAATGAA GGGGGTGTCT GTGGGGTCAT CCACCTTGCT CTGTGAAGAC GGCTGCCTGG
H Pepsin: CACTGGAAGT CTGAACTGGG TGCCTGTTAC CGTCGAGGGT TACTGGCAGA TCACCGTGGA CAGCATCACC ATGAACGGAG AGGCCATCGC CTGCGCTGAG GGCTGCCAGG
B Rennin: CACAGGGTCC CTGCACTGGG TGCCCGTGAC AGTGCAGCAG TACTGGCAAT TCACTGTGGA CAGTGTCACC ATCAGCGGTG TGGTTGTAGC CTGTGAGGGT GGCTGTCAGG
                780        790        800        810        820        830        840        850        860        870        880

                      220                        230       234236     240
          V V D T G S S F I S A P T S S L K L I M Q A L G A K E K R I D E Y V V N
M Renin : TAGTGGTGGA CACTGGTTCA TCCTTTATCT CGGCTCCTAC GAGCTCCCTG AAGTTGATCA TGCAAGCCCT GGGAGCCAAG GAGAAGAGAA TAGATGAATA TGTTGTGAAC
H Renin : CATTGGTAGA CACCGGTGCA TCCTACATCT CAGGTTCTAC CAGCTCCATA GAGAAGCTCA TGGAGGCCTT GGGAGCCAAG AAGAGGCTGT TTGATGTATA TGTCGTGAAG
H Pepsin: CCATTGTTGA CACCGGCACC TCTCTGCTGA CCGGCCCAAC CAGCCCCATT GCCAACATCC AGAGCGACAT CGGAGCCAGC GAGAACTCAG ATGGCGACAT GGTGGTCAGC
B Rennin: CCATCCTGGA CACGGGCACC TCCAAGCTGG TCGGGCCCAG CAGCGACATC CTCAACATCC AGCAGGCCAT TGGAGCCACA CAGAACCAGT ACGATGAGTT TGACATCGAC
                890        900        910        920        930        940        950        960        970        980        990

          250                   260                   270                       a b c d 280
          C S Q V P T L P D I S F D L G G R A Y T L S S T D Y V L Q Y P Y R R D K L
M Renin : TGTAGCCAGG TGCCCACCCT CCCCGACATT TCCTTTGACC TGGGAGGCAG GGCCTACACA CTCAGCAGTA CGGACTACGT GCTACAGTAT CCCTACAGGA GAGACAAGCT
H Renin : TGTAACGAGG GCCCTACACT CCCCGACATC TCTTTCCACC TGGGAGGCAA AGAATACACG CTCACCAGCG CGGACTATGT ATTTCAGGAA TCCTACAGTA GTAAAAAGCT
H Pepsin: TGCTCAGCCA TCAGCAGCCT GCCCGACATC GTCTTCACCA TCAATGGAGT CCAGTACCCC GTGCCACCCA GTGCCTACAT CCTGCAGAGC GAG------- -----GGGAG
B Rennin: TGCGACAACC TGAGCTACAT GCCCACTGTG GTCTTTGAGA TCAATGGCAA AATCTACCCA CTGACCCCCT CCGCCTATAC CGGCCAGGAC CAA------- -----GGCTT
               1000       1010       1020       1030       1040       1050       1060       1070       1080       1090       1100

                      290                   300                   310
          C T L A L H A M D I P P P T G P V W V L G A T F I R K F Y T E F D R H N N
M Renin : GTGCACACTG GCTCTCCATG CCATGGACAT CCCACCACCC ACTGGGCCTG TCTGGGTCCT GGGTGCCACC TTCATCCGCA AGTTCTATAC AGAGTTTGAT CGGCATAACA
H Renin : GTGCACACTG GCCATCCACG CCATGGATAT CCCGCCACCC ACTGGACCCA CCTGGGCCTT GGGGGCCACC TTCATCCGAA AGTTCTACAC AGAGTTTGAT CGGCGTAACA
H Pepsin: CTGCATCAGT GGCTTCCAGG GCATGAACCT CCCCACCGAA TCTGGAGAGC TTTGGATCCT GGGTGATGTC TTCATCCGCC AGTACTTTAC CGTCTTCGAC AGGGCAAACA
B Rennin: CTGTACCACT GGCTTCCAGA GTGAAAATCA TTCC------ ------GAGA AATGGATCCT GGGGGATGTT TTCATCCGAG AGTATTACAG TGTCTTTGAC CGGGCCAACA
               1110       1120       1130       1140       1150       1160       1170       1180       1190       1200       1210

           320
           R I G F A L A R
M Renin : ATCGCATTGG ATTCGCCTTG GCCCGC
H Renin : ACCGCATTGG CTTCGCCTTG GCCCGC
H Pepsin: ACCAGGTCGG CCTGGCCCCC GTGGCT
B Rennin: ACCTCGTGGG GCTGGCCAAA GCCATC
               1220       1230
```

Figure 5. Alignements of mouse renin, human renin (20), human pepsin (21) and bovine rennin (26). Only coding sequences are shown. Dashes are introduced to maximise homology. The single letter amino acid sequence of mouse kidney renin is shown above the DNA sequence and is numbered with reference to porcine pepsin (3). The DNA sequences are numbered starting from the ATG initiation codons. Vertical bars indicate the position of intron/exon boundaries. ∇ represent proteolytic cleavage points that produce the signal and activation peptides.

Evolutionary rates between the aspartyl protease genes.

We have presented evidence for gene duplication, followed by sequence divergence as support for the model of aspartyl protease gene evolution. Using the alignments given in figure 5, we can estimate the rates of evolution and times of divergence between these genes. The three substitution type (3-ST) model of Kimura (27) was used to correct for multiple base substitutions to estimate the number of base substitutions per site at each of the three codon positions (Table 4). For the comparison of mouse and human renin a time of divergence of 80 million years (24) was used to estimate the rates of evolution (base substitutions per site per year) at the first (k1), second (k2) and third (k3) codon positions. These estimates were $k1=1.38\times10^{-9}$, $k2=0.84\times10^{-9}$ and $k3=2.71\times10^{-9}$. When all mutations are selectively neutral the rate of evolution is approximately equal to the mutation rate. The synonymous component at the third codon position (k_s) approximates to this and consequently, is nearly equal in all genes (27). For the mouse/human renin comparison $k_s=2.23\times10^{-9}$, a value similar to that found for other genes (27). Assuming that this rate is constant and the same for all aspartyl proteases, we may estimate the times of divergence (T_d, years) between pairs of genes ($T_d=K_s/2\times k_s$). The estimated times of divergence are summarised (Table 5) and suggest pepsin and rennin to be more closely related to each other than to the renins. The data would suggest that these genes are the result of a gene duplication approximately 121 million years ago. Since divergence, these genes now respond to different developmental controls, rennin being expressed as a fetal pepsin (26).

The data suggest that all the mammalian aspartyl protease genes were derived from a common ancestral sequence approximately 160 million years ago, by a process of gene duplication and sequence divergence. However, due to gene correction mechanisms such as gene conversion, there would be a tendency to preserve homology. The times of divergence could therefore indicate the times of either gene duplication or the last round of gene correction.

	human renin	human pepsin	bovine rennin
mouse renin	K1=0.22±0.03 K2=0.13±0.02 K3=0.43±0.05 K_s=0.36±0.05	K1=0.66±0.07 K2=0.50±0.05 K3=1.10±0.13 K_s=0.69±0.08	K1=0.73±0.08 K2=0.52±0.06 K3=1.38±0.27 K_s=0.69±0.08
human renin		K1=0.78±0.08 K2=0.49±0.05 K3=1.08±0.12 K_s=0.69±0.09	K1=0.76±0.08 K2=0.52±0.06 K3=1.30±0.24 K_s=0.99±0.23
human pepsin			K1=0.43±0.05 K2=0.31±0.04 K3=0.75±0.08 K_s=0.54±0.06

Table 4. The number of base substitutions per site at the first (K1), second (K2) and third (K3) codon positions estimated from the alignment of codons given in figure 5. For each comparison an estimate of the silent or synonymous component at the third position (K_s) is given together with an estimate of the standard error. The three substitution type model of Kimura (27) was used to calculate these values.

	human renin	human pepsin	bovine rennin
mouse renin	80	155±18	164±18
human renin		155±20	222±54
human pepsin			121±14

Table 5. Estimates of times of divergence between pairs of aspartyl protease genes. Divergence times in millions of years.

Gene Fusion Model.

The prediction of the gene-fusion hypothesis (8) is that aspartyl protease genes should display internal sequence homology. The structural homology observed by Tang *et al* (8) would sugest that exons II-V and VI-IX are the result of a gene duplication and fusion (Figure 6). Dot-plots of aspartyl protease genes with themselves may reveal this internal sequence homology (Figure 7).

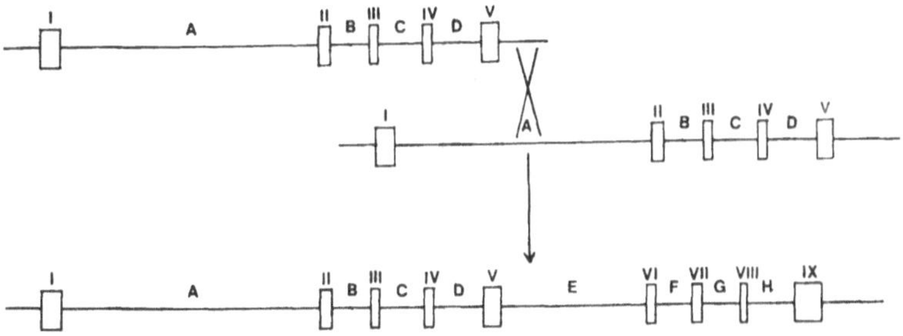

Figure 6. Gene-Fusion Model.

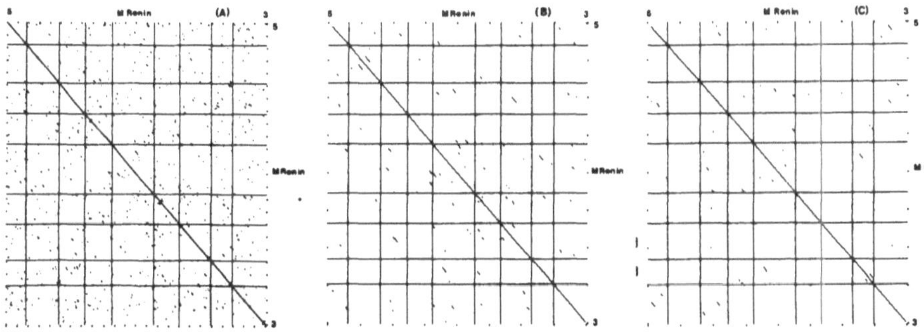

Figure 7. Dot-plot analysis of mouse kidney renin using DNA and amino acid sequence data for internal sequence homologies. Two kinds of analysis were used; (a) base identity (b) amino acid identity (c) score matrix. Window size is 9 residues and stringencies (a) 77 % (b) 44% (c) 110.

The dot-plots using DNA sequences did not reveal any significant regions of internal homology in any of the genes examined (example mouse renin, Figure 7a). This result is not really surprising, since evidence for such an ancient duplication event would probably be masked by many sequence changes since that time. However, it might be expected that the type of amino acids used would be conserved. Initially, to examine this possibility the amino acid sequences were compared in much the same way as DNAs, and again no significant homologies were found (Figure 7b). Using a nine residue window revealed homology between active sites, but even these localised regions of homology disappeared with a larger window of 29 residues. For many distantly related proteins many significant alignments of sequences contain almost no identities, but are formed from chemically and structurally similar amino acids. A score matrix was calculated (28) by looking at accepted point mutations in 71 families of closely related proteins and was found to be the most powerful score matrix for finding distant relationships between amino acid sequences. This matrix was used to calculate scores for each comparison in a dot-plot analysis in much the same way as described earlier for DNAs. Again, no internal homologies were found with small or large search windows. The sensitivity of this last method is illustrated in figure 8, where pepsin sequences from human and *Penicillium jantinellum* are compared. These proteins are clearly homologous and this conservation has persisted since their divergence some 1000 million years ago.

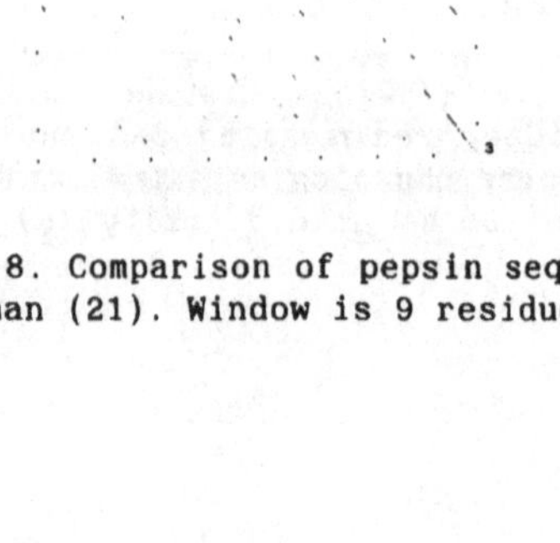

Figure 8. Comparison of pepsin sequences from *Penicillium jantinellum* (6) and human (21). Window is 9 residues and stringency score 110.

The lack of large stretches of internal homology within the aspartyl proteases indicates either that the gene duplication/fusion hypothesis is wrong and the homology between exon III and VII is perhaps due to a requirement dictated by some repeating feature in the tertiary structure or alternatively, the long evolutionary times have caused the internal homology to be less obvious. The tertiary structure may then show an evolutionary connection that has long been lost in the primary sequence. The isolation and characterisation of further genes of the aspartyl protease gene family within and between different species, will provide a clearer picture of the evolutionary history of these genes.

Functional significance of renin coding regions conserved in evolution.

The DNA sequences derived from the nine exons of the *Ren*-1 gene are shown in figure 9. This DNA sequence shows 97 % DNA sequence homology with that of the cDNA sequence derived from SMG mRNA of the SWR mouse (29 and 30). A partial kidney cDNA sequence derived from SWR mouse kidney mRNA (18) is identical to sequences within exon VII of the *Ren*-1 gene (Figure 10). Preliminary analysis of exons of the *Ren*-2 gene shows the identity of these sequences with those of the cDNAs from SMG mRNA (29 and 30). The data show that *Ren*-1 codes for a mouse kidney renin and *Ren*-2 is the major determinant of SMG renin (*Ren*-1 is probably expressed in the latter tissue, but at a low level as in C57BL/10).

Figure 9. Nucleotide sequence of the *Ren*-1 gene and its 3' flanking sequence. The amino acid sequence of kidney renin, derived from the exons of the *Ren*-1 gene is indicated above the DNA sequence. The amino acid residues are numbered according to the numbering system of porcine pepsin (3). Differences between *Ren*-1 genomic and SMG cDNA sequences are shown below the *Ren*-1 sequence. Polyadenylation and potential glycosylation signals are underlined. Potential proteolytic cleavage sites are indicated with arrows.

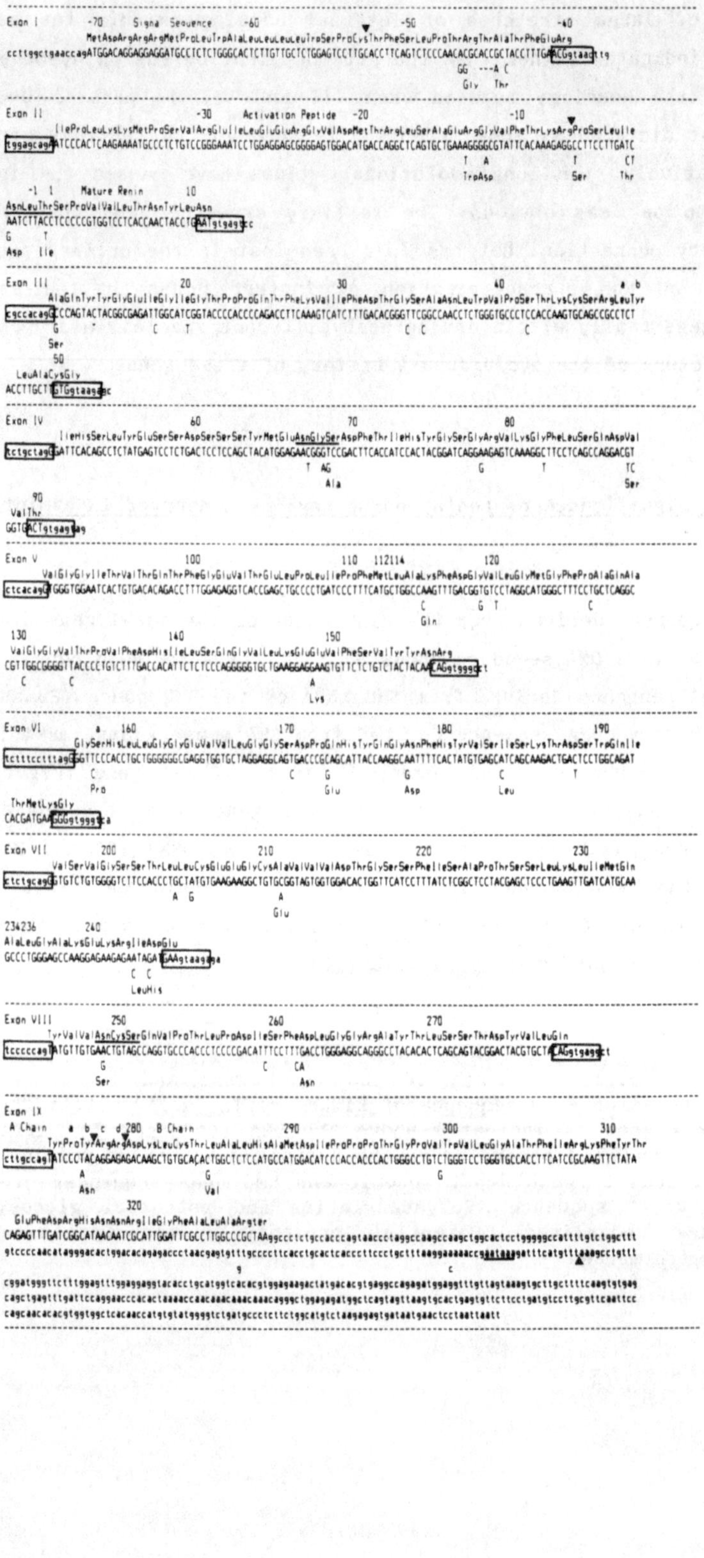
Exon I
Signal Sequence
Exon II
Activation Peptide
Mature Renin
Exon III
Exon IV
Exon V
Exon VI
Exon VII
Exon VIII
Exon IX
A Chain
B Chain

```
Ren1   : CTCTGCAGGGTGTCTGTGGGGTCTTCCACCCTGCTATGTGAAGAAGGCTGTGCGGTAGTGGTGGACACTGGTTCATCCTTT
Kidney :       GlyValSerValGlySerSerThrLeuLeuCysGluGluGlyCysAlaValValValAspThrGlySerSerPhe
mRNA   :         GGUGUCUGUGGGGUCUUCCACCCUGCUAUGUGAAGAAGGCUGUGCGGUAGUGGUGGACACUGGUUCAUCCUUU
Ren2   : ---------------------------------A--G----------------A----------------------------
SMG    :       GlyValSerValGlySerSerThrLeuLeuCysGluGluGlyCysGluValValValAspThrGlySerSerPhe
mRNA   :         -------------------------A--G----------------A----------------------------

Ren1   : ATCTCGGCTCCTACGAGCTCCCTGAAGTTGATCATGCAAGCCCTGGGAGCCAAGGAGAAGAGAATAGATGAAGTAAGAGAT
Kidney : IleSerAlaProThrSerSerLeuLysLeuIleMetGlnAlaLeuGlyAlaLysGluLysArgIleAspGlu
mRNA   : AUCUCGGCUCCUACGAGCUCCCUGAAGUUGAUCAUGCAAGCCCUGGGAGCCAAGGAGAAGAGAAUAGAUGAA
Ren2   : ----------------------------------------------------------------C--C--------------
SMG    : IleSerAlaProThrSerSerLeuLysLeuIleMetGlnAlaLeuGlyAlaLysGluLysArgLeuHisGlu
mRNA   : ----------------------------------------------------------------C--C-----
```

Figure 10. Comparison of *Ren*-1 and *Ren*-2 exon VII sequences with kidney and SMG cDNAs. The amino acid sequences are predicted from the nucleotide sequences. The upper three lines give the nucleotide sequence of *Ren*-1 exon VII (DBA/2), the amino acid and nucleotide sequences of kidney renin and mRNA. The three lower lines give the nucleotide sequence of *Ren*-2 exon VII (DBA/2), the amino acid and nucleotide sequences of SMG renin and mRNA. Only nucleotides which differ from the *Ren*-1 sequences are shown. Homologous positions are indicated by dashes. Amino acid differences are underlined. A partial kidney cDNA sequence (18) is identical to the first twenty six codons for the *Ren*-1 mRNA.

The *Ren*-2 gene appears to function only in SMG tissue and the basis of this tissue specificity is unknown. The primary amino acid sequences of the SMG and kidney renins also predict potential glycosylation sites in kidney renin (positions -2, 67 and 249), but none in SMG renin. Such a pattern is found for kidney renin isolated *in vivo* (31), further support for identifying the product of *Ren*-1 as kidney renin.

	Signal Peptide	Activation Peptide	Mature Enzyme
mouse renin	18	46	340
human renin	18	46	338
human pepsin	15	47	326
bovine rennin	16	42	323

Table 6. Comparison of the polypeptide sizes produced from the precursors of aspartyl proteases by proteolytic cleavage. Sizes in kd, kilodaltons.

Comparison of the sizes of the polypeptides produced by proteolytic processing of the aspartyl proteases indicate similarly sized products

(Table 6). Exon I in all aspartyl protease genes codes for a typical signal sequence (32). For mouse kidney renin these characteristics are as follows: positively charged groups (Arg -68, Arg -67 and Arg -66) within the first five amino acid residues, followed by a core of hydrophobic residues and a helix-breaking group (Pro -54) two residues before the cleavage site. At the carboxyl terminus there is a characteristic small, uncharged amino acid residue (Cys -53). From dot-plots it was shown earlier that there was no apparent DNA homology between exon I of the two renin genes and the pepsins, rennin and pepsin (Figure 4). This could indicate that these sequences were derived from different sources rather than by gene duplication, probably an example of "exon-shuffling" (33). Alternatively, the distant relationship between renins and pepsins may have destroyed the homology between signal sequences: consistent with this view are the number of codons deleted/substituted to make the alignments (Figure 5). All are exact multiples of three (42 rennin and 48 pepsin), a result expected if these genes shared a common ancestor. Rennin and pepsin have similar-sized signal peptides and show some sequence homology, further evidence for a recent ancestor to these genes.

Exon II in each case codes for an activation or pro-peptide, that is cleaved from the pro-form of an aspartyl protease to produce an active mature enzyme. Consequently, the peptide itself is discarded in the process and it is possible that the only functional constraints on this region are those to ensure correct processing. This would predict that these sequences would be hypervariable and the rates of base substitution at each of the three positions in any codon would be similar. The calculations for the mouse/human renin comparison give $k1=3.19\times10^{-9}$, $k2=1.69\times10^{-9}$ and $k3=4.11\times10^{-9}$ (base substitutions per site per year) values twice that found in the rest of the gene (Table 4). The C peptide of the pro-insulin molecule seems to be an analagous case. It forms a peptide bridge between the A and B peptide regions of pro-insulin and ensures that these chains pair correctly during the folding of the nascent protein; the C peptide is then removed when active insulin is formed (34). It also shows a high degree of base substitution, $k=4.8\times10^{-9}$ (27) because its role does not depend greatly on its structure. This may

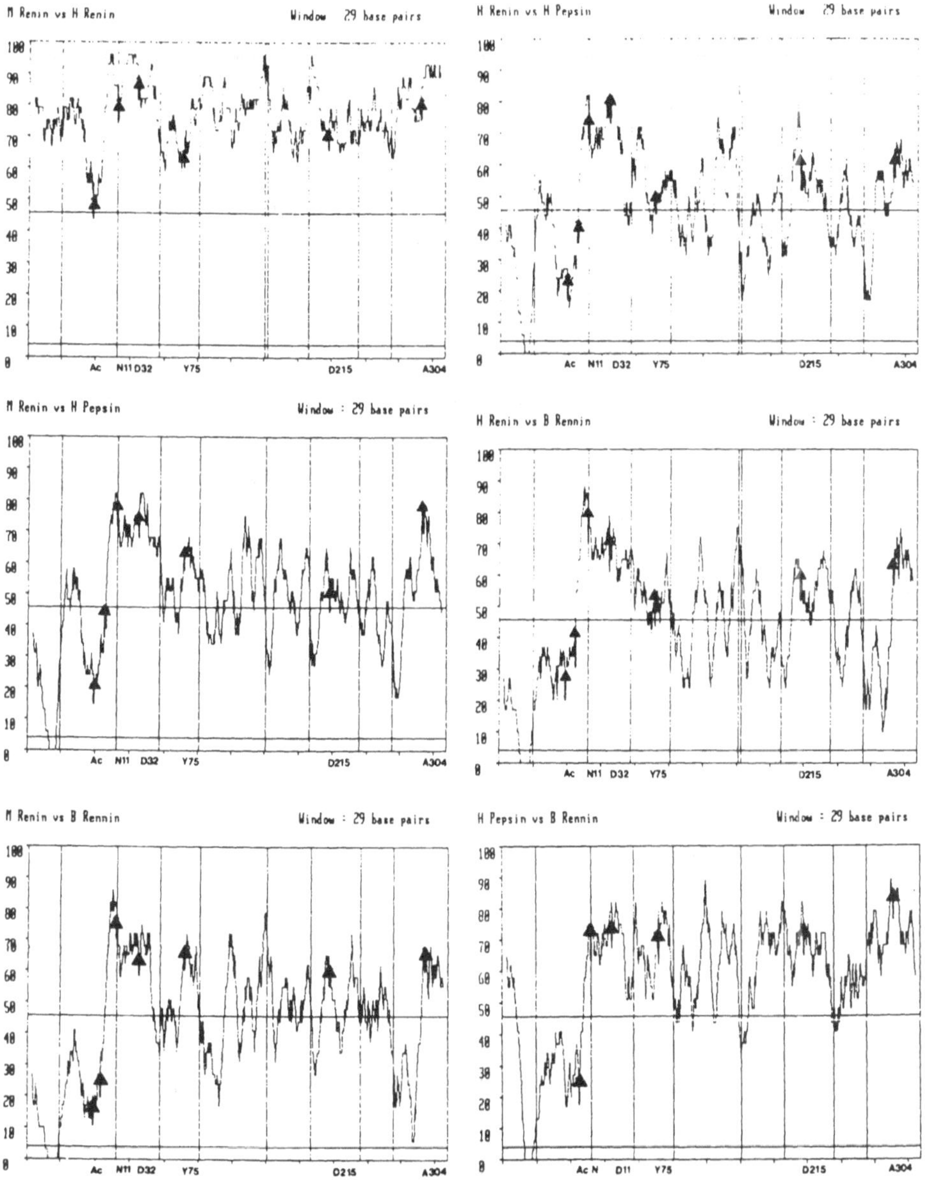

Figure 11. Plots of homology between pairs of aspartyl proteases. The plots use the sequence alignments shown in figure 5 and plot the percentage homology found in a window centred at a given base. The window size was 29 base pairs. The vertical scale shows the percentage homology and the horizontal scale the base position (scale in intervals of 100 base pairs). Arrows point to specific codons discussed in the text and the corresponding amino acid is shown below the horizontal axis.

also be true for the activation peptides of the aspartyl proteases, where the high degree of base substitution is shown as a region of very low homology in all plots of sequence homology (Figure 11).

In all aspartyl protease genes a highly conserved region (greater than 75 %) is found flanking the cleavage point between the activation peptide and the mature enzyme (Figure 11). The transition is abrupt in all pairs examined. This region may contain residues essential for specific cleavage at this site.

Residues shared by renin and other aspartyl proteases may reflect amino acids important in the general structure and function of aspartyl proteases. Many regions of homology are seen between mouse renin and human pepsin, in particular exon III (80 %). This region codes for Asp 32, a key residue in the catalytic mechanism of all aspartyl proteases. Interestingly, the region in exon VII that codes for the second aspartate residue, Asp 215, is not so well conserved (60 %), possibly indicating that this region is not so important as the first in the general catalytic mechanism of aspartate proteases. This pattern was found in all comparisons except between pepsin/rennin, where 80 % homology was found at each active site (Figure 11).

Comparison of human and mouse kidney renins together with their comparison with other aspartyl proteases should identify coding regions specific to the renin molecules. The extremely high substrate specificity and neutral pH optima of renins clearly distinguish them from other aspartyl proteases. The recent model of SMG renin (35) suggests that Asn 11 and Ala 304 are responsible for the neutral pK. Two highly conserved regions (95 %) containing these residues are found in the mouse/human renin comparison, but not between other pairs of aspartyl proteases (Figure 11).

The pattern of homology seen between pairs of aspartyl protease sequences shows many other striking similarities, particularly in the position and frequency of peak homologies. The significance of these conserved regions awaits further knowledge on the structure and function of these proteins. However, the types of analyses shown here highlight regions within proteins that are conserved and possibly important for the function of these enzymes. The DNA sequences encoding these regions would be suitable targets for *in vitro* mutagenesis, to produce mutant proteins that could be used to explore the molecular mechanisms of aspartyl protease action. The powerful approach of "Protein Engineering" can now be brought to bear on an interesting and important group of proteolytic enzymes.

Acknowledgements.

We are very grateful for the technical assistance of J. Bridges, to Dr. M.D. Edge for synthesis of an oligonucleotide probe, to Dr. J.D. Windass for discussion and to I.C.I for financial support.

References.

1. Burris, J.F.: Lancet 249, (1983).
2. Cowley, A.W., Miller, J.P.: Circ. Res. 28, 568-581 (1971).
3. Sepulveda, P., Maraniszyn, J.Jr., Liu, D and Tang, J.: J. Biol. Chem. 250, 5082-5088 (1975).
4. Foltmann, B., Pedersen, V.B., Kauffman, D and Wybrandt, G.: J. Biol. Chem. 254, 8447-8456 (1979).
5. Barrett, A.J.: in Acid Proteinases, Structure, Function and Biology (ed. Tang.) pp 291-300 (Plenum, New York, 1977).
6. Hsu, I-N., Delbaere, L.T.J., James, M.N.G and Hofmann, T.: Nature (London) 266, 140-145 (1977).
7. Subramanian, E., Swan, I.D.A., Liu, M., Davies, D.R., Jenkins, J.A., Tickle, I.J and Blundell, T.L.: Proc. Natl. Acad. Sci. (USA) 74, 556-559 (1977).
8. Tang, J., James, M.N.G., Hsu, I.N., Jenkins, J.A and Blundell, T.L.: Nature (London) 271, 618-621 (1978).

9. Tanaka, T., Gresik, E.W., Michelakis, A.M. and Barka, T.: J. Histochem. Cytochem. 28, 1113-1118.

10. Wilson, C.M., Erodos, E.G., Dunn, J.F and Wilson, J.D.: Proc. Natl. Acad. Sci. (USA) 74, 1185-1189.

11. Wilson, C.M., Cherry, M., Taylor, B.A. and Wilson, J.D.: Biochem. Genet. 19, 509-523.

12. Wilson, C.M., Myhre, M.J., Reynolds, R.C. and Wilson, J.D.: Endocrinology 110, 982-989 (1982).

13. Wilson, C.M., Erodos, E.G., Wilson, J.D and Taylor, B.A.: Proc. Natl. Acad. Sci. (USA) 75, 5623-5626.

14. Piccini, N., Knopf, J.L and Gross, K.W.: Cell 30, 205-213.

15. Panthier, J.J., Holm, I and Rougeon, F.: EMBO J. 1, 1417-1421.

16. Mullins, J.J., Burt, D.W., Windass, J.D., McTurk, P., George, H and Brammar, W.J.: EMBO J. 1, 1461-1466.

17. Holm, I., Ono, R., Panthier, J-J and Rougeon, F.: EMBO J. 3, 557-562 (1984).

18. Panthier, J.J and Rougeon, F.: EMBO J. 2, 675-678.

19. Rougeon, F., Chambraud, B., Foote, S., Panthier, J_J., Nageotte, R., and Corvol, P.: Proc. Natl. Acad. Sci. (USA) 75, 6367-6371 (1981).

20. Hobart, P.: Personal Communication (1984).

21. Sogawa, K., Fujii-Kuriyama, Y., Mizukami, Y., Ichihara, Y and Takahashi, K.: J. Biol. Chem. 258, 5306-5311 (1983).

22. Mount, S.M.: Nucleic Acids Res. 10, 459-472 (1982).

23. Jeffreys, A.J.: in Genetic Engineering 2, (ed. Williamson, R) pp 1-48 (1981).

24. Romero-Herrera, A.E., Lehmann, H., Joysey, K.A and Friday, A.E.: Nature (London) 246, 389-395 (1973).

25. White, C.T., Hardies, S.C., Hutchison III, C.A and Edgell, M.H.: Nucleic Acids Res. 12, 751-766 (1984).

26. Moir, D., Mao, J-I., Schumm, J.W., Vovis, G.F., ALford, B.L and Taunton-Roby, A.: Gene 19, 127-138 (1982).

27. Kimura, K.: Proc. Natl. Acad. Sci. (USA) 78, 454-458.

28. Staden, R.: Nucleic Acids Res. 10, 2951-2961.

29. Panthier, J-J., Foote, S., Chambraud, B., Strosberg, P.V., Corvol, P and Rougeon, F.: Nature (London) 298, 90-92 (1982).

30. Soubrier, F., Panthier, J-J., Corvol, P and Rougeon, F.: Nucleic Acids Res. 11, 7181-7190 (1983).

31. Inagami, T., Murakami, K., Yokosawa, H., Matoba, T., Yokosawa, N., Takii, Y and Misono, K.: in Enzymatic Release of Vasoactive Peptides, eds. Gross, F and Vogel, H.G. (Raven Press, NY), pp. 7-30.

32. Watson, M.E.E.: Nucleic Acids Res. 12, 5145-5164 (1984).

33. Gilbert, W.: Nature (London) 271, 501 (1978).

34. Perler, F., Efstratiadis, A., Lomedico, P., Gilbert, W., Kolodner, R. and Dodgson, J.: Cell 20, 555-566 (1980).

35. Blundell, T., Sibanda, B.L and Pearl, L.: Nature (London) 304, 273-275 (1983).

PEPSTATIN-INSENSITIVE ACID PROTEINASES

Sawao Murao, Kohei Oda

Department of Agricultural Chemistry, University of Osaka Prefecture, Sakai, Osaka 591, Japan

Introduction

It is well known that acid proteinases (calf rennin) have been used for making cheese since the period before Christ. The association between man and acid proteinase is thus very old; however, research on this enzyme has progressed very slowly compared with that on other proteinases. In 1960, when Hartley (1) classified proteinases based on the catalytic residue, no information was available on the catalytic residue of any acid proteinase. The term "acid proteinase" was, therefore, given to the group of proteinases having optimum pH in the acid region.

From such a background, we attempted to obtain a microbial inhibitor of acid proteinases, and obtained S-PI (Pepstatin AC) in 1970 (2). The finding of S-PI by us and the development of such active-site-directed reagents as DAN (diazoacetyl-DL-norleucine methylester) (3) and EPNP [1,2-epoxy-3-(*p*-nitrophenoxy)propane] (4) have made remarkable progress in studying acid proteinases (Fig. 1). Now, it is believed that two aspartic acid residues constitute the active site of acid proteinase, and that the enzyme is specifically inactivated by Pepstatin, DAN and EPNP (5). Therefore, acid proteinase is called also carboxyl or aspartic proteinase.

Since 1970, we have paid attention to the inhibitory spectrum of S-PI inhibiting all the acid proteinases tested. We considered that a S-PI-insensitive acid proteinase, if there was any, should be distinct from known acid proteinases, as is

Aspartic Proteinases and their Inhibitors

trypsin from chymotrypsin among serine proteinases. From such a standpoint, we started screening for a particular microorganism producing S-PI-insensitive acid proteinase and isolated in 1972 Scytalidium lignicolum, later designated ATCC 24568 (6,7).
In this paper, we describe Pepstatin-insensitive acid proteinases that should be classified in a new subclass.

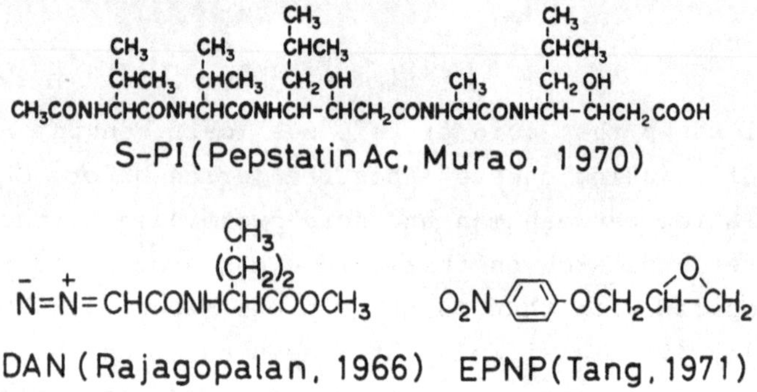

Fig.1 Chemical Structures of S-PI, DAN and EPNP

I. Finding Pepstatin -Insensitive Acid Proteinases and Their Properties

1. Isolation of a strain of microorganisms producing Pepstatin-insensitive acid proteinases (6,7)

Screening for such a microoganism producing S-PI-insensitive acid proteinases was carried out with casein plates, containing 10 μg/ml of S-PI. A number of soil samples collected from all over Japan were tested. We succeeded in isolating M-133 strain, that produced S-PI-insensitive acid proteinase, from soil collected on the farm of the University of Osaka Prefecture. The acid proteinase in culture filtrate was not inactivated by S-PI even in an amount of 100 μg, nor by DAN in the presence of cupric ions. By taxonomical studies, the M-133 strain was identified as Scytalidium lignicolum and deposited in ATCC as Scytalidium lignicolum ATCC 24568. Its photograph is shown in Fig. 2. For production of the acid proteinase, optimal culture conditions were looked for. A largest amount of acid proteinase, about 250 mg/liter, was produced by cultivation of the strain for about 80 hr under the conditions shown in the legend to Fig. 3.

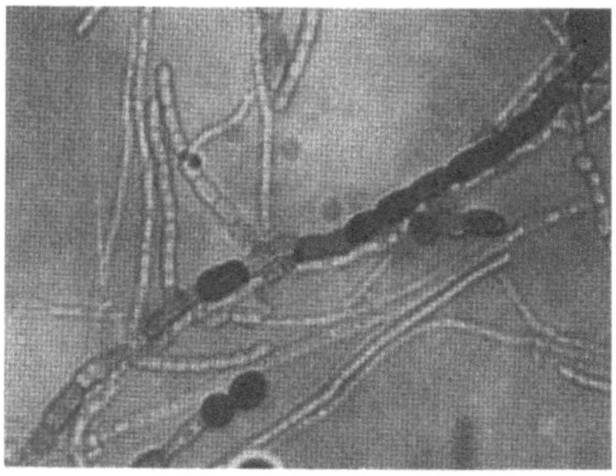

Fig.2 Photomicrograph of Scytalidium lignicolum ATCC 24568

Cells were cultured on potato-dextrose agar for 9 days at 25°C

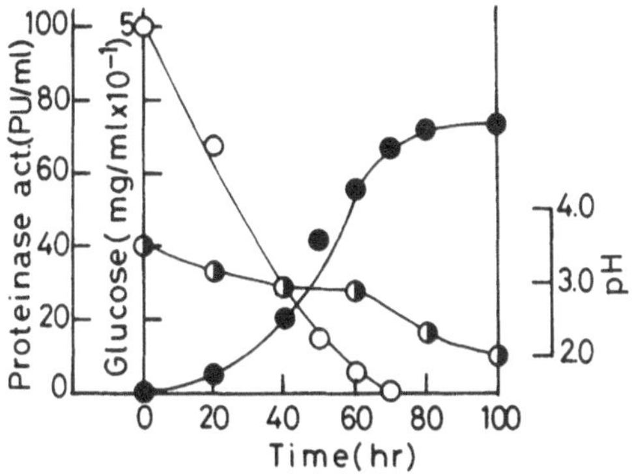

Fig.3 Time Course of Acid Proteinase Production by S.lignicolum

Medium: 5% glucose, 1.5% meat extract, 0.2% KH_2PO_4, 0.05% $MgSO_4 \cdot 7H_2O$, 0.1% Yeast extract, pH 3.5.
25°C, 350 rpm, 15 l medium/20 l Jar
—○—: glucose, —◐—: pH,
—●—: proteinase activity.

2. Purification(8)

The acid proteinase was purified by the procedures shown in Fig. 4. Four distinct enzymes were obtained. About 0.3 g of A-1, 0.6 g of A-2, 6.4 g of B and 0.7 g of C were obtained from 100 liters of culture filtrate. All the enzymes were monodisperse in ultracentrifugal analysis and disc electrophoresis.

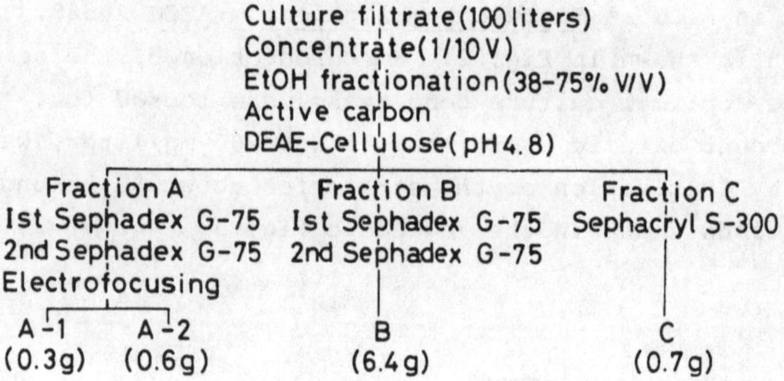

Fig.4 Purification of Acid Proteinases of Scytalidium lignicolum

3. General properties of acid proteinases A-1, A-2, B and C (8,9,10,11)

Some of their properties are summarized in Table 1. A-1 and A-2 were very similar to each other in the enzymic properties, though slightly different in the isoelectric point. They were active at pH 3.0-3.5 towards casein and stable at pH 2.5-5.5 for 20 hr at 37 °C. B was most active at pH 2.2 and stable at pH 1.5-5.0. C was most active at pH 2.0 and stable at pH 3.0-5.0.

Table 1 Some Properties of S.lignicolum Acid Proteinases

	S.lignicolum			Pepsin	R.chinensis
	A-1 (A-2)	B	C		
Optimum pH	3.0-3.5	2.2	2.0	1.8-2.0	2.9-3.3
Molecular weight	40,000	21,969	360,000	34,500	35,000
Isoelectric point	3.6 (3.8)	3.2	2.3	1.0	5.0-5.2
Amino acid composition	His;none	His;none	His;yes	Ser;rich	His;none
Carbohydrate content	≒10%	0	≒30%	0	0
Inhibition with S-PI	−	−	−	+	+
DAN	−	−	−	+	+
EPNP	−	+	−	+	+

The molecular weights of A-1, A-2, B and C were 40,000, 40,000, 21,969 and 360,000, respectively. A-1, A-2, B enzymes did not contain any histidine residue in the molecules as shown in Table 2. Both A-1 and A-2 are glycoproteins, containing 3 moles of glucosamine and 10-15 moles of mannose (11). Effects of S-PI, DAN and EPNP on these acid proteinases were examined, comparing with porcine pepsin. As shown in Fig. 5, A-1, A-2, and C were not inactivated by S-PI or DAN. EPNP inactivated B like pepsin, but A-1, A-2 or C. Furthermore, none was inactivated by any of such proteinase inhibitors as PCMB, EDTA and DFP nor by such a microbial proteinase inhibitor (12,13,14,15) as S-SI or Talopeptin. Such behavior towards inhibitors suggests that the active site or its adjacent area of acid proteinases A-1, A-2, B and C may differ from that of known acid proteinases.

Table 2 Amino Acid Composition of _S.lignicolum_ Acid Proteinases

Amino acid	A-1	A-2	B	C
Lys	4	4	2	21
His	0	0	0	21
Arg	2	3	1	63
$CysSO_3H$	9	8	6	10
Asp	44	39	28	250
Thr	31	28	22	173
Ser	41	36	32	121
Glu	23	26	20	183
Pro	22	21	4	269
Gly	37	37	19	204
Ala	25	25	15	130
Val	20	19	16	62
Met	1	1	1	45
Ile	13	12	12	45
Leu	20	19	6	131
Tyr	16	16	5	329
Phe	16	16	10	78
Trp	1	1	5	48
Total	325	311	204	2183
Carbohydrate	≒10%	≒10%	0	≒30%

4. Substrate specificity (9,10,16,17)

Morihara (18) studied the substrate specificity of Pepstatin-sensitive acid proteinases of microbial origin and showed that they each possess an active site with a size correspond-

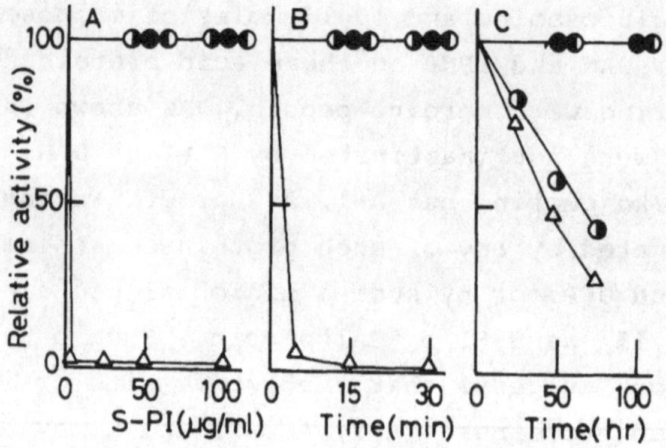

Fig.5 Inhibition of S.lignicolum Acid Proteinases with S-PI, DAN and EPNP
A : [S-PI] The reactions were carried out at 37°C and optimal pH for 30 min.
B : [DAN] The reactions were carried out at 14°C, pH 5.5, according to the method of Rajagopalan. Enzyme:DAN:Cu(II)= 1:40:33
C : [EPNP] The reactions were carried out with 0.5 ml of 0.1% enzyme solution in 0.05M sodium citrate buffer, pH 4.5, and 5 mg of solid powder of EPNP at 25°C, according to the method of Tang. —●— : Acid proteinase A-1 and A-2, —◐— : Acid proteinase B, —◑— : Acid proteinase C, —△— : Pepsin.

ing to an oligopeptide containing at least six amino acid residues (P_1-P_3 and P_1'-P_3') as does pepsin. The acid proteinases from S. lignicolum behaved differently from the above enzymes towards inhibitors. From these results, we considered that the substrate specificity of acid proteinases A-1, A-2, B and C might differ from those of known acid proteinases. To confirm this, we examined acid proteinases A-1, A-2 and B for the substrate specificity with such synthetic peptides as N-acylated di-, tri-, tetra-, penta-,and hexapeptides in comparison with Pepstatin-sensitive ones (18).
The results of these studies may be summarized as follows: (1) the influence of the molecular size of the peptide substrate upon acid proteinases A-1, A-2 and B is completely different from that of Pepstatin-sensitive ones. As summarized in Table 3, maximum hydrolysis was observed with N-acylated tetrapeptides by enzymes A-1 and A-2, in which the presence of amino acid residues at the P_2 and P_2' positions was essential. The enzymes A-1 and A-2 may have a more compact active site than known Pepstatin-sensitive acid proteinases. Enzyme B showed, on the other hand, very low activity

Table 3 Comparison of Acid Proteinases from Scytalidium lignicolum with Those from Aspergillus niger and Rhizopus chinensis, which are Sensitive to Pepstatin.

Substrate[a]	Acid proteinase ($kcat/Km$, $M^{-1} \cdot s^{-1}$)			
	Scytalidium		Asp.niger[b]	Rh.chinensis[b]
	A-1 and A-2	B		
Z-Phe-Leu-Ala	10	0	0	10
Z-Phe-Leu-Ala-Ala	10,700	negligible[c]	50	33
Z-Lys-Ala-Ala-Ala	140	negligible[c]	16	10
Z-Lys-Leu-Ala-Ala	2,090	30	156	111
Z-Ala-Lys-Ala-Ala-Ala	7,970	15	200	63
Z-Ala-Ala-Lys-Ala-Ala-Ala	2,600	45	10,500	4,750

a ↑ shows the bond split by the Scytalidium enzymes and ▽ that by Asp.niger and Rh.chinensis ones. b References (18). c The bond split was the C-terminal peptide bond (Ala-Ala).

Table 4 Hydrolysis of Z-X-Leu-Ala-Ala and Z-Phe-Y-Ala-Ala (X, Y=various amino acid residues) by Acid Proteinases

Acid proteinase	Z—X—Leu-Ala-Ala Z-Phe—Y—Ala-Ala		$kcat/Km$ ($M^{-1} \cdot s^{-1}$)
Pepsin Rhizopus chinensis Mucor pusillus Cladosporium Rhodotorula glutinis Aspergillus saitoi Aspergillus niger B etc.	↑	X=Phe, Leu Y=Tyr, Leu	1,000-200
S.lignicolum A-1, A-2	↑	X=Phe, Ala, Tyr Y=Leu	11,000-8,000
S.lignicolum B	↑	X=Lys, Leu, Phe Y=Tyr, Lys	40-30

towards the peptides tested, although there was a tendency that the larger the molecular size of the peptide substrate, the higher the activity. It is not clear whether the active site of enzyme B is also larger than known Pepstatin-sensitive acid proteinase, because the peptide substrates tested may have been inadequate.

In comaring acid proteinases A-1, A-2 and B with Pepstatin-, DAN- and EPNP-sensitive ones, a marked difference was further found (Table 4). When such a Z-tetrapeptide as Z-X-Leu-Ala-Ala or Z-Phe-Y-Ala-Ala (X, Y = various amino acid residues) was used as substrate, acid proteinases A-1 and A-2 split the Leu-Ala or Y-Ala bond (Z-X-Y↑Ala-Ala), and acid proteinase B split the carboxyl terminal bond (Z-X-Y-Ala↑Ala), whereas known acid proteinases split the X-Leu or Phe-Y bond (Z-X↑Y-Ala-Ala), as reported by Morihara (18).

It can be concluded that the substrate specificity of *Scytalidium* acid proteinases is distinguishable from that of known microbial acid proteinases sensitive to Pepstatin, DAN and EPNP.

5. The active center and its structure

Acid proteinases A-1, A-2, B and C are insensitive to Pepstatin, DAN and EPNP, except B to EPNP. They behave characteristically towards synthetic substrates. These facts suggest that the active sites or the vicinities of acid proteinases A-1, A-2, B and C are considerably different in the structure from those of known acid proteinases. To confirm this, the following studies were undertaken with Pepstatin-sensitive acid proteinases as control.

a) Effect of Pepstatin Ac (8). As shown in Table 5, acid proteinases A-1, A-2, B and C were not inhibited by S-PI at such a high concentration as 100-fold molar excess. Pepsin was completely inactivated by an equal mole of S-PI; acid proteinases isolated from *Rhodotorula glutinis* K-24, *Cladosporium* sp. No. 45-2, *Trametes sanguinea*, *Rhizopus chinensis* and *Aspergillus saitoi* were also inactivated, though the rates of inhibition were different from one to another. We examined

Table 5 Effect of S-PI (Pepstatin Ac)

Acid proteinase	Remaining activity(%) Molar ratio (S-PI/Acid proteinase) 1	2	3	100
Scytalidium lignicolum A-1				100
" " A-2				100
" " B				100
" " C				100
Pepsin	0			
Cladosporium sp.No.45-2	80	65	55	0
Trametes sanguinea	65	50	42	0
Rhodotorula glutinis K-24	40	25	20	0
Rhizopus chinensis	42	25	22	0
Aspergillus saitoi	55			0

The enzyme solutions were each incubated at 37°C for 10 min with an equal volume of S-PI at optimal pH and the remaining activity was assayed.

further by ultraviolet difference spectrophotometry the interaction between S-PI and each of acid proteinases A-1, A-2 and B. A-1 or A-2 did not show any difference spectrum in the presence of S-PI at such a high concentration as 20 molar excess, which indicates that S-PI does not bind to any of them (Fig.6). On the other hand, B showed difference spectrum, with absorption maxima at 273, 285 and 293 nm, but the spectrum was very different (blue shift) from those of S-PI-sensitive acid proteinases (red shift).

b) Effect of synthetic pepsin inhibitors (8). The four acid proteinases were modified chemically with DAN and EPNP, which react specifically with different aspartic acid residues at the active center of porcine pepsin. These results are summarized in Table 6. A-1, A-2 or C was not inactivated by either modifier. B was not inactivated by DAN, but inactivated by EPNP. From these results, it was expected that these enzymes differ from known acid proteinases and that acid proteinases A-1 and A-2 differ from B in the structure at or near the active site.

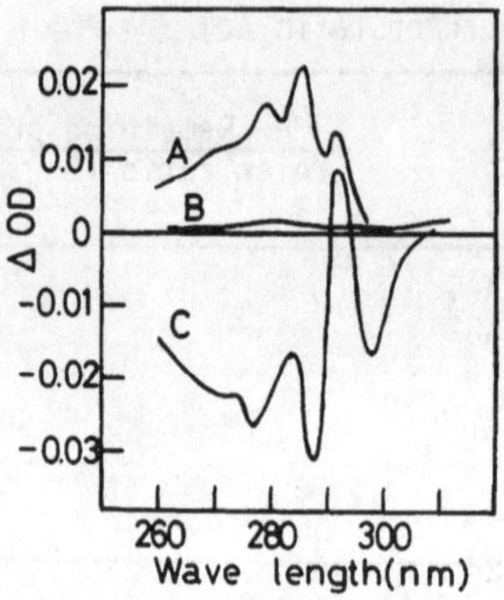

Fig.6 Difference Spectra of Acid Proteinases A-2 and B induced by the the Addition of S-PI
A : Pepsin (S-PI/E=1,M/M)
B : Acid proteinase A-2(S-PI/E=20,M/M)
C : Acid proteinase B (s-PI/E=20,M/M)

Table 6 Effects of DAN and EPNP

Acid proteinase	Remaining activity(%)	
	DAN	EPNP
Scytalidium lignicolum A-1	83	95
〃 〃 A-2	90	100
〃 〃 B	96	32
〃 〃 C	100	100
Pepsin	0	34
Rhodotorula glutinis K-24	0	0
Cladosporium sp. No.45-2	0	0

DAN : The reactions were carried out at 14°C, pH 5.5 for 1 hr, according to the method of Rajagopalan.
EPNP: The reactions were carried out with 0.5 ml of a 0.1% enzyme solution in 0.05M sodium citrate buffer, pH 4.5, and 5 mg of solid powder of EPNP at 25°C for 71 hr, according to the method of Tang.

c) Behavior towards pyridine-2-azo-*p*-dimethylaniline (PAD)

The active sites of acid proteinases A-1, A-2 and B were examined for the structure by the method of Nakatani using PAD (Fig. 7) (19). The method is based on the following facts: (1) the catalytic activity of pepsin and other acid proteinases is functioned by two carboxyl groups interacting

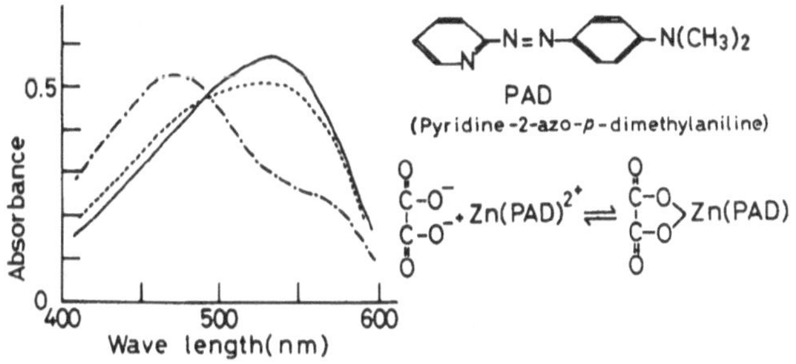

Fig.7 Chemical Structure of PAD and Its Reaction with Oxalic Acid
Absorption spectra were determined at pH 5.0.
-·-·: PAD (2.9×10^{-5}M)
------: PAD + $ZnSO_4$ (2×10^{-3}M)
——: PAD + $ZnSO_4$ + Oxalic acid (2×10^{-3}M)

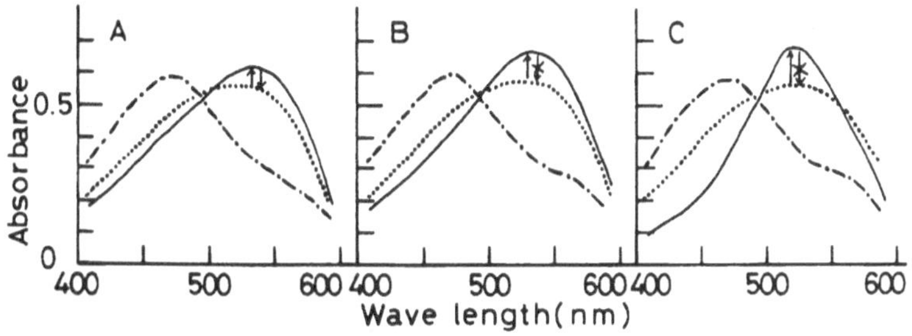

Fig.8 Absorption Spectra of Zn(II)-PAD-Acid Proteinases
Absorption spectra were determined at pH 5.0. -·-·: PAD (2.9×10^{-5}M), ----- : PAD + $ZnSO_4$ (2×10^{-3}M), ——: PAD + $ZnSO_4$ + Acid proteinase (2.9×10^{-5}M). A: Pepsin, B: Acid proteinase A-1, C: Acid proteinase B.

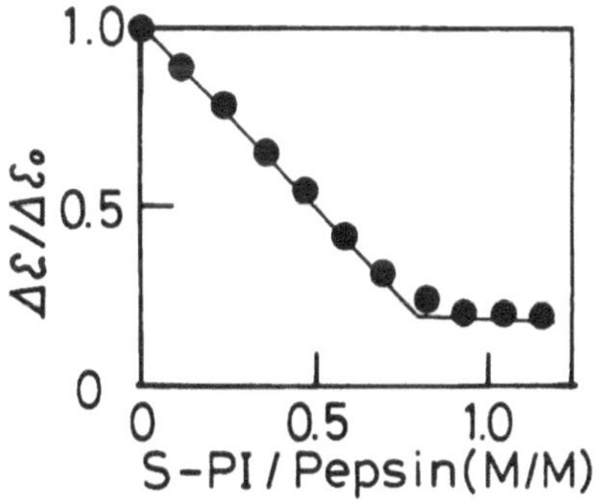

Fig.9 Titration of Zn(II)-PAD-Pepsin with S-PI

each other; (2) S-PI binds to the active site, probably the substrate-binding site, of acid proteinases.

As shown in Fig.8, A-1 and B each bound to zinc(II)-PAD, forming a zinc(II)-PAD-enzyme complex, as did pepsin. For Pepstatin Ac-sensitive acid proteinases, it was speculated that zinc(II)-PAD would bind to the two catalytic carboxyl groups in the active site, for zinc(II)-PAD binding to the enzyme was replaced by S-PI (Fig.9). On the other hand, with S. lignicolum acid proteinases, unlike pepsin, the absorption spectrum characteristic for the zinc(II)-PAD-enzyme complex did not change by the addition of S-PI; the initial absorption spectrum of zinc(II)-PAD was not restored. Thus the zinc(II)-PAD binding to the Scytalidium enzymes was not replaced by the addition of S-PI.

It can, therefore, be said that acid proteinases A-1, A-2 and B have two carboxyl groups, located very closely each other (about 3 A), like the active carboxyl groups of Pepstatin-sensitive acid proteinase. It is not clear, however, whether or not the carboxyl groups are involved in their catalytic action. Figure 10 shows schematically the results obtained.

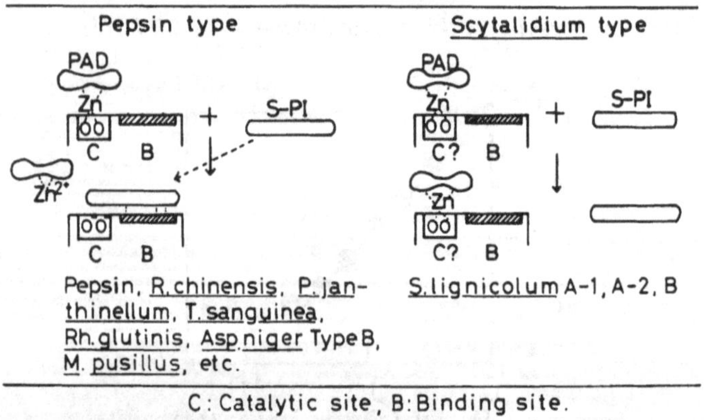

Fig.10 Schematic Diagram showing the Active Sites of Pepsin and Scytalidium-type Acid Proteinases based on the Release of Zn(II)-PAD from the Complex by the Addition of S-PI.

d) Complete amino acid sequence of acid proteinase B (20)

It is well known that carboxyl proteinases of gastric and microbial origin are highly homologous in the primary and tertiary structures and in the structure of the catalytic site. Penicillopepsin, an acid proteinase from Penicillium janthinellum shows about 40% homology with animal pepsin in the primary structure.

We attempted to determine the amino acid sequences of acid proteinases from S. lignicolum and to compare the structures of their active sites with those of Pepstatin-sensitive ones.

We first determined the amino acid sequence of acid proteinases B, since its molecular weight was the smallest.

Its complete amino acid sequence is as shown in Fig. 11.

Thr-Val-Glu-Ser-Asn-Trp-Gly-Gly-Ala-Ile(10)-Leu-Ile-Gly-Ser-Asp-Phe-Asp-Thr-Val-Ser(20)-
Ala-Thr-Ala-Asn-Val-Pro-Ser-Ala-Thr-Gly(30)-Ala-Ser-Gly-Gly-Ser-Ser-Ala-Ala-Trp-Val(40)-
Gly-Ile-Asp-Gly-Asp-Thr-Cys-Gln-Thr-Ala(50)-Ile-Leu-Gln-Thr-Gly-Phe-Asp-Trp-Tyr-Gly(60)-
Asp-Gly-Thr-Tyr-Asp-Ala-Trp-Tyr-Glu-Trp(70)-Tyr-Pro-Glu-Val-Ser-Asp-Asp-Phe-Ser-Gly(80)-
Ile-Thr-Ile-Ser-Glu-Gly-Asp-Ser-Ile-Gln(90)-Met-Ser-Val-Thr-Ala-Thr-Ser-Asp*-Thr-Ser(100)-
Gly-Ser-Ala-Thr-Leu-Glu-Asn-Leu-Thr-Thr(110)-Gly-Gln-Lys-Val-Ser-Lys-Ser-Phe-Ser-Asn(120)-
Glu-Ser-Ser-Gly-Leu-Cys-Arg-Thr-Asn-Ala(130)-Glu-Phe-Ile-Ile-Glu-Asp-Phe-Glu-Glu-Cys(140)-
Asn-Ser-Asp-Gly-Ser-Asp-Glu-Phe-Val-Pro(150)-Phe-Ala-Ser-Phe-Ser-Pro-Ala-Val-Glu-Phe(160)-
Thr-Asp-Cys-Ser-Val-Thr-Ser-Asp-Gly-Glu(170)-Ser-Val-Ser-Leu-Asp-Asp-Ala-Gln-Ile-Thr(180)-
Gln-Val-Ile-Ile-Asn-Asn-Gln-Asp-Val-Thr(190)-Asp-Cys-Ser-Val-Ser-Gly-Thr-Thr-Val-Ser(200)-
Cys-Ser-Tyr-Val(204)

Fig.11 Complete Covalent Structure of Scytalidium lignicolum Acid Proteinase B

The enzyme is a single polypeptide, with a molecular weight of 21,969, composed of 204 amino acid residues. Three intramolecular disulfide bridges were found at Cys^{47-126}, $Cys^{140-163}$, and $Cys^{192-201}$. The amino acid sequence of acid proteinase B is quite different from those of such Pepstatin- and DAN-sensitive enzymes as pepsin and penicillopepsin; however only slight sequence homology was observed between acid proteinase B and other known acid proteinases. The sequence from Asp^{98} to Leu^{105} in acid proteinase B showed some homology to those around Asp^{32} of pepsin and other known acid proteinases (Fig. 12). Interestingly, Asp^{32} of pepsin is reactive to EPNP and it is considered to be one of the active site amino acids in other known acid proteinases. Acid proteinase B is also inactivated by EPNP. The sequence from Asp^{98} to Leu^{105} in acid proteinase B is also somewhat homologous to the sequences around DAN-reactive Asp^{215} of pepsin-type acid proteinases . It remains unclear whether or not Asp^{98} of acid proteinase B is one of the active site amino acids and reactive to EPNP. These problems are now under investigation. Thus, the complete amino acid sequence of acid proteinase B is quite different from those of known acid proteinases. More marked differences would be observed between each of acid proteinases A-1, A-2 and C, that are insensitive to DAN and EPNP, and known Pepstatin-sensitive acid proteinases.

<u>S.lignicolum</u> B	(96) Thr-Ser-Asp-Thr-(100) Ser-Gly-Ser-Ala-Thr-(105) Leu-Glu-Asn-Leu-(109) Thr-
Porcine pepsin	(30) Ile-Phe-Asp*-Thr- -Gly-(35) Ser-Ser-Asn-Leu-Trp-(40) Val-Pro-Ser-
Cow pepsin	Ile-Phe-Asp-Thr- -Gly-Ser-Ser-Asp-Leu-Trp-Val-Pro-Ser-
Human pepsin	Val-Phe-Asp-Thr- -Gly-
Calf chymosin	Leu-Phe-Asp-Thr- -Gly-Ser-Ser-Asp-Phe-Trp-Val-Pro-Ser-
Penicillopepsin	Asn-Phe-Asp-Thr- -Gly-Ser-Ala-Asp-Leu-Trp-Val-Phe-Ser-
<u>Rhizopus</u> pepsin	Asx-Phe-Asp-Thr- -Gly(Ser)-Thr
<u>P. roqueforti</u>	Asn-Phe-Asp- X - -Gly(Ser)

Fig.12 Homology in the Amino Acid Sequence among <u>Scytalidium</u> Acid Proteinase B and Pepstatin-sensitive Acid Proteinases
Identical residues are shown in boxes. *EPNP-reactive aspartic acid.

6. Catalytic residues (to be published)

Acid proteinases A-1, A-2, B and C are all insensitive to S-PI, DAN and EPNP, except B to EPNP, which suggests that no carboxyl residue participates in the action. On the other hand, the zinc(II)-PAD experiments suggest the presence of two carboxyl groups located very closely each other in these enzymes. To explain the discrepancy, we carried out two distinct experiments; one was by kinetics and the other by the zinc(II)-PAD method.

a) Kinetics. Figure 13-A indicates the pH-dependence of ko/Km of hydrolysis of Z-Phe-Leu-Ala-Ala catalyzed by acid proteinase A-2, which suggests that two residues each having an apparent dissociation constant at pH 1.53 or 3.77 are involved in the action of acid proteinase A-2. Acid proteinase A-2 does not contain any histidine residue in the molecule. Therefore, both of them are considered to be carboxyl residues. Similar experiments were carried out with acid proteinase C with Suc-Phe-Arg-Ala-Phe-$NPhNO_2$ as substrate. The pH dependence of ko/Km indicates that the two residues, each having an apparent dissociation constant at pH 2.89 or 5.43, are involved in the action. Based on the data at 30 °C and 40 °C, ΔHe was calculated by the equation of Van't Hoff. Both of them were regarded as carboxyl residues. We could not examine acid proteinase B since no good substrate was available.

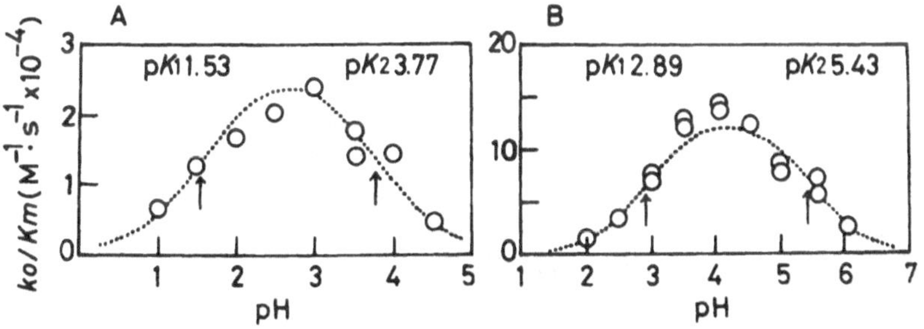

Fig.13 pH-Dependence of *ko/Km* of Hydrolysis of Synthetic Substrates catalysed by S.lignicolum Acid Proteinases

A : Acid proteinase A-2, Z-Phe-Leu-Ala-Ala, 30°C.

B : Acid proteinase C, Suc-Phe-Arg-Ala-Phe-$NPhNO_2$, 30°C.

b) Zinc(II)-PAD. By the method of zinc(II)-PAD, two carboxyl groups located very closely each other were demonstrated in acid proteinases A-1, A-2 and B. We were unable, however, to find whether they are involved in the action, although such is the case in Pepstatin-sensitive acid proteinases. To answer this question, we attempted screening for such substances that release zinc(II)-PAD from the complex of zinc(II)-PAD-Pepstatin-insensitive acid proteinases. Pepstatin derivatives, synthetic substrates, Angiotensin I, etc. were examined. Among them, angiotensin I alone released zinc(II)-PAD from the complex. Angiotensin I was effective also for Pepstatin-sensitive acid proteinases as shown in Fig. 14. As reported by Ichishima (21), angiotensin I is a good substrate for these enzymes regardless of the Pepstatin-sensitivity. Therefore, the two carboxyl groups capable of binding to zinc(II)-PAD were considered to be located

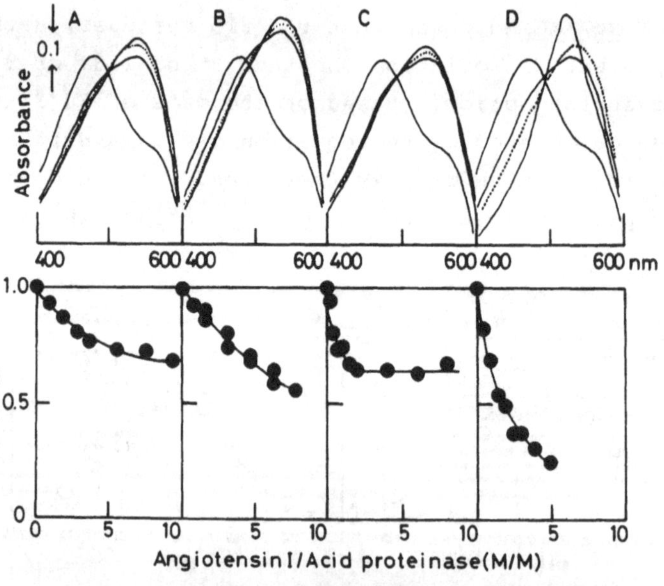

Fig.14 Absorption Spectra of Zn(II)-PAD-Acid Proteinases
A:Pepsin (1.2 x 10^{-5}M), B:Cladosporium acid proteinase (1.4 x 10^{-5}M), C:Acid proteinase A-2 (1.1 x 10^{-5}M), D:Acid proteinase B (1.5 x 10^{-5}M). Other experimental conditions are the same as those shown in Fig.8.

at the active-site regions of S. lignicolum enzymes (Fig. 15). From these results, it can be concluded that Scytalidium enzymes have two active carboxyl residues participating in the action, as do the Pepstatin-sensitive acid proteinases. It remains unclear whether the carboxyl residues are of aspartic acid or glutamic acid residues.

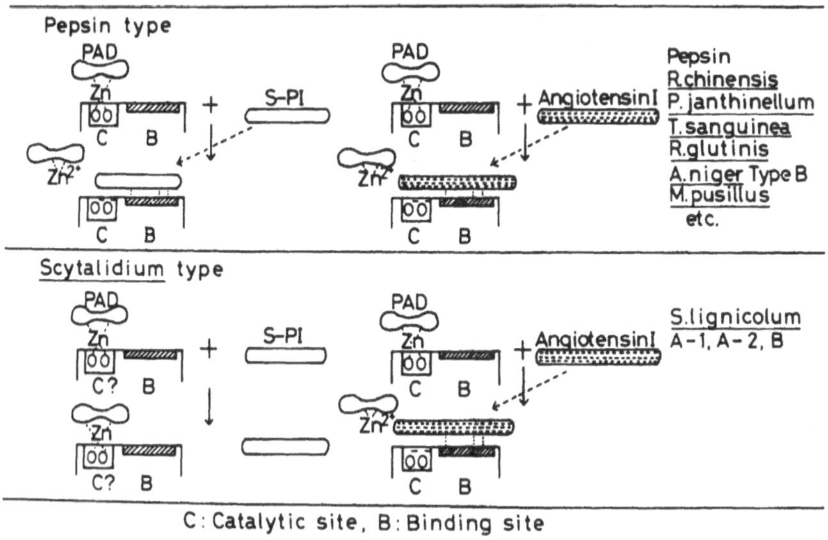

Fig.15 Schematic Diagram showing the Active Sites of Pepsin and Scytalidium-type Acid Proteinases based on the Release of Zn(II)-PAD from the Complex by the Addition of S-PI or Angiotensin I

II. Occurence of Pepstatin-Insensitive Acid Proteinases in Basidiomycetes

In 1972, we, using S-PI, isolated Scytalidium lignicolum ATCC 24568, that produced new carboxyl proteinases (6).
In our subsequent studies, we found that such enzymes resembling Scytalidium acid proteinase are widely distributed in Basidiomycetes, especially in edible mushrooms; Lentinus edodes TMI-563 (Shii-take), Ganoderma lucidum IFO 4912 (Man-

nen-take), Pleurotus cornucopiae (Tamogi-take), Pleurotus ostreatus IFO 7051 (Hira-take), Flammulina velutipes IFO 7046 (Enoki-take), and Lentinus edodes IFO 4902 (Shii-take), etc. (22). Some of their enzymes were purified and characterized (23,24) as summarized in Table 7.

Table 7 Some Properties of Pepstatin-Insensitive Acid Proteinases from Basidiomycetes

	L.edodes	G.lucidum	F.velutipes	P.ostreatus
Molecular weight	40,000	36,000	43,000	33,000
Isoelectric point	4.2	5.3	n.d	5.8
Amino acid composition	His;none	Met;none	n.d	n.d
Optimum pH (casein)	2.7	2.0	3.0	3.5
pH Stability (37°C, 3 hr)	3.2-5.2	3.5-5.2	3.3-5.5	3.8-6.0
Inhibition				
with S-PI	—	—	—	—
DAN	—	—	—	—
EPNP	—	—	—	—
Substrate specificity				
Insulin B-chain	Tyr(16)-Leu ↑ 35%	Phe(24)-Phe ↑ 49%	n.d	n.d

n.d : not determined

Some of their characteristics are: (1) The proteolytic activity was not inhibited by such a carboxyl proteinase inhibitor as S-PI, DAN or EPNP. (2) The enzyme was very labile. (3) The hydrolytic activity towards casein was lower than towards hemoglobin (the relative hydrolysis rate of G. lucidum acid proteinase was 0.15). The relative hydrolysis rates of pepsin and other known microbial acid proteinases were 0.5-1.0. (4) The enzyme showed a high substrate specificity towards oxidized insulin B-chain. For example, G. lucidum acid proteinases split preferentially (50% of total hydrolysis) the Phe(24)-Phe(25) bond of oxidized insulin B-chain. These results suggest that the low caseinolytic activity and the high substrate specificitiy towards insulin B-chain are common properties of Pepstatin-insensitive acid proteinases.

III. A Proposal for Classification of Acid Proteinases

It is known that two aspartic acid residues constitute the active site of acid proteinases and that the enzyme is specifically inactivated with Pepstatin, DAN or EPNP. Therefore, acid proteinase is called also carboxyl or aspartic proteinases.

We isolated Scytalidium lignicolum ATCC 24568 in 1972 (6,7). This strain produces four distinct new acid proteinases A-1, A-2, B and C (8). We studied these enzymes in detail, especially on the behavior towards inhibitors (8), the substrate specificities (9,10,16,17,21), the natures of active sites (19), and the complete amino acid sequence (acid proteinase B) (20). We concluded that the Scytalidium enzymes are new-type carboxyl proteinases. We further demonstrated that such enzymes having properties similar to those of Scytalidium enzymes are widely distributed among Basidiomycetes (22,23, 24). Chang (25) reported that acid proteinase Type A of Aspergillus niger var. macrosporus is also insensitive to Pepstatin and EPNP.

Table 8 Classification of Acid Proteinases

Type	Acid proteinase	Optimum pH	Inhibitor			Substrate specificity		Active center (catalytic residue)
			S-PI	DAN	EPNP	Z-X-Y-Ala-Ala	casein/hemoglobin	
Scytalidium	S.lignicolum							
	A-1, A-2	3.0-3.5	−	−	−	↑	0.2	
	B	2.2	−	−	+	↑	0.3	-COOH (Asp or Glu)
	C	2.0	−	−	−		1.3	
	Others[1]	2.0-3.3	−	−	−		0.2-0.4	
Pepsin	Pepsin	1.8	+	+	+	↑	0.4	-COOH (Asp)
	Others[2]	1.8-3.3	+	+	+	↑	0.5-1.0	

1) Lentinus edodes, Ganoderma lucidum, Pleurotus ostreatus, Flammulina velutipes, Aspergillus niger type A, etc.
2) Chymosin, Gastricsin, Microbial acid proteinases reported so far.

We propose that Pepstatin- and DAN-insensitive caroboxyl proteinases should be classified in a new subclass, the Scytalidium type, of carboxyl proteinases, as shown in Table 8.

References

1. Hartley, B.S.: Annu. Rev. Biochem. 29, 45-72 (1960).
2. Murao, S., Satoi, S.: Agric. Biol. Chem. 34, 1265-1267 (1970).
3. Rajagopalan, T.G., Stein, W.H., Moore, S.: J. Biol. Chem. 241, 4295-4297 (1966).
4. Tang, J: J. Biol. Chem.: 246, 4510-4517 (1971).
5. Tang, J.: Molecular and Cellular Biochemistry, 26, 93-109 (1979).
6. Murao, S., Oda, K., Matsushita, Y.: Agric. Biol. Chem. 36, 1647-1650 (1972).
7. Murao, S., Oda, K., Matsushita, Y.: Agric. Biol. Chem. 37, 1417-1421 (1973).
8. Oda, K., Murao, S.: Agric. Biol. Chem. 38, 2435-2444 (1974).
9. Oda, K., Murao, S., Oka, T., Morihara, K.: Agric. Biol. Chem. 40, 859-866 (1976).
10. Oda, K., Murao, S., Oka, T., Morihara, K.: Agric. Biol. Chem. 39, 477-484 (1975).
11. Murao, S., Morita, N., Yamamoto, Y., Oda, K.: Agric. Biol. Chem. 42, 1343-1349 (1978).
12. Murao, S., Katsura, M., Fukuhara, K., Oda, K.: Agric. Biol. Chem. 44, 701-703 (1980).
13. Murao, S., Kasai, N., Kimura, Y., Oda, K., Fukuhara, K.: Agric. Biol. Chem. 46, 855-857 (1982).
14. Murao, S., Watanabe, T.: Agric. Biol. Chem. 41, 1313-1314 (1977).
15. Murao, S., Sato, S.: Agric. Biol. Chem. 36, 160-163 (1972).
16. Morihara, K., Tsuzuki, H., Murao, S., Oda, K.: J. Biochem. 85, 661-668 (1979).
17. Oda, K., Murao, S.: Agric. Biol. Chem. 40, 1221-1225 (1976).
18. Morihara, K.: Proteinases and Their Inhibitors (Turk, V., Vitale, Lj. eds.) pp. 213-222, Mladinska Knjiga-

Pergamon Press, Ljubljana, Oxford.

19. Nakatani, H., Hiromi, K., Satoi, S., Oda, K., Murao, S., Ichishima, E.: Biochim. Biophys. Acta. 391, 415-421 (1975).

20. Maita, T., Nagata, S., Matsuda, G., Maruta, S., Oda, K., Murao, S., Tsuru, D.: J.Biochem. 95, 465-475 (1984).

21. Ichishima, E., Majima, E., Emi, M., Hayashi, K., Murao, S.:Agric. Biol. Chem. 45, 2391-2393 (1981).

22. Oda, K., Terashita, T., Kono, M., Murao, S.: Agric. Biol. Chem. 45, 2339-2340 (1981).

23. Terashita, T., Oda, K., Kono, M., Murao, S.: Agric. Biol. Chem. 45, 1937-1943 (1981).

24. Terashita, T., Oda, K., Kono, M., Murao, S.: Agric. Biol. Chem. 48, 1029-1035 (1984).

25. Chang, W. J., Horiuchi, S., Takahashi, K., Yamasaki, M., Yamada, Y.: J. Biochem. 80, 975-981 (1976).

RENIN INHIBITORS. DESIGN OF ANGIOTENSINOGEN TRANSITION-STATE ANALOGS CONTAINING STATINE

Conformationally Restricted Inhibitors and a Model for the Bound Conformation of Renin Substrate

Joshua Boger

Department of Medicinal Chemistry, Merck Sharp & Dohme Research Laboratories, West Point, Pennsylvania 19486 USA

Introduction

The proteinase renin (EC 3.4.99.19) is the first enzyme in the renin-angiotensin system, a proteolytic cascade responsible for the formation of the potent pressor substance angiotensin II and the aldosteronogenic angiotensin III (see Figure 1). Renin cleaves the plasma protein substrate angiotensinogen, removing the N-terminal decapeptide angiotensin I, transformed further by converting enzyme into the active angiotensin II. In addition to its action in plasma, a renin-angiotensin system may be present in vascular wall, in kidney, and in brain. Although the contribution of this pathway to the maintenance of normal blood pressure is unclear, it appears to be a major factor in various hypertensive states (1). Inhibition of this system at the converting enzyme reaction, with captopril and with enalapril maleate, is a major advance in antihypertensive therapy (2). Inhibition of renin would provide an alternative means of interruption of this system.

Unlike the angiotensin converting enzyme, renin is an extraordinarily specific enzyme. Its only known natural substrate is

Aspartic Proteinases and their Inhibitors

the protein angiotensinogen, and the minimum synthetic substrate is the octapeptide segment 6-13 around the cleavage site (3, see Figure 2). Renin is an aspartic proteinase, in the same family as pepsin, cathepsin D, and the fungal enzymes rhizopuspepsin, penicillopepsin, and endothiapepsin. These aspartic proteinases have a pair of aspartic acid residues in their active sites, as determined either by structural studies or by inhibition with specific reagents. The most potent competitive inhibitor of this enzyme class is the natural product pepstatin **8**, isovaleryl-Val-Val-Sta-Ala-Sta, which contains the unusual amino acid statine (Sta). The 3S-hydroxyl group of the central Sta of pepstatin is thought to function in pepsin as a transition-state or intermediate analog of the substrate scissile carbonyl (4, see Figure 2).

While pepstatin **8** is a subnanomolar competitive inhibitor of most aspartic proteinases (pepsin, K_i = 0.046 nM, 5), it is a comparatively weak inhibitor of renin (human renin, K_i = 13000 nM). The high substrate specificity of renin led us to propose that incorporation of Sta into sequences more closely resembling the renin minimum substrate might give more potent renin inhibitors (6,7). Reviewed here are computer modelling

```
1   2   3   4   5   6   7   8   9   10  11  12  13
Asp-Arg-Val-Tyr-Ile-His-Pro-Phe-His-Leu-Leu-Val-Tyr-R Porcine
                                       |-Val-Ile-His-R Human
                                       |          Angiotensinogen
                                       ↓ RENIN
Asp-Arg-Val-Tyr-Ile-His-Pro-Phe-His-Leu           Angiotensin I
                                ↓ CONVERTING ENZYME
Asp-Arg-Val-Tyr-Ile-His-Pro-Phe                   Angiotensin II
   ↓ AMINOPEPTIDASE
    Arg-Val-Tyr-Ile-His-Pro-Phe                   Angiotensin III
```

FIGURE 1. Renin-Angiotensin System. R = Remainder of the α_2-globulin. MW of angiotensinogens = 50-60,000. Numbering begins from the N-terminus of the protein.

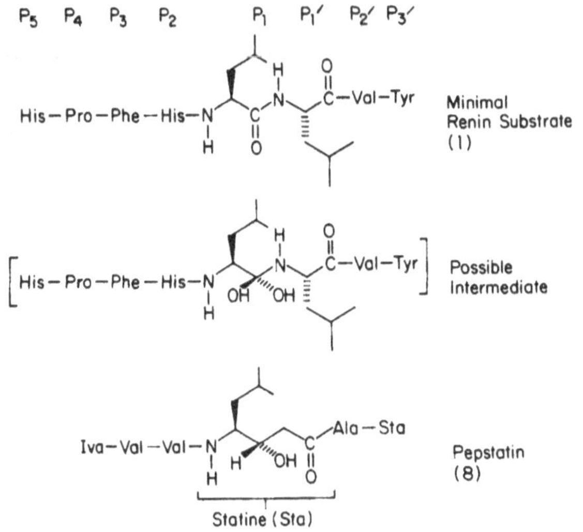

FIGURE 2. Minimal fully active porcine renin substrate octapeptide **1**, sequence 6-13 (3). The renin substrate residues are designated using the nomenclature of Schechter & Berger (15), in which the scissile dipeptide is P_1-$P_{1'}$. Possible intermediate from renin-catalyzed attack of water on the scissile Leu-Leu amide bond. Pepstatin **8**, with the central Sta aligned to show its proposed role in an analog of the intermediate.

studies, using conformationally restricted renin inhibitors containing Sta and an X-ray structure of rhizopuspepsin containing pepstatin **8**, from which can be constructed a conformation for the octapeptide renin substrate as bound in renin.

Results and Discussion

<u>Design of renin inhibitors.</u> Considering the high substrate

specificity of renin, it is surprising that pepstatin **8** is an inhibitor of renin. Despite the lack of structural homology between pepstatin **8** and the minimum renin substrate octapeptide **1** (see Figure 2 and Table 1), pepstatin **8** must provide some of the recognition and binding sites demanded of a renin substrate. One of these is the Sta side chain itself, which is identical to the Leu-10 side chain of the renin substrate. With Sta substituted for Leu-10, the analogy with substrate

TABLE 1. Inhibition of Renin by Statine-Containing Substrate Analogs.

COMPOUND	STRUCTURE	INHIBITION	
	P_5 P_4 P_3 P_2 P_1 $P_{1'}$ $P_{2'}$ $P_{3'}$	K_i, 10^{-9}M	
NO.	6 7 8 9 10 11 12 13	PORCINE	HUMAN
1*	His-Pro-Phe-His-Leu-Leu-Val-Tyr	55000	ND
2	His-Pro-Phe-His-Sta-----Leu-Phe-NH_2	6	130
3	Iva-His-Pro-Phe-His-Sta-----Leu-Phe-NH_2	10	8.8+
4	Iva-His-Pro-Phe-His-Sta-----Ile-Phe-NH_2	48	1.9
5	Ibu-His-Pro-Phe-His-Sta-----NH_2	11000	870
6	Ibu-His-Pro-Phe-His-Sta-----Leu-NH_2	260	165
7	Ibu-His-Pro-Phe-His-Sta-----Leu-Phe-NH_2	16	7.7+
8 (Pepstatin)	Iva-Val-Val-Sta-----Ala-Sta	380	13000

Porcine kidney renin (pH 7.3, 30°C) and human kidney renin (pH 7.4, 37°C) assays as described (6). Human renin K_i values consistent with competitive inhibition. Porcine renin K_i values calculated from IC_{50} values assuming competitive inhibition (14). Compounds characterized by TLC, HPLC, ^1H NMR, amino acid analyses, and elemental analyses. Iva = isovaleryl; Ibu = isobutyryl; Sta = statine, (3<u>S</u>,4<u>S</u>)-4-amino-3-hydroxyl-6-methyl heptanoyl; ND = not determined. The numbering schemes above the compound list are both the Schechter & Berger scheme (scissile bond = P_1-$P_{1'}$, 15) and that of the porcine renin substrate and reflect the hypothesized analogy between substrate and inhibitors. *Minimum substrate sequence, K_m value (3). +Revised from Ref. 6 and 7.

can be made exact on the N-terminal side of the substrate as in **5** (Table 1), giving a µM inhibitor of porcine renin and a somewhat better inhibitor of human renin. The continuation of this inhibitor toward the C-terminus is problematical, since Sta is two carbon atoms longer than an α-amino acid, which should introduce a "frame-shift" into the subsequent amino acids. On the basis of molecular modelling using the X-ray conformation of pepstatin in rhizopuspepsin (**8**), we proposed (6,7) that Sta could be a dipeptide analog (Figure 3). Thus Sta could replace Leu-10 (P_1) and Leu-11 ($P_{1'}$), allowing correction of the frame shift and restoration of the analogy with substrate. Modelling indicated that the restoration of the analogy could be approximate by the Val-12 ($P_{2'}$) and almost exact by the important Tyr-13 ($P_{3'}$). Thus extension of the pentapeptide **5** to **6** and **7** gives potent competitive inhibitors of porcine renin with K_i values 3000-9000 times smaller than a comparable substrate K_m value (Table 1, compare **2, 3, 7** and **1**), as expected for a transition-state analog inhibitor.

Leu was found to be the optimal substitution to mimic Val-12, while Phe, Tyr, or Leu were equally good for Tyr-13. Extension further, past position 13, gives no added inhibition. Substitution of the human sequence-like Ile at position 12 does give more potent inhibitors of human renin, such as **4**. As expected, inhibitors designed from the porcine substrate sequence (Figure 2) are potent inhibitors of human renin as well, since human renin, a less fastidious enzyme, cleaves the porcine substrate and its own human sequence equally well (9).

Inhibitors deleting His-6 at the N-terminus (7) show activity comparable to the full heptapeptides, provided that a free amino terminus is acylated. Thus there appears to be little enzyme recognition of the His-6 side chain, although modifications at this P_5 position and at P_6 show small and often different effects on the inhibitions of various renins.

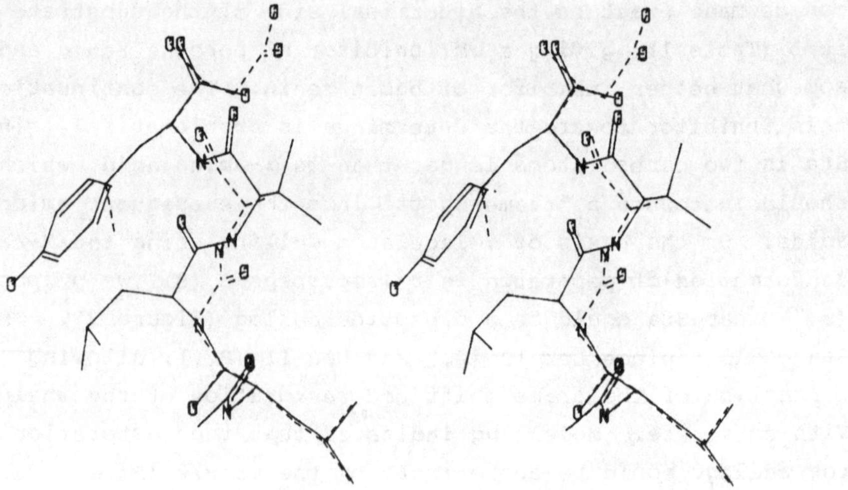

FIGURE 3. Ac-Leu*Leu-Val-Tyr (solid line, substrate fragment 10-13), where * denotes a tetrahedral carbonyl (-CHOH-NH-), matched onto the ...Sta-Ala-Sta portion (dashed line) of iso-valeryl-Val-Val-Sta-Ala-Sta **8** (pepstatin) as in the complex with Rhizopus chinensis aspartic proteinase (22, refitted to 2.5Å data, see also Ref. 8). Matched were N, C_α, and C(OH) of Leu-10 to corresponding atoms of the central Sta of pepstatin, then adding the C_α and C_β of Tyr-13 matched onto the terminal Sta of pepstatin. The intervening eight dihedral angles in the substrate fragment backbone were free to rotate, with the two moving amide bonds restrained near 180°. Termini backbone dihedrals and side chain dihedral angles for Leu-10 and Tyr-13 were adjusted to match onto pepstatin. Side chains of Leu-11 and Val-12 are in an energy-minimized conformation.

Aspartic proteinase specificity. Statine-containing heptapeptide analogs of the renin octapeptide substrate are specific for renin versus other aspartic proteinases (see **3**, **4**, and **9** in Table 2), in contrast to pepstatin **8**. Of special note is the unique importance for renin of Phe-8, which is worth a factor of 4000 in K_i versus Ala, as in **10**, a substitution of only small effect on other aspartic proteinases. The P_3 site (Phe-8) is a primary renin-specific substrate recognition site.

TABLE 2. Aspartic Proteinase Specificity of Renin Inhibitors

COMPOUND	STRUCTURE	K_i, 10^{-9}M			
		HUMAN	PEPSIN	CAT.	RHIZ.
NO.	6 7 8 9 10 12 13	RENIN		D	PEPSIN
4	Iva-His-Pro-Phe-His-Sta-Ile-Phe-NH_2	1.9	43	134	24
3	Iva-His-Pro-Phe-His-Sta-Leu-Phe-NH_2	8.8*	27	210	200
9	Boc-His-Pro-Phe-His-Sta-Leu-Phe-NH_2	2.3	56	1600	110
10	Boc-His-Pro-Ala-His-Sta-Leu-Phe-NH_2	9300	21	15200	73
8	Iva-Val-Val-Sta-Ala-Sta	13000	0.05**	$\leq 0.5^{+}$	$\leq 17^{\#}$

Human kidney renin assay as described in Table 1. Porcine pepsin (pH 4.0) and rabbit liver cathepsin D (Cat. D, pH 4.0) assays as described previously (6, 16). Rhizopuspepsin assay was performed as for porcine pepsin, but with E_t = 60 nM, S_o = 0.1 mM, and K_m = 50 μM. Abbreviations: see Table 1; Boc = <u>tert</u>-butyloxycarbonyl. *Data revised from Ref. 6. **Ref. 5. +Ref. 17. #Upper limit estimated from IC_{50} values close to E_t.

<u>Inhibitors containing 2-alkyl-statine.</u> Statine acting as a dipeptide analog has no second side chain. Modelling based upon the bound conformation of pepstatin in rhizopuspepsin indicated that an added 2-<u>R</u>-alkyl side chain would be bound in a large unfilled hydrophobic cleft in rhizopuspepsin, the proposed $P_{1'}$ binding site (10). Heptapeptide renin inhibitors containing either 2-<u>R</u> or 2-<u>S</u> alkyl-statines (**12** and **13**, Table 3) confirmed that the 2-<u>R</u> isomer **12** is the more active renin inhibitor, but only slightly more active when compared to the reference unsubstituted compounds **7**, **9**, and **11**. The role of the $P_{1'}$ site of renin substrate in binding is unknown, but the successful attachment to Sta of a large side chain in the predicted stereochemistry is consistent with the hypothesis that Sta is a dipeptide analog and strengthens the justification for the use of the rhizopuspepsin structure for renin modelling. The rhizopuspepsin inhibition results for **12** and **13** are puzzling, but may indicate that the 2-isobutyl group

TABLE 3. Inhibition of Renin by Substrate Analogs Containing 2-Alkyl-Statines

COMPOUND NO.	STRUCTURE 6 7 8 9 10 11 12 13	K_i, 10^{-9}M HUMAN RENIN	RHIZOPUS PEPSIN
7	Ibu-His-Pro-Phe-His-**Sta**---------Leu-Phe-NH_2	7.7*	230
9	Boc-His-Pro-Phe-His-**Sta**---------Leu-Phe-NH_2	2.3	110
11	Boc-His-Pro-Phe-Phe-**Sta**---------Leu-Phe-NH_2	0.78	220
12	Ibu-His-Pro-Phe-Phe-**Sta**(2<u>R</u>-<u>i</u>bu)-Leu-Phe-NH_2	1.7	1900
13	Ibu-His-Pro-Phe-Phe-**Sta**(2<u>S</u>-<u>i</u>bu)-Leu-Phe-NH_2	120	1900

Human kidney renin and rhizopuspepsin assays as described in Tables 1 and 2. Syntheses of **13** and **14** reported in Ref. 10. Abbreviations: See Tables 1 and 2; ibu = isobutyl (i.e., 2-methyl-1-propyl). *Value revised from Ref. 7.

itself is not optimal for rhizopuspepsin. From the modelling, this 2-**R**-isobutyl group would resemble most closely a Val in the $P_{1'}$ position.

<u>Cyclic inhibitors: C-terminal cycle.</u> Modelling based on the pepstatin structure in rhizopuspepsin indicated that a disulfide cycle could be constructed from the P_2 position toward the C-terminus, as shown in Figure 4. Either a homocysteine (Hcy) or Cys substituted for His-9 was found to allow completion of a cycle to an added C-terminal Cys side chain (7). The renin inhibition results for these cyclics, **15** and **16** (Table 4) show that, although there is some loss in potency due to replacement of the His-9 residue, inhibition is comparable to that of the linear analog **14**. The replacement of His-9 with Ala or the linear Cys or Hcy chains leads to increased binding by pepsin and rhizopuspepsin, showing that His-9 is a renin-specific residue.

TABLE 4. C-Terminal Cyclic Inhibitors: Aspartic Proteinase
6 7 8 9 10 12 13 14
Specificity of Ibu-His-Pro-Phe-Y-Sta-Leu-Phe-Z

COMPOUND NO.	POSITION 9 Y	POSITION 14 Z	K_i, 10^{-9}M PORCINE RENIN	HUMAN REININ	PEPSIN	RHIZOPUS PEPSIN
7	His	NH_2	16	7.7*	23	230
14	Ala	NH_2	110	99	3.7	$\leq 17^{+}$
15	Cys	$NH-CH_2-CH_2-S$	48	380	$\leq 1.0^{+}$	$\leq 17^{+}$
16	Hcy	$NH-CH_2-CH_2-S$	71	200	$\leq 1.0^{+}$	40

Assays and abbreviations as described in Tables 1 and 2. Hcy = L-homocysteinyl. *Revised from value reported in Ref. 7. ⁺Upper limit estimated from IC_{50} values close to E_t.

TABLE 5. N-Terminal Cyclic Inhibitors: Aspartic Proteinase
6 7 8 9 10 12 13 14
Specificity of H-X-Pro-Phe-Y-Sta-Leu-Phe-Z

COMPOUND NO.	POSITION 6 X	POSITION 9 Y	POSITION 14 Z	K_i, 10^{-9}M PORCINE RENIN	PEPSIN	CATHEPSIN D	RHIZOPUS PEPSIN
2	His	His	NH_2	7.7	34	900	1200
17	Abu	Abu	OCH_3	9.2	≤ 1.0*	85	≤ 17*
18	Hcy	Hcy	OCH_3	19	≤ 1.0*	63	≤ 17*

See notes for Table 4. Abu = L-α-aminobutyryl. *Upper limit estimated from IC_{50} values close to E_t.

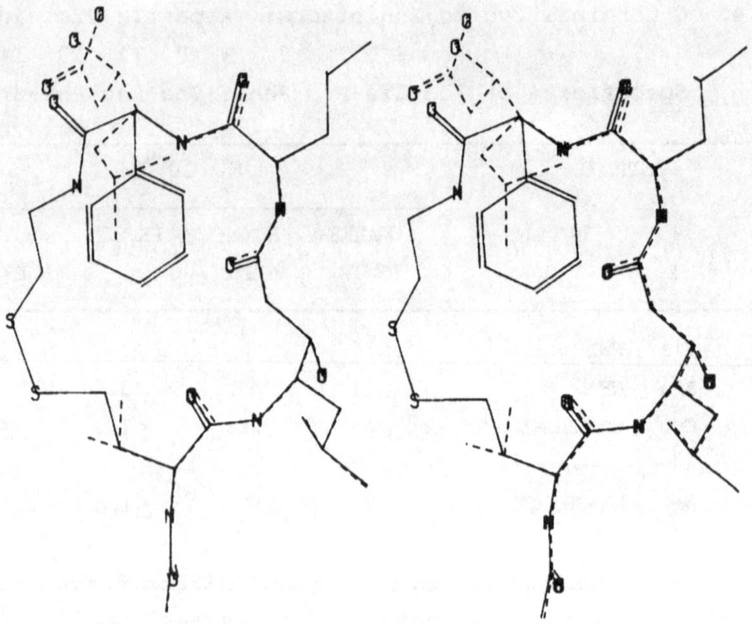

FIGURE 4. Ac-Hcy-Sta-Leu-Phe-NH-CH_2-CH_2-S (solid line) matched onto Val-Sta-Ala-Sta portion of pepstatin **8** (dashed line), which is in the conformation found in rhizopuspepsin-pepstatin structure (see Figure 3). Hcy = L-homocysteinyl.

Cyclic inhibitors: N-terminal cycle. Paiva et al. (11) reported a cyclic disulfide constrained renin substrate analog with Cys substitutions at positions 5 and 10, enclosing a proposed Pro-Phe turn in a 20-membered ring. Our modelling indicated (7) that a similar constraint could be made by forming a disulfide from positions 6 to 9. The optimal size for such a disulfide restrained turn for renin was found to be a 16-membered ring, using a Hcy-6, Hcy-9 disulfide (**18**, Table 5). This disulfide ring provides a strong constraint on the position of the Phe-8 required by renin, and thus this disulfide is a sensitive probe of renin bound conformation. As with the C-terminal cycles, replacement of the His-9 with a

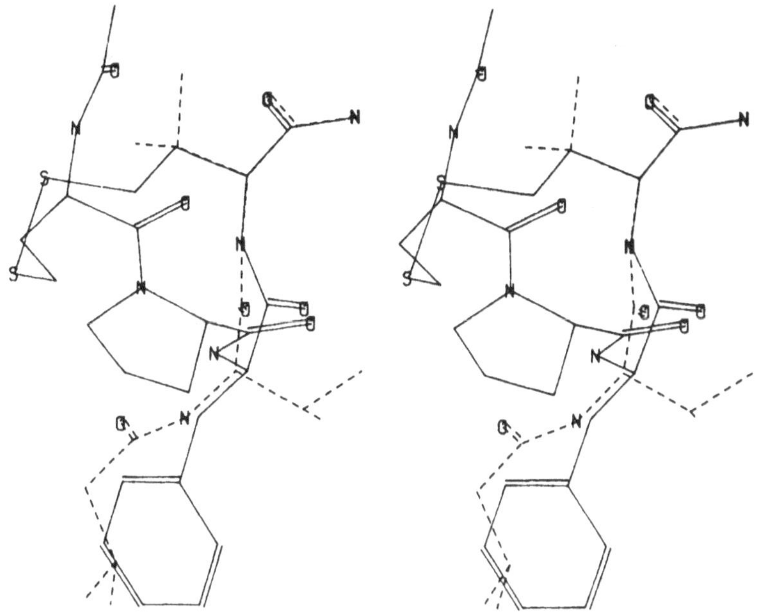

FIGURE 5. Ac-Hcy-Pro-Phe-Hcy-NH-CH_3 (solid line), type I β-turn, energy minimized conformation, fit onto the Iva-Val-Val portion (dashed line) of pepstatin **8**. The fit used the C_β, C_α, and C=O of the C-terminal Hcy onto the C-terminal Val of the pepstatin fragment.

linear side chain, as in **17** or **18**, leads to decreased aspartic proteinase specificity.

Calculations suggest that the disulfide constrained Pro-Phe is most likely in a Type I beta turn, a turn structure common in proteins. An energy-minimized Type I Hcy-Pro-Phe-Hcy cycle is shown in Figure 5 superimposed onto the pepstatin structure. Although the overall fit is poor, since the structures differ greatly, it is of interest to note that the renin inhibitor Phe (P_3) is near the pepstatin isovaleryl group, suggesting pepstatin may thus provide for a P_3 mimic, allowing for the modest renin inhibition demonstrated by pepstatin.

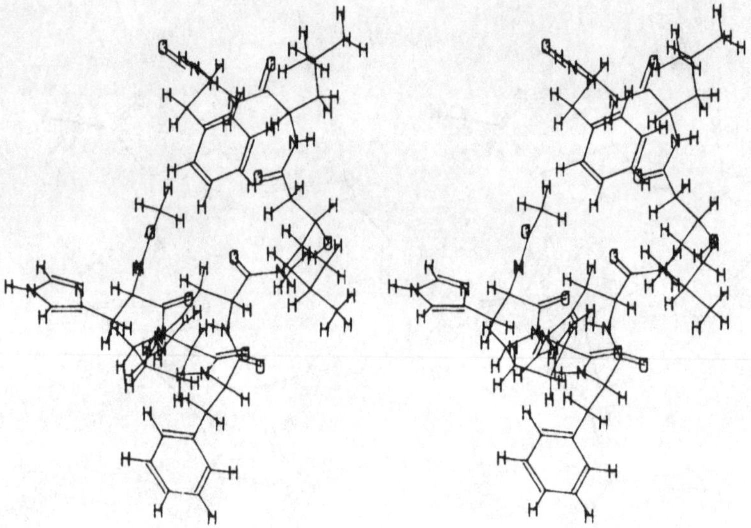

FIGURE 6. Hypothetical conformation of the renin inhibitory heptapeptide sequence Ac-His-Pro-Phe-His-Sta-Leu-Phe-NH_2 based on the rhizopuspepsin bound conformation of pepstatin **8** and the cyclic analogs in Tables 4 and 5 and Figures 4 and 5.

Bound conformation of statine-containing renin inhibitors. The modelling and inhibition data on the N- and C-terminal cyclic inhibitors, along with the X-ray conformation of pepstatin bound in rhizopuspepsin, allow construction of a model for the renin-bound conformation of statine-containing renin inhibitors, shown in Figure 6. The conformation at the C-terminus is identical to that seen for pepstatin, while the N-terminus is fixed in the Type I beta turn structure. This pepstatin-based conformation, while not constructed using the native rhizopuspepsin structure explicitly as a guide, can be slipped into the rhizopuspepsin active site (with pepstatin removed), without serious unfavorable interactions with the enzyme. Moreover, the inhibitor residues found to be sensitive to substitution (positions 7-13) all are within the influence of the active site residues, while the non-required position 6 extends into solvent.

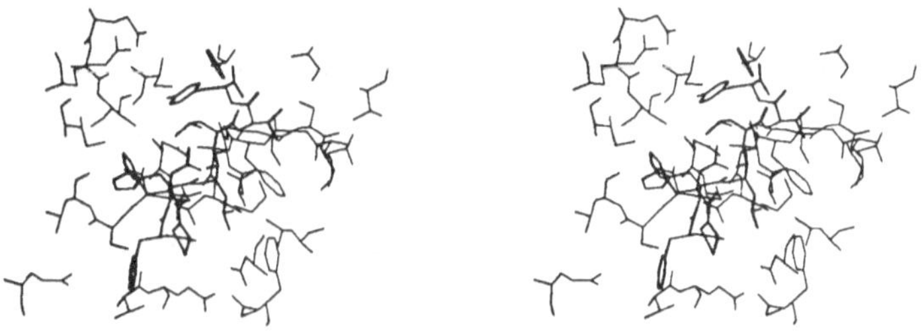

FIGURE 7. Hypothetical conformation of the renin substrate octapeptide intermediate analog sequence Ac-His-Pro-Phe-His-Leu*Leu-Val-Tyr (bold line) in the rhizopuspepsin active site (1.8Å native structure, R. Bott, unpublished results), where * denotes a tetrahedral amide,-CHOH-NH-. The substrate conformation is based on the pepstatin-rhizopuspepsin X-ray structure (22, refitted 2.5Å data, see also Ref. 8), and the conformations implied in Figures 3-5. Rhizopuspepsin residues shown are those within 6Å of the bound renin substrate. See Table 6 for the identity of those residues within 4Å.

Bound conformation of renin substrate octapeptide. The hypothesis that statine functions as a dipeptide analog, the starting point for this work, has received support from the structure-activity results at positions following Sta-10, which have been found generally to parallel the results from substitutions in renin substrate at positions 12 and 13. Additional support is provided by the 2-alkyl-substituted statines, which show that a large aliphatic side chain is allowed in the position analogous to $P_{1'}$. Together these results suggest that the renin-bound conformation of the C-terminus of renin substrate resembles the model shown in Figure 3. The renin-bound conformation of the N-terminus is suggested to resemble the model shown in Figure 5. Combining these models generates a model for the renin-bound conformation of renin substrate octapeptide, shown in Figure 7 bound in rhizopuspepsin.

TABLE 6. Binding of Proposed Renin Substrate Conformation in Rhizopus Pepsin: 4Å Neighbors

SUBSTRATE		ENZYMES				
POSITION	RESIDUE	RHIZOPUS PEPSIN	PORCINE PEPSIN	CATHEPSIN D	MOUSE RENIN	HUMAN RENIN
P_4	Pro	Asp (77)	Thr	Ser	Ser	Thr
P_3	Phe	Val (12)	Thr	Ala	Ser	Thr
		Thr (218)	Thr	Thr	Ser	Ala
		Ser (219)	Ser	Ser	Ser	Ser
		Ile (222)	Thr	Val	Ser	Ser
P_2	His	Thr (218)	Thr	Thr	Ser	Ala
		Ile (222)	Thr	Val	Ser	Ser
P_1	Leu	Asp (30)	Ile	Val	Ile	Val
		Asp (32)	Asp	Asp	Asp	Asp
		Phe (75)	Tyr	Tyr	Tyr	Tyr
		Asp (77)	Thr	Ser	Ser	Thr
		Phe (111)	Phe	Thr	Pro	Pro
		Leu (120)	Ile	Ile	Val	Val
		Gly (217)	Gly	Gly	Gly	Gly
$P_{1'}$	Leu (Val)	Ile (213)	Ile	Ile	Val	Leu
		Asp (215)	Asp	Asp	Asp	Asp
		Thr (218)	Thr	Thr	Ser	Ala
		Ile (301)	Ile	Ile	Val	Ala
$P_{2'}$	Val (Ile)	Ser (35)	Ser	Ser	Ser	Ser
		Ile (73)	Ile	Ile	Ile	Leu
		Phe (75)	Tyr	Tyr	Tyr	Tyr
$P_{3'}$	Tyr (His)	Tyr (189)	Tyr	Tyr	Ser	Val
		Ile (213)	Ile	Ile	Val	Leu
		Ile (301)	Ile	Ile	Val	Ala

TABLE 6. (Adjacent page). Binding of Proposed Renin Substrate Conformation in rhizopuspepsin: 4Å Neighbors. See notes to Figure 7 and text. Aspartic proteinase sequences: porcine pepsin (18), porcine cathepsin D (19), mouse submaxillary gland renin (20), and human kidney renin (21). Numbering for rhizopuspepsin residues is based on homology with the porcine pepsin sequence. The renin substrate residues are designated using the nomenclature of Schechter & Berger (15), in which the scissile dipeptide is P_1-$P_{1'}$. Substrate residues in parenthesis are those which differ in the human substrate sequence (23).

The construction of an enzyme three-dimensional structure using its primary structure and the known X-ray of a related enzyme is a perilous task. Where the primary homology is strong, replacements of a few differing side chains can be made with some confidence, perhaps especially for conservative changes, such as Glu to Asp, or Ile to Val. Where the homology is weaker, such replacement is difficult to perform, even with the aid of various energy calculation techniques, and is only a smaller version of the unsolved problem of a priori protein structure prediction. Nevertheless, it may be useful to examine the X-ray structure of rhizopuspepsin (8) along with the proposed renin substrate conformation, with regard to variations in the complex which might characterize the renin substrate-renin interaction.

Table 6 lists the residues contacted, within 4Å, between the side chains of the proposed renin substrate octapeptide bound conformation and the rhizopuspepsin X-ray structure. Listed also are the likely homologous residues in several other aspartic proteinases, including mouse and human renins. For none of these, beside rhizopuspepsin, are X-ray structures available. It should be emphasized that in these other enzymes, these homologous residues may not be in the same three-dimensional structure, due to changes elsewhere in the chain. More importantly, these other enzymes may, and probably do, have main and side chain differences which would lead to additional interactions with this proposed substrate conformation. Nevertheless, it may be instructive to examine these proposed binding sub-sites.

The rhizopuspepsin active site, where pepstatin binds, is a long, deep cleft in one enzyme face, formed at the nearly symmetrical pairing of two enzyme lobes (8). At the middle of the site, at its deepest point, are the two catalytically important aspartic acid residues, which give the enzyme class its name. Running down one side of the site is a wall of largely hydrophobic residues. The other side is characterized by the flap, a β-hairpin loop on the surface, which may move at a hinge region to open and close on the P_1 binding site (8).

At the top of the active site (see Figure 7) are the residues which may interact with the $P_{3'}$ residue (= the $S_{3'}$ enzyme site). Three residues of this site in rhizopuspepsin, Tyr-189, Ile-213, and Ile-301, are conserved in pepsin and cathepsin D (see Table 6). The binding site appears to open up in the renins, yet retains its generally hydrophobic character, appropriate for binding Tyr or Phe. In human renin, the Ile-301 of rhizopuspepsin is an Ala, accounting perhaps for the relatively weak binding of the $P_{3'}$ residue (for example, compare **6** and 7 in Table 1). Additionally, the more open accessible $S_{3'}$ site in human renin may allow for the binding of the more polar His, the $P_{3'}$ residue of human substrate.

On the other side of the active site, parts of three largely-conserved residues form an $S_{2'}$ site. This site is rather shallow, consistent with the observation that this $P_{2'}$ position for renin substrate is not strongly recognized by renins (unpublished results). Moving back to the other side of the active site, several residues, including the carbon side chain portion of Asp-215, form a hydrophobic pocket identified as the $S_{1'}$ site. This pocket enlarges somewhat in the renins.

Many residues form the S_1 site of rhizopuspepsin, all making hydrophobic contacts with the P_1 side chain of Leu. Although parts of this site change somewhat unpredictably in the renins (see below), the flap region which covers the site is retained. Two residues of the hydrophobic wall interact with the P_2 (His) residue. As has been noted (see Tables 4 and 5), only renins prefer the polar His residue. The very hydrophobic S_2 sites of the non-renin enzymes give way to Ser and Ala in the renins.

Other than P_1, the most important substrate residue for renin is P_3 (Phe). An area of poor homology in rhizopuspepsin sequence 274 to 282 compared to the renins makes it likely that the residues identified in Table 6 as the P_3 binding site are not the only ones interacting with P_3 in renin. The few good contacts these residues make with the substrate Phe are consistent with the unimportance of this position for binding to rhizopuspepsin and the other non-renin aspartic proteinases (see **9** and **10** in Table 2). Something not explained by the homologous replacements listed must be important in renin. Model building of the renin sequence in the regions 108 to 117 and 274 to 282 has been inconclusive, but suggests that residues from these areas may interact with renin substrate P_3 (Phe).

Similarily, the proposed binding of P_4 (Pro) may be described inadequately by the single contact noted in rhizopuspepsin. For rhizopuspepsin, the Asp-77 makes bad contacts with substrate Pro, but these can be relieved in the other enzymes. For renins, residues in the 108 to 117 area may interact with P_4. Consistent with the structure activity results, there are no residues which make contact with P_5 (His), although a P_6 residue could lie across the outside of the site.

Conclusions

The hypothesis that Sta can act as a dipeptide analog of the scissile residues for aspartic proteinases led to the design of potent, selective, competitive inhibitors of renin (6, 7). These compounds, transition-state or intermediate analogs of the substrate, have been shown to be effective in vivo inhibitors of the renin-angiotensin system (12), and may provide a novel site for the control of hypertension. Design and syntheses of conformationally-restricted renin inhibitors (7) and the availability of an X-ray structure of a related enzyme, rhizopuspepsin (8), have allowed construction of a model for the bound conformation of these inhibitors. Consideration of the role of Sta enables the elaboration of the model to include the bound conformation of the minimum renin substrate octapeptide. This model conformation can be placed back into the rhizopuspepsin enzyme active site, and favorable binding contacts can be seen and extrapolated to the renin structures, based upon primary sequence homology. Confirmation of this bound renin substrate conformation and examination of the renin interactions in detail await completion of the mouse renin-statine renin inhibitor X-ray structure, in progress (13).

Acknowledgements. I wish to thank my collaborators for their contributions to the ideas and experimental work presented here: Linda S. Payne, Debra S. Perlow, Nancy S. Lohr, Mark G. Bock, Edgar H. Ulm, Bruce I. LaMont, Tsau-Yen Lin, Martin Poe, and Daniel F. Veber. I am grateful to Drs. Richard Bott and David R. Davies for provision of the pepstatin-rhizopuspepsin coordinates prior to publication and for the gift of rhizopuspepsin for assay, to Dr. Graham M. Smith for aid with the molecular modelling, to Professors Jeremy R. Knowles and Daniel H. Rich for helpful consultations, and to Drs. Paul S. Anderson and Ralph Hirschmann for encouragement and support.

References

1. Peach, M.J.: Physiol. Rev. 57, 313-370 (1977).

2. Abrams, W. B., Davies, R. O., Ferguson, R. K.: Federation Proc. 43, 1314-1321 (1984).

3. Skeggs, L.T., Lentz, K.E., Kahn, J.R., Hochstrasser, H.J.: J. Exp. Med. 128, 13-34 (1968).

4. Marciniszyn, J., Hartsuck, J.A., Tang, J.J.: J. Biol. Chem. 251, 7088-7094 (1976).

5. Workman, R., Burkitt, D.W.: Archs. Biochem. Biophys. 194, 157-164 (1979).

6. Boger, J., Lohr, N.S., Ulm, E.H., Poe, M., Blaine, E.H., Fanelli, G.M., Lin, T.-Y., Payne, L.S., Schorn, T.W., LaMont, B.I., Vassil, T.C., Stabilito, I.I., Veber, D.F., Rich, D.H., Boparai, A.S.: Nature 303, 81-84 (1983).

7. Boger, J.: in Peptides: Structure and Function, Proceedings of the Eighth American Peptide Symposium, ed. Hruby, V.J., Rich, D.H., 569-578, Pierce Chemical, Rockford, Illinois 1983.

8. Bott, R., Subramanian, E., Davies, D.R.: Biochem. 21, 6956-6962 (1982).

9. Braun-Menendez, E., Fasciolo, J.C., Leloir, L.F., Munoz, J.M., Taquini, A.C.: in Renal Hypertension, 113-117, Charles C. Thomas, Springfield, Illinois 1946.

10. Veber, D.F., Bock, M.G., Brady, S.F., Ulm, E.H., Cochran, D.W., Smith, G.M., LaMont, B.I., DiPardo, R.M., Poe, M., Freidinger, R.M., Evans, B.E., Boger, J.: Biochem. Soc. Trans., in press (1984).

11. Nakaie, C.R., Oliveira, M.C.F., Juliano, L., Paiva, A.C.M.: Biochem. J. 205, 43-47 (1982).

12. Blaine, E.H., Schorn, T.W., Boger, J.: Hypertension 6 Suppl. I I-111-118 (1984).

13. Navia, M.A., Springer, J.P., Poe, M., Boger, J., Hoogsteen, K.: J. Biol. Chem., in press (1984).

14. Cha, S.: Biochem. Pharmacol. 24, 2177-2185 (1975).

15. Schechter, I., Berger, A.: Biochem. Biophys. Res. Commun. 27, 157-162 (1967).

16. Lin, T.-Y., Williams, H.R.: J. Biol. Chem. 254, 11875-11883 (1979).

17. Knight, C.G., Barrett, A.J.: Biochem, J. 155, 117-125 (1976).

18. Tang, J., Sepulveda, P., Marciniszyn, J., Jr., Chen, K.C.S., Huang, W.-Y., Tao, N., Liu, D., Lanier, J.P.: Proc. Natl. Acad. Sci. USA 70, 3437-3439 (1973).

19. Shewale, J.G., Tang, J.: Proc. Natl. Acad. Sci. USA 81, 3703-3707 (1984).

20. Panthier, J.-J., Foote, S., Chambraud, B., Strosberg, A.D., Corvol, P., Rougeon, F.: Nature 298, 90-92 (1982).

21. Imai, T., Miyazaki, H., Hirose, S., Hori, H., Hayashi, T., Kageyama, R., Ohkubo, H., Nakanishi, S., Murakami, K.: Proc. Natl. Acad. Sci. USA 80, 7405-7409 (1983).

22. Bott, R.R., Davies, D.R.: in Peptides: Structure and Function, Proceedings of the Eighth American Peptide Symposium, ed. Hruby, V.J., Rich, D.H., 531-540, Pierce Chemical, Rockford, Illinois 1983.

23. Tewksbury, D.A., Dart, R.A., Travis, J.: Biochem. Biophys. Res. Commun. 99, 1311-1315 (1981).

CHEMISTRY OF RENIN INHIBITORS

Michael Szelke
Department of Chemical Pathology, Royal Postgraduate Medical School, London, W12 OHS, U.K.

Introduction

Renin is the aspartic proteinase that catalyses the first and rate-limiting step in the release of the pressor octapeptide angiotensin-II from the glycoprotein substrate angiotensinogen present in blood (1). It is a very specific enzyme: it only cleaves the 10,11 peptide bond in the N-terminal region of angiotensinogen to produce the biologically inactive decapeptide angiotensin-I. Hydrolysis of the latter at the 8,9 bond by the carboxydipeptidase angiotensin converting enzyme (ACE) gives rise to angiotensin-II which plays an important role in the regulation of blood pressure, water and electrolyte balance. Figure 1 shows the formation of angiotensins I and II from the N-terminal tetradecapeptide of horse renin substrate 1 (1). Human angiotensinogen 1a differs from the horse substrate at positions 11, 12 and 13 (2,3).

Cloning and sequence analysis of the cDNA coding for rat angiotensinogen indicates that the mature protein contains 453 amino acid residues and three glycosylation sites (4). At present only the N-terminal tridecapeptide sequence of human renin substrate is known (2).

Human renin has been isolated (5,6,7,8) and its amino acid sequence has been determined by cloning and sequence analysis of the cDNA coding for it (9). The mature enzyme contains 340

Aspartic Proteinases and their Inhibitors

residues, three intra-chain disulphide bridges and two glycosylation sites.

```
                                             Renin
                                               ↓
H-Asp-Arg-Val-Tyr-Ile-His-Pro-Phe-His-Leu-Leu-Val-Tyr-Ser-  1
   1   2   3   4   5   6   7   8   9  10  11  12  13  14
                                               ↓
H-Asp-Arg-Val-Tyr-Ile-His-Pro-Phe-His-Leu-Val-Ile-His-      1a
                                   ACE
                                    ↓
H-Asp-Arg-Val-Tyr-Ile-His-Pro-Phe-His-Leu-OH    Angiotensin-I

H-Asp-Arg-Val-Tyr-Ile-His-Pro-Phe-OH            Angiotensin-II
```

Figure 1. Formation of angiotensin-I and angiotensin-II from the N-terminal tetradecapeptide of horse renin substrate 1. ACE = angiotensin converting enzyme.

Inhibitors of Renin

Current interest in renin inhibitors has been stimulated by the expectation that they will prove useful in the treatment of hypertension. Inhibitors of ACE are already in clinical use for the same purpose (10).

Although the activity of renin can be blocked by antibodies to the enzyme (11) or by peptides related to the N-terminal sequence of prorenin (12), this paper will only discuss inhibitors derived from the substrate sequence. In particular, it will focus on inhibitors based on synthetic isosteres of the transition state.

(A) Substrate analogues

Much of our fundamental knowledge on the biochemistry of the renin-angiotensin system comes from the classical work of

Skeggs and his co-workers that began in the early fifties (for a review see ref. 1). It was Skeggs who first demonstrated in 1964 that the (6-13)-octapeptide 2 of horse angiotensinogen, though still hydrolysed slowly by renin, was a weak competitive *in vitro* inhibitor (IC_{50}=0.2mM) of the enzyme (13). Later Kokubu and his co-workers found that the minimum sequence still capable of acting as an inhibitor ($IC_{50} \simeq$ 1mM) was the (10-13)-tetrapeptide in the form of its methyl ester 3 or the corresponding Phe^{13}-analogue 4 (14,15).

Skeggs' octapeptide 2 has formed the basis for the development of more effective substrate-based inhibitors with inhibitory constants in the micromolar range. Burton, Poulsen and Haber replaced the leucine residues at positions 10 and 11 in 2 with the corresponding D-amino acid, or with phenylalanine and found that these modifications blocked hydrolysis of the 10-11 bond and increased binding affinity to the enzyme, although the resulting analogues were rather insoluble in water (16). Historically, the most important compound of this series was the decapeptide 5 (also known as renin inhibitory peptide or RIP; K_i=2μM), because it was the first renin inhibitor to be investigated in man (17). Structures of the substrate analogue inhibitors 2 - 5 are shown in Figure 2.

```
H-Asp-Arg-Val-Tyr-Ile-His-Pro-Phe-His-Leu-Leu-Val-Tyr-Ser-    1
   1   2   3   4   5   6   7   8   9  10  11  12  13  14
                    H-His-Pro-Phe-His-Leu-Leu-Val-Tyr-OH      2
                                    H-Leu-Leu-Val-Tyr-OMe     3
                                    H-Leu-Leu-Val-Phe-OMe     4
                H-Pro-His-Pro-Phe-His-Phe-Phe-Val-Tyr-Lys-OH  5
```

Figure 2. Substrate analogue inhibitors of renin.

(B) Transition-state analogues

Transition state in a chemical reaction is defined as the intermediate structure of highest free energy content along the pathway from substrate to product. By definition, therefore the transition state represents a very labile, reactive structure. Transition-state theory postulates that acceleraction of a reaction by a catalyst such as an enzyme is due to the much higher binding affinity of catalyst for the transition state than for the ground state of the substrate (18-20).

Kinetic and structural studies indicate that aspartic proteinases (unlike the serine proteinases) do not form a covalent intermediate with their substrate. Current ideas on the hydrolytic mechanism of aspartic proteinases are discussed in other parts of this volume (21). One of the proposed mechanisms is illustrated in Figure 3: it consists of attack at the amide carbonyl carbon of substrate A by a hydroxide ion generated via proton abstration from the molecule of water present between the active site aspartic carboxyl groups (22). This results in the transformation of the planar trigonal carbonyl

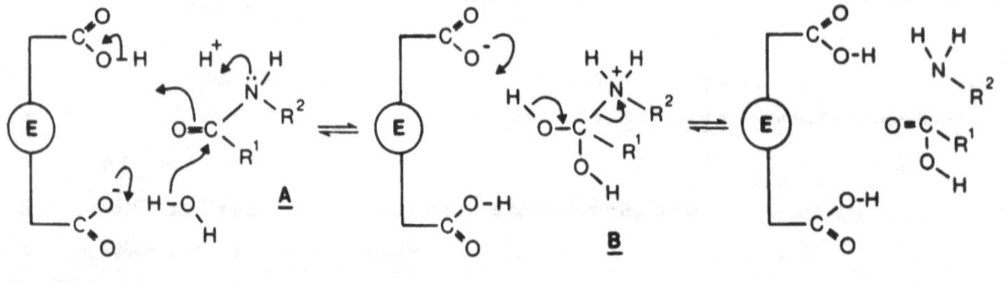

Figure 3. Mechanism of hydrolysis of substrate A by an aspartic proteinase E. Structure B is the transition state.

carbon in A into a tetrahedral one in the transition state B (Figure 4). Non-hydrolysable analogues of this transition state, i.e. stable structures which resemble B in their stereochemistry or electron distribution will, according to transition state theory, bind to the enzyme more tightly than the substrate and will act as competitive inhibitors.

There is a limited number of chemically accessible structures that fulfil the criteria for a peptide transition state ana-

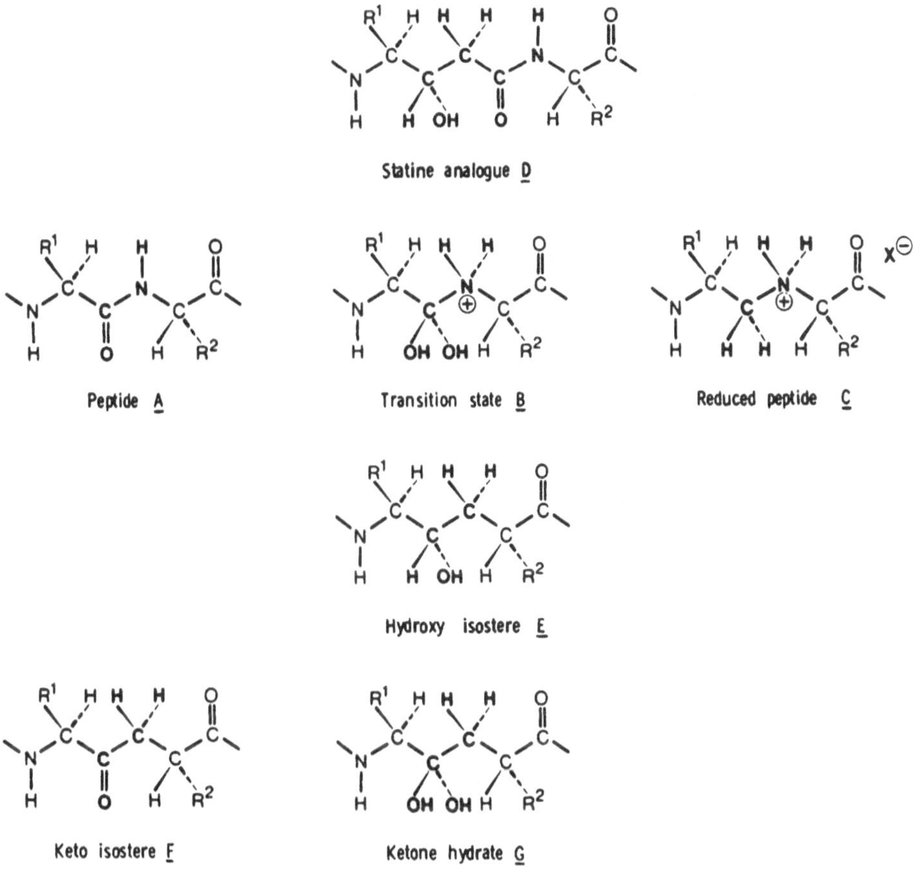

Figure 4. Stereochemical relationship of the peptide bond A and its transition state B to the reduced peptide C, statine analogue D, hydroxy isostere E, keto isostere F and ketone hydrate G.

logue. In the course of our search for renin inhibitors, we have synthesised, in a suitably protected form, dipeptide analogues C, E and F, and incorporated them into the (6-13)-octapeptide sequence of human renin substrate in place of the scissile (10,11)-dipeptide. Structures C and E were chosen because of their stereochemical resemblance to the tetrahedral transition state B, and the keto analogue F because of the possibility that it may form the tetrahedral hydrate G when bound to the active site of renin.

Another type of analogue was also synthesised: that containing the unusual amino acid statine (Sta) in place of Leu^{10} (structure D in Figure 4). Statine is responsibile for the activity of the general aspartic proteinase inhibitor pepstatin (Figure 5) where it is thought to play the role of a transition-state analogue (29).

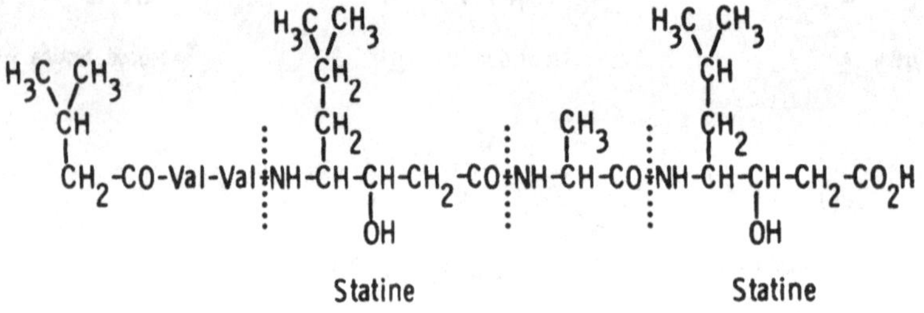

Figure 5. The structure of pepstatin A.

The substrate specificity of renin depends on the tight binding of the side-chains R^1 and R^2 of the transition state B to complementary hydrophobic pockets in the enzyme binding cleft. Specific inhibitors should therefore be similar to the transition-state B in the spatial geometry of their side-chains R^1 and R^2. While the reduced peptide C and the hydroxy analogue E fulfil this criterion (i.e. they are isosteric with B), the statine analogue D does not: its side-chains R^1 and R^2 are separated by seven instead of five interatomic distances and therefore they are likely to be misaligned with the hydrophobic binding pockets of the enzyme.

Several other groups have also made renin inhibitors containing statine. For these the reader is referred to the original papers (30-33) or to J. Boger's review in this volume (34). The remaining sections of the present paper will describe the chemistry of renin inhibitors based on structures C, E and F (Figure 4) with a brief reference only to the essential in vitro activity data. A full discussion of their in vitro and in vivo biological activities is presented in this volume by B.J. Leckie (35).

Reduced analogues. The first transition-state analogue to be investigated by us was the secondary amine C (Figure 4) which can be obtained by reduction of the amide bond in the appropriately protected dipeptide (23,24) and is therefore referred to as a reduced isostere.

The currently favoured synthesis of reduced peptides is shown in Figure 6. It is based on the reductive alkylation of an amino acid trichloroethyl ester with a protected alpha-aminoaldehyde using sodium cyanoborohydride as the reducing agent. After protection of the secondary amine with 3,4-dichlorobenzyloxycarbonyl, the trichloroethyl group is removed by treatment with zinc and acetic acid to give a protected reduced dipeptide which is ready for incorporation into the peptide sequence by either solid-phase or solution synthesis (26). The protected alpha-aminoaldehyde can be prepared either by reduction of the corresponding amino ester with di-isobutyl aluminium hydride (27) or by treatment of the N-methoxy-N-methylamide with lithium aluminium hydride (28). Chromatographic and spectroscopic evidence indicates that optical integrity at the asymmetric centres is retained during synthesis, and therefore the side-chains R^1 and R^2 possess the same orientation in the reduced dipeptide as they did in the parent amino acids.

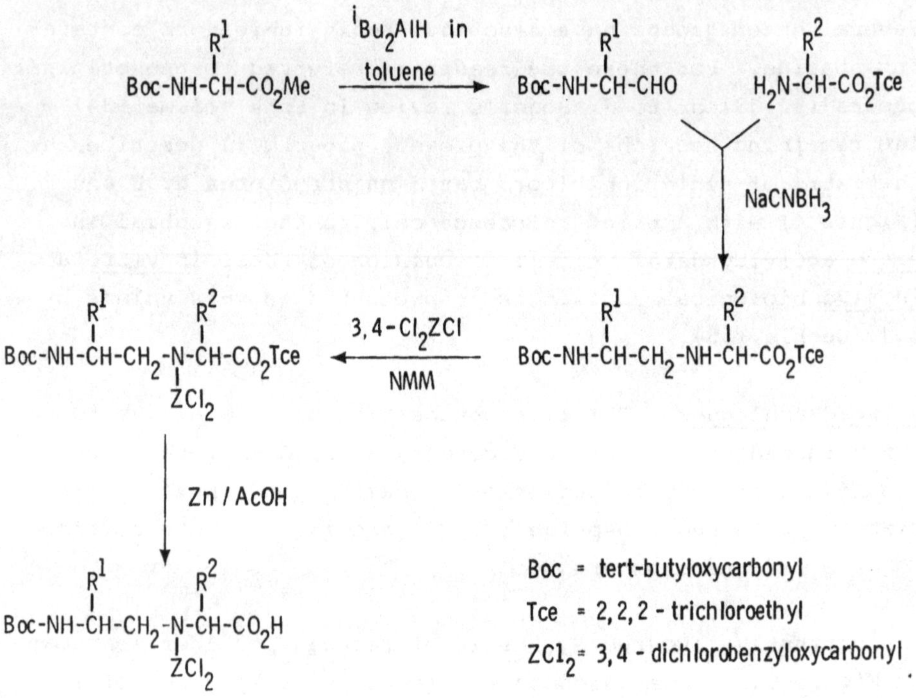

Figure 6. A general synthesis of protected reduced dipeptides.

A reduced analogue _6_ of Kokubu's tetrapeptide inhibitor _4_ was reported by us in 1972 (25). It was the first effective _in vivo_ inhibitor of renin which gave significant falls of blood pressure in Goldblatt hypertensive rats. However, _6_ lacked some of the important secondary binding sites and consequently had an IC_{50} of only 0.4mM against porcine renin.

H-Leu$\underline{R}$Leu-Val-Phe-OMe _6_

(_R_ indicates $-CH_2-NH-$ in place of -CO-NH-)

A dramatic 10,000-fold increase in binding affinity to dog renin was observed when the reduced isostere Leu$\underline{R}$Leu was introduced into Skeggs' octapeptide 2 to give the analogue bearing the Hammersmith code number H-76, and the corresponding DHis^6-derivative H-77 (Table I). However, these inhibitors based on the equine substrate sequence (which is, presumably, identical or similar to the canine sequence) were found to be 30-50 times less active against human renin. It was only when the N-terminal sequence of human substrate was elucidated (2) that it became possible to synthesise inhibitors (such as H-113 and H-142 in Table I) which showed a marked preference for human renin.

The decapeptide analogue H-142 containing a reduced 10-11 scissile bond has proved to be a non-toxic and effective inhibitor of human renin in vivo and has given us valuable information on the likely clinical effects of renin inhibitors (35,36).

Keto isosteres of peptides (see structure F in Figure 4) were first prepared in our laboratory in 1976 (37) as part of a programme aimed at metabolically stable hormone analogues, and were first used by us in the synthesis of mechanism - based inhibitors of renin in 1980 (38). Figure 7 shows our currently preferred synthesis of dipeptide keto isosteres (38) based on the alkylation of di(trichloroethyl)malonate with the bromoketone derived from the left-hand amino acid residue.

The resulting Boc-protected dipeptide keto analogue is, of course, racemic at position 2. Although optical integrity at position 5 is retained during synthesis, this asymmetric centre is extremely susceptible to racemisation in the presence of base.

Table I

Code No.	Structure 6 7 8 9 10 11 12 13	IC_{50} (µM) vs. human plasma renin	dog plasma renin
2	H-His-Pro-Phe-His-Leu—Leu-Val-Tyr-OH	200	200
H-76	H-His-Pro-Phe-His-Leu^R^Leu-Val-Tyr-OH	1.0	0.03
H-77	H-DHis-Pro-Phe-His-Leu^R^Leu-Val-Tyr-OH	1.0	0.02
H-112	H-His-Pro-Phe-His-Leu—Val-Ile-His-OH	313	1000
H-113	H-His-Pro-Phe-His-Leu^R^Val-Ile-His-OH	0.19	155
H-142	H-Pro-His-Pro-Phe-His-Leu^R^Val-Ile-His-Lys-OH	0.01	10
H-272	H-His-Pro-Phe-His-Leu^K^Val-Ile-His-OH	0.5	-
H-176	H-His-Pro-Phe-His-Sta—Val-Ile-His-OH	0.017	0.22
H-194	H-His-Pro-Phe-His-Leu^OH^Val-Ile-His-OH	0.003	0.02
H-261	Boc-His-Pro-Phe-His-Leu^OH^Val-Ile-His-OH	0.0007	0.02

R = $-CH_2-NH-$ K = $-CO-CH_2-$

Sta = 3S - statine OH = $-\overset{*}{C}H(OH)-CH_2-$ * S-configuration

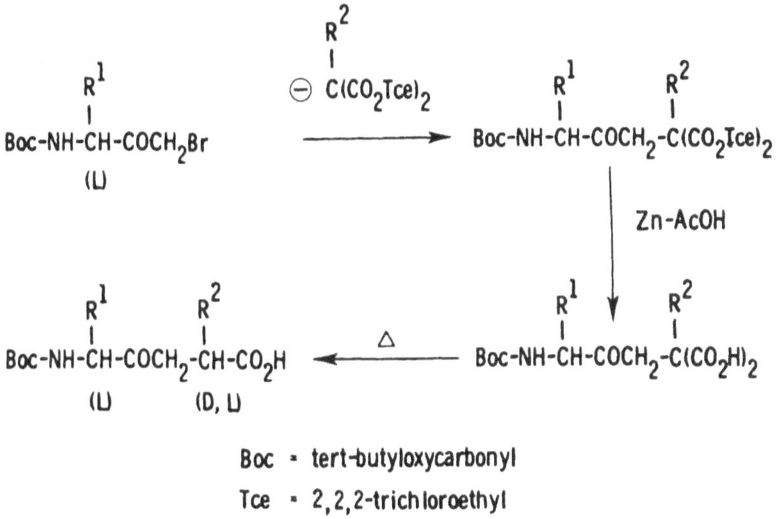

Figure 7. Synthesis of protected dipeptide keto isosteres.

Replacement of the scissile bond in the human renin substrate (6-13)octapeptide with a keto isostere yields a relatively poor inhibitor of renin (compound H-272 in Table I), suggesting that hydration of the ketone F to the geminal diol G (Figure 4) is not favoured in this case. In contrast, a keto analogue of pepstatin was shown by Rich and his coworkers to be an effective inhibitor of pepsin and to form a tetrahedral species on binding to the enzyme (39).

Hydroxy isosteres of peptides (see structure E in Figure 4) were first prepared in 1977 (40,41) and were first applied to renin inhibitors in 1980 (38). A practical synthesis of protected leucyl-valine hydroxy isostere is shown in Figure 8 (42). It involves construction of a five-carbon backbone containing three asymmetric centres. Two of these, C-2 and C-5 bearing, respectively, the iso-propyl (^{i}Pr) and iso-butyl (^{i}Bu) groups representing the side-chains of leucine and valine, are required in the same configuration as in the corresponding natural L-amino acid. Statine has the 3-hydroxy function in the S-configuration. Pepstatin analogues containing 3R-statine show binding affinities for pepsin which are 2-3 orders of magnitude smaller than those of the natural S-epimers (43). By analogy, our aim was to obtain the 4S-hydroxy isostere.

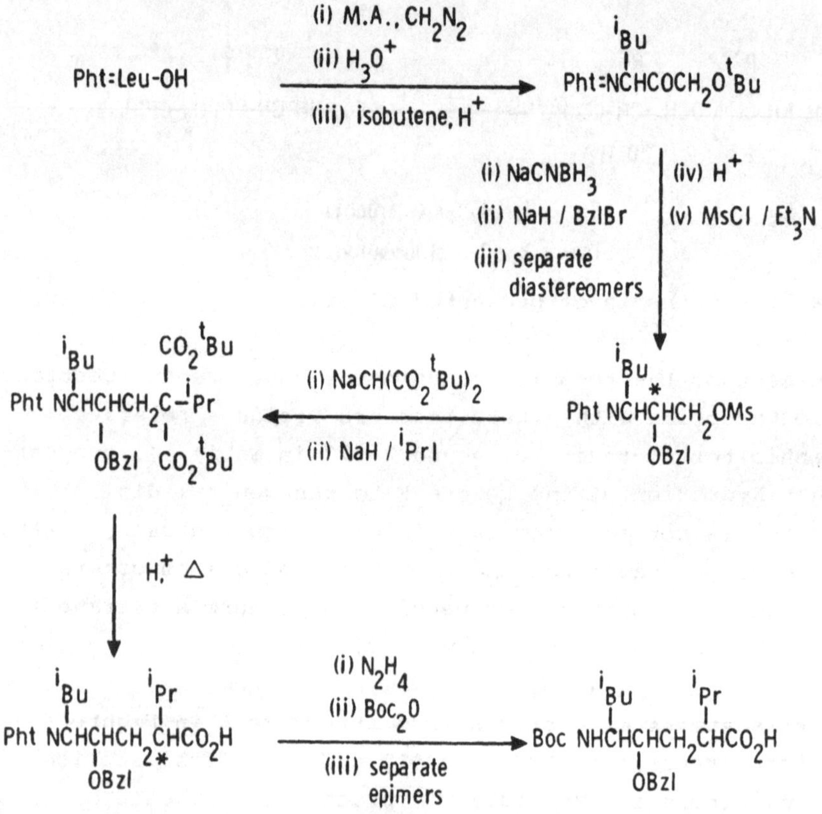

Figure 8. Synthesis of the protected leucyl-valine hydroxy isostere.
Boc = tert-butyloxycarbonyl; M.A. = mixed anhydride
Ms = methanesulphonyl; Pht = phthaloyl. Asterisks indicate new asymmetric centres introduced during synthesis.

The orientation of the ^{i}Bu side-chain was retained during synthesis and was therefore the same as in the starting phthaloyl-L-leucine. The next asymmetric centre (marked with an asterisk in Figure 8) was formed during reduction of the protected hydroxy-ketone with $NaCNBH_3$ which gave predominnantly the desired epimer (configuration confirmed by x-ray crystallography). The third asymmetric carbon at position 2 was introduced during extension of the carbon chain via malonic ester synthesis and yielded an equal mixture of the two epimers which were separated by chromatography. A stereospecific synthesis of both keto and hydroxy isosteres of dipeptides has been reported (44).

Incorporation of the protected leucyl-valine hydroxy isostere into the (6-13)octapeptide of human angiotensinogen was carried out by conventional peptide coupling methods according to the fragment condensation scheme shown in Figure 9.

Comparison of transition-state analogues C, D, E and F (Figure 4) has been carried out by incorporating each one in place of the scissile Leu^{10}-Val^{11} dipeptide, and measuring inhibition of human plasma renin in vitro by the resulting octapeptide analogues. The results are shown in Table I.

Least effective is the keto analogue H-272, giving a modest 600-fold increase over the activity of the parent octapeptide H-112. The reduced analogue H-113 (1,650 x H-112) is about three times more potent than H-272. Addition of Burton's terminal Pro^5 and Lys^{14} modifications in H-142 brings a further twenty-fold increase (31,000 x H-112), emphasizing the necessity of secondary binding sites for reduced analogues (cf. also the reduced tetrapeptide 6 discussed earlier). The Sta^{10}- compound H-176 (18,400 x H-112) is ten times more active than the reduced analogue H-113 which indicates the importance of functionality (and not only geometry) at the tetrahedral carbon atom. However, most potent of all the octapeptide analogues studied here is the S-hydroxy isostere H-194 (104,000 x H-112), being almost six times more active than the statine compound H-176, and clearly representing the closest structural approximation to the transition state within the above series.

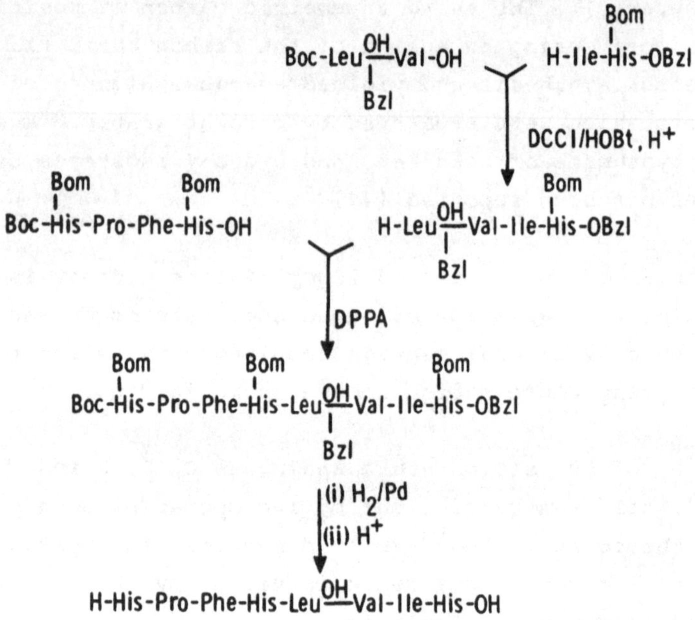

Figure 9. Synthesis of compound H-194, the (10-11)hydroxy isostere of human renin substrate (6-13)octapeptide. Bom = pi-benzyloxymethyl; DPPA = diphenylphosphoryl azide.

A further increase in potency can be achieved by adding a bulky lipophilic group at the N-terminus to give H-261 (450,000 x H-112) which is one of the most potent inhibitors of human renin both in vitro and in vivo (35).

Peptide aldehydes, like the keto analogues already discussed, are capable of forming a tetrahedral hydrate which is isosteric with the transition state (Figure 10).

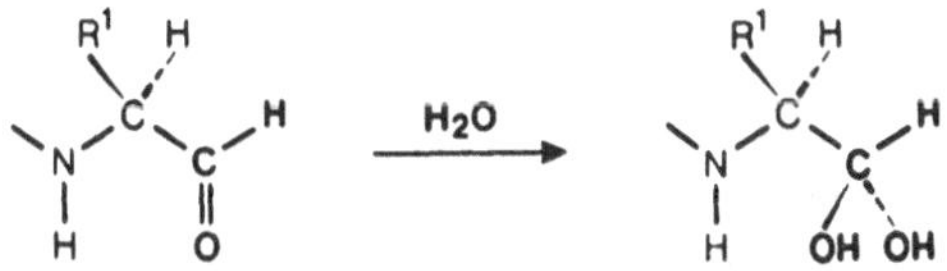

Figure 10. Conversion of peptide aldehyde into the tetrahedral hydrate.

Peptide aldehyde inhibitors of serine and thiol proteinases have been known for the past fifteen years. Some of these (e.g. leupeptin, antipain) were isolated from microbial sources (45), others were synthesised (e.g. 46) on the basis of a known substrate sequence.

More recently, several new peptide aldehyde inhibitors of renin have been described (47-49). They all share a C-terminal L-leucinal (Table II) corresponding to Leu^{10} of the renin substrate sequence, and contain lipophilic groups at their N-termini.

Table II

6 7 8 9 10 ↓ His Pro Phe His Leu	IC_{50} (μM) vs. human renin	Ref.
Z-His-Pro-Phe-His-Leu-H	50.0	47,48
Z-Pro-Phe-His-Leu-H	0.32	48
Z-Nal-His-Leu-H	0.08	48
Boc-Phe-Leu-Leu-H	1.1	49

-Leu-H represents C-terminal leucinal
Nal = 3-(1′-naphthyl)alanine; Z = benzyloxycarbonyl

In general, their <u>in vitro</u> potencies are 2-3 orders of magnitude smaller than those of transition-state analogue inhibitors based on a hydroxy isostere and containing secondary binding sites on both sides of the scissile linkage.

Summary

The original work presented here serves to illustrate the rational development, from first principles, of potent and specific renin inhibitors that have already proved valuable in human studies. In the human renin substrate (6-13)octapeptide, hydroxy isosteres offer the best approximation to the transition state at present. Empirical approaches based on naturally occurring transition-state analogues have also yielded effective inhibitors of renin. Time will only tell which, if any, of these substances will have the attributes required to compete successfully with ACE inhibitors as clinically useful antihypertensive agents.

Acknowledgements

The development of transition-state analogue inhibitors reported in this paper has been the result of close collaboration between two academic laboratories: the Peptide Unit at the Royal Postgraduate Medical School, Hammersmith Hospital, London, and the Medical Research Council Blood Pressure Unit, Western Infirmary, Glasgow. Compounds were designed and synthesised at Hammersmith by Drs. D. Michael Jones, Allan Hallett, Butrus Atrash and Javier Sueiras-Diaz. All in vitro, in vivo and clinical evaluation was carried out at Glasgow by Drs. Brenda J. Leckie, Sheila Beattie, Malcolm Tree, James J. Morton and David Webb under the direction of Drs. Anthony F. Lever and J. Ian S. Robertson. Financial support was provided by Ferring Pharmaceuticals Ltd., England and Ferring AB, Sweden.

The author is indebted to Dr. D. Michael Jones for agreeing, at short notice, to present this paper on his behalf at the Prague meeting.

References

1. Skeggs, L.T., Dorer, F.E., Levine, M., Lentz, K.E., Kahn, J.R. in: The Renin-Angiotensin System (Johnson, J.A., Anderson, R.R., editors), Plenum Press, New York, 1980, p.1.

2. Tewksbury, D.A., Dart, R.A., Travis, J.: Biochem. Biophys. Res. Comm. 99, 1311 (1981).

3. Tewksbury, D.A.: Fed. Proc. 42, 2724 (1983).

4. Ohkubo, H., Kageyama, R., Ujihara, M., Hirose, T., Inayama, S., Nakanishi, S.: Proc. Nat. Acad. Sci. USA. 80, 2196 (1983).

5. Yokosawa, H., Holladay, L.A. Inagami, T., Haas, E., Murakami, K.: J. Biol. Chem. 255, 3498 (1980).

6. Slater, E.E., Strout, H.V. Jr.: J. Biol. Chem. 256, 8164 (1981).

7. Inagami, T., Chang, J.J., Dykes, C.W., Takii, Y., Kisaragi, M., Misono, K.S.: Fed. Proc. 42, 2729 (1983).

8. McIntyre, G.D., Leckie, B., Hallett, A., Szelke, M.: Biochem. J. 211, 519 (1983).

9. Imai, T., Miyazaki, H., Hirose, S., Hori, H., Hayashi, T., Kageyama, R., Ohkubo, H., Nakanishi, S., Murakami, K.: Proc. Nat. Acad. Sci. USA. 80, 7405 (1983).

10. Sweet, C.S., Blaine, E.H. in: Handbook of Hypertension 3, Pharmacology of Antihypertensive Drugs (Van Zweiten, P.A., editor), Elsevier, Amsterdam 1984, p. 343.

11. Dzau, V.J. et al: Clin. Exp. Hypertension A5, 1207 (1983)

12. Evin, G., Devin, J., Castro, B., Ménard, J., Corvol, P.: Proc. Nat. Acad. Sci. USA. 81, 48 (1984).

13. Skeggs, L.T., Lentz, K.E., Hochstrasser, H., Kahn, J.R.: Canad. Med. Assocn. J. 90, 185 (1964).

14. Kukubu, T., Ueda, E., Fujimoto, S., Hiwada, K., Kato, A., Akutsu, H., Yamamura, Y.: Nature, 217, 456 (1968).

15. Kokubu, T., Hiwada, K., Ito, T., Ueda, E., Yamamura, Y., Mizoguchi, T., Shigezane, K.: Biochem. Pharmacol. 22, 3217 (1973).

16. Haber, E., Burton, J.: Fed. Proc. 38, 2768 (1979).

17. Haber, E.: J. Hypertension 2, 223 (1984).

18. Wolfenden, R.: Nature 223, 704 (1969).

19. Lienhard, G.E.: Science 180, 149 (1973).

20. Wolfenden, R., in: Transition States of Biochemical Processes (Gandour, R.D., Schowen, R.L., editors) Plenum Press, New York 1978, p. 555.

21. See especially the papers in this volume by Antonov, V.K., Blundell, T.L., James, M.N.G. and Pearl, L.:pp 203-220, 151-161, 163-177, and 189-195, resp.

22. Pearl, L., Blundell, T.: FEBS Letters 174, 96 (1984).

23. Szelke, M., Hudson, D., Sharpe, R., MacIntyre, I., Fink, G., Pickering, A.M.C. in: Molecular Endocrinology (MacIntyre, I., Szelke, M., editors) Elsevier, Amsterdam 1977, p. 57.

24. Szelke, M., Leckie, B., Hallett, A., Jones, D.M., Sueiras, J., Atrash, B., Lever, A.F.: Nature 299, 555 (1982).

25. Parry, M.J., Russell, A.B., Szelke, M. in: Chemistry and Biology of Peptides (Meienhofer, J., editor) Ann Arbor Science Publishers, Ann Arbor 1972, p. 541.

26. Szelke, M., Jones, D.M., Hallett, A.: United States Patent 4,424,207 (1984).

27. Ito, A., Takahashi, R., Miura, C., Baba, Y.: Chem. Pharm. Bull. 23, 3106 (1975).

28. Fehrentz, J.A., Castro, B.: Synthesis 676 (1983).

29. Marciniszyn, J. Jr., Hartsuck, J.A., Tang, J.: J. Biol. Chem. 251, 7088 (1976).

30. Evin, G., Castro, B., Devin, J., Ménard, J., Corvol, P., Guegan, R., Diaz, J., Demarne, H., Cazaubon, C., Gagnol, J.-P. in: Peptides: Structure and Function (Hruby, V.J., Rich, D.H. editors), Pierce Chemical Co., Rockford, Ill. 1983, p. 583.

31. Boger, J., Lohr, N.S., Ulm, E.H., Poe, M., Blaine, E.H., Fanelli, G.M., Lin, T.-Y., Payne, L.S., Schorn, T.W., LaMont, B.I., Vassil, T.C., Stabilito, I.I., Veber, D.F., Rich, D.H., Bopari, A.S.: Nature 303, 81 (1983).

32. Boger, J. in: Peptides: Structure and Function (Hruby, V.J., Rich, D.H., editors), Pierce Chemical Co., Rockford, Ill. 1983, p. 569.

33. Wood, J.M., Fuhrer, W., Bühlmayer, P., Riniker, B., Hofbauer, K.G.: This volume, pp 463-466 (1984).

34. Boger, J.: This volume, pp 401-420 (1984).

35. Leckie, B.J.: This volume, pp 443-461 (1984).

36. Webb, D.J., Cumming, A.M.M., Leckie, B.J., Lever, A.F., Morton, J.J., Robertson, J.I.S., Szelke, M., Donovan, B.: The Lancet ii, 1486 (1983).

37. Sharpe, R., Szelke, M.: U.S. Patent 4,242,256. U.K. Priorty: Application No. 13192/76 (1976).

38. Szelke, M., Jones, D.M., Hallett, A.: Eur. Pat. Appl. 0-045-665. U.K. Priority: Application No. 80-25579 (1980).

39. Rich, D.H., Bernatowicz, M.S., Schmidt, P.G.: J. Amer. Chem. Soc. 104, 3535 (1982).

40. Hudson, D., Sharpe, R., Szelke, M.: U.S. Patent 4,198,398 (Re. 30,731). U.K. Priority: Application No. 29207/77 (1977).

41. Hudson, D., Sharpe, R., Szelke, M.: U.S. Patent 4,304,715 U.K. Priority: Application No. 20124/79 (1979).

42. Szelke, M., Jones, D.M., Atrash, B., Hallett, A., Leckie, B.J. in: Peptides: Structure and Function (Hruby, V.J., Rich, D.H., editors), Pierce Chemical Co., Rockford, Ill. 1983. p. 579.

43. Rich, D.H., Sun, E.T.O.: J. Med. Chem. 23, 27 (1980).

44. Holladay, M.W., Rich, D.H.: Tetrahedron Letters 24, 4401 (1983).

45. Umezawa, H., Aoyagi, T. in: Proteinases in Mammalian Cell and Tissues (Barrett, A.J., editor) Elsevier, Amsterdam 1979, p. 637.

46. Bajusz, S., Barabás, E., Tolnay, P., Széll, E., Bagdy, D. Int. J. Peptide Protein Res. 12, 217 (1978).

47. Ito, A., Miura, C., Horikoshi, H., Miyagawa, H., Baba, Y. in: Peptide Chemistry 1977 (Shiba, T., editor) Protein Research Foundation, Osaka 1978. p. 165.

48. Kokubu, T., Hiwada, K., Sato, Y., Iwata, T., Imamura, Y., Matsueda, R., Yabe, Y., Kogen, H., Yamazaki, M., Iijima, Y., Baba, Y.: Biochem. Biophys, Res. Com. 118, 929 (1984)

49. Fehrentz, J.-A., Heitz, A., Castro, B., Cazaubon, C., Nisato, D.: FEBS Letters 167, 273 (1984).

HUMAN RENIN INHIBITORS

B.J. Leckie
MRC Blood Pressure Unit, Western Infirmary, Glasgow G11 6NT, U.K.

Introduction

Renin is an aspartate proteinase produced mainly in the kidney, although extrarenal sources of the enzyme have been identified (1). Human renin has been purified (2-5) and the amino-acid sequence has been deduced from cloning of cDNA (6,7). Certain regions of the sequence are homologous with other aspartate proteinases, namely the residues 31 to 35 and 215 to 217 around the active site aspartyls corresponding to asp-32 and asp 215 in the porcine pepsin sequence (8). However, renin shows differences in both structure and properties when compared with most other aspartate proteinases. The prosequence varies from that of pepsinogen and prochymosin, and this may be related to the fact that conversion of prorenin to renin appears to be mediated by a serine proteinase rather than by intramolecular cleavage (9,10). Renin is highly specific. Unlike pepsin and cathepsin D it is not a general proteinase, and is only capable of splitting one peptide bond in its substrate, angiotensinogen. The product is the decapeptide angiotensin 1 which is then converted to the vasoactive octapeptide angiotensin 11 by a carboxydipeptidase: converting enzyme. The pH optimum for the action of human renin on human angiotensinogen is around neutrality, although higher or lower pH optima can be obtained with other substrates (11). The amino-acid sequence of the N-terminus of horse angiotensinogen was published by Skeggs and his colleagues (12) and that of human angiotensinogen by Tewksbury (13) and that of the rat by Bouhnik (14).

Aspartic Proteinases and their Inhibitors

The scissile bond is leucyl-valine in man compared with leucyl-leucine in rat and horse. There are variations in the P2´and P3´ positions as well between species but the sequence on the N-terminal side of the scissile bond, corresponding to angiotensin I, is highly conserved. Human renin will cleave human and animal angiotensinogen, but some animal renins show relatively poor activity against human substrate (15).

Renin is the rate-controlling step of the renin-angiotensin system. The effector hormone, angiotensin II, constricts smooth muscle, stimulates release of aldosterone from the adrenal gland, stimulates vasopressin production but suppresses ACTH and has direct effects on the kidney, the sympathetic nervous system and the brain (16). The direct pressor effect maintains blood pressure to a small extent in normal people and to a greater extent in normals during sodium depletion. Angiotensin II has a second, "slow" effect in maintaining blood pressure and this is as important as the immediate pressor effect in normal people (16). Plasma renin concentration and circulating levels of angiotensin I and angiotensin II are raised in only a small proportion of hypertensive patients. However, angiotensin II maintains blood pressure in a far larger proportion of the hypertensive population, as shown by the study of agents which block the renin-angiotensin system.

The first renin-angiotensin antagonists were analogues of angiotensin II, such as saralasin, which competed with the hormone at the receptor. They showed partial agonist activity and also had other disadvantages such as short duration of action. Inhibitors of converting enzyme were developed as a result of work on naturally occuring peptide inhibitors of the enzyme (17). Non-peptide, orally active converting enzyme inhibitors have been developed. These have proved valuable not only in the treatment of renovascular hypertension with high levels of circulating angiotensin II, but also in hypertension with apparently normal concentrations of circulating renin and angiotensin, and in other diseases such as cardiac failure

(17). Converting enzyme inhibitors also inhibit the enzymes which destroy kinins and it has been suggested that part of the antihypertensive action of converting enzyme inhibitors might be due to their action as potentiators of bradykinin. However, the evidence for this is inconclusive. The slow reduction in blood pressure seen during long-term administration of inhibitors such as Captopril to normal people and hypertensives could be due to their action in blocking the slow pressor effect of angiotensin II.

The therapeutic benefits of converting enzyme inhibition have stimulated interest in renin inhibitors. Various naturally occurring renin inhibitors have been described but some of them are of low potency and unknown specificity. Pepstatin is a much better inhibitor of pepsin and cathepsin D tha n of renin (18). Antibodies to renin have been used in physiological studies in animals (19). Work on synthetic renin inhibitors that are analogues of renin substrate began almost at the same time as work on converting enzyme inhibitors. The first inhibitors were amino-acid substituted modifications of the (6-13)-octapeptide sequence of horse angiotensinogen (12,20,21). A different approach was used by Szelke and his colleagues who modified the scissile peptide bond in the (6-13)-octapeptide (22,23,24). The resulting compounds were powerful renin inhibitors both in vitro and in vivo.

Results

Inhibition of the Renin-Substrate Reaction

Effect of substituting the scissile peptide bond

Figure 1 shows the structure of the (6-13)-octapeptide of human, horse and rat angiotensinogen. Figure 2 shows the "reduced", "keto" and "hydroxy" isosteres that were substituted for the scissile peptide bond in the (6-13)-octapeptide of

	1	2	3	4	5	6	7	8	9	10	11	12	13	
										Renin ↓				
Human	Asp	Arg	Val	Tyr	Ile	His	Pro	Phe	His	Leu	Val	Ile	His	-
	├── Angiotensin I ──┤													
	├── Angiotensin II ──┤													
					*									
Horse	Asp	Arg	Val	Tyr	Ile	His	Pro	Phe	His	Leu	Leu	Tyr	Ser	-
Rat	Asp	Arg	Val	Tyr	Ile	His	Pro	Phe	His	Leu	Leu	Tyr	Tyr	-

* Val^5 substituted for Ile^5 in some animals e.g. Ox

Fig. 1 The amino-acid sequence of the N-terminus of human, horse and rat angiotensinogen showing sites of cleavage by renin and converting enzyme.

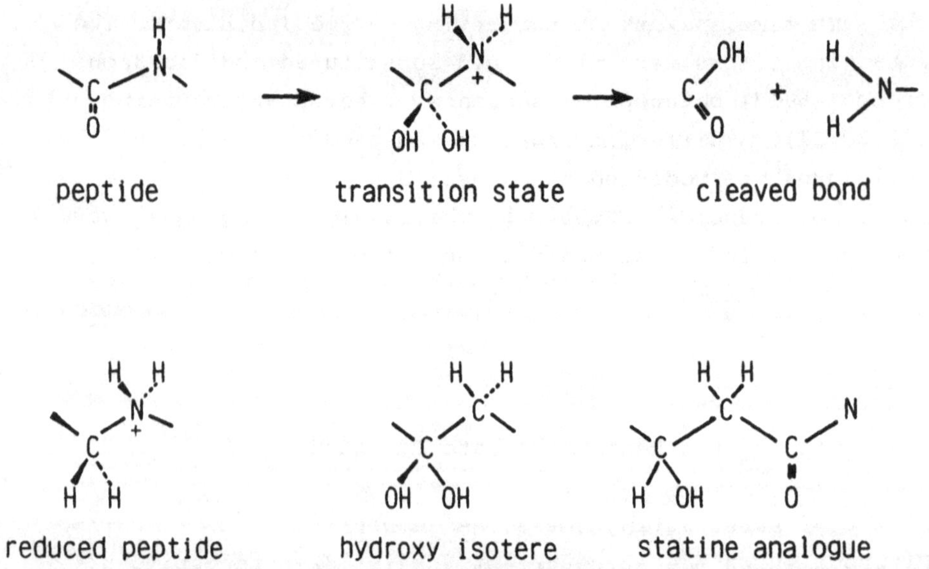

Fig. 2 Structure of isosteres compared with the peptide bond and the hypothetical transition state.

human angiotensinogen (24). Also shown is the structure of the peptide backbone that results when statine, a naturally occurring amino acid,is substituted for Leu-10 at the P1 position. The resulting compounds H113, H194 and H272 were tested for their ability to inhibit the production of angiotensin I by renin in human plasma at pH 7.0 (25). Figure 3 shows the concentration of inhibitor in the incubation mixture plotted against renin enzymatic activity. The (6-13)-octapeptide is a weak inhibitor of renin. The reduced isostere is 1600-fold more powerful as an inhibitor, and the hydroxy isostere shows an increase in potency of 100,000-fold compared with the unmodified peptide. Also shown in Figure 3 and Table 1 is a peptide, H-176, in which statine was substituted for Leu-10 in the P1 position of human renin substrate. This peptide was a better inhibitor of human renin than the reduced isostere, but it was less powerful than the hydroxy isostere.

Table 2 shows the action of four of these compounds on dog and rat renin. The human (6-13)-octapeptide gave little detectable inhibition of the animal renins. The reduced isostere was extremely species-specific with a human/dog potency ratio of 833. The hydroxy isostere gave a human/dog ratio of 6.7 compared with 12.5 for the statine-substituted peptide.

Effect of substitutions of the amino-acid sequence around the reduced isostere

The species-specificity of the reduced isostere was investigated by substituting "equine" for "human" residues in the P1´, P2´ and P3´ positions (26). Substitution of His-13 (human) by Tyr-13 (horse) led to little change in the potency of the inhibitor against human renin but to a 14-fold improvement in the action against dog renin. Substitution of Ile-12 by Val-12 led to an approximately three-fold loss of potency against human renin but to a six-fold increase against dog renin, while substitution of Val-11 by Leu-11 led to a further three-fold loss of potency against human renin but to a sixty-

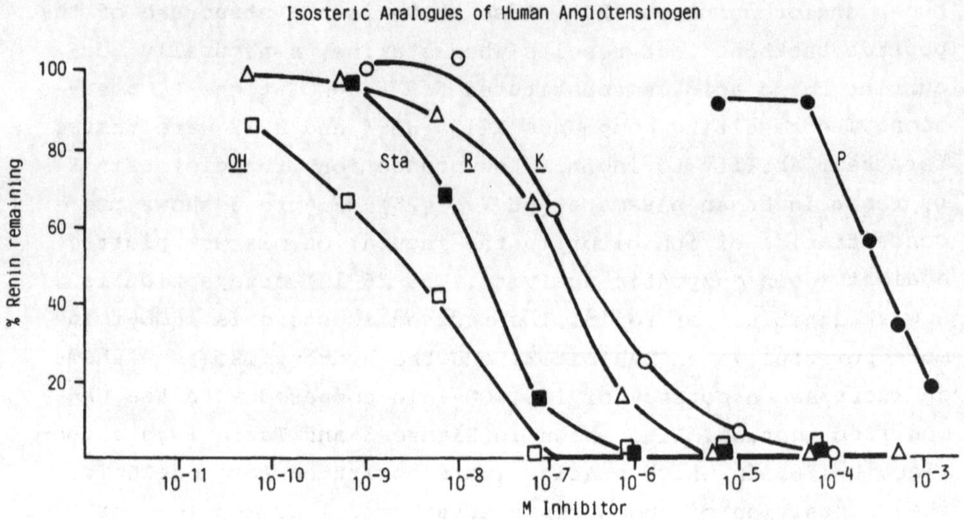

Fig. 3 Inhibition of renin in human plasma by the human (6-13)-octapeptide and by peptides with a reduced (R), keto (K), or hydroxy (OH) isostere at the scissile bond. A peptide with Sta-10 substituted for Leu-10 is also shown.

Table 1

Substitution of the Leu-10, Val-11 scissile bond in the (6-13)-octapeptide of human angiotensinogen by a reduced (R), keto (K) or hydroxy (OH) isostere, and the resulting IC50 against human renin, where the IC50 is the concentration of peptide needed to give 5o percent inhibition of renin in a standard assay system (25).

	6	7	8	9	10		11	12	13	IC^{50} µM
H112	His	Pro	Phe	His	Leu		Val	Ile	His	313
H113	His	Pro	Phe	His	Leu	R	Val	Ile	His	0.19
H272	His	Pro	Phe	His	Leu	K	Val	Ile	His	0.5
H194	His	Pro	Phe	His	Leu	OH	Val	Ile	His	0.003
H176	His	Pro	Phe	His	Sta		Val	Ile	His	0.017

fold improvement against dog renin. Thus the P1´ and the P3´ positions appear to be important in the determination of species-specificity, although dog renin seems to be more exacting in its requirement for the "correct" residue than human renin does. The results are summarized in Table 3.

Burton, Poulsen and Haber (21) and Burton, Cody, Herd and Haber (27) reported that replacing both Leu-10 and Leu-11 with Phe in the equine sequence gave a good inhibitor of human renin. Table 2 shows the results that we obtained by making a similar peptide, and also shows that replacing the Phe-Phe peptide bond by a reduced isostere improved the inhibition some thirty--fold against human renin. However this compound (H110) was a less powerful inhibitor of human renin than the reduced isostere of the natural human substrate. Substitution of Phe-11 for Leu-11 caused an approximately 100-fold reduction of potency against dog renin.

Table 2

Species-specificity of the inhibition of renin by the compounds shown in Table 1.

		IC^{50} μM		
		Human	Dog	Rat
H112	Human octapeptide	313	>1000	>1000
H113	Reduced isostere	0.19	155	194
H194	OH isostere	0.003	0.02	1.0
H176	Statine substituted	0.017	0.22	0.15

Burton and his colleagues also showed that the addition of Pro to the N-terminus and Lys to the C-terminus of their peptide improved its capacity to inhibit human renin. We confirmed this (Table 3) and also showed that the same modification applied to several reduced isosteres also improved the inhibition. The effect is due to the addition of proline, giving a residue in the P6 position, rather than to the presence of lysine at P4´ (Table 3). The same modification had little effect on compounds containing the hydroxy isostere or the statine substitution (Table 4). However, when a BOC group was placed at the N-terminus in an analogue of human angiotensinogen containing the hydroxy isostere, some improvement of the inhibition was noted. This compound, H261, is one of the most potent renin inhibitors synthesized (28).

Inhibition of Renin in Dogs, Primates and Man

The first series of experiments was carried out using an analogue of horse renin substrate with a reduced isostere at the scissile bond and D-His substituted for L-His at the N-terminus (23). This compound, H-77, was infused into conscious, trained dogs which were sodium-depleted in order to raise the plasma concentration of renin and thus make it easier to detect inhibition of the renin-angiotensin system. A dose-dependent fall in the concentrations of circulating blood angiotensin I and plasma angiotensin II occurred. Mean arterial pressure fell during infusion of the inhibitor. Figure 4 summarizes the results. It shows the change in blood angiotensin l concentration, plasma angiotensin ll concentration, mean arterial pressure and heart rate after a 30-minute infusion of H-77 at doses between 0 and 10 mg/kg/h, (9.9 µmole/kg/h) compared with basal values. Infusion of dextrose or of H-77 at 0.01 mg/kg/h did not change pulse or blood pressure. However, when H-77 was infused at 0.1, 1.0 or 10 mg/kg/h, a dose--dependent fall in blood angiotensin I concentration, plasma angiotensin II concentration and mean arterial pressure occurred. Heart rate increased during the two highest infusion rates. After the infusion had stopped, the angiotensin concentrations and the blood pressure returned to basal levels within 2.5 hours.

Inhibitors based on human angiotensinogen were tested in so-

Table 3

Substitution of amino-acids in reduced isosteres

Compound	Residue										IC50, µm	
		6	7	8	9	10	11	12	13		Human	Dog
H112+		HIS	PRO	PHE	HIS	LEU	VAL	ILE	HIS		313	1000
H113		HIS	-	-	-	-	rVAL	-	-		0.19	155
H116		HIS	-	-	-	-	rVAL	-	TYR		0.10	11
H117		HIS	-	-	-	-	rVAL	VAL	TYR		0.37	1.8
H76		HIS	-	-	-	-	rLEU	VAL	TYR		1.0	0.03
H77		DHIS	-	-	-	-	rLEU	VAL	TYR		1.0	0.02
Horse		HIS	-	-	-	-	LEU	VAL	TYR		200	200
Burton		HIS	-	-	-	PHE	PHE	VAL	TYR		31	500
H110		HIS	-	-	-	PHE	rPHE	VAL	TYR		0.82	3.9
H115		HIS	-	-	-	LEU	rPHE	VAL	TYR		1.4	0.26
Burton	PRO	HIS	PRO	PHE	HIS	PHE	PHE	VAL	TYR	LYS	5.9	140
H108	PRO	-	-	-	-	PHE	rPHE	VAL	TYR	LYS	0.04	0.6
H142	PRO	-	-	-	-	LEU	rVAL	ILE	HIS	LYS	0.01	10
H155	PRO	-	-	-	-	LEU	rVAL	ILE	TYR	LYS	0.016	0.23
H156	PRO	-	-	-	-	LEU	rLEU	ILE	TYR	LYS	0.03	0.02
H158	PRO	-	-	-	-	LEU	rVAL	ILE	TYR		0.03	x
H153		HIS	-	-	-	-	rVAL	ILE	TYR	LYS	0.05	22.0
H154		HIS	-	-	-	-	rLEU	ILE	TYR	LYS	0.15	0.20
H164		HIS	-	-	-	-	rVAL	ILE	HIS	LYS	0.05	32.0
H148	PRO	HIS	-	-	-	-	rVAL	ILE	HIS		0.016	x

+Human (6-13)-octapeptide
r = reduced isostere
- = residue the same as in the preceding compound

Table 4

Compound	Residue									IC50, μm
		6	7	8	9	10	11	12	13		Human renin
H113		HIS	PRO	PHE	HIS	LEU	rVAL	ILE	HIS		0.19
H142	PRO	HIS	PRO	PHE	HIS	LEU	rVAL	ILE	HIS	LYS	0.01
H176		HIS	PRO	PHE	HIS	STA	VAL	ILE	HIS		0.017
H189	PRO	HIS	PRO	PHE	HIS	STA	VAL	ILE	HIS	LYS	0.012
H194		HIS	PRO	PHE	HIS	LEU	ohVAL	ILE	HIS		0.003
H263	PRO	HIS	PRO	PHE	HIS	LEU	ohVAL	ILE	HIS	LYS	0.0042
H261	BOC	HIS	PRO	PHE	HIS	LEU	ohVAL	ILE	HIS		0.0007

r = reduced isostere
oh = hydroxy isostere

dium-deplete anaestetized baboons (26). In the first series of experiments we compared the effect of H142 (reduced isostere) with H189 (statine-substituted peptide). H142 had an IC50 of 0.01 μM in human plasma and 0.02 μM in baboon plasma, H189 had an IC50 of 0.01 μM in human plasma and 0.015 μM in baboon plasma. Each peptide was infused for 10 minutes at 10 mg/kg/h which was equivalent to a dose rate of 7.13 μmol/kg/h for H142 and 6.65 μmol/kg/h for H189 (26). Table 5 shows that H142 and H189 were equally effective at lowering blood angiotensin 1 concentration and plasma angiotensin II concentration.

In a second series of experiments on the baboon we tested the effect of the hydroxy isostere H261, compared with captopril. Animals were sodium-depleted and anaesthesia was induced with thiopentone sodium (8 mg/kg) and maintained with halothane.

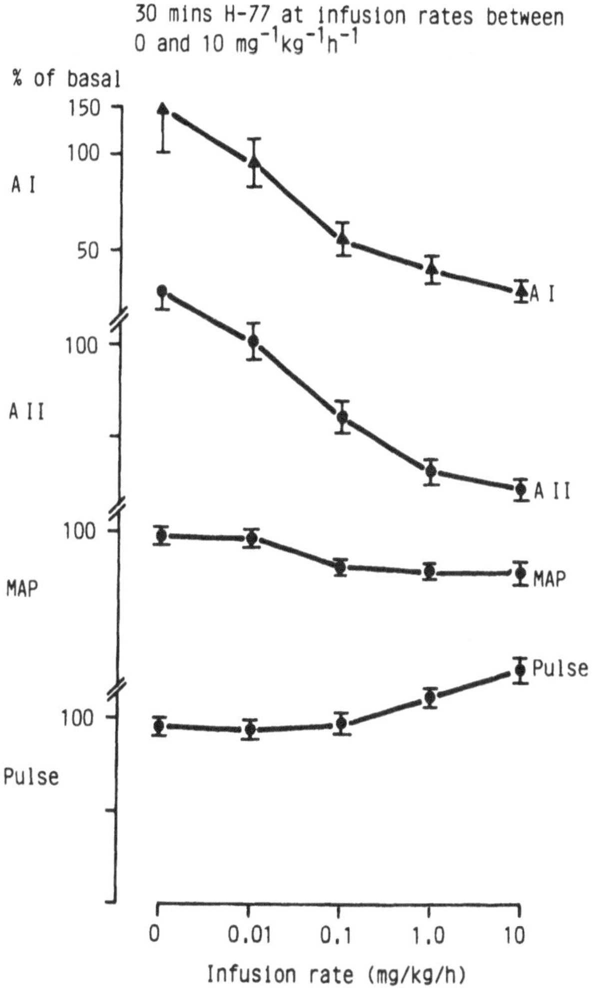

Fig. 4 Circulating levels of angiotensin I, angiotensin II and the mean arterial pressure and heart rate after a 30-minute infusion of H-77 into 6 conscious sodium--deplete dogs at various dose rates. Mean and standard error as a percentage of the basal pre-infusion value is shown (23).

Suxamethonium was given as a muscle relaxant. Animals were infused with dextrose for an hour, followed by H261 at 0.1 and 1.0 µmol/kg/h for 15 minutes. Then further 2 hours of dextrose was given, followed by 15 minutes of captopril at 0.025 mmol/kg/h. Figure 5 shows the results. H261 at 0.1 µmol/kg/h produced significant falls in renin concentration, blood

Table 5

Circulating concentrations of renin enzymatic activity, blood angiotensin I and plasma angiotensin II after infusion of the renin inhibitors H142 and H189 in baboons

Sample	Renin µU/ml plasma	AI pg/g blood	AII pg/ml plasma
Basal	636 ± 112	35.0 ± 8	154 ± 31
H189	4.3 ± 0.5	6.3 ± 1	16.3± 2.4
H142	17.0 ± 1.5	7.8 ± 2.5	14.8± 0.6

Mean and standard error of four infusions.

angiotensin 1 and plasma angiotensin II concentrations and in mean arterial pressure. A further fall occurred during infusion at 1.0 µmol/kg/h. The fall in plasma angiotensin II concentration and in mean arterial pressure was similar to that obtained with captopril. H261 reduced plasma angiotensin II concentration to 3.0 + 0.5 % and mean arterial pressure to 51 + 3 % of the basal value. Captopril caused a reduction in plasma angiotensin II concentration to 2.8 + 1 % of basal and a reduction in mean arterial pressure to 51 + 5 % of basal when the 210-minute sample (Fig. 5) was taken as the basal value. At this time, 120 minutes after the infusion had ceased, blood concentration of angiotensin I, plasma concentration of angiotensin II, and renin concentration, were not significantly different from basal values. Thus a large overshoot of the renin-angiotensin system did not occur when infusion of the renin inhibitor was stopped.

The reduced isostere H142 was tested in man (29) by infusion into 9 sodium-deplete normal volunteers. Three studies were done in each subject. An intravenous infusion of dextrose for one hour was followed by an infusion of either dextrose or H142 at 1 or 2.5 mg/kg/h for 30 minutes, followed by dextrose

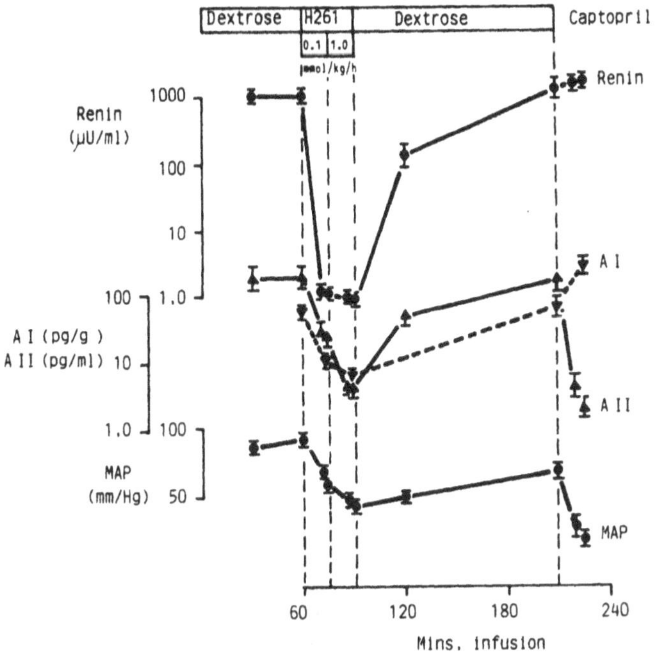

Fig. 5 Infusion of H261 into 6 anaesthetised baboons. The time-course of the experiment is shown on the abscissa and the plasma concentrations of renin, angiotensin I, angiotensin II and the mean arterial pressure are shown on the ordinate.

for 1.5 hours. Figure 6 shows the results for the experiment with the highest dose of H142. Infusion of H142 resulted in a fall in circulating angiotensin 1 and angiotensin II concentrations, in renin enzymatic activity and in diastolic blood pressure. Heart rate increased slightly. After the infusion at 2.5 mg/kg/h ended, blood angiotensin I had increased to 158 % of basal, plasma angiotensin II to 286 % of basal and renin to 389 % of basal. However, there was no significant increase of blood pressure above basal values.

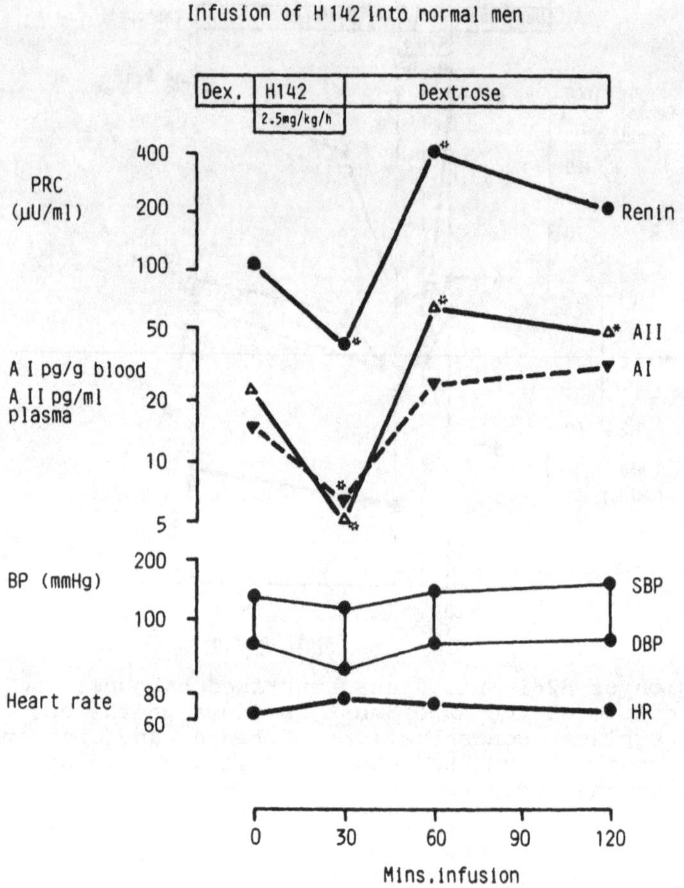

Fig. 6 Infusion of H142 at 2.5 mg/kg/h into 9 sodium-deplete male subjects. The stars show significant ($p < 0.05$) change from basal. The fall in diastolic blood pressure was significant but not that in systolic blood pressure.

Conclusion

Inhibitors of renin were synthesized by replacing the scissile peptide bond with non-hydrolysable isosteres. The "hydroxy isostere" was more potent than the "reduced" isostere, presumably because its structure was closer to that of the transition-state of the enzyme. Substitution of the Leu-10- Val-11 scissile bond in the (6-13)-octapeptide sequence of human angiotensinogen by a Leu-Val hydroxy isostere produced a compound that was approximately 100,000-fold more potent than the parent peptide. The efficiency of the isosteric renin inhibitors could also be modified by altering the amino-acid sequence in the peptide. Position P1´, followed by P3´ was important in determining species-specificity. Dog renin seemed to be more stringent in its requirement for the "correct" residue than human renin. Addition of proline to the N-terminus and lysine to the C-terminus of reduced isosteres improved their inhibitory capacity, the greatest effect being due to the addition of a proline residue at the N-terminus of the peptide (P6). However, the same substitution had little effect on the hydroxy isostere. Possibly, this compound bound more strongly at the active site of the enzyme and therefore changes at distant portions of the peptide had less effect than with the reduced isostere. Alternatively, the conformation of the peptide chain in the binding cleft of the enzyme might be different with the two types of isostere, bringing different residues in contact with the binding sites on the enzyme.

It was interesting that the addition of an N-terminal BOC group to the hydroxy-isosteric analogue of human angiotensinogen improved the inhibition to give one of the most powerful renin antagonists so far synthesized. Part of the effect of the BOC group may be due to the fact that could block aminopeptidase action in plasma, however, it also occupies P6 and may bind tightly to S6 of the enzyme, possibly to the same site that accepts proline in H142. If this is the case, however, it is surprising that proline has such a small effect when H194 is compared with H263 (Table 4).

Peptides with statine substituted for Leu-10 in the human angiotensinogen sequence were also synthesized. These were good inhibitors, but less powerful than the hydroxy isosteres. Statine-containing compounds have been synthesized by Boger and his colleagues and shown to be good renin inhibitors in vitro and in vivo (30,31). Since statine introduces an extra carb on atom into the peptide chain it is likely that the binding of the statine-hydroxyl is slightly different to that of the hydroxy isostere. The active site of renin can accommodate a bulky benzenesulphonyl group and therefore it is not surprising that it may accept some distortion of the peptide chain in the statine-substituted analogues.

A series of isosteric renin inhibitors was tested for their ability to inhibit renin in animals and man. With all inhibitors a dose-dependent reduction in circulating blood angiotensin I and plasma angiotensin II occurred together with a reduction in mean arterial pressure (except when ketamine was used as an anaesthetic (26)). An increase in heart rate was noted as the b lood pressure fell. When converting enzyme inhibitors are given acutely, an increase in heart rate is not always noted, but more experiments are required to establish if there is a genuine difference between converting enzyme inhibitors and renin inhibitors in this respect. The fall in blood pressure obtained with the renin inhibitors was similar to that obtained with the converting enzyme inhibitor captopril. This suggests that under acute conditions, captopril has no additional effect on blood pressure other than its action in blocking angiotensin II formation. When infusion of the inhibitor was stopped, angiotensin concentration, blood pressure and heart rate returend to normal within two hours. It is interesting that in humans an increase in renin enzymatic activity, angiotensin I concentration and angiotensin II concentration was noted immediately after the infusion had stopped, but this did not result in an increase in blood pressure above basal values. No adverse side-effects were seen when H142 was infused into human volunteers. Thus a more stable, longer-acting isosteric compound may prove useful in reducing blood pressure in man.

Acknowledgements

The work described in this report was a collaborative study between the Medical Research Council Blood Pressure Unit and the group of Dr M. Szelke at the Hammersmith Hospital,London, and was carried out by Drs S. Beattie, A.F. Lever, J.J. Morton, J.I.S. Robertson, M. Tree and D. Webb (MRC) and M. Szelke, B. Atrash, A. Hallett, D.M. Jones and J. Sueiras (Royal Postgraduate Medical School, Hammersmith Hospital).

References

1. Ganten, D., Schelling, P., Vecsei, P., Ganten, U.: Am. J. Med. 60, 760-772 (1976).
2. Galen, F.X., Devaux, C., Guyenne, T.T., Menard, J., Corvol, P.: J. Biol. Chem. 254, 4848-4855 (1979).
3. Yokosawa, H., Holladay, L.A., Inagami, T., Haas, E., Murakami, K.: J.Biol. Chem. 255, 3498-3502 (1980).
4. Slater, E.E., Cohn, R.C., Dzau, V.J., Haber, E.: Clin. Sci. Mol. Med. 55, 117s-119s (1978).
5. McIntyre, G.D., Leckie, B., Hallett, A., Szelke, M.: Biochem. J. 211, 519-522 (1983).
6. Imai, T., Miyazaki, H., Hirose, S., Hori, H., Hayashi, T., Kageyama, R., Ohkubo, H., Nakanishi, S., Murakami, K.: Proc. Natl. Acad. Sci. USA 80, 7405-7409 (1983).
7. Soubrier, F., Panthier, J-J., Corvol, P., Rougeon, F.: Nucleic Acid Res. 11, 7181-7190 (1983).
8. Sibanda, B.L., Blundell, T., Hobart, P.M., Fogliano, M., Bindra, J.S., Dominy, B.W., Chirgwin, J.M.: FEBS 174, (In Press)
9. Poulsen, K., Jacobsen, J.: J. Hypertens. 1, 3-5 (1983).
10. McIntyre, G.D., Pau, B., Hallett, A., Leckie, B.J., Szelke, M.: J. Hypertens. 2, 305-310 (1984).
11. Inagami, T., Murakami, K., Misono K., Workman, R.J., Cohen, S., Suketa, Y.: In Acid Proteases,Structure Function and Biology (Tang, J., ed.), pp. 225-248, Plenum Press, New York 1977.
12. Skeggs, L.T., Kahn, J.R., Lentz, K., Shumway, N.P.: J.Exp. Med. 106, 439-453 (1957).

13. Tewksbury, D.A., Frome, W.L., Dumas, M.L.: J. Biol. Chem. 253, 3817-3820 (1978).

14. Bouhnik, J., Clauser, E., Strosberg, D., Frenoy, J.P., Menard, J., Corvol, P.: Biochemistry 20, 7010-7015 (1981).

15. Poulsen, K., Haber, E., Burton, J.: Biochim. Biophys. Acta 452, 533-537 (1976).

16. Brown, J.J., Leckie, B., Lever,A.F., McIntyre, G., Morton, J.J., Semple, P.F., Robertson, J.I.S.: In Handbook of hypertension (Robertson, J.I.S., ed.), vol. 1, pp. 278-323, Elsevier Science Publishers, Amsterdam 1983.

17. Cushman, D.W., Cheung, H.S., Sabo, E.F., Ondetti, M.A.: Am. J. Cardiology 49, 1390-1394 (1982).

18. Evin, G., Gardes, J., Kreft, C., Castro, B., Corvol, P., Menard, J.: Clin. Sci. Molec. Med. 55, 167s-169s (1978).

19. Haber, E.: Clin. Sci. 59, 7s-19s (1980).

20. Kokubu, T., Ueda, E., Fujimoto, S., Hiwada, K., Kato, A., Akutsu, H., Yamamura, Y., Saito, S., Mizoguchi, T.: Nature 217, 456-457 (1968).

21. Burton, J., Poulsen, K., Haber, E.: Biochemistry 14, 3892-3898 (1975).

22. Parry, M.J., Russell, A.B., Szelke, M.: In Chemistry and Biology of Peptides (Meienhofer, J., ed.), pp. 541-544, Ann Arbor Science Publications, Ann Arbor 1972.

23. Szelke, M., Leckie, B., Tree, M., Brown, A., Grant, J., Hallett, A., Jones, D.M., Lever, A.E.: Hypertension 4, 59-69 (1982).

24. Szelke, M., Leckie, B., Hallett, A., Jones, D.M., Sueiras, J., Atrash, B., Lever, A.F.: Nature 299, 555-557 (1982).

25. Millar, J.A., Leckie, B.J., Morton, J.J., Jordan, J., Tree, M.: Clin. Chim. Acta 101, 5-15 (1980).

26. Leckie, B., Szelke, M., Hallett, A., Hughes, M., Lever, A.F., McIntyre, G., Morton, J.J., Tree, M.: Clin. and Exper. Hyperten. A5 788, 1221-1236 (1983).

27. Burton, J., Cody, R.J., Herd, J.A., Haber, E.: Proc. Natl. Acad. Sci. USA 77, 5476-5479 (1980).

28. Szelke, M., Jones, D.M., Atrash, B., Hallett, A., Leckie, B.: In: Peptides, Structure and Function (Hruby, V.J. and Rich, D.H., eds), pp. 579-582, Pierce Chemical Company, Rockford, IL 1983.

29. Webb, D.J., Cumming, A.M.M., Leckie, B.J., Lever, A.F., Morton, J.J., Robertson, J.I.S., Szelke, M., Donovan, B.: Lancet 2, 1486-1487 (1983).

30. Boger, J., Lohr, N.S., Ulm, E.H., Poe, M., Blaine, E.H., Fanelli, G.M., Lin, T-S., Payne, L.S., Schorn, T.W.,

LaMont, B.I., Vassil, T.C., Stabilito, I.I., Veber, D.F., Rich, D.H., Bopari, A.S.: Nature 303, 81-84 (1983).

31. Blaine, E.H., Schorn, T.W., Boger, J.: Hypertension 6, I-111-I-118 (1984).

PROTECTION GROUPS INCREASE THE IN VIVO STABILITY OF A STATINE-CONTAINING RENIN INHIBITOR

Jeanette M. Wood, Walter Fuhrer, Peter Bühlmayer, Bernhard Riniker and Karl G. Hofbauer

Research Department, Pharmaceuticals Division,
CIBA-GEIGY Ltd., Klybeckstr. 141,
4002 Basle, Switzerland

Introduction

Renin inibitors of the substrate or transition state analogue type (1, 2, 3) are interesting pharmacological tools for investigating the renin-angiotensin system (RAS). However, the extremely short biological half-life of these peptidic inhibitors limits studies _in vivo_. In our experiments we evaluated whether substituents used to protect such peptides during their synthesis, might also protect the compounds from degradation _in vivo_. Statine was incorporated into the N-terminal sequence of human angiotensinogen to produce CGP 29 291, the unprotected compound (Fig. 1a), and CGP 29 287, the same compound with protection groups at both terminals (Fig. 1b). Another limitation of peptidic inhibitors can be their poor aqueous solubility. Due to the 2 Arg residues, both CGP 22 291 and CGP 29 287 are readily soluble. Therefore both compounds could be administered intravenously and their effects _in vivo_ evaluated.

Methods

Renin activity was measured by incubation of human or marmoset plasma at 37°C and pH 7.4 in the presence of angiotensinase inhibitors (BAL, EDTA). The angiotensin I formed was measured by radioimmunoassay. The effects of varying concentrations of CGP 29 291 and CGP 29 287 on renin activity were determined. The concentration of compound inducing 50 % inhibition of

Aspartic Proteinases and their Inhibitors

renin activity (IC_{50}) was calculated.

The effects of these peptides on plasma renin activity (PRA) and blood pressure (BP) were evaluated in conscious, catheterized marmosets after mild volume depletion (2). The compounds were infused for 30 min. Saline (0.9 %) was infused (40 µl/min) for 15 min before and 30 min after infusion of a compound. CGP 29 291 and CGP 29 287 were infused in a dose of 0.1 mg/kg/min (n = 4 and n = 5 respectively). CGP 29 287 was also infused in a dose of 0.01 mg/kg/min (n = 5). BP was recorded continuously (2). Blood samples for measurement of PRA were taken before infusion, after 30 min of infusion and 30 min after stopping infusion of the compound. Values given in the text and the figure are mean ± SEM.

Results

In vitro CGP 29 291 was a potent inhibitor of human (IC_{50} = 0.006 µM) and marmoset (IC_{50} = 0.02 µM) renin. CGP 29 287 had a slightly increased potency (IC_{50}: human = 0.001 µM, marmoset = 0.005 µM).

In vivo, BP was lowered (-15 ± 5 mmHg) and PRA inhibited (92 ± 2 %) during infusion of CGP 29 291 (Fig. 1a). After stopping the infusion of CGP 29 291, BP and PRA returned to preinfusion values within 30 min. During infusion of CGP 29 287 in the same dose, a similar fall in BP was observed (-16 ± 1 mmHg) and PRA was inhibited (100 %) (Fig. 1b). After stopping the infusion of CGP 29 287, BP remained lowered and PRA inhibited at 30 min. The 10 fold lower dose of CGP 29 287 also reduced BP (-11 ± 2 mmHg) and inhibited PRA (100 %) during infusion with the response persisting 30 min after infusion (Δ BP = -12 ± 3 mmHg; inhibition of PRA = 84 ± 11 %).

Conclusions

Addition of protection groups at both terminals of a statine-containing human angiotensinogen analogue slightly increased its apparent potency *in vitro*.

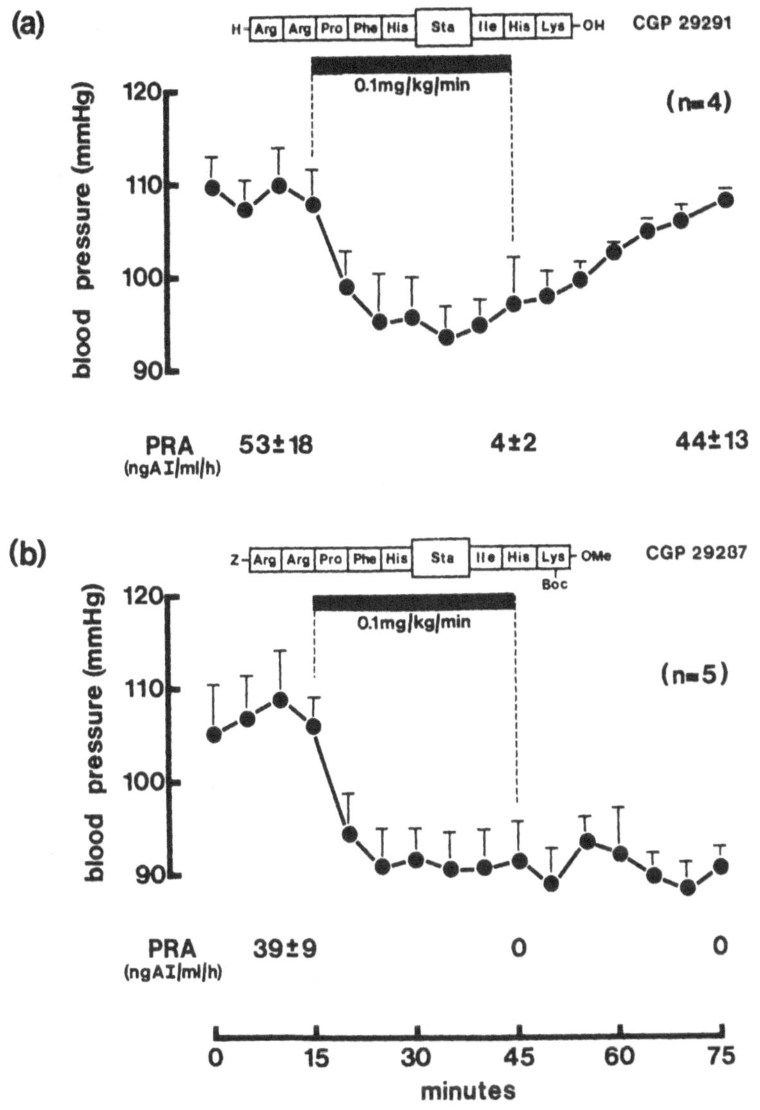

Fig. 1: Effects of infusion of 0.1 mg/kg/min of CGP 29 291 (a) or CGP 29 287 (b) on blood pressure and plasma renin activity (PRA) in conscious, volume-depleted marmosets.

In vivo, the protected and unprotected compound induced a similar fall in BP during intravenous infusion of equal doses. Previous experiments have shown that this is the maximum fall in BP obtained in this experimental model after inhibition of the RAS (2). Both compounds also inhibited PRA during infusion. However, there was a considerable difference in the duration of these responses after stopping the infusions. Both BP and PRA had recovered within 30 min after the unprotected compound whereas there was no recovery at 30 min after the protected compound. The approximately 4 fold increase in potency against marmoset renin may contribute to some extent to the increased duration of action *in vivo*. However, after infusion of a 10 fold lower dose of CGP 29 287 the response still persisted for up to 30 min. Moreover, we have shown that CGP 29 287 is effective after a single intravenous injection of 0.1 mg/kg. It appears, therefore, that the protection groups prevent the peptidic inhibitor from being rapidly degraded *in vivo*.

In conclusion, substituents used to protect peptidic renin inhibitors during their synthesis may significantly prolong their biological half-life.

References

1. Haber, E.: J. Hypertension 2, 223 (1984).
2. Wood, J.M., Forgiarini, P., Hofbauer, K.G.: J. Hypertension 1 (suppl. 2), 189 (1983).
3. Tree, M., Atrash, B., Donovan, B., Gamble, J., Hallet, A., Hughes, M., Jones, D.M., Leckie, B., Lever, A.F., Morton, J.J., Szelke, M.: J. Hypertension 1, 399 (1983).

INHIBITION OF ASPARTIC PROTEINASES BY TRANSITION STATE SUBSTRATE ANALOGUES

X-Ray Studies of the Complex of Endothiapepsin with the Renin Inhibitor H-142

A. Hallett, D. M. Jones, B. Atrash, M. Szelke

Department of Chemical Pathology, Royal Postgraduate Medical School, London W12 0HS, UK

B. J. Leckie, S. Beattie

MRC Blood Pressure Unit, Western Infirmary, Glasgow G11 6NT, UK

B. M. Dunn

Department of Biochemistry and Molecular Biology, J. Hillis Miller Health Centre, University of Florida, Gainsville, Florida 32610, USA

M. J. Valler, C. E. Rolph, J. Kay

Department of Biochemistry, University College, PO Box 78, Cardiff CF1 1XL, UK

S. I. Foundling, S. P. Wood, L. H. Pearl, F. E. Watson, T. L. Blundell

Laboratory of Molecular Biology, Department of Crystallography, Birkbeck College, University of London, Malet Street, London WC1E 7HX, UK

Introduction

Many previous investigations including several X-ray crystallographic studies have indicated the extended nature of the active site cleft of aspartic proteinases with at least seven identifiable subsites occurring along its length (1). By analysis of complexes of logically designed synthetic peptide substrates and inhibitors with these enzymes, it is possible to gain considerable insight into the interactions which occur at each of these subsites and which contribute to the overall position of the substrate so that the scissile peptide bond is in the correct orientation in close juxtaposition to the catalytic apparatus of the two active site aspartic acid residues (2).

Kinetic Parameters of Substrates and Inhibitors

Introduction of the chromophoric amino acid, p-nitro-phenylalanine (Nph) into position P_1' of the peptide substrates permits the cleavage of the P_1-P_1' peptide bond by aspartic proteinases to be followed spectrophotometrically (3,4). On the basis of a statistical survey of the amino acid residues found to occur most frequently in the other subsites adjacent to pepsin-susceptible bonds in proteins (5), the following peptide was synthesised by solid phase methods (4) to contain sufficient residues to occupy each of the seven subsites :

$$\begin{array}{ccccccc} S_4 & S_3 & S_1 & S_1' & S_2' & S_2' & S_3' \\ \text{Pro} - & \text{Thr} - & \text{Glu} - & \text{Phe} - & \text{Nph} - & \text{Arg} - & \text{Leu} \end{array} \qquad \underline{1}$$

Kinetic parameters (K_M,k_{cat}) for the hydrolysis of this substrate by a number of mammalian and microbial aspartic proteinases were determined at pH3.1 and a representative selection of these is shown in Table 1.

From the specificity constant, k_{cat}/K_M, it can be seen that 1 is an excellent substrate for fungal enzymes such as those from Penicillium janthinellum (penicillopepsin) and Endothia parasitica (endothiapepsin), a good substrate for most of the mammalian enzymes, and it is still a useful substrate for human gastricsin, despite the very high K_M value, since the peptide is readily soluble at high concentrations in aqueous buffers. Further derivatives of 1 were synthesised with systematic replacement of the residue occupying the P_3 position and the kinetic parameters and pH dependence of the interactions of these with several aspartic proteinases are described, together with the molecular interpretation of the results, in the chapter by Dunn et al (6).

Table 1

Enzyme	k_{cat} (sec^{-1})	K_M (mM)	k_{cat}/K_M ($sec^{-1} mM^{-1}$)
Human pepsin	72	0.17	420
Human gastricsin	9	2.0	4
Calf chymosin	51	0.46	110
Human cathepsin D	173	0.55	310
Penicillopepsin	24	0.005	4,670
Endothiapepsin	12	0.003	4,000
Yeast proteinase A	29	0.30	100

Kinetic parameters for the hydrolysis of 1 by several aspartic proteinases as determined at pH3.1 (in 0.1M sodium formate buffer) at 37°. Absolute concentrations of each enzyme were determined by active site titration against isovaleryl-pepstatin and s/v against S plots were used to derive the kinetic constants.

We have previously shown in the development of renin inhibitors (7) that the substrate of an aspartic proteinase can be converted into a potent inhibitor by modification of the scissile bond so that it resembles stereochemically the tetrahedral transition state formed during amide hydrolysis. Analogues of the transition state bind more tightly to the enzyme than the substrate (8) and if they are stable to hydrolysis, they act as potent competitive inhibitors.

This approach has now been applied to synthetic substrates in order to produce effective inhibitors of other aspartic proteinases. The following compounds were synthesised by the incorporation of protected dipeptide analogues (see paper in this volume by M. Szelke (9)) into peptides by standard methods of solid-phase peptide synthesis.

H-256	Pro-Thr-Glu-Phe$\overset{R}{-}$Phe-Arg-Glu
H-297	Pro-Thr-Glu-Phe$\overset{R}{-}$Nle-Arg-Leu

where R represents $-CH_2NH-$ in place of -CONH-.

Table 2

Enzyme	H-256	H-297	Substrate 1
	K_i (nM)		K_M (nM)
Pig pepsin	41	14	85,000
Calf chymosin	2,600	3,200	460,000
Human cathepsin D	-	120	550,000
Pig cathepsin D	110	-	240,000
Endothiapepsin	60	18	3,000
Penicillopepsin	10	not detd.	5,000

Kinetic constants (K_i) for the inhibition of several aspartic proteinases by two synthetic inhibitors, H-256 and H-297 as determined at pH3.1 and 37°.

Kinetic constants (K_i) for the interaction of these peptides with several aspartic proteinases are shown in Table 2. These inhibitors show an increased affinity for all the enzymes of several orders of magnitude as indicated by the ratio of the K_i values to the K_M values obtained with the parent substrate 1. Neither compound is an effective inhibitor of calf chymosin or a good inhibitor of cathepsin D despite a larger increase in affinity. Both appear to be potent inhibitors of the microbial enzymes penicillopepsin and endothiapepsin. Alteration of the residue occupying the P_3' position from a hydrophilic Glu residue to a hydrophobic Leu residue seemed to have very little influence on the extent of interaction.

Since the design logic for the original synthetic substrate was based on information obtained with pig pepsin, it was to be expected that the derived inhibitors were considerably more effective towards pig pepsin than towards, for example, calf chymosin. Thus, using this approach it should be possible to design specific inhibitors for individual aspartic proteinases based on the knowledge of their preferences in each subsite.

This information can be gleaned from examination of the interactions of each enzyme with synthetic and naturally-occurring protein substrates. For example, the details of the interaction of renin with the (6-13) octapeptide sequence of the protein substrate angiotensinogen are now well mapped out in a number of species. Thus it was possible to produce a synthetic inhibitor centred around this sequence but containing the reduced $-CH_2-NH-$ isostere in place of the scissile -CO-NH- of the -Leu-Val- peptide bond (7). The decapeptide analogue

H-142 Pro-His-Pro-Phe-His-Leu$\overset{R}{-}$Val-Ile-His-Lys

is a highly potent inhibitor of human renin (9) and has been shown to be effective in lowering blood pressure in human volunteers (10).

The selectivity of this inhibitor can be assessed from the kinetic constants given in Table 3 for its interaction with several other aspartic proteinases. It has no significant effect on any of the other aspartic proteinases known to be present in the human body so that drugs based on this type of inhibitor may well have significant value as antihypertensive agents with no side effects due to the inhibition of other potentially competitive enzymes. Another significant point to emerge from Table 3 is that, of all the aspartic proteinases tested so far against H-142, only renin and endothiapepsin were inhibited effectively by this peptide.

Table 3

Enzyme	K_i (nM)
Human pepsin	60,000
Human gastricsin	15,000
Human cathepsin D	40,000
Human renin	10*
Endothiapepsin	290
Penicillopepsin	24,000

Kinetic constants (K_i) for the inhibition of several aspartic proteinases by H-142 as determined at pH3.1 (in 0.1M sodium formate buffer) and 37°. *The value for human renin was an IC_{50} determined at pH7.0.

These results suggest that endothiapepsin may be the enzyme of choice for modelling human renin (11,12) and that a three-dimensional X-ray structure analysis of endothiapepsin complexed with the inhibitor H-142 may be of considerable interest.

X-Ray Difference Fourier Studies

The crystal structure of endothiapepsin has been investigated by X-ray techniques at 2.1Å resolution and refined to an agreement factor (R) of 0.16 (13). This allows X-ray analysis using difference Fourier techniques of enzyme-inhibitor complexes provided the crystals are isomorphous with those of the native enzyme.

In order to achieve this 600μM of inhibitors H-142 and H-256 were stirred together with 60μM of endothiapepsin at 4°C for one hour in 0.1M sodium acetate at pH4.5. Then ammonium sulphate was added to 55% saturation and the resultant solution millipore filtered into clean vials. A little acetone was added to clear turbidity. Crystals of the complexes grew after 4 - 6 months in habits identical to those of the native crystals. The spacegroup is $P2_1$. The cell dimensions of the complex of endothiapepsin with H-142 are compared with those of the native crystals in Table 4 and can be seen to be isomorphous.

Table 4

Cell Dimension	Native	Complex with H-142
a (Å)	53.6	53.7
b (Å)	74.1	74.0
c (Å)	45.7	45.5
β (°)	110.0	109.2

Crystallographic data for endothiapepsin and its complex with the inhibitor H-142.

X-ray data can be collected to a resolution of 2.1Å. At the time of writing, two equivalent reflections for data to 2.8Å have been collected but difference Fouriers have been calculated only to 3.0Å resolution. The molecule of H-142 has been fitted to the difference electron density using FRODO on an Evans and Sutherland Picture System. There is no density for the COOH-terminal lysine (P_4') indicating a disordering of this residue. Nevertheless adequate density exists to give approximate positions for the other residues (P_6 to P_3') although residues P_4 to P_1 are most clearly shown. We have subjected these coordinates and those of the enzyme to restrained least-squares and the agreement value is presently ∿0.20 although refinement is still proceeding. The medium resolution precludes a detailed description of the interactions between H-142 and the endothiapepsin; this must await the refinement at high resolution. Nevertheless the general arrangement of the inhibitor with respect to the enzyme can be discerned with some confidence.

Figures 1 and 2 show the general arrangement of the inhibitor in the enzyme active site and the conformation of the inhibitor. The inhibitor adopts an extended conformation from P_6 to P_1' while residues P_2' and P_3' are gently turned towards the mouth of the cleft. The mainchain appears to make a number of hydrogen bond interactions. These include :

P_3	C=O-------H-N	Thr 219
P_1	N-H-------O=C	Gly 217

which comprise an antiparallel ß-sheet between enzyme and inhibitor mainchain. The hydrogen bonding appears to be extended by :

P_3	N-H-------Oγ	Thr 219

as found in the inhibitor complexes between penicillopepsin and various tetrapeptides (14). There is a similar hydrogen bond on the COOH-terminal side of the reduced bond :

P_1'	C=O-------H-N	Gly 76

as predicted from the previous modelling studies (for example see refs. 11, 12 and 14 and the references cited therein). From the present model it appears that P_2 CO, P_1' CO and P_3' NH make interactions with mainchain groups on the flap (residues 74 - 77).

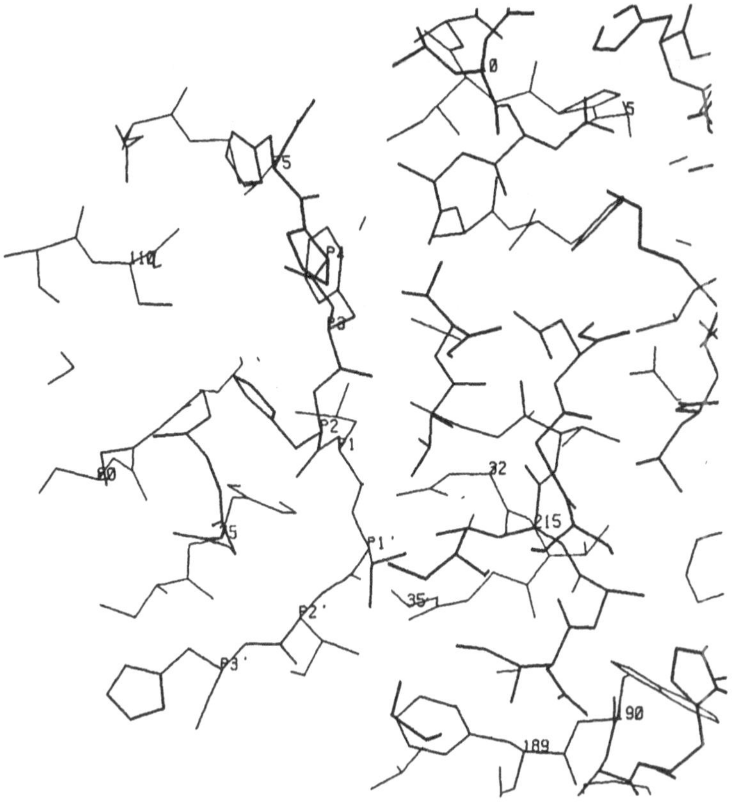

Figure 1

The binding site of H-142 (P_6 to P_3') in the active site cleft of endothiapepsin. The flap containing residue Tyr 75 is on the left-hand side while the two active site aspartates are on the right-hand side close to the reduced bond (P_1 - P_1') of H-142.

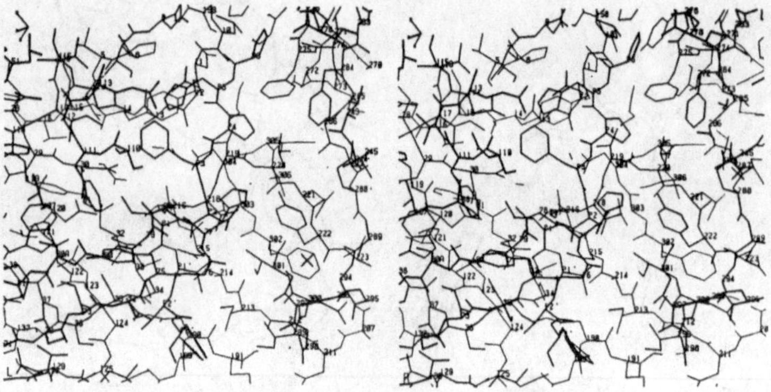

Figure 2a (above)

A stereo view of the active site with H-142 bound from a direction orthogonal to that of Figure 1.

Figure 2b (below)

The conformation of H-142 viewed from the same direction.

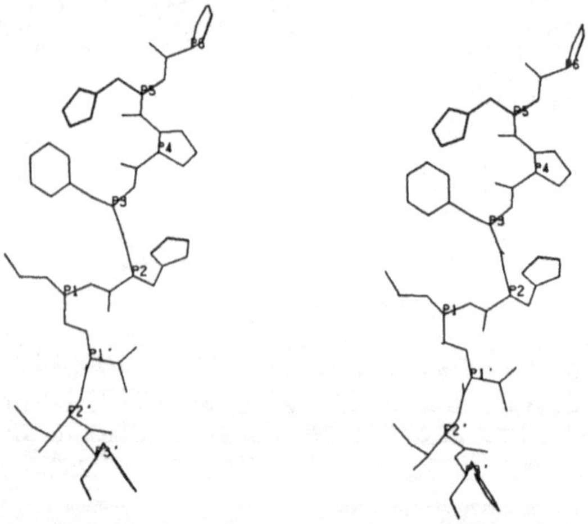

The sidechains at P_1 and P_1' are oriented towards the large hydrophobic pockets that make up S_1 and S_1' as suggested in earlier studies (12,13,15). The inhibitor sidechains (P_4 to P_2) pack closely into the cleft making variable numbers of intermolecular contacts as indicated in Table 5. It is interesting that P_5 and P_6 also interact with the enzyme although the residues lie on the surface at the end of the active site cleft.

Table 5

Subsite	Number of Contacts	Residue in H-142
S_6	9	Pro P_6
S_5	11	His P_5
S_4	8	Pro P_4
S_3	22	Phe P_3
S_2	11	His P_2
S_1 }	35	Leu P_1
S_1' }		Val^R P_1'
S_2'	17	Ile P_2'
S_3'	10	His P_3'

Numbers of contacts between enzyme and different residues of H-142.

The close correlation between the interactions defined here and those defined previously both by X-ray studies and predictive techniques implies that at least the interactions of P_4 - P_1' are well defined and largely preserved in different enzyme complexes. This may not be true of P_5 and P_6 where the substrate or inhibitor may optimise interactions with residues on the surface of the enzyme where there is room for movement. Also the positions of P_2' and P_3' seem unlikely to be preserved in all complexes. It is probable that the presence of an aromatic residue (tyrosine or phenylalanine) at position 189 in aspartic proteinases other than renins demands a different conformation in the polypeptide substrate in this region.

References

1. Subramanian, E., Swan, I.D.A., Liu, M., Davies, D.R., Jenkins, J.A., Tickle, I. J., Blundell, T.L.: Proc. Natl. Acad. Sci. (USA) 74, 556-559 (1977).

2. Fruton, J.S.: Adv. Enzymol. 44, 1-36 (1976).

3. Hofmann, T., Hodges, R.S.: Biochem. J. 203, 603-610 (1982).

4. Dunn, B.M., Kammermann, B., McCurry, K.: Anal. Biochem. 138, 68-73 (1984).

5. Powers, J.C., Harley, A.D., Myers, D.V.: Acid Proteases (J. Tang, ed.) Adv. Exp. Biol. Med. 95 pp.141-157, Plenum Press, New York 1977.

6. Dunn, B.M., Parten, B., Jiminez, M., Rolph, C.E., Valler, M.J., Kay, J.: this volume pp.,221-243 (1984).

7. Szelke, M., Leckie, B.J., Hallett, A., Jones, D.M., Sueiras-Diaz, J., Atrash, B., Lever, A.F.: Nature 299, 555-557 (1982).

8. Lienhard, G.D.: Science 180, 149-254 (1973).

9. Szelke, M.: this volume pp. 421-441 (1984).

10. Webb, D.J., Cumming, A.M.A., Leckie, B.J., Lever, A.F., Morton, J.J., Robertson, J.I.S., Szelke, M., Donovan, B.: Lancet ii, 1486-1487 (1983).

11. Blundell, T.L., Sibanda, B.L., Pearl, L. H.: Nature 304, 273-275 (1983).

12. Sibanda, B.L., Blundell, T.L., Hobart, P.M., Fogliano, M., Bindra, J.S., Dominy, B.W., Chirgwin, J.M.: FEBS Letts. 174, 102-111 (1984).

13. Pearl, L.H., Blundell, T.L.: FEBS Letts. 174, 96-101 (1984).

14. James, M.N.G., Sielecki, A., Salituro, F., Rich, D.H., Hofmann, T.: Proc. Natl. Acad. Sci. (USA) 79, 6137-6141 (1982).

15. Blundell, T.L., Jones, H.B., Khan, G., Taylor, G.L., Sewell, B.T., Pearl, L.H., Wood, S.P.: Enzyme Regulation and Mechanism of Action (P. Mildner and B. Ries, eds.) pp.281-288, Pergamon, Oxford, 1980

DESIGN AND SYNTHESIS OF STATINE-CONTAINING INHIBITORS OF CHYMOSIN

M.J.Powell, R.J.Holdsworth, T.S.Baker
Boots-Celltech Diagnostics Ltd, 240, Bath Road, Slough SL1 4ET. U.K.

R.C.Titmas, C.C.Bose, A.Phipps, M.Eaton
Celltech Ltd, 244, Bath Road, Slough SL1 4DY, U.K.

C.E.Rolph, M.J.Valler and J.Kay
Dept. of Biochemistry, University College, P.O.Box 78, Cardiff CF1 1XL.
Wales, U.K.

Acidified extracts of animal stomachs have been used for many years for the coagulation of milk to produce cheese. The major ingredient of this extract or rennet is the aspartic proteinase, chymosin. This acts upon certain proteins in milk at pH 6.0-6.5 to initiate a cascade reaction which results ultimately in the separation of solid curd from the liquid whey (1). Coagulants such as chymosin begin this process by the selective hydrolysis of a Phe-Met bond in the following sequence in κ-casein:-

101 105 106 111

Pro-His-Leu-Ser-Phe-Met-Ala-Ile-Pro-Pro-Lys-Lys-

There is currently a chronic world shortage of calf chymosin for cheese-making so that recombinant DNA technology has been harnessed to introduce and express the gene for calf (pro) chymosin in E.Coli (2,3) and yeast(4).

It was also of interest for commercial purposes to develop a tight-binding inhibitor of chymosin that would be effective at pH values as high as 6.0. The naturally-occurring compound, pepstatin, is a good inhibitor of several aspartic proteinases including pepsin and cathepsin

Aspartic Proteinases and their Inhibitors

D (5) but it is relatively ineffective towards calf chymosin (Table 1). Previous reports have indicated that it is possible to prepare synthetic inhibitors of renin by introduction of the unusual amino acid, statine (4(S)-amino-3(S)hydroxy-6 methylheptanoic acid) as a dipeptide analogue into peptides containing the sequence of residues found on either side of the scissile bond in the substrate, angiotensinogen (6). A similar strategy was adopted to develop the peptide inhibitors for chymosin described herein.

The decapeptide I (Table 1) containing statine as a replacement for the dipeptide Phe-Met of the scissile bond of κ-casein was synthesised by the polyamide solid-phase method using N-α-fluorenylmethoxy carbonyl (FMOC)-protected amino acids and purified by HPLC. The inhibition constant (K_i) for the interaction of this substrate analogue (I) with calf chymosin was determined at pH 3.1 and pH 6.0 using Lys-Pro-Val-Glu-Phe-(NO_2)Phe-Arg-Leu as substrate (7). However, as a competitive inhibitor of chymosin, it was found to be no more effective than pepstatin although its water solubility (probably due to the C-terminal-Pro-Pro-Lys-Lys sequence) was almost infinitely greater.

Aspartic proteinases are generally considered to have seven subsites (designated S_4, S_3, S_2, S_1-S_1', S_2', S_3') in their active site cleft. By inspection of a computer graphic representation of the chymosin molecule (based on the known structure of the aspartic proteinase from Endothia parasitica (8)) with the appropriate κ-casein sequence modelled in to occupy these seven subsites (by courtesy of Prof.T.L.Blundell and associates, Birkbeck College, London), it was reasoned that deletion of the histidine residue from the N-terminal end of the decapeptide should not introduce any gross steric constraints on interaction with the enzyme. Accordingly, the des-His-N-acylated nonapeptide (II) was synthesised and purified and was found to have K_i values for chymosin almost 20 fold better than the histidine-containing direct substrate analogue. Treatment of II with carboxypeptidase B to remove the two C-terminal lysine residues did not affect the ability of the peptide to inhibit chymosin (K_i at pH 6.0 = 1.4×10^{-8}M).

Compound	Structure	K_i ($\times 10^8$) M pH 6.0	pH 3.1
Pepstatin	Iva-Val-Val-Sta-Ala-Sta	22	3.2
I	R-C(=O)-NH-His-Leu-Ser-Sta-Ala-Ile-Pro-Pro-Lys-Lys	16	3.7
II	R-C(=O)-NH-Leu-Ser-Sta-Ala-Ile-Pro-Pro-Lys-Lys	1.1	0.3
III	R-C(=O)-NH-Leu-Asp-Sta-Ala-Ile-Pro-Pro-Lys-Lys	2.9	0.7

Table 1 Kinetic constants (K_i) for the inhibition of calf chymosin were determined at pH 3.1 (in 0.1M sodium formate buffer) and pH 6.0 (0.1M MES buffer) at 37° using Lys-Pro-Val-Glu-Phe-(NO_2)Phe-Arg-Leu as substrate. Plots of 1/v against inhibitor concentration at different substrate concentrations were used to derive the K_i values. All inhibitions were of the competitive type.

A further analogue (III) was synthesised in which the serine residue occupying the S_2 sub-site was replaced by an aspartic acid since it was considered possible for the Lys residue in position 220 of the chymosin sequence to approach close enough to the side-chain of the substrate residue occupying the S_2 sub-site to form an electrostatic interaction. This inhibitor had a K_i value at pH 3.1 virtually identical to that of the serine-containing analogue (II) but the value at pH 6.0 (where the Asp should have been fully ionised) was no better than that obtained with II(Table 1).

Both II and III were as effective at pH 6.0 in inhibiting recombinant chymosin obtained from E.Coli (K_i values 1.1 and 2.8×10^{-8}M respectively) as they were towards the naturally-occurring calf chymosin (K_i=1.1 and 2.9×10^{-8}M - see Table 1).

Thus, it would appear that with the sub-sites of chymosin occupied as follows:

$$\begin{array}{ccccccc} S_4 & S_3 & S_2 & S_1 & S_1' & S_2' & S_3' \\ \text{R-}\underset{\displaystyle \|\atop \text{O}}{\text{C}}\text{-NH-} & \text{Leu -} & \text{Ser} \text{ —— } & \text{Sta} \text{ —— } & \text{Ala -} & \text{Ile -} & \end{array}$$

effective inhibition of the enzyme can be achieved.

Peptide II was also a potent inhibitor of the ability of chymosin to clot milk at pH 6.0 and since this peptide has a high aqueous solubility, it may find application in a new method for the detection of hormones in milk.

References

1. Fox, P.A.: Biochem.Soc.Trans. 10, 282-284 (1982).
2. Emtage, J.S., Angal, S., Doel, M.T., Harris, T.J.R., Jenkins, B., Lilley, G., Lowe, P.A.: Proc.Natl. Acad.Sci (USA) 80, 3671-3675(1983).
3. Marston, F.A.O., Lowe, P.A., Doel,M.T., Schoemaker, J.M., White, S., Angal, S.: Biotechnology. 2, No.9, September (1984).

4. Mellor, J., Dobson, M.J., Roberts, N.A., Tuite, M.F., Emtage, J.S., White, S., Lowe, P.A., Patel, T., Kingsman, A.J., Kingsman, S.M.: Gene 24, 1-14 (1983).

5. Kay, J., Valler, M.J., Dunn, B.M. Proteinase Inhibitors, Medical & Biological Aspects (N.Katunuma, H. Umezawa, H.Holzer, eds.) pp.201-210. Springer-Verlag, Berlin (1983)

6. Boger, J.S., Lohr, N.S., Ulm, E.H., Poe, M., Blaine, E.H., Fanelli, G.M., Lin, T-Y., Payne, L.S., Schorn, T.W., Lamont, B.I., Vassil, T.C., Stabilito, I.I., Veber, D.F., Rich, D.H., Bopari, A.S. : Nature 303, 81-84 (1983).

7. Dunn, B.M., Parten, B., Jimenez, M., Rolph, C.E., Valler, M.J., Kay, J. : this volume, pp 221-243 (1984).

8. Subramanian, E., Swan, I.D.A., Liu, M., Davies, D.R., Jenkins, J.A., Tickle, I.J., Blundell, T.L. Proc.Natl.Acad Sci (USA) 74, 556-559 (1977).

INTERACTION OF CATHEPSIN D AND PEPSIN WITH α_2-MACROGLOBULIN

Tamara Lah, Maja Vihar, Vito Turk

Department of Biochemistry, J.Stefan Institute, E.Kardelj University,
Jamova 39, 61000 Ljubljana, Yugoslavia

Introduction

Alpha$_2$-macroglobulin (α_2M) is a high-molecular-weight inhibitor, present in plasma and in connective tissue generally (1), where it modulates the activity and clearance of proteinases of various types, and might, in addition to this, play an important role in immunological response (2,3). The clearance of α_2M from the bloodstream or from biological fluids is a receptor-mediated process (3, 4), leading to the former's final degradation in lysosomes (2), where cathepsins play a major role. Very few investigations have been carried out on the interaction of aspartic proteinases with α_2M, although it is commonly believed that α_2M generally inhibits all four classes of proteinases. Some doubt, however, has been cast on this hypothesis by two findings: firstly, non-typical behaviour of α_2M-cathepsin D complex (5) and secondly, the binding of Candida albicans aspartic proteinase to α_2M, which occurred in a narrow pH interval, with only slight inhibition of the enzyme (6). The present studies were designed to elucidate the mechanism of the interaction of cathepsin D and pepsin with α_2M. It was found that both enzymes caused a change in α_2M conformation and degraded the inhibitor in an acidic environment, while their enzymatic activity was not inhibited.

Aspartic Proteinases and their Inhibitors

Results

The stability of pure human α_2M, isolated by a combination of Cibacron affinity (7), gel permeation and antihaptoglobin-Sepharose chromatography, was investigated in the pH range 2.5-8.0 by far ultraviolet (UV) circular dichroism (CD) spectroscopy and the activity of trypsin-α_2M complex on N-a-benzoyl-L- Arg- p- nitroanilide (BAPNA)in the presence of soy -bean trypsin inhibitor (STI). After being incubated for 24 hours at 25° C or for 30 min at 37° C, in a buffered solution of different pH values, the inhibitor was brought to pH 8.0 and its residual inhibitory potential on trypsin was measured: it was found that at pH value of 4.0, 40% of the inhibitory potential of α_2M was irreversibly lost at 25° C, and about 80% at 37° C. A small, but irreversible increase of ellipticity ([θ]) was observed in far UV CD spectra (Fig. 1). Spectra were analysed by the method of Chang et al. (8) which included β-turns (Table 1). A non-significant increase in the helical structure and a similar decrease in the unordered structure was observed, but significant rearrangement in the β-structure followed the acidification of the inhibitor solution.

Interaction with the proteinases pepsin (SIGMA, USA) and cathepsin D, isolated from bovine spleen (9), was observed by means of far UV CD spectroscopy, polyacrylamide gel electrophoresis (PAGE) and sodium dodecysulphate-PAGE (SDS-PAGE). Simultaneously, proteolytic activity was measured on hemoglobin substrate at pH 4.0 and 3.0 by Anson's method (10) for cathepsin D and pepsin, respectively. No influence of α_2M on the proteolytic activity of cathepsin D, preincubated with the inhibitor at pH values from 8.0-4.0 was observed at any molar ratio of enzyme: inhibitor (ranging from 3:1 to 0.5:1) and any T (25° C, 37° C). Similarly, pepsin activity was not affected when the enzyme was preincubated with α_2M under varying conditions such as pH, molar ratio and T.

Far UV CD spectra of α_2M and proteinase- α_2M complexes are shown on Fig. 1. The interaction of α_2M with aspartic proteinases led to a red shift of negative ellipticity maximum (210 nm → 214 nm) and an increase of $[\theta]_{mrw}$, which was higher with increasing enzyme: inhibitor ratio. Cathepsin D- α_2M and pepsin- α_2M complexes contained significantly less helical and ordered β-sheet structure compared to native or partially inactivated α_2M(Table 1). The electrophoretic mobility of α_2M at pH 4.0 is shown on Fig. 2. The middle bands represent the S-form of the inhibitor. Association with cathepsin D at 25° C and 37° C obviously changed the inhibitor's structure and increased mobility was observed in lanes 1,2 and 4,5, respectively (Fig. 2). No change of electrophoretic mobility was observed, when α_2M was incubated with cathepsin D at pH above 6.0. SDS-PAGE experiments were designed to show pH dependence and proteolytic fragmentation of α_2M in the presence of molar excess of cathepsin D, 20:1 molar ratio (Fig. 3). Limited proteolysis of 185 kDa subunits at pH 7.0 by cathepsin D produced about 95 kDa proteolytic fragments , similarly to trypsin hydrolysis under the same conditions (lanes 4 and 5, respectively). At lower pH nearly complete splitting of 185 kDa subunit was observed (lanes 2 and 3).

Fig. 1. Far UV CD spectra

Spectra were measured on Jobin-Yvon Dichrograph III, 25° C, 0.1 mm cell. $[\theta]_{mrw}$ was calculated from the difference between α_2M/ buffer or proteinase- α_2M/proteinase CD absorbance.

—— α_2M, pH 8.0;
- - - α_2M, pH 4.0;
--- cathepsin D- α_2M, pH 4.0, 1:1;
-×- pepsin- α_2M, pH 4.0, 2:1.

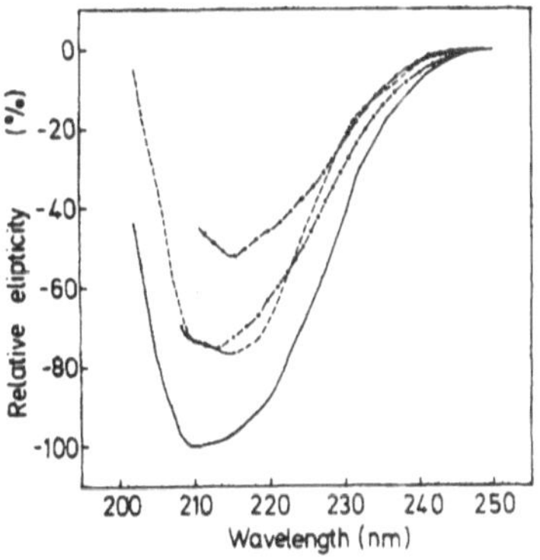

Table 1. Secondary Structure of α_2M and Proteinase- α_2M complex (in %)

Incubation	pH	T	t	helix	β-sheet	β-turn	β-total	random
α_2M	8.0	25°C	24 h	14	46	10	56	30
Cath.D- α_2M^{++}	8.0	25°C	24 h	15	42	13	55	30
α_2M	4.0	37°C	30 m	15	41	19	60	25
Cath.D- α_2M^{+}	4.0	37°C	30 m	5	45	16	61	34
Cath.D- α_2M^{+++}	4.0	37°C	30 m	0	30	20	50	50
Pepsin- α_2M^{++}	4.0	37°C	30 m	0	24	24	48	52

$^{+}$molar ratio 1:1; $^{++}$molar ratio 2:1; $^{+++}$molar ratio 10:1.

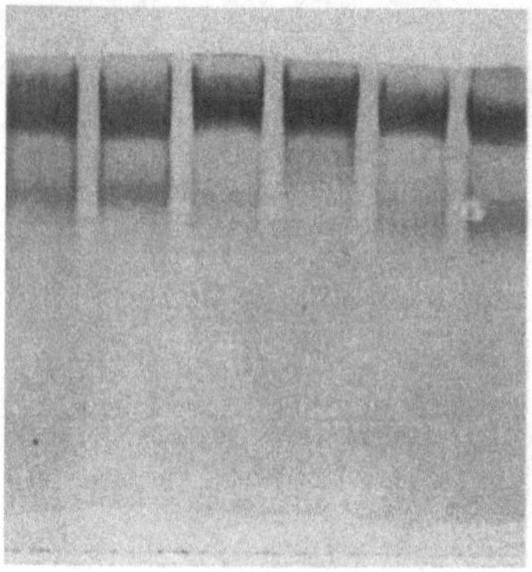

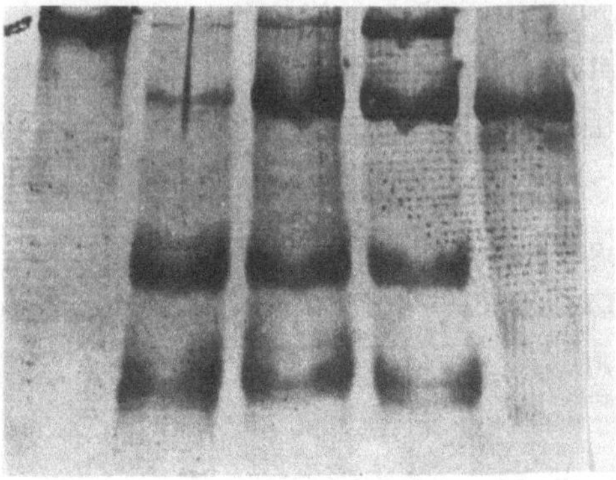

Fig. 2. PAGE of α_2M and cathepsin D-α_2M complex was performed as described by Barrett et al. (11). α_2M and cathepsin D- α_2M complexes (2:1 molar ratio) were preincubated at pH 4.0 and 25°C for 24 hrs (lanes 1,2,3) or 37°C for 30 min (lanes 4,5).

Fig. 3. SDS-PAGE with reduction and boiling (4 min) of the samples, was performed in gradient gels (11). Lanes represent 1) α_2M; 2) cathepsin D- α_2M, pH 5.0; 3) cathepsin D- α_2M, pH 6.0; 4) cathepsin D- α_2M, pH 7.0; 5) trypsin- α_2M, pH 7.0.

Discussion

Taken together, the above results provide evidence for the binding of the aspartic proteinases, cathepsin D and pepsin to α_2M. This process is strongly pH dependent as indicated by the conformational change of α_2M, observed by far UV CD, PAGE and SDS-PAGE which occurred at a pH $\leqslant 6.0$, and is fairly similar to the conformational change described upon serine proteases binding (12). The interaction, studied in greater detail with cathepsin D at 37° C, a molar ratio varying from 1:1 to 1:10 and the pH range from 4.0 to 6.0, resulted in increased electrophoretic mobility of the complex (Fig. 2) and an extensive degradation of the inhibitor (Fig. 3). These changes, however, could not involve effective "entrapment" of cathepsin D or pepsin, as both retained their proteolytic activity on the hemoglobin substrate under the described experimental conditions of enzyme-inhibitor interaction. Our results are not in disagreement with reported observations on adherence and cleavage of α_2M by some aspartic proteinases, cathepsin D (13), Candida albicans proteinase (6) and calf chymosin (14). They complement previous studies with respect to pepsin interaction as well as provide measurements of activity of the α_2M-cathepsin D and α_2M-pepsin complexes on hemoglobin, confirming non-inhibitory binding between the enzyme and the inhibitor. Although considerable inactivation of α_2M occurred on acidification at 37° C, due probably to a conformational change in α_2M, non-inhibitory binding and limited proteolysis of the inhibitor was observed also under the conditions where both α_2M and the aspartic proteinase were in the native, active state. These observations lead to two main conclusions :

- conformational change, observed in α_2M upon the proteolytic attack by aspartic proteinases might be similar to the changes observed in the case of other proteinases. "Entrapment", however, is not a necessary consequence and it is suggested, that, generally, the activities of proteinases in the α_2M-complexes might also be regulated by the conformational distortion of enzymes' active sites.
- α_2M is probably not a physiological modulator of the proteolytic activity of aspartic proteinases in contrast to what has been suggested for other classes of proteinases. Its inactivation and rapid degradation by cathepsin D under conditions which mimic intralysosomal pH, T and a molar excess of proteinase, permits us to conclude that α_2M, rather than being an

inhibitor, the physiologic substrate for cathepsin D (and pepsin). Such a conclusion could be of considerable importance in some pathological conditions, such as inflammation, where cathepsin D is released extracellularly (15).

We thank dr. B. K. Shenton, dr. P.K. Donnelly and dr. A. Alomran (University of Newcastle upon Tyne, UK) for help in isolation of α_2M and dr. A.J. Barrett (Strangeways Lab., UK) for part of the α_2M sample and for fruitful discussions. This work was financed by the Research Council of Slovenia and NSF Grant No. F7F030Y.

References

1. Cassiman,J.J.,Van Leuven,F.,Van der Schueren,B.,Van den Berghe,H.: Cell Tissue Res. 213, 301-310 (1980).
2. James,K.: TIBS 5, 43-47 (1980).
3. Borth,W.,Susani,M.: Proteinases:Potential Role in Health and Disease (Hori,H.,Heidland,A.,eds.)Plenum Publ.Corp.pp 371-377(1984).
4. Van Leuven,F.: TIBS 5, 185-187 (1982).
5. Rüchel,R.,Böning,B.: J.Immun.Meth.61,107-116 (1983).
6. Feldman,S.R.,Nevy,K.A,Gonias,S.L.,Pizzo,S.V.: BBRC 114,757-762(1983).
7. Virca,G.D.,Travis,J.,Hall,P.K.,Roberts,R.C.:Anal.Biochem.89,274-278 (1978).
8. Chang,C.T.,Wu,C.S.C.,Yang,J.T.: Anal.Biochem. 91, 13-31 (1978).
9. Smith,R.,Turk,V.: Eur.J.Biochem. 48,245-254 (1974).
10. Anson,M.L.: J.Gen.Physiol. 22, 79-89 (1939).
11. Barrett,A.J.,Brown,M.A.,Sayers,C.A.: Biochem.J. 181,401-418(1979).
12. Björk,I.,Fisch,W.W.: Biochem.J. 207, 347-356(1982).
13. Barrett,A., Starkey,Ph.: Biochem.J. 133, 709-724 (1973).
14. Mortensen,B., Sottrup-Jensen,L., Hansen,H.F.,Petersen,T.E., Magnusson,S.: FEBS Letters 135, 295-300 (1981).
15. Marzella,L., Glaumann,H.: Int. Rev. Exp.Pathol. 25, 239-279 (1983).

METHODS FOR DETECTION OF PROTEINASES: ELECTROPHORETIC AND IMMUNOLOGICAL COMPARISON OF ASPARTIC PROTEINASES OF DIFFERENT ORIGINS

Bent Foltmann, Nadezhda I. Tarasova, Pal B. Szecsi.
Institute of Biochemical Genetics, University of Copenhagen,
Ø. Farimagsgade 2A, (DK) 1353 Copenhagen K, Denmark.

Introduction

In many preparations of aspartic proteinases the purifications have primarily been monitored by determinations of the proteolytic activity in test tubes. In our recent investigations on gastric proteinases and their zymogens (1,2) we have found monitoring by gel electrophoresis and immunoelectrophoresis most useful.

In this communication we summarize some of our experiences with different methods and we describe a new and sensitive technique for the detection of proteinases after gel electrophoresis. In the latter method a second skim-milk containing gel is "sandwiched" against the first gel after electrophoresis has been performed. After incubation, the proteinases produce precipitates in the skim-milk, and since the precipitates consist of casein mainly, we use the term caseogram for this type of detection.

Furthermore we illustrate the use of those techniques by the detection of aspartic proteinases or their zymogens in physiological fluids and extracts of different origins.

Materials

Cheese rennets. The samples were gifts from Chr. Hansens Laboratory Ltd. Copenhagen, Denmark.
Blood was outdated human blood from Blodbanken, Rigshospitalet, Denmark.

Aspartic Proteinases and their Inhibitors

Seminal liquid was a pool of samples used for fertility investigations at Praktiserende Lægers Laboratorium, Copenhagen, Denmark.
Specimens of tissues were obtained at autopsy from Bispebjerg Hospital, Copenhagen, Denmark. Only specimens without any sign of cadaverosis were used.
Antisera against chromatographically purified pepsinogens and partly purified membrane aspartic cathepsin (for purification, see below) were raised in rabbits at Statens Seruminstitut, Copenhagen, Denmark, by courtesey of Dr. N.H. Axelsen.
Human pepsins and pepsinogens. In the electrophoretic comparison of pepsinogens from different origins we included a reference of partly purified and partly activated extract of gastric mucosa. Like other adult mammals man has two types of pepsins in the gastric juice: Pepsin A (EC 3.4.23.1) generally called just pepsin, and pepsin C (EC 3.4.23.3) also called gastricsin (3,4). The corresponding zymogens in this paper are designated pepsinogen A and pepsinogen C. In the clinical litterature the designations pepsinogen I and II are often used, e.g.(5). The two types show no immunochemical cross reaction; within each type isoforms may occur, and these can not be distinguised with the use of polyclonal antisera. The isoforms are numbered according to their electrophoretic mobilities with number 1 having the highest mobility towards the anode. The fast moving number 1 components occur in minor amounts only and they were not detectable in our reference. The human gastric mucosa further contains minor amounts of some electrophoretically slow moving proteinases (abbreviated SMP (5)). These were not detectable in the reference.
Chymotrypsin (EC 3.4.21.1) was bovine from Worthington, Bedford, USA.
Trypsin (EC 3.4.21.4) was porcine from NOVO Industry, Bagsværd, Denmark.

Methods

Extractions of tissue samples were carried-out in an ice-bath with a Potter-Elvehjem homogenizer, (1 g of wet weight tissue and 2 ml of 0.05 M sodium phosphate pH 6). Cell debris was removed by centrifugation at 20000xg for 30 min (5°C).

Agar gel electrophoresis was carried-out in gels of 1% agar (Noble, Difco Laboratories, Detroit, USA). For analyses of samples that contained active proteinases only, 0.05 M sodium acetate pH 4.3, pH 5.0 or pH 5.5 were used. For samples that contained active enzymes and zymogens, 0.05 M sodium acetate pH 5.5 or 0.05 M sodium phosphate pH 6.0 were used; samples that were tested for zymogens only were run in 0.02 M Tris/Barbital pH 8.2. Electrophoresis took place in an apparatus with a horizontal water cooled plate (1).

Immunoelectrophoresis was carried-out in gels of 1% agarose (Litex, Glostrup, Denmark) made up in 0.02 M Tris/Barbital pH 8.6. Conditions for electrophoresis and composition of buffers were those described in (6). The following versions were used: crossed (7), crossed tandem (8), rocket (9), and line rocket (10).

Disc electrophoresis was performed in slab gels of polyacrylamide. The system was that of Maurer (11) with a stacking gel at pH 6.7 and a separation gel at pH 8.7; concentrations of polyacrylamide from 5 to 12.5% were tested in the separation gel.

Isoelectric focusing was made in gels of polyacrylamide (1 mm thick, T = 7.5% C = 3%) containing 5% Pharmalyte pH 2.5 - 5 (Pharmacia, Sweden) and 1% Triton-X100. Anode solution 0.1 M H_2SO_4, cathode solution 0.1 M NaOH.

Detection by digestion of hemoglobin (hemogram). After gel electrophoresis proteinases may be detected by incubation with a hemoglobin solution followed by precipitation and staining of undigested protein. The principle of this method was introduced by Uriel (12) and it has been used in several investigations on pepsins (13) and pepsinogens from gastric mucosa (5) or urine (14). Results of investigations from our institute are presented in (1) and (2). In this paper we shall present a broad outline and some general comments on our experience with the method.

After electrophoresis, the gel was immersed in an acid solution of hemoglobin for 20 min at room temperature. The gel was removed from the hemoglobin solution, and digestion was continued for 2 to 24 h at 37°C in a moist chamber. Undigested hemoglobin was precipitated e.g. with ethanol, acetic acid, water 2:1:2 v/v. Gels of polyacrylamide could be stained directly with Coomassie Brilliant Blue; gels of agar were pressed under filterpaper and dried before staining.

The sensitivity of the method may be increased with decreasing concentrations of hemoglobin in the substrate solution, but with concentrations less than 0.3% it is often difficult to obtain an even blue background. Consequently we recommend a substrate solution of 0.3% hemoglobin.

It should be observed that the impregnation with hemoglobin in different gels shows great variation. In gels that contain more than 8% of polyacrylamide it is difficult to obtain sufficient impregnation of the hemoglobin in the gels, and with gels of polyacrylamide we have obtained the best result with 5% gels. The agarose (Litex, type HSA) which we normally use for immunoelectrophoresis behaves very badly for the detection of proteinases with hemoglobin. We have made several unsuccessful experiments in which we tried to combine immunoelectrophoresis with hemograms. Agarose of Sigma type II is better for such experiments, but we have obtained the best results by combining immunoelectrophoresis with caseograms as described below.

For the adjustment of the pH before digestion we have used preincubation of the gels in 0.05 or 0.1 M HCl if the electrophoretic fractionation was carried-out at a pH above 6. With the other buffers used in our experiments it was sufficient to equilibrate for 20 min with the same buffer as used for the subsequent incubation with the hemoglobin solution.

Detection by clotting of casein (caseogram). Clotting of casein depends on a limited proteolysis of κ-casein. Most known aspartic proteinases are able to produce such a limited proteolysis, and clotting of skim-milk at pH 5.3 is a sensitive test for aspartic proteinases. For a rapid monitoring of chromatographic fractionation of proteinases we have used radial-diffusion from wells in a skim-milk containing gel (15). This semiquantitative evaluation was later supplemented with gel electrophoresis and detection of the proteinases by casting a casein containing gel on top of the first gel, after electrophoresis had been performed. The method was subsequently standardized as summarized below.

After electrophoresis the gel was immersed in 0.05 or 0.1 M HCl in order to activate zymogens (this step was omitted if the aim was test for active enzymes only). The gel was subsequently equilibrated with 0.15 M sodium acetate pH 5.3. A 1% gel of agarose (due to its mechanical strength, preferably Sigma type II) was made up to 1% with skim-milk powder in 0.1 M sodium acetate pH 5.3. The skim-milk gel was cast to a thickness of 1 mm on a glass plate of the same dimensions as the electrophoresis gel. The skim-milk gel was then placed upon the electrophoresis gel and after incubation at 37°C in a moist chamber, the proteinase containing zones produced distinct precipitates in the skim-milk gel. The precipitates were easily seen in diffuse side light, and were copied on a clear plastic foil placed on the gel. For permanent documentation, the skim-milk gel was removed from the electrophoresis gel, and stained with 0.02% amido black in water. Destaining and elution of non-precipitated proteins was carried out with 5% glycerol in 0.01 M ammonium acetate pH 5.8. The gel was finally dried between two layers of cellophane.

The only difficulty in the method is that of permanent documentation. By visual inspection it is easy to see the precipitated casein, but the contrast is not good for photography, therefore tracings are used for most illustrations in this communication.

The caseogram print works equally well after electrophoresis on agar gel, polyacrylamide gel (disc or isoelectric focusing) and after different kinds of immunoelectrophoresis (in the latter case, preincubation at pH 2.8 is advisable). The only condition which one must strictly adhere to, is that the surface of the electrophoresis gel must have a pH between 5.1 and 5.3 before contact with the casein gel is established. If pH is too low, non-specific acid precipitation of casein may occur, and with too high pH the sensitivity of the method is diminished.

The detection limit is 5 ng for pepsin A against 25 ng by a hemogram at pH 2.8. For chymosin it is 1 ng against 100 ng by a hemogram at pH 3.5 (all values are for 2 h of incubation). In addition to the great sensitivity, the caseogram has the advantage that inert proteins in the sample do not interfere with detection. Caseograms are therefore the method of choice for analyses of crude preparations of aspartic proteinases.

Results

Cheese rennets

The caseogram detection offers an excellent opportunity to characterize different types of cheese rennets, that otherwise are difficult to analyse by gel electrophoresis due to the weak general proteolytic activity of the milk-clotting proteinases used in cheese making.

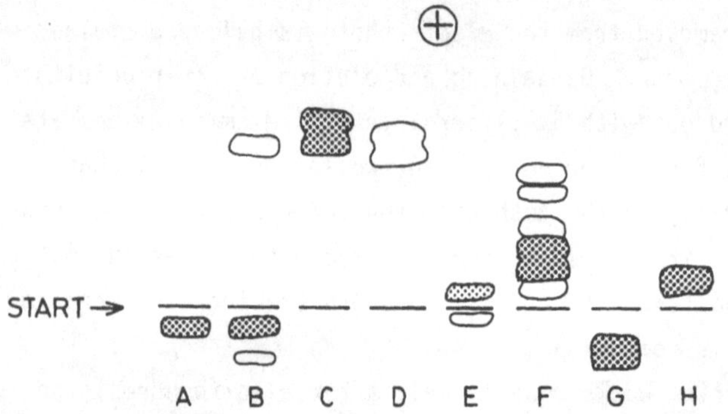

Fig.1: Tracings of a caseogram with different rennets. All samples were diluted to a milk-clotting activity of approx. 0.1 CU per ml; 5 µl was used in each slit. Electrophoresis was carried out in a 1% agar gel, 0.05 M sodium acetate pH 5.0 at 15 V/cm for 1.5 h. For comments to the samples see the text.

Fig.1 shows the results of electrophoretic analyses of different rennets. The milk-clotting unit (CU) is that defined in (16), 0.1 CU per ml is equivalent to the milk-clotting activity of 0.7 µg of calf chymosin per ml. Due to the electroosmotic flow in the system, calf chymosin has an apparent cathionic mobility (lane A). In calf rennet (lane B) a faint zone is observed below the main zone of chymosin. This faint zone corresponds to partial degradation products which are eluted before the main components of chymosin by chromatography on columns of DEAE cellulose(16). The anionic component corresponds to traces of bovine pepsin A that are also present in calf rennet. The bovine rennet (lane C) shows pepsin A only. Under these conditions bovine pepsin A has almost the same electrophoretic mobility as porcine pepsin A (lane D). If the electrophoretic fractionation is performed at pH 4.3, the bovine pepsin A may be resolved into 3 components of which the 2 predominant move ahead of porcine pepsin A. The individual zones of pepsin A may represent components with a different degree of phosphorylation. Pepsin C was not present in these preparations; other experiments with crude and activated extracts of bovine or porcine gastric mucosa have shown that their components of pepsin C have mobilities that are about half of the mobility of pepsin A. Lamb rennet (lane E) shows two components, one has a mobility that is slightly larger than that of calf chymosin, and like calf rennet, minor amounts of a component with a smaller electrophoretic mobility is also observed, presumably also due to partial degradation products. Chicken pepsin (lane F) is resolved into about 6 components, it is unknown to which extend these represent genetic differences or post-translational modifications with different degrees of sulfatation in the glycosylated protein (17). The microbial rennets from Mucor miehei (lane G) and Mucor pusillus (lane H) are easily separated in this system.

One electrophoretic fractionation does not allow a complete identification of all the commonly used rennets, but the method provides a simple and rapid way for the qualitative characterization of the preparations.

Pepsinogens outside the gastric mucosa

Serum pepsinogen has been subject to quantitative determinations by radioimmunoassay, but this method does not give information about the distribution of isopepsinogens in the samples. In spite of the high sensitivity of the caseogram, it was not sufficient for the direct analysis of serum pepsinogen. The samples were therefore concentrated and partially purified: 20 ml of serum was diluted with 200 ml of distilled water, pH was adjusted to 6 by a slow addition of 2 M acetic acid, and the precipitate was removed by centrifugation (2000xg for 20 min at 5°C). DEAE cellulose was equilibrated with 0.05 M sodium phosphate pH 6; after suction with a Buchner funnel, 0.5 g of wet weight cellulose was suspended in the diluted serum,

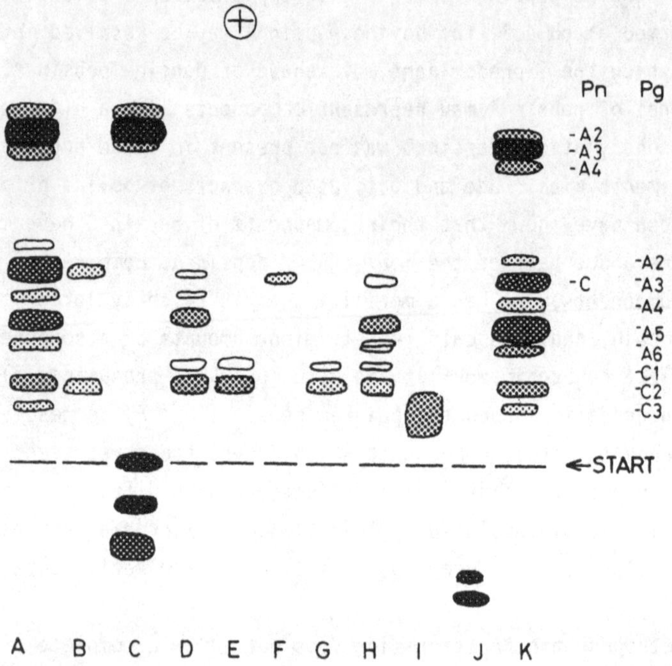

Fig.2: Tracings of a caseogram with physiological fluids and tissue extracts. Electrophoresis was carried-out in 1% agar gel, 0,05 M sodium phosphate pH 6.0 at 15 V/cm for 2 h. Abbreviations; Pn: pepsin; Pg: pepsinogen. Note that PgA3 and PgA5 both give rise to a pepsin with mobility as PnA3 (14). Isoforms of PnC were not observed, PnC has a mobility like PgA3, and an activation intermediate of PgC2 has a mobility like PgA5 (1). Lane A - H and K contain the following samples of human origin: A and K: reference of partly purified and partly activated extract of gastric mucosa. B: extract of striated muscle (psoas). C: extract of pancreas (during the preparation all pepsinogens have been converted into pepsins, the slow moving and cathionic components are pancreatic serine proteinases).
D: concentrated sample of serum pepsinogen. E: seminal liquid. F: urine. G: extract of vesicular glands. H: extract of prostatic gland. I: reference of bovine chymotrypsin (1 mg per ml). J: reference of porcine trypsin (1 mg per ml).

and adsorption of the zymogens took place by gentle stirring overnight at 5°C. The DEAE cellulose was collected by centrifugation and elution took place by suspension in 1.5 ml 1 M NaCl and stirring for 30 min. The DEAE cellulose was again removed by centrifugation and the supernatant was dialysed overnight against 500 ml of 0.005 M sodium phosphate pH 6.5. The sample was freeze-dried and dissolved in 0.3 ml distilled water, 5 µl of this solution was used for the analyses as shown in Fig.2. In addition to the zones of pepsinogen A all samples showed appreciable amounts of pepsinogen C.

It should be added that pepsinogen A in the concentrated serum samples could be detected by hemograms after 20 h of digestion at 37°C; the samples, however, still contained a great deal of other serum proteins, that have almost the same electrophoretic mobility as pepsinogen C and covered this completely in the hemograms. This example illustrates the advantage that inert proteins do not interfere with detection by caseograms.

The method described here allows a direct comparison between the electrophoretic pattern of serum pepsinogens and that observed in urine. This investigation is not yet completed but we may report that we have not observed pepsinogen C in the samples of urine.

Seminal pepsinogen C. In 1952 Lundquist and Seedorff (18) demonstrated the presence of a pepsinogen-like proteinase in the seminal fluid. In subsequent investigations Hirsch-Marie and Conte (19), and Samloff and Liebman (20) showed that the seminal pepsinogen had immunochemical properties that were closely related to if not identical with gastric pepsinogen C.

Fig.2 includes the electrophoretic pattern of a pool of seminal liquid that had been concentrated with DEAE-cellulose. Pepsinogen C2 is the predominant component but minor amounts of pepsinogen C1 are also observed. We have confirmed the immunochemical relationships by tandem crossed immunoelectrophoresis with authentic gastric pepsinogen C and seminal liquid (Fig. 3). The precipitates fuse completely without any sign of spurs. We have not yet had the opportunity to carry-out a chemical analysis of seminal pepsinogen C, but the high degree of immunochemical identity raises some questions concerning the amino acid compositions of the seminal zymogen and the active proteinase that have been reported previously (21). These amino acid compositions differ so much from those found for gastric pepsinogen C and pepsin C (1) that it seems incompatible with the apparent complete immunochemical identity.

In a discussion of the origin of the seminal zymogens it should be recalled that the testes produce about 10% of the volume of the seminal liquid only. The bulk of the fluid is secreted by a pair of glands, the seminal vesicles, and by the prostate gland. From histochemical investigations, Lundquist and Seedorff suggested that the seminal zymogen was formed in the seminal vesicles, whereas Chiang et al. (22) have prepared the zymogen from prostate glands. We have analysed extracts of all three types of glands and examples of the electrophoretic patterns are shown in Fig.2. In glands from 4 individuals we observed pepsinogen C in all. Quantitative determinations were not carried-out; the intensities of the precipitates showed considerable variation, and with this material we are not able to decide whether the seminal vesicles or the prostate gland produce most of the pepsinogen C in the seminal liquid. It is noteworthy that pepsinogen C1 was observed in all samples, and two individuals showed minor amounts of pepsinogen A in the testes and prostate gland.

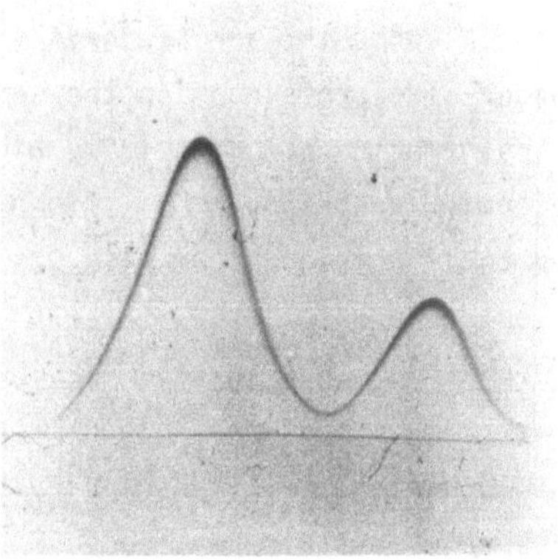

Fig.3: Crossed tandem immunoelectrophoresis of pepsinogen C from seminal liquid and from gastric mucosa. First dimension, anode to the right. Second dimension run into a gel containing anti-(gastric pepsinogen C), anode at the top. Right peak: seminal liquid. Left peak: gastric mucosal extract.

Other tissue extracts. We have screened other tissue extracts for the presence of pepsins and pepsinogens, with the following preliminary results. Duodenum: 4 samples examined, pepsinogen C was present in all. Jejunum: 1 sample, pepsinogen C was observed, together with a proteinase with an electrophoretic mobility like cathepsin D. Pancreas: 6 samples, 5 showed the presence of pepsinogen A and 4 of pepsinogen C. Large amounts of trypsin and chymotrypsin were observed in all cases. Striated muscle (psoas): 5 samples, 4 contained pepsinogen C and 2 of these also pepsinogen A.

The immunological identity of the tissue pepsinogens and those of gastric origin were tested by crossed tandem immunoelectrophoresis and rocket line immunoelectrophoresis. Quantitative determinations have not been carried-out, but visual inspection of the caseograms indicated considerably inter-individual variations in the intensity of precipitates.

Membrane aspartic cathepsin

In order to obtain a cathepsin D-like proteinase from a well defined cell type we made preparations of the membrane bound cathepsin from human erythrocytes. This enzyme has previously been partly purified e.g. (23), and the name membrane cathpsin D has been used. Both our preparation and that of (23) were inhibited by pepstatin, so the enzyme does belong to the group of aspartic proteinases. But it must be borne in mind that the delimitation between cathepsin D and E is not yet sufficiently clarified, and the names cathepsin D, types H and L have been used for membrane bound cathepsins from lymphocytes (24), for review see (25). As our experiments went on, doubt has arisen about the right name. Thus, until the enzyme is better characterized, we will use the neutral designation, membrane aspartic cathepsin.

One volume of washed and packed erythrocytes was lysed with 10 volumes of 0.005 M sodium phosphate pH 7.5 and standing for 20 h at 5°C. The membranes were collected by centrifugation (6000xg for 2.5 h at 5°C). The membranes were subsequently washed with a phosphate buffer of the same composition as the lysis buffer and collected by centrifugation (58000xg for 30 min at 5°C). After 4 to 6 washings the supernatant was colourless. One volume of packed membranes was mixed with 2 volumes of ice-cold water and 3 volumes of ice-cold 1-butanol. The suspension was centrifuged for 2 h at 58000xg (5°C). The water phase was made 0.1 M with respect to sodium acetate pH 4.0 and applied to a column of bacitracin-silochrom (26), 10 ml of extract was used per ml of silochrom containing 15 mmole of bacitracin. Elution took place with 0.1 M sodium phosphate pH 7.5. Some non-specific adsorption may occur on the silochrom, and a second affinity chromatography was carried-out with bacitracin-sepharose. For the second adsorption the solution was made 0.5 M with respect to sodium chloride and 0.2 M with respect to sodium acetate pH 4.0; 1 ml of sorbent was used per 100 ml of solution. Elution took place with 0.1 M sodium phosphate pH 8.2.

Determinations of proteolytic activities against hemoglobin at pH 3.5 relative to the absorbancies at 280 nm indicated about a 150 fold purification relative to the dialyzed butanol extract. The preparation was analysed by SDS-polyacrylamide electrophoresis; 3-4 minor bands were seen together with 2 predominant bands with apparent molecular weights of about 35000 and 15000, these correspond to the molecular weights of heavy and light chains from lysosomal spleen cathepsin D (27). This preparation was used for immunization of rabbits. The antiserum was used for comparison of aspartic cathepsins from other tissues, and the membrane aspartic cathepsin was used as a reference in the electrophoretic comparisons.

Spleen has been used in several experiments as a source for preparations of cathepsin D. We have compared the proteinases from the phosphate extract of human spleen with membrane aspartic cathepsin, (butanol was required to extract cathepsin from erythrocytes therefore a contamination of membrane aspartic cathepsin was excluded in the phosphate extracts).

By agar gel electrophoresis and development as a caseogram the crude spleen extract showed minor amounts of pepsin A and C only. Spleen cathepsin D was concentrated and partly purified on bacitracin-silochrom with the same conditions used for membrane aspartic cathepsin. During the elution at pH 7.6 the active pepsins were denatured.

The two preparations were compared and partly characterized by isoelectric focusing. After development as caseograms both showed two components. Those of membrane had pI at 4.1 and 4.2, whereas those from spleen had pI at 4.1 and 4.5. Furthermore the preparations were tested by affinity electrophoresis in presence of Con A-Sepharose (28), and both were retained to the same extend. It is also noteworthy that precipitation on caseograms was not possible without preincubation of the electrophoresis gel in 5% 1-O-methyl-α-D-glucopyranoside i.e. without dissociation of the proteinase-Con A complex. These findings show that both cathepsins are glycosylated.

The antiserum against membrane aspartic cathepsin was used for rocket line immunoelectrophoresis with an intermediate gel containing membrane aspartic cathepsin and wells with tissue extracts below. The extracts of liver and striated muscle (psoas) did not contain detectable amounts of membrane aspartic cathepsin. Detectable amounts were observed in extracts of testes, seminal vesicles and prostate glands. Moderate amounts were seen in the crude extract of spleen, and the precipitate from gastric mucosal extracts indicated a content of a membrane aspartic cathepsin that was at least 10 times larger than that of our spleen extract. This observation prompted us to further investigations on of the proteinases of the gastric extract.

Fig.4: Caseogram print of crossed tandem immunoelectrophoresis of a partially purified preparation of gastric SMP (left peak) and membrane aspartic cathepsin (right peak). First dimension, anode to the right. Second dimension run into a gel containing anti-(membrane aspartic cathepsin), anode at the top.

Extract of gastric mucosa was analysed by agarose gel electrophoresis at pH 6.0; 4 samples of the same extract were run in parallel and a sample of erythrocyte membrane aspartic cathepsin was included as a reference. Detection was carried-out by caseograms on the agarose gel and by crossed immunoelectrophoresis against anti-(membrane aspartic cathepsin), anti-pepsinogen C and anti-pepsinogen A. In all experiments the precipitations were detected by caseogram prints. The electrophoretic mobilities in gels of agar and agarose are not the same but the relative order of migration is the same in the two systems. The 3 antisera produced precipitates without any cross-reactions, and the reference of membrane aspartic cathepsin showed an electrophoretic mobility that was within the range of the mobilities of the electrophoretically slow moving proteinases from the gastric mucosa.

The immunological relationship between the gastric slow moving proteinases and membrane aspartic cathepsin was further tested by tandem crossed immunoelectrophoresis of partly purified preparations of both enzymes against anti-(membrane aspartic cathepsin). In this experiment the amounts of protein in the precipitates were so small that direct protein staining was impossible. The immunoprecipitates were visualized with a caseogram print (Fig.4). This way the precipitates appear more blurred than by direct protein staining, but the two precipitates are clearly fused without any major spurs.

Discussion

The object of this communication is dual. First, to demonstrate the applicability of a simple and versatile method for detection the of proteinases, and second to report some results obtained with this method.

The zymogens for the aspartic proteinases are easy to deal with since their activation is promoted by a conformational change induced by hydrogen ions only (4). The only problem one might expect during detection of pepsinogen and similar zymogens is an inhibition of the milk-clotting activity by the propart peptides which are liberated during activation. Such inhibition has not hampered the detection of zymogens in our experiments; the reason is twofold. First, the inhibition by equimolar amounts of propart peptides and active enzyme is far from complete. Second, a considerable degradation of the peptides probably takes place during changes of pH, from that of the electrophoresis buffer down to a pH of about 2 and up again to a pH 5.3 for the milk-clotting system.

The physiological relevance of the preliminary results remains unclear. The consistent observations of pepsinogen C in the seminal fluid and the production of pepsinogen C in the glands of the genital tract indicate that pepsin C is important for fertility. One may expect that activation takes place in the acidic conditions of the vagina, and that pepsin C may induce activation of other proteinase zymogens during fertilisation.

The small amounts of pepsinogens which we have observed in other tissues show large inter-individual variations. Such pepsinogens may represent a fortuitous gene expression, but the finding of extra gastric pepsinogens throws light on two other observations. First, this explains that persons who have undergone total gastrectomy still have some circulating serum pepsinogen. Second, Pals et al. (29) have compared the electrophoretic pattern of pepsinogens from mucosa obtained by biopsy with pepsinogens from urine of the same patients; 39% of the individuals showed a difference in the two electrophoretic patterns, and this is presumably due to pepsinogens that are synthesized outside the gastric mucosa.

In this connection attention should be drawn to investigations by Moriyama et al. (30), who isolated an aspartic proteinase zymogen from monkey lungs. This was called procathepsin D-II, but in all analysed properties it was indistinguisable from pepsinogen C. Thus, in keeping with the terminology used here, we feel that the name pepsinogen C could be used without reservations. We take these results as a confirmation of our own, that extra gastric pepsinogens are more widespread than previously thought.

As regards the immunological relationships among cathepsin D's from different tissues, identities have been reported among the enzymes from different organs of chicken (31) and rat (32) respectively. The latter investigation also reported an immunological difference between cathepsin D and E. Comparison of human cathepsins have shown immunological identity between cathepsin D from liver extract and that of rheumatoid synovia (33). The designations gastric cathepsin or gastric cathepsin D-like proteinase have been used for a proteinase that presumably corresponds to one of the electrophoretically slow moving proteinases (34,35), but no immunochemical comparison with other human cathepsin D was reported.

The immunochemical relationship which we have observed supports the designation gastric cathepsin for the electrophoretically slow moving proteinases, but the question is whether they may be called cathepsin D. By gelfiltration molecular weigths of around 90000 have been observed for the gastric cathepsins (35,36) and unpublished results from our institute. Similar molecular weigths are reported for cathepsin E. By SDS-polyacrylamide electrophoreses of the gastric cathepsins, molecular weights of approx. 45000 were observed, and these properties correspond to those of membrane bound cathepsin D type H (24). One may speculate if some hydrophobic parts of the membrane bound cathepsins also may cause formation of dimers.

Our knowledge is still incomplete. We may, however, conclude that the gastric proteinases have closely related counterparts outside the stomach.

Note added after the paper was presented: By the courtesy of dr. A. J. Barrett (Cambridge, UK) we have obtained a sample of human cathepsin D. Neither gastric SMP nor membrane aspartic proteinase gave any immunological cross-reaction with the human cathepsin D.

References:

1. Foltmann, B., Jensen, A.L.: Eur.J.Biochem. 128, 63-70 (1982).
2. Jensen, T., Axelsen, N.H., Foltmann, B.: Biochim.Biophys.Acta 705, 249-256 (1982).
3. Tang, J.: Methods Enzymol. 19, 406-421 (1970).
4. Foltmann, B.: Essays in Biochem. 17, 52-84 (1981).
5. Samloff, I.M.: Gastroenterology 60, 586-604 (1971).
6. Axelsen, N.H.(ed.): "Handbook of immunoprecipitation in gel techniques" Scand.J.Immunol. 17 suppl.no.10 (1983).
7. Grubb, A.O.: Scand.J.Immunol. 17 suppl.no.10, 113-124 (1983).
8. Krøll, J.: Scand.J.Immunol. 17 suppl.no.10, 135-139 (1983).
9. Weeke, B.: Scand.J.Immunol. 2 suppl.no.2, 37-46 (1973).
10. Krøll, J.: Scand.J.Immunol. 17 suppl.no.10, 165-169 (1983).

11. Maurer, H.R.: "Disc electrophoresis and related techniques of polyacrylamide electrophoresis" Walter de Gruyter, Berlin-New York (1971).

12. Uriel, J.: Nature (London) 188, 853-854 (1960).

13. Etherington, D.J., Taylor, W.H.: Biochem.J. 113, 663-668 (1969).

14. Korsnes, L., Gedde-Dahl, T.Jr.: Ann.Hum.Genet.Lond. 43, 199-212 (1980).

15. Lawrence, R.C., Sanderson, W.B.: J.Dairy Res. 36, 21-29 (1969).

16. Foltmann, B.: Methods Enzymol. 19, 421-436 (1970).

17. Baudys, M., Kostka, V.: Eur.J.Biochem. 136, 89-99 (1983).

18. Lundquist, F., Seedorff, H.H.: Nature (London) 170, 1115-1116 (1952).

19. Hirsch-Marie, H., Conte, M.: Bull.Soc.Chim.Biol. 49, 147-155 (1967).

20. Samloff, I.M., Liebman, W.M.: Clin.exp.Immunol. 11, 405-414 (1972).

21. Chulavatnatol, M., Ruenwongsa, P.: Biochim.Biophys.Acta 452, 525-532 (1976).

22. Chiang, L, Contreras, L., Chiang, J., Ward, P.H.: Arch.Biochem. Biophys. 210, 14-20 (1981).

23. Pontremoli, S., Salamino, F., Sparatore, B., Melloni, E., Morelli, A., Benati, U., De Flora, A.: Biochem.J. 181, 559-568 (1979).

24. Yago, N., Bowers, W.E.: J.Biol.Chem. 250, 4749-4754 (1975).

25. Barrett, A.J.: in "Proteinases in Mammalian Cells and Tissues", (Barrett, A.J. ed.) Elsevier/North-Holland. Amsterdam-London 209-247 (1977).

26. Stepanov, V.M., Rudenskaya, G.N., Gaida, A.V., Osterman, A.L.: J.Biochim.Biophys.Meth. 5, 177-186 (1981).

27. Huang, J.S., Huang, S.S., Tang, J.: J.Biol.Chem. 254, 11405-11417 (1979).

28. Bøg-Hansen, T.C.: Scand.J.Immunol. 17 suppl.no.10, 243-253 (1983).

29. Pals, G., Defize, J., Meuwissen, S., Westerveld, B., Verhoeven, H., Oms, E., Kreuning, J., Frants, R., Eriksson, A.: Gastroenterology 80, 1247 (1981).

30. Moriyama, A., Kageyama, T., Takahashi, K.: Eur.J.Biochem. 132, 687-697 (1983).

31. Weston, P.D., Poole, A.R. in "Lysosomes in biology and pathology" vol. 3 (Dingle, J.Tr. ed.) North-Holland Publ.Comp.Amsterdam-London, 425-464 (1973).

32. Yamamoto, K., Kamata, O., Katsura, N.: J.Biochem.(Tokyo) 87, 511-516 (1980).

33. Dingle, J.T., Barrett, A.J., Weston, P.D.: Biochem.J. 123, 1-13 (1971).

34. Rapp, W., Lehmann, H.E.: J.Clin.Chem.Clin.Biochem. 14, 569-576 (1976).

35. Kageyama, T., Takahashi, K.: J.Biochem.(Tokyo) 87, 725-735 (1980).

36. Muto, N., Arai, K.M., Tani, S.: Biochim.Biophys.Acta 745, 61-69 (1983).

APPARENT INHIBITION OF PEPSIN BY AN EXCESS OF HAEMOGLOBIN SUBSTRATE

Baldur Símonarson, Institute for Experimental Pathology, Keldur, University of Iceland, PO Box 8540, Reykjavík, Iceland

Introduction

The Anson assay (1) is familiar to most proteinase biochemists. It was originally introduced as a general assay for proteolytic enzymes, but specific synthetic substrates with chromogenic or fluorogenic groups have largely limited its use to the assay of enzymes where such synthetic substrates are lacking, like pepsin and cathepsin D, or newly isolated enzymes.

When studying ruminant pepsins, it was observed that maximum enzyme activity was obtained with a substrate concentration of 0.5% haemoglobin. When the final haemoglobin concentration was 1.67%, as in the original Anson assay, the activity was 35-40% less (2). A similar inhibition by haemoglobin substrate above a concentration of 1.25% has been reported for cathepsin D (3). Barth and Afting (4) failed to observe this, but the highest haemoglobin concentration they used was only 50 µM or 0.33%.

Materials and methods

This phenomenon was studied with commercial pig pepsin (Sigma P 7012), and its activity assayed both with the Anson method as well as by stopping the reaction with alkali and then measuring the newly liberated amino groups by a modification of the trinitrobenzenesulphonic acid (TNBS) method of Fields (4).

Aspartic Proteinases and their Inhibitors

Results

The action of pepsin on different haemoglobin concentrations, measured by the Anson assay, is shown in Table 1. Rather than expressing pepsin activity in the customary units, the final concentration of tyrosine-like substances in the assay mixture at the end of the incubation period is shown. It is also shown how many tyrosyl residues are produced per each chain of the haemoglobin molecule initially present. There are three tyrosyl residues per chain in the intact haemoglobin molecule. Inhibition of proteolytic activity is seen at high haemoglobin concentrations.

Table 1 also shows the action of pepsin on different haemoglobin concentrations, when the increase in amino groups liberated is measured by TNBS. No inhibition of proteolytic activity is seen at higher haemoglobin concentrations.

Discussion

The inhibition of pepsin by high haemoglobin concentrations is an apparent effect, which somehow must be explained by the nature of the Anson assay.

The following points should be considered:

1) The specificity and the extent of endopeptidase cleavage.
2) The order of proteinase attack on susceptible peptide bonds: "one-by-one", "zipper", or intermediate.
3) The number and distribution of aromatic amino acid residues in the protein substrate.
4) The complexity of the hydrolysis products. Some of the peptides produced may act as inhibitors.
5) The molecular weight of the peptides soluble in trichloroacetic acid (TCA) is related to the concentration of TCA used for precipitation.

From Table 1 it is seen that the number of tyrosyl residues produced per chain increases as the haemoglobin concentration is lowered. It is also seen from Table 1 that shorter peptides are produced as the haemoglobin concentration is lowered.

At low haemoglobin concentrations one runs the risk of exhausting the supply of substrate. If the supply of tyrosine-containing peptides which are insoluble in TCA is adequate at higher concentrations of haemoglobin, the Anson assay should still be valid. Only at the highest haemoglobin concentration are 10% or less of the tyrosyl residues liberated. It is possible that some of the peptides produced by the initial peptic hydrolysis of haemoglobin are further hydrolysed by a "zipper" mechanism rather than "one-by-one".

It is unlikely that these results invalidate the Anson assay, but they highlight some of its limitations. Anson's original procedure has often been modified, but it is clearly risky to do so, both with respect to the concentration of protein substrate used, as well as the concentration of TCA used for precipitation. A high concentration of TCA will permit only the analysis of very small peptides and free amino acids, and thus make the assay less suitable for assaying endopeptidases. Apparent k_M values obtained for proteolytic enzymes using the Anson assay have limited physical meaning, although such values have been reported in the literature.

In the absence of suitable synthetic substrates, proteinases are better assayed by their action on N,N-dimethyl proteins (6) or succinyl proteins (7), omitting the precipitation step, and measuring the amino groups liberated with TNBS or fluorescamine. Such modified proteins are easily prepared and the detection methods are very sensitive.

Table 1. The action of pepsin on different haemoglobin concentrations. 10 minutes incubation at 37°C, pepsin 0.049 μM.

	Anson assay		Increase in amino groups	
% Hb	mM Tyr in assay	Tyr produced per chain	Δ-NH_2 (mM) in assay	Δ-NH_2 per chain
1.67	0.35	0.34	2.67	2.58
1.25	0.38	0.49	2.31	2.98
1.00	0.42	0.67	2.24	3.61
0.75	0.44	0.95	1.70	3.66
0.62	0.47	1.21	1.42	3.66
0.50	0.48	1.55	1.37	4.40
0.40	0.46	1.87	1.24	5.0
0.33	0.45	2.16	1.06	5.1
0.20	0.36	2.94	0.83	6.7
0.10	0.22	3.58	0.43	6.9
0.05	0.13	4.16	0.20	5.5
0.03	0.09	4.34	0.16	7.7

References

1. Anson, M.L.: J. Gen. Physiol. 22, 79-89 (1938).
2. Magnússon, S., Símonarson, B.: Hoppe-Seyler's Z. Physiol. Chem. 363, 983-984 (1982).
3. Yamamoto, K., Katsuda, N., Kato, K.: Eur. J. Biochem. 92, 499-508 (1978).
4. Barth, R., Afting, E.-G.: Biochem. J. 219, 899-904 (1984).
5. Fields, R.: Biochem. J. 124, 581-590 (1971).
6. Lin, Y., Means, G.E., Feeney, R.E.: J. Biol. Chem. 244, 789-793 (1969).
7. Schwabe, C.: Anal. Biochem. 53, 484-490 (1973).

DETERMINATION OF CHYMOSIN BY ROCKET IMMUNOELECTROPHORESIS

Rolf Kleine

High School for Engineering, Köthen, Section of Biology and Institute of Physiological Chemistry, University of Halle/S., GDR

Introduction

A reliable and specific determination of chymosin activity in crude or partially purified rennet preparations is a prerequisite of optimal and economical application of this acid gastric aspartic proteinase in cheesemaking and related fields. Furthermore, the detection of extremely small amounts of chymosin during the production by recombinant DNA-techniques in genetically engineered microorganisms, e.g. E. coli or yeasts is essential for an efficient assay of the expression rate. Of the current enzymatic chymosin assays the milk-clotting test of Berridge with slight modifications, such as determination of the activity ratio (clotting activity at pH 6.0/clotting activity at 6.5) (2) or the micro-titer plate assay (3) are generally preferred. All these enzymatic coagulation tests, however, codetermine pepsin contaminations in the chymosin samples. The only possibility to circumvent this disadvantage offer highly specific immunological methods, such as the immunodiffusion assay or rocket immunoelectrophoresis (4).

The aim of this work was to develop an immunological micromethod of selective detection of minute amounts of chymosin in presence of pepsin using anti-chymosin antisera.

Aspartic Proteinases and their Inhibitors

Materials and Methods

Antisera. Monospecific anti-chymosin antibodies were raised in rabbits by repeated subcutaneous injections of freshly purified chymosin in complete (first injection) and incomplete (the following injections) Freund´s adjuvant (Difco Laboratories, Detroit, Mich., USA), respectively, in ten to twelve day intervals (total amount: 2.5 mg chymosin per animal). Chymosin was isolated either from Chr. Hansen´s rennet (Denmark) or from chymosin (Sigma, USA) by gel filtration (Sephadex G-50 or G-75, fine), ion exchange chromatography on DEAE-cellulose and ultrafiltration (Amicon UM-10 or PM-10). Blood was collected from the ear vein starting two weeks after the last injection. Control serum was obtained from rabbits injected with 0.9 % NaCl. The sera were stored at -25 °C.

Immune techniques

Immune double diffusion (5) was performed in 1.5 % Noble Agar gel (pH 8.6; 0.038 M barbital buffer). Samples (5 or 10 μl) of antigen (chymosin, pepsin or protein mixtures) were applied in serial dilution into wells surrounding the center well containing the anti-chymosin antiserum (20 μl or 10 μl). After 24 to 48 h of diffusion in a moist chamber the precipitation lines were stained with Coomassie Blue R-250 (0.05 %) after exhaustive washing with 0.15 M NaCl and air drying.

Quantitative immunoelectrophoresis (rocket technique) (4) was carried out in 1 % agarose gel (0.18 ml/cm^2 of plate) in 0.038 M barbital buffer (pH 8.6). Agarose with standard electroendosmosis (EEO) (Serva, Heidelberg) or agarose C with low EEO (Pharmacia Fine Chemicals) were used. The respective gel suspension (4.5 ml) containing 200 to 400 μl of monospecific anti-chymosin antiserum was cast on slides (5 x 5 cm). Usually five samples (1 to 10 μl) containing 0.01 to 1.5 μg of purified chymosin (for calibration curve, see Fig. 1) or samples

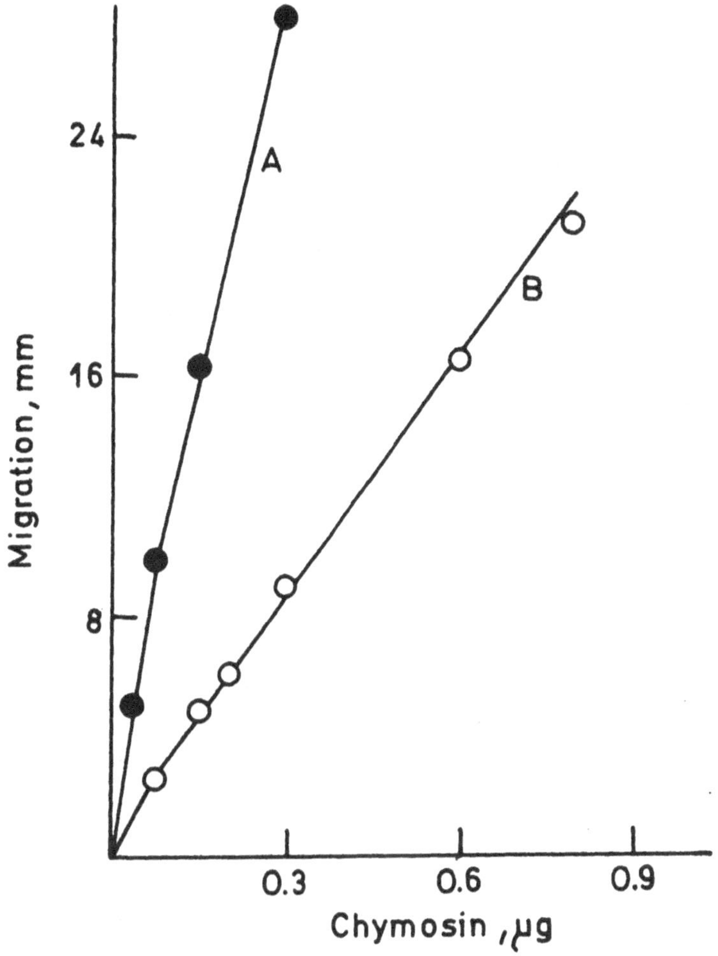

Fig. 1 Relationship between antigen (chymosin) concentration and rocket height. Each point on the calibration curves corresponds to a rocket obtained by electrophoretic migration of chymosin (0.037 µg to 0.8 µg) into anti-chymosin antiserum (300 µl in A, 250 µl in B) -containing agarose gel (1 %, pH 8.6; 4.5 ml gel per 25 cm^2). A: agarose with low electroendosmosis (EEO); B: agarose with standard EEO.

containing unknown amounts of chymosin, contaminated with pepsin and other proteins, were pipeted into the wells, localized at the cathodic side of the plate. Electrophoresis was run in a horizontal chamber at pH 8.6 and 10-12 °C for 18 h at a constant voltage of 187 V and at a current of 6 mA. Before staining the plates were treated as described above.

Results

Using the Ouchterlony double diffusion technique it was possible to detect chymosin semiquantitatively in crude rennet preparations even at concentrations of 90 to 100 ng per sample. For selective and quantitative determination of chymosin in the nanogram range rocket immunoelectrophoresis was used. However, we found that the height of the rockets, i.e. the sensitivity of the method, strongly depends on the type of agarose used. In agarose gels showing EEO, the migration distance of the chymosin antigen was significantly decreased when compared with the height of rockets in agarose showing low or no EEO. As shown in Fig. 1 the lowest detection level for chymosin in agarose with EEO lies between 30 and 50 ng, compared with 5 to 10 ng of chymosin in agarose with low EEO, corresponding $2 \cdot 10^{-13}$ mol of enzyme. Agarose free of EEO gave the same results.

Moreover, the shape and height of the rockets depends on the concentration of antiserum in the gel. Runs with different amounts of antiserum (200 to 400 µl per 4.5 ml gel) showed that 250 to 300 µl (= 10 to 12 $\mu l/cm^2$ gel) of anti-chymosin antiserum were optimal for the chymosin assay.

The procedure described was used to assay the chymosin content in commercial samples of calf rennet (Ostralab, Dresden, GDR, 150 to 200 mg/100 ml extract). Bovine and pig pepsin (both from Serva, Heidelberg) did not react with the anti-chymosin antiserum. The occurrence of pepsin in chymosin samples had no effect on the chymosin determination.

Discussion

Although it is possible to determine the approximate content of chymosin in the presence of pepsin by the modified Berridge test (the so-called activity ratio) (2) the immunological technique using anti-chymosin antibodies is the most specific and at the same time most sensitive assay of chymosin. The reported procedure gives reproducible results in the range of 10 to 500 ng and 50 to 1200 ng of chymosin using agarose with or without EEO, respectively. A technique superior to immunoelectrophoresis in sensitivity is the radioimmunoassay using ^{125}J protein A or radiolabelled anti-chymosin antibody or the enzyme-linked immunoadsorbent assay. However, these methods are costly, highly specialized and time-consuming, whereas rocket immunoelectrophoresis is simple and can be carried out routinely in any laboratory.

Acknowledgements

The author wishes to appreciate gratefully the skillful technical assistance of Ms. Ursula Goder.

References

1. Berridge, N.J.: Methods Enzymol. 2, 69-77 (1955).
2. Harboe, M.: This volume, pp. 537-550 (1984).
3. Emtage, J.S., Angal, S., Doel, M.T., Harris, T.J.R., Jenkins, B., Lilley, G., Lowe, P.A.: Proc. Nat. Acad. Sci. USA 80, 3671-3675 (1983).
4. Laurell, C.-B.: Anal. Biochem. 15, 45-52 (1966).
5. Ouchterlony, O.: Progr. Allergy 5, 1-78 (1958).

ASPARTIC PROTEINASES IN GASTRIC CARCINOMAS

William A. Reid,
Department of Pathology, University of Leeds, Leeds LS2 9JT, England, U.K.

Martin J. Valler and John Kay,
Department of Biochemistry, University College, Cardiff CF1 1XL, Wales, U.K.

Many cancer patients present with metastatic disease and determination of the location of the primary tumour is a common problem. Investigation of such a patient normally includes biopsy of the secondary tumour. While in some cases it is possible to identify from this the site of the primary carcinoma, in many instances an adequate diagnosis cannot be made so that the patient must be subjected to further investigations. It would thus appear useful to have specific markers to apply to secondary tumours in order to identify the primary site. For gastric carcinoma and its metastases, previous work has suggested that some of the tumours contain (and presumably therefore synthesise) gastric enzymes (1,2). We have therefore attempted to establish whether aspartic proteinases are detectable in the formalin-fixed, paraffin embedded tissue routinely used for diagnostic purposes on surgical biopsy specimens.

Materials and Methods

Human pepsin and human gastricsin were generous gifts from Dr. A.P. Ryle, University of Edinburgh and Professor J. Tang, Oklahoma Medical

Aspartic Proteinases and their Inhibitors

Research Foundation respectively. Cathepsin D was purified to homogeneity from human spleen as described by Afting and Becker (3). Antisera were raised in rabbits by serial injection of purified antigens in complete Freund's adjuvant and characterised as described previously (4,5). All antisera were used on 6 micron paraffin sections at a dilution of 1 in 100, apart from antipepsin (1/200). The sections were stained by the immunoperoxidase (PAP) method and counterstained with haemotoxylin. Absorption experiments were performed by incubating antiserum with antigen for one hour at room temperature before use. Tumour blocks studied were from the Surgical Routine File, Glasgow Royal Infirmary, Scotland and had been fixed in 10% unbuffered formalin and paraffin embedded. Wherever possible, blocks of gastric carcinoma were chosen so as to include adjacent normal mucosa as a positive control. A series of adenocarcinomas from various other sites was also studied.

Results

In the normal stomach, pepsin (ogen) was readily detected in the mucosa of the gastric body with staining mainly confined to the chief cells (Fig. 1a) although some parietal cells were also positively stained. Similar results were obtained for the cells in the gastric body mucosa when the antiserum to gastricsin was used, but this enzyme was also present in the gastric antrum in the deep glands and in mucous neck cells (Fig. 1b)

In addition, the Brunner's glands of the duodenum gave a positive result for (pro) gastricsin. Absorption experiments with pepsin and gastricsin showed complete abolition of staining by their respective antisera but not cross absorption with the other antiserum. The antiserum to cathepsin D stained a number of cells in the gastric mucosa of normal stomach. In the lamina propria cells, probably of macrophage origin, were positive. In the gastric body the parietal cells and mucous neck cells were strongly stained. In the antrum, the deep antral glands and mucous neck cells were positive, together with the Brunner's glands and certain mucosal cells of the duodenum. In all cases, staining was strong and granular.

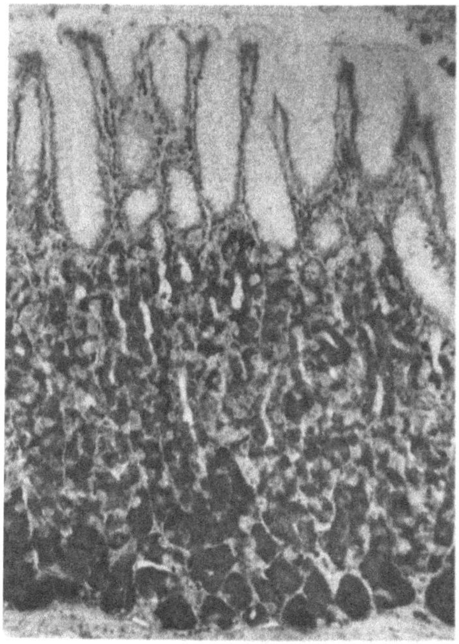

Fig. 1a.

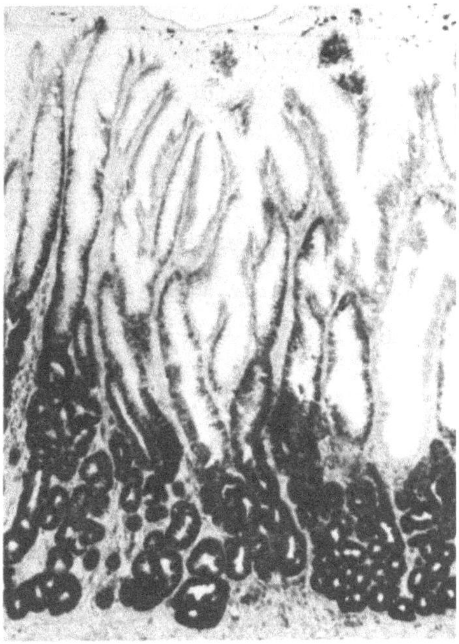

Fig. 1b.

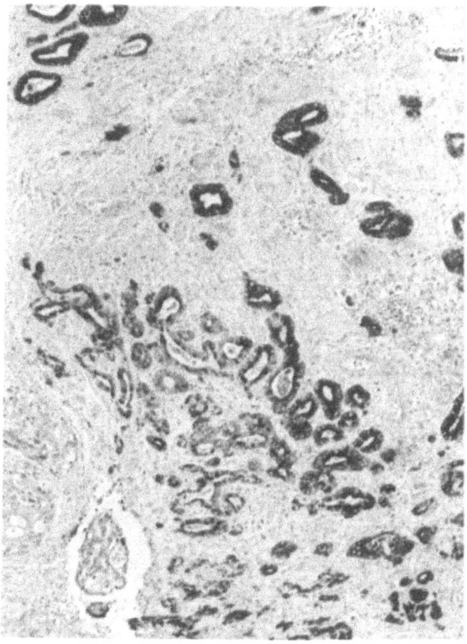

Fig. 1a. Normal gastric body mucosa stained with antipepsin. Note staining of chief cells.

Fig. 1b. Normal gastric antral mucosa stained with antigastricsin. Note staining of antral glands.

Fig. 2. Gastric adenocarcinoma stained with antigastricsin. Note focal positive staining in tumour cells.

Staining of a series of 125 gastric adenocarcinomas for pepsin gave a positive reaction in only 7 cases (5.6%). The staining was abolished by prior absorption of antiserum with pepsin or pepsinogen. However, in a series of 114 cases of gastric carcinoma stained for gastricsin, 33 showed positive granular staining in the cytoplasm of the tumour cells (Fig.2). In most cases, staining was focal. Twelve of the positive cases had metastatic tumour in local lymph nodes; ten of these cases also gave positive staining in the tumour cells of these deposits. Absorption experiments on these positive cases showed that staining was abolished by prior incubation of antiserum with gastricsin. Staining of a series of carcinomas of colon, breast, ovary, endometrium, pancreas and bile duct gave negative results for pepsin and gastricsin.

Staining of gastric carcinomas by antiserum to cathepsin D has only been carried out in a small series, but in the 12 cases so far tested, there was strong positive staining in the tumours. In most instances, staining was present both in the tumour cells and in cells, presumably of macrophage origin, in the tumour stroma.

Discussion

Our studies have revealed that aspartic proteinases are readily demonstrable in routine histological sections of gastric carcinomas. Gastricsin is present in almost a third of cases and is also found in lymph node secondary deposits. By contrast, pepsin is a rare finding. We have previously suggested (5) that since most gastric carcinomas arise from the antrum of the stomach, tumour production of gastricsin rather than pepsin may be a reflection of this. An alternative explanation might be that gastricsin is the more primitive enzyme and more likely to occur in carcinomas as an oncofetal antigen.

The significance of the Cathepsin D findings is not clear. Cathepsin D is of course a lysosomal enzyme and is found in cells of the mononuclear phagocyte system. This may explain its presence in stromal cells of the normal stomach and of tumours. Its localisation in the mucous neck cells of the stomach may be of interest, but its presence in the parietal cells is remarkable.

The value of these enzymes as tumour markers has not yet been worked out fully. Pepsin occurs too rarely in tumours to have any significance, whereas gastricsin may be of value as a marker, especially as it is generally absent fromcarcinomas of other sites. Its presence in carcinoma of the prostate (6,7) is a possible difficulty, although primaries in this location can be diagnosed readily by clinical examination and biopsy. Cathepsin D was observed in all gastric carcinomas so far examined, but preliminary studies have shown that it also occurs in carcinoma of colon, so that its specificity remains to be determined.

References

1. Hirsch-Marie H., Loisillier F., Touboul J.P., Burton P. :Lab. Invest. 34 (6), 623-32 (1976).

2. Kalinowsky V.P. and Parshin A.H. :Vopr. Onkol. 26, 31-6 (1975.)

3. Afting E.G. and Becker M.L. :Biochem. J. 197, 519-522 (1982).

4. Reid W.A., Thompson, W.D., Kay J.: J.Clin. Path. 36, 137-139 (1983).

5. Reid, W.A., Kay J.: Dis. Markers 1, 263-269 (1983).

6. Reid, W.A., Vongsorasak L., Svasti J., Valler M.J., Kay J.: Cell Tiss. Res., 236, 597-600 (1984).

7. Reid, W.A., Liddle C.L., Kay J.: in preparation.

GASTRIC ACID PROTEINASES IN VARIOUS DISEASES

Zdena Kučerová, Libuše Korbová, Jiřina Čížková
Department of Pathological Physiology

Jiří Kohout
Ist Department of Surgery

Josef Marek
IInd Department of Pathological Anatomy, Faculty of Medicine, Charles University, Prague, Czechoslovakia

Introduction

Little attention only has been paid so far to the secretion and content of individual pepsinogens and acid gastric proteinases in different gastric diseases. The results obtained by various authors often differ considerably (1-3) and various numbers of enzymatically active fractions have been detected by different methods; therefore a common nomenclature is still missing. In our Laboratory we adhered to the widely used classification of acid proteinases introduced by Samloff (1). The cleavage of synthetic substrates by acid proteinases in various gastric diseases has not yet been assayed to our knowledge either. Substrate Ac-Phe-Tyr(I_2) is specifically cleaved by group I gastric pepsinogens (PG I). We have found that For-Tyr-Tyr and For-Tyr-Phe are hydrolyzed by group II pepsinogens (PG II) (4,5).

In this study electrophoretic separation of gastric mucosa extracts of patients with gastric (UV) and duodenal ulcer (UBD) and with gastric carcinoma (Ca) as well as hydrolysis of the above synthetic substrates by these extracts were carried out.

Aspartic Proteinases and their Inhibitors

Material and Methods

The resected parts of stomachs were either deep-frozen (at -20 °C) or processed immediately. After washing with 0.1 M phosphate buffer, pH 7.3, the stomach was divided into several transversal strips. A part of each was taken for histological examination. Dissected mucosa was homogenized in 0.1 M phosphate buffer, pH 7.3 (4 ml per 1 g of tissue). The supernatant separated by centrifugation at 12 000 r.p.m./min was analyzed by agar (Difco) gel electrophoresis according to Mangla et al. (6). Enzymatic activity determined on the agar gel electropherograms was expressed semiquantitatively (3 degrees, +, ++, +++). Cleavage of synthetic substrates: the homogenate of gastric mucosa was dialyzed against Ringer´s solution, activated at pH 2 and diluted 1:20 by 0.05 M HCl. The substrate (0.15 ml of 0.002 M solution in 0.1 M acetate buffer, pH 5.3) was added to 0.5 ml samples and the mixtures were incubated 5 h at 37 °C. Free amino acids were determined by ninhydrin, protein content according to Lowry, proteolytic activity according to Anson and Mirsky.

Results

The assays (375 electrophoreses and 384 hydrolyses of synthetic substrates) were carried with samples obtained from 77 patients with gastric and duodenal ulcer and with gastric carcinoma (Table 1).

Table 1

Number of examinations	UBD	UV	Ca
Patients with	34	18	25
Electrophoreses	141	85	119
Hydrolyses of substrates	158	107	119

We found that all pepsinogens (except Pg 1) isolated from patients with gastric carcinoma have a higher electrophoretic mobility (a higher negative charge, Fig. 1) than the pepsinogens of patients with gastric and duodenal ulcer.

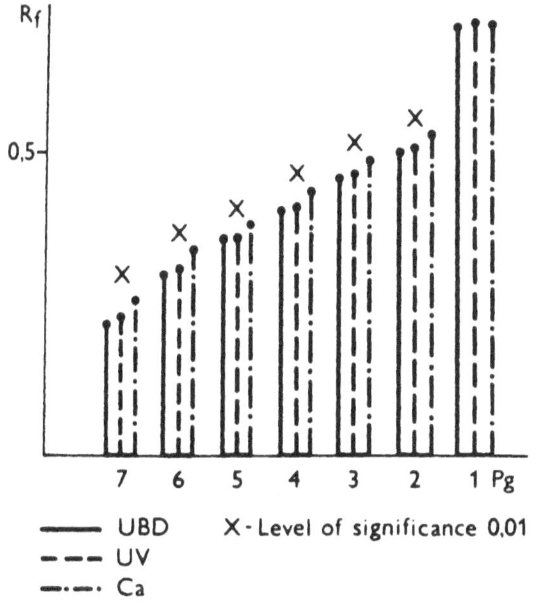

Fig. 1 Mobility of pepsinogens 1-7 during electrophoresis in agar gel
UBD - patients with duodenal ulcer, UV - patients with gastric ulcer, Ca - patients with gastric carcinoma

The experiments with the hydrolysis of synthetic substrates showed that the cleavage of Ac-Phe-Tyr(I_2) proceeds in accordance with the activ ity values which had been determined on the agar gel elctropherograms for PG I from patients with all gastric diseases studied (Fig. 2). The cleavage of For-Tyr-Tyr and For-Tyr-Phe, however, gave values which correspond rather to activities found for samples of PG II. This holds only for pepsinogens isolated from the stomachs of patients with gastric and duodenal ulcer. Extracts of homogenates of samples from patients with gastric carcinoma hydrolyze more intensively the substrate For-Tyr-Tyr and to some extent also For-Tyr-Phe. Proteolytic activity assays yielded the ratios of PG I to PG II shown in Table 2. There is no significant difference in the amount of PG I in antral mucosa in patients with duodenal ulcer and stomach cancer. In patients with stomach ulcer, however, the amount of PG I is decreased.

Table 2
Ratio of pepsinogens of group I to pepsinogens of group II

	Antrum	Corpus
UBD	1.56	4.66
UV	1.09	4.29
Ca	1.75	3.36

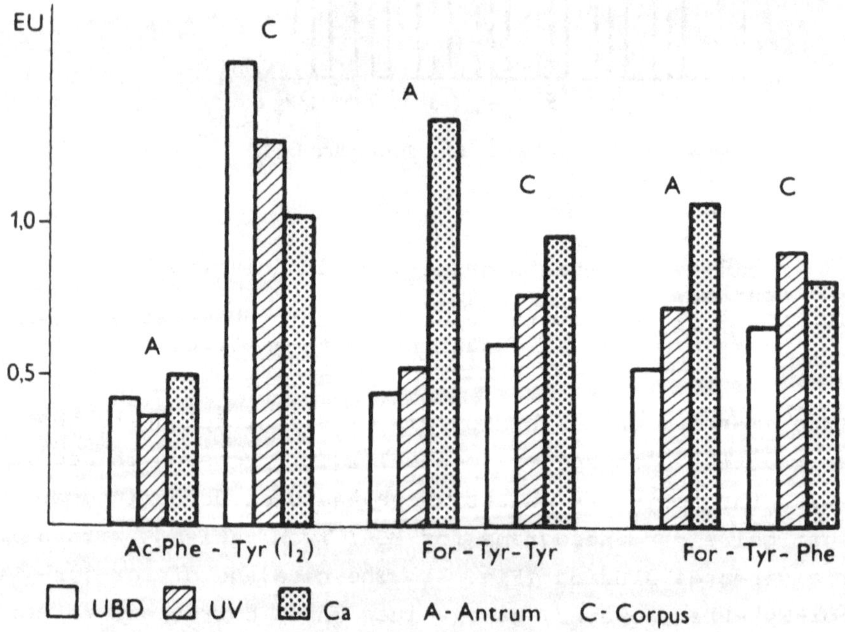

Fig. 2 Rate of hydrolysis of Ac-Phe-Tyr(I_2), For-Tyr-Tyr, and For-Tyr-Phe by homogenized gastric antrum and corpus mucosa of patients with duodenal ulcer (UBD), gastric ulcer (UV), and gastric carcinoma (Ca).

The quantity of PG I in the corpus mucosa of patients with gastric cancer is also decreased when compared with the two other groups.

We conclude that patients with gastric carcinoma seem to produce PG I and PG II of different nature. These pepsinogens have a higher electrophoretic mobility and a different affinity for synthetic substrates.

References

1. Samloff, I.M.: In Pathophysiology of carcinogenesis in digestive organs. (Farber, E., Kawachi, T., Nagayo, T., Sugano, H., Sugimura, T., Weisberger, J.H., eds), pp. 65-72, Japan Scient. Society Press, 1977.
2. Furihata, H., Tatematsu, M., Shirai, T., Yokochi, K., Takahashi, M., Sugimura, T.: ibid., pp. 49-61.
3. Hirsch-Marie, H., Loisillier, F., Trouboul, J.P., Burtin, P.: Lab. Invest. 34, 623-632 (1976).
4. Kohout, J., Korbová, L., Krejčík, L.: Prakt. Lék. 49, 882-885 (1969).
5. Korbová, L., Kohout, J., Kasafírek, E., Čížková, J.: Collect. Czech. Chem. Commun. 42, 2577-2585 (1977).
6. Mangla, J.C., Guarasci, G., Turner, M.D.: Amer. J. Dig. Dis. 18, 857-864 (1973).

ACTIVITIES OF SOME PROTEOLYTIC ENZYMES IN THE CARTILAGE AND SUBCHONDRAL BONE OF OSTEOARTHROTIC RABBITS

D. Rohožková, B. Tesárek, K. Trnavský
Research Institute for Rheumatic Diseases
Na slupi 4, Prague 2, Czechoslovakia

Introduction

The pathogenesis of osteoarthrosis is regarded either as a consequence of mechanical damage or as a result of enzymatic processes. Even though the primary cause of this disease has not been established as yet most workers ascribe the damage of the cartilage during osteoarthrosis predominantly to enzymatic processes (1-4). Of the large number of enzymes which may be involved a major role in development of the disease play proteolytic enzymes produced by cartilage cells, the chondrocytes.

In the present study changes in enzyme activities during the development of experimental osteoarthrosis in laboratory animals were investigated and the effect of administration of antirheumatics on the level of the enzymes was examined.

Results

Female rabbits, aged six months, were used throughout our experiments. Experimental osteoarthrosis was developed either by surgical intervention (5) or by intraarticular injections of papain (6). Specimens of articular cartilage and of subchondral bone, and in some cases also of synovial effusion, were obtained from the knee joints of hind legs. The determinations were carried out with fresh specimens or after their lyophil-

Aspartic Proteinases and their Inhibitors

ization. The substrates used for the determination of enzymatic activity were h emoglobin for cathepsin D (7), glycyl--phenylalanyl-p-nitranilide for cathepsin C (dipeptidyl peptidase) (8) and the synthetic pentapeptide prepared according to Wünsch and Heidrich (9) for collagenase (PZ-peptidase). The effect of two antirheumatics, Solu-cortef (hydrocortisone sodium succinate, Upjohn, Belgium) in doses of 5 mg/rabbit and day and Ketazon (ketophenylbutazone, Spofa, Czechoslovakia) in doses of 100 mg per rabbit and day, was investigated in rabbits with surgical osteoarthrosis after daily administration for one month from the first postoperative day onwards.

A statistically significant increase of the activity of the individual enzymes was observed during the early phases of surgical osteoarthrosis, i.e. during the second and third week after the surgery. The levels of the enzymes in the synovial fluid showed a marked increase during the first postoperative week already.

Similar results were obtained with the model of papainosteoarthrosis. This shows that neither the joint surgery performed in the former model, nor the subsequent wound healing had any substantial effects on the activity of the enzymes investigated. (See Table 1)

Long-term administration of antirheumatics resulted in a statistically significant decrease of the activity of both cathepsins in the articular cartilage, as well as in the subchondral bone. (See Table 2 and 3)

Our results are in good agreement with the data reported by other authors (10) who have also described an increase of enzyme activities during the early stages of the development of experimental osteoarthrosis.

A long-term administration of the two antirheumatics resulted in the expected inhibition of activity of the enzymes investigated. However, from this study it is not possible to draw any final conclusions on the mechanism of these inhibition pro-

Table 1

Enzyme	Week	Sample	Surgical OA $\bar{x}_{cont.}$	Surgical OA $\bar{x}_{exper.}$	Papain OA $\bar{x}_{cont.}$	Papain OA $\bar{x}_{exper.}$
cath. D	1	cartil.	0.92	1.75$^+$	1.46	1.70
	2	cartil.	1.04	2.58$^+$	0.70	1.86$^+$
	3	cartil.	0.94	1.54$^+$	1.20	2.70$^+$
	4	cartil.	0.83	1.67$^+$	1.98	1.90
	12	cartil.	0.68	0.79		
cath. C	1	cartil.	77.7	93.7$^+$	113	105
	2	cartil.	65.7	104.1$^+$	80	167$^+$
	3	cartil.	99.7	159.4	86	90
	4	cartil.	104.0	126.4	90	84
	12	cartil.	106.0	130.0		
	1	synov. fluid	69	341$^+$		
	2	synov. fluid	70	1000$^+$		
	3	synov. fluid	90	446$^+$		
	4	synov. fluid	98	110		
Pz-ase	1	cartil.	11.2	28.1$^+$		
	2	cartil.	7.9	21.0$^+$		
	3	cartil.	11.2	25.0$^+$		
	4	cartil.	17.1	20.7		
	1	bone	24.5	27.1		
	2	bone	9.1	20.9$^+$		
	3	bone	14.4	45.7$^+$		
	4	bone	19.6	20.8		

cath. D = cathepsin D (nmol hemoglobin/1 h)
cath. C = cathepsin C (nmol p-nitroaniline)
Pz-ase = Pz-peptidase (nmol Pz-Pro-Leu/1 h)
OA = osteoarthrosis
$\bar{x}^+$ = values statistically significant

Table 2

Enzyme	Drug	Sample	Week	$\overline{x}_{cont.}$	$\overline{x}_{exper.}$
cath. D	Solu-cortef	cartil.	1	3.66	3.81
			2	4.18	2.52^{+}
			4	5.33	2.02^{+}
		bone	1	1.92	1.77
			2	3.19	2.85
			4	3.15	2.26^{+}
cath. C	Solu-cortef	cartil.	1	77.5	51.7^{+}
			2	82.8	83.5
			4	98.1	43.5^{+}
		bone	1	84.1	47.5^{+}
			2	106.2	100.7
			4	154.3	101.2^{+}

cath. D = cathepsin D (nmol hemoglobin/l h)
cath. C = cathepsin C (nmol p-nitroaniline)
$\overline{x}^{+}$ = values statistically significant

Table 3

Enzyme	Drug	Sample	Week	$\bar{x}_{cont.}$	$\bar{x}_{exper.}$
cath. D	Ketazon	cartil.	1	3.67	3.77
			2	4.19	3.27^{+}
			4	5.20	3.62^{+}
		bone	1	1.92	1.77
			2	3.17	2.45^{+}
			4	3.15	1.92^{+}
cath. C	Ketazon	cartil.	1	76.8	52.5^{+}
			2	82.8	103.7
			4	98.1	43.5^{+}
		bone	1	86.0	51.2^{+}
			2	104.2	108.7
			4	155.0	97.5^{+}

cath. D = cathepsin D (nmol hemoglobin/l h)
cath. C = cathepsin C (nmol p-nitroaniline)
$\bar{x}^{+}$ = values statistically significant

cesses. The similar effect of both drugs on the enzymes, i.e. on an aspartic endopeptidase cathepsin D and on an exopeptidase of the cathepsin C type suggests that - unlike the inhibition brought about by specific inhibitors - the effect of the two drugs on these enzymes is not selective.

References

1. Mohr, W.: Akt. Rheum. 3, 161-182 (1978).

2. Lindenhayn, K., Heilmann, H.H.: Wiss. Z. Ernst-Moritz--Arndt-Univ. Greifswald 31, 91-96 (1982).

3. Bollet, A.J.: Arth. Rheum. 12, 152-163 (1969).

4. Glynn, L.E.: Lancet 1, 574-582 (1977).

5. Colombo, C., Butler, M., O´Byrne, E., Hickman, L.,Selwyn, M., Schwartzendruber, D., Steiner, B.: Arth. Rheum. 26, 875-886 (1983).

6. Farkas, T., Bihari-Varga, M., Biró, T.: Ann. Rheum. Dis. 35, 23-26 (1976).

7. Bjelle, A., Osterlin, S.: J. Rheumat. 3, 400-408 (1976).

8. Tuppy, H., Wiesbauer, U., Wintersberger, E.: Z. physiol. Chem. 329, 278-288 (1962).

9. Wünsch, E., Heidrich, H.G.: Z. physiol. Chem. 332, 300-304 (1963).

10. Pelletier, J.P., Martel-Pelletier, J., Altman, R.D., Chandur-Mnaymneh, L., Howell, H.S., Woessner, J.F., jr.: Arth. Rheum. 26, 866-874 (1983).

COMMERCIAL ASPECTS OF ASPARTIC PROTEASES

Marianne K. Harboe
Chr. Hansen's Laboratorium A/S
DK-1250 Copenhagen

Introduction

The present paper will be limited to those of the aspartic proteases (EC 3.4.23) which are used for cheese making. The name milk-clotting enzymes or milk coagulants will be used for all enzyme preparations which are able to clot milk, while the name rennet will be reserved for commercial products containing aspartic proteases of mammalian origin. The milk-clotting enzymes constitute a considerable part of the total market for enzymes. The individual milk-clotting enzymes have very similar structures but rather different enzymic properties. Analysis of the milk-clotting enzymes is therefore important for estimating their value for cheese making. Even though cheese has been made for thousands of years new enzymes are still found and new processes are developed. The use of the recombinant DNA technique for making chymosin and the application of UF milk for cheese will probably contribute to the most important changes in the use of milk-clotting enzymes which have been seen in many years.

Market for milk-clotting enzymes

The total market for milk clotting enzymes is estimated to be 50-80 mill. dollars. This is only a rough estimate because very little exact knowledge of the market has been collected. The following does therefore only intend to give an idea about

Aspartic Proteinases and their Inhibitors

the size and distribution of the market. The milk-clotting enzymes constitute one of the big markets for enzymes probably accounting for around 15% of the total market for enzymes. There has been an increase in the market value for milk clotting enzymes during the last 10 years. This is both due to rising prices and to growth in the volume of milk coagulants. The production of cheese in Europe and the USA is increasing slightly every year.

The dominating milk coagulant on the market is the animal rennet constituting approximately 80% based on prices. The microbial coagulants account for approximately 20% and less than 1% is derived from plants.

The major enzyme in animal rennet is chymosin which always naturally exists in mixtures with more or less bovine pepsin. Calf-rennet containing 70-90% chymosin has traditionally been used for many years because it is generally suitable for all types of cheese. Bovine pepsin is the second most common enzyme for cheese. It has the disadvantage of being more sensitive to variations in pH and Ca than chymosin, but rennets containing 50-90% bovine pepsin can often replace chymosin without any change in the cheese technology. Porcine pepsin is also used but not as frequently as bovine pepsin. It is not likely that any other animal rennets will be used in larger amounts. Lamb and kid chymosin seem very similar to calf chymosin but relatively little rennet is obtained because the stomachs are smaller and the animals live more sparsely. Chicken pepsin is too proteolytic and the stomachs are too small. It is used mainly in Israel.

Mucor miehei protease is the dominating enzyme among the microbial coagulants. It is used especially in the United States where approximately 40% of the rennet is Mucor miehei protease. Mucor pusillus is not used to the same extend and Endothia parasitica is used even less. Many countries have restrictions on the use of microbial enzymes. Most countries do not allow mixtures of microbial and animal coagulants, and microbial coagulants are not allowed in Norway, Sweden and Holland at all. Another coagulant, Endothia parasitica, is mainly used in the production of emmenthaler cheese, as the too proteolytic Endothia parasitica protease (Epp) is destroyed during the high production temperature. Epp is used for emmenthaler especially in Finland and West Germany.

Milk coagulants from plants are only rarely used. The most important which is derived from Cardo (1), has been used in Portugal since ancient times.

The use of other milk-clotting enzymes for cheese making is of no importance at all.

Coagulants are produced by a great number of companies. Chr. Hansen's Laboratory, a Danish company, is the largest manufacturer of animal rennet and Novo, also Danish, is the main company on microbial milk coagulants. Novo covers a considerable part of the market for microbial coagulants. So, Denmark has quite a strong position on the market for milk-clotting coagulants. Some other manufacturers of animal rennet are Miles, Dairyland, Pfizer, Kemikalia AB, Clerici, Granday, Hauser, and New Zealand Rennet Corporation. Some important manufacturers of microbial coagulants are Meito, Gist-Brocades, Miles, Pfizer, and Dairyland.

The cheese maker has to make his choice among the above mentioned coagulants. Chymosin is generally the preferred enzyme in spite of its high price. Many markets quote a price for Bovine rennet twice as high as for Mucor miehei coagulant and a price approximately three and a half times as high for calf rennet as for Mucor miehei coagulant. The price of a coagulant is not the total dominating factor because it constitutes only approximately 0.5% of the cheese price. However, the lower price for microbial coagulants is almost always the main reason for changing from calf rennet. Except for the price, calf and bovine rennet is equal or better in all other aspects. The choice of milk coagulant is often based on the price combined with one or several of the following requirements for a good milk coagulant. A high ratio between milk-clotting activity and proteolytic activity is of importance for obtaining a high quality cheese with suitable texture and without bitter taste. The cheese makers are also very concerned about cheese yields. Even as little as 0.1% additional yield often means quite a significant increase in profit as many dairies produce several tons of cheese daily. And a low sensitivity of the clotting activity to pH and calcium is desirable otherwise small variations in the milk will disturbe the very tight time schedule for cheese-making. Most dairies avoid changes in the production procedure because they are afraid of getting the large batches of cheese which can only be used for processed cheese.

Analysis of milk coagulants

As many coagulants of various prices are offered for sale it is obvious that there is a need for comparing the coagulants. The value of coagulants for cheese-making is evaluated by analysis of milk-clotting activity, enzymic composition, and by various other characterizations.
Measurement of the milk-clotting activity (strength) is still the most important part of the analysis. The three methods most often used are the Soxhlet method (2), the Berridge method (3), the Chr. Hansen Laboratory method (4). I will only comment on the most important difference which concerns choice of pH and the degree of standardization. The Berridge method uses reconstituted skim milk of the for cheese-making artificially low pH 6.3. This low pH gives rise to a too high clotting strength and it does not say much about the clotting strength at the pH of cheese-making, usually 6.5 - 6.7. Further no standards are used in the Berridge method. The CHL method uses reconstituted skim milk at pH 6.5, but it actually measures at pH 6.6 approximately by using standards for each type of animal rennet which are all standardized to have the same clotting strength in milk at pH 6.6. By using such stan-

dards minor variations in the test conditions are eliminated and it is secured that the dairy can always use the same amount of rennet for the cheese no matter which type is used. The activity is expressed in Chr. Hansen units (CHU). The Soxhlet method has the advantage of simulating real cheese making, by using cheese milk and by using standard conditions for cheese-making. The disadvantages are the facts that it does not use a standard, that cheese milk varies, and that it is too time consuming. The activity is here expressed as the relationship between the amount of rennet and milk which results in clotting after 40 min.

Many methods have been suggested for analysing the enzymic composition of the milk coagulants (5-15). The three most important ones are Rocket immunoelectrophoresis (10, 11, 16), activity ratio (4) and the Garnot method (12-15). Most of the methods are designed for milk coagulants containing only two specific aspartic proteases, and none of the known methods are perfect for quantitative analysis of any type of milk coagulants.

Rocket immunoelectrophoresis (RIE) is however ideal for identification. By RIE a sample is run into a gel containing monospecific antiserum which reacts with the aspartic proteases corresponding to the antiserum. The quantification is performed by comparing the height of the sample precipitates with that of the standards. One plate has to be run for each enzyme to be estimated. That the enzyme in the sample and in the standard is completely identical is a very important assumption to the analysis. In case of partial identity the quantification will be wrong. This is often the case when the rennet contains denatured enzymes. We have found that denatured enzymes react with the antibodies in some cases but not in all. It depends on how the enzymes have been denatured. This means that some rennets contain more activity as determined by RIE than found by the milk-clotting test. We have found denatured enzymes in all types of rennet. This is in contradiction with the results obtained by A. Andrén (17).Andrén found that the decrease in amount of chymosin follows the decrease in milk-clotting activity. The discrepancy is probably due to differences in the denaturations.

The activity ratio method (4) is another valuable method for determination of enzymic composition. The method is based on the different effect of pH changes on the milk clotting of the enzymes. The ratio between the strength relative to a calf rennet standard (90% chymosin/10% bovine pepsin) at pH 6.0 and 6.5 is calculated. The activity ratio is thus the increase in activity relative to the calf rennet standard when the pH of the milk is changed from 6.5 to 6.0. The big difference between the values for chymosin (0.85) and bovine pepsin (2.20) can be used for determination of the composition of rennets containing only chymosin and bovine pepsin. The values of activity ratio for chymosin and for bovine pepsin are plotted against the percentage content of chymosin and a line is drawn between the two points. For calf or bovine rennets to be investigated the activity ratio is measured and the % chymosin is read from the curve. The remaining proportion of the enzyme is then bovine pepsin. The advantage of the method is that it is fast, the disadvantage that the method is indirect and without any identification of the enzymes, and also that each individual enzyme component in some rennets can have different values of activity ratio.

The third important method for determining the enzyme composition is the Garnot method named after the French researcher, Pascaline Garnot who first published it (13). The principle is as follows: A sample of dialyzed rennet is applied to an ion exchange column. The elution is performed in two steps. First, chymosin and the microbial coagulants, if any, are eluted by 0.2 M NaCl in 0.025 M piperazine buffer pH 5.3 and second bovine and/or porcine pepsin are eluted by 0.7 M NaCl in the same buffer. The clotting strength is measured in each fraction and the composition is calculated from these. The method has been modified in several ways (13-15). I will comment on IDF's version (International Dairy Federation) which is based on measurement in modified Berridge units and Chr. Hansen's which is measured in Chr. Hansen's units (CHU). The main difference between the two versions is caused by lack of standard in the IDF version and by the different pH values used for the milk-clotting tests. The IDF method uses pH around 6.3 - 6.4 whereas CHL uses approximately pH 6.6. The clotting time varies very much with the pH and the changes are different for chymosin and bovine pepsin. A decrease in pH results in a decreased clotting time or increased clotting activity. The activity of bovine pepsin increases more than that of chymosin. Because of the lower pH in the IDF version and the following higher activity of bovine pepsin 1 CHU chymosin corresponds to approximately 1.8 RU whereas 1 CHU bovine pepsin corresponds to approximately 2.7 RU BpA. The effect on the enzymic composition is that the IDF Garnot method will show a higher % bovine pepsin than Chr. Hansen's method, which uses a pH value close to real cheese-making conditions and the results therefore express the enzyme composition as it will work in the cheese-milk. This means that if a rennet contains 50% Ch and 50% BpA by Chr. Hansen's method then half of the clotting activity during cheese making will come from chymosin and the other half from bovine pepsin. The IDF Garnot method will by analysis of the same sample show 31% chymosin and 69% bovine pepsin which does not reflect the enzyme composition as it will act during clotting of cheese milk. The IDF method is however only provisional, changes are likely to occur. The fact that calf rennet costs about twice as much as bovine rennet stresses the importance of knowing the enzymic composition under cheese-making conditions.

Milk coagulants are often further characterized by measuring possible bacterial contamination, the stability of the coagulants, the ability of the products to form a curd and by measuring the proteolytic activity.

The bacteriological control of milk coagulants is very important. Total bacterial counts, yeast and mould, coli and gas producing anaerobic spore formers (e.g. Clostridium tyrobutyricum), which may cause late blowing in the cheese, are probably the most important tests.

Another important characteristic is storage stability and heat lability. Fairly heat lable products are desired in order to recover whey proteins free of active clotting enzymes which may prevent the application of the whey for e.g. food. Calf rennet is as heat labile as needed for later application of the whey whereas the conventional Mucor miehei coagulant is too stable. Several heat labile products of Mucor miehei coagulants have therefore been developed by oxidation of the conventional products (18-20). The heat lability of the latest products is as high as that of calf rennet.

The proteolytic activity of milk coagulants is important for the development of texture and flavour in cheese as well as for the cheese yield. Numerous papers have been published on methods for determination of proteolytic activity (e.g. 21-23). Each paper describes a new method or a modification of previous methods. None of the methods seems to have been generally adopted without being modified and probably none of the methods measures the proteolytic activity as it will behave in cheese. This is because artificial substrates and non-cheese conditions often are used for the assays, together with the fact that the proteolytic activity varies much with the condition for the proteolysis (e.g. substrate, pH, temperature, sodium chloride, calcium etc.). Generally it is therefore advisable to be as close as possible to the cheese making conditions in order to obtain usable results. The most reliable results are probably obtained by measuring the actual degradation in each type of cheese e.g. by measuring amount of soluble protein or number of peptide bonds split. This is however not the only factor being important. Also the character e.g. flavour especially bitternes of the peptides is important together with the yield and texture of the cheese. Based on our experience on measuring proteolytic activity the main milk coagulants can be listed in the following order of increasing proteolytic activity: lamb and calf chymosin, bovine pepsin, Mucor miehei protease, Mucor pusillus protease, porcine pepsin, Endothia parasitica protease, and chicken pepsin.

Milk coagulants may also be characterized by their ability to form a curd. This can be measured on e.g. a formagraph (N.Foss and Co., Denmark) by following the changes in viscosity of the milk (24). The principle is as follows. A sample cuvette containing milk and milk coagulant is moved back and forth ($\pm$ 0.7 mm). A pendulum immersed into this non-coagulated milk is not affected by the movement and a straight line is drawn on the paper. Gradually as the milk begins to coagulate the movement is transmitted to the pendulum. The higher curd tension the bigger oscillation of the pendulum. These swings are transmitted via a mirror attached to the hanging of the pendulum to a chart paper where a curve is drawn. Ten samples can be analyzed simultaneously. This measurement of curd tension is performed in order to compare the ability of various rennets to form coagulum of a suitable strength and to find the best time for cutting the curd. The curd tension obtained is very important for the consistency and the yield of the cheese. If the curd is too soft when it is cut, too much dry matter is lost during the whey draining and subsequently low cheese yield is obtained.

It is also of interest to know the amount of milk coagulant retained in cheese because the coagulant influences on the cheese ripening. The amount of rennet in cheese is unknown since the distribution of the enzyme between the cheese and the whey and the survival of rennet activity depend on the type of rennet and the production condition. A possible method for identification of the milk coagulant by Rocket immunoelectrophoreses (RIE) does not function yet, because the great amount of casein peptides interferes with the formation of rockets. A method for extraction and quantification of the amount of rennet activity has been developed at Chr. Hansen's Laboratory. The analysis takes about 4 hrs. The method is modified from that of Stadhouders (25) which takes approximately one week. The principle is as follows: 2 g. cheese is homogenized in 15 ml 1 M acetate buffer pH 4.7, and centrifugated. The supernatant which contains the rennet is desalted on Pharmacia PD-10 columns and the clotting activity is measured directly by a sensitive milk clotting test ($\geq$10 ng chymosin per/ml). 1 ml sample is added to 10 ml reconstituted skim-milk containing 1.1% skim-milk powder and 0.2% $CaCl_2$. A standard curve is drawn by plotting the clotting time versus the reciprocal clotting units. A straight line is obtained over a large range. The clotting time of the sample is compared to a standard curve made of a standard of the same enzymic composition. The clotting activity is found and multiplied by the dilution.

Future aspects

Of interest in the future is mainly the application of ultrafiltrated milk for cheese (26), the preparation of chymosin by recombinant DNA technique and accellerated cheese ripening (28) by which the ripening time is shortened through addition of proteolytic enzymes. Only the two first items will be discussed here as only these are expected to affect the use of milk-clotting enzymes.

The traditional hard cheese is produced by addition of a starter and then milk-clotting enzyme to warm milk to form a curd. The curd is cut into cubes, the whey is drained and the curd is pressed to make a cheese. Finally the cheese is ripened. By this procedure approximately 15-20% of the milk dry matter, mainly whey proteins, are lost in the whey. Furthermore, this traditional process requires large vats because only 1 kg cheese is obtained from 10 kg of milk. By the new procedure the milk is concentrated often 2-5 times by ultrafiltration before the starter and milk-clotting enzyme is added. The whey proteins are retained in the milk and the proportion between the casein and whey are changed often from 16:1 in the traditional milk to 4:1 in the UF milk. Further the concentration of Calcium and Phosphate is increased. Great effort has been made to use this method for cheese-making because the retained whey proteins increase the yield of cheese by up to 15-20% and because the reduced volume makes the production method more rational. However, due to the changed composition UF milk has properties quite different from those of the conventional milk. This has led to some problems concerning the use of milk-clotting enzymes for UF-cheese and to several studies on the clotting and coagulation of UF milk. In four times concentrated UF milk only 50% of the κ casein has to be hydrolyzed before the milk coagulates whereas 90% has to be hydrolyzed in conventional milk (28). In many cases UF-milk does not clot at all, it only coagulates. Both the clotting and the coagulation time depends on the concentration factor, pH, and several other factors. Some investigations have found an increase in clotting time with increasing protein concentration while others have found the opposite result (26). The inconsistency is at least partly due to different experimental conditions. The coagulation in UF milk depends very much on pH (29). Practical cheese experiments indicate that the clotting time decreases with increasing concentration degree. Further investigations have to be performed before the right type and amount of rennet can be chosen for each type of cheese. Problems like these and several others which are all connected to the changed substrate for the milk clotting enzymes and the starter cultures are the reason for UF until now mainly being used for Feta and semisoft cheeses (Camenbert).

One of the most interesting future commercial aspects of the aspartic proteases is the production of chymosin by the recombinant DNA technique. Many laboratories are trying to make chymosin in this way (30-36) but no product is on the market yet. Genencor, Inc. in collaboration with Chr. Hansen's Laboratorium A/S is one of the main companies on this project. Production of chymosin has been developed at Genencor by cloning of the prochymosin gene in E.Coli. The N-terminal and C-terminal halves of the molecule, were cloned in two separate experiments and assembled with a small synthetic piece of DNA, to complete the N-terminus, in a three way ligation to form met-prochymosin. High levels of expression in E.Coli were obtained, but the prochymosin was stored in E.Coli as insoluble refractile particles containing approximately 50% prochymosin. The particles were difficult to solubilize and the molecules were easily oxidized upon isolation which resulted in both correct and incorrect disulphide bonds. The best results were obtained by dissolving the particles in e.g. urea followed by oxidative sulfitolyses, diafiltration, ion exchange chromatography, reformation of the correct disulphide bridges by a glutathione redoxbuffer and a final diafiltration. The isolated prochymosin was contaminated with a 32,000 molecular weight protein which was part of the prochymosin molecule. The contaminant could not be activated but could easily be removed by precipitation. The remaining part was monomers of prochymosin. This prochymosin appeared to be identical to that of calf after thorough analysis. The main methods used were amino acid analysis, N and C-terminal sequence analysis, gelfiltration, SDS-polyacrylamide gelelectrophoresis, rocket and crossed immunoelectrophoresis, activation compared to that of calf prochymosin, and by measurement of milk-clotting activity, proteolytic activity and stability of the activated sample at various pH values. The product has been further characterized by evaluation of its value for cheese. Large scale production (37) of cheddar showed excellent flavour and texture equal to cheese prepared by commercial calf rennet.

References

1. Barbosa, M., Corradini, C., Battistotti, B.: Neth. Milk Dairy J. 35, 307-312 (1981).
2. Soxhlet, F. von: Milchzeitung 6, 495-513 (1877).
3. Berridge, N.J.: Biochemistry 39, 179-186 (1945).
4. Rothe, G.A.L., Axelsen, N.H., Jøhnk, P.: J. Dairy Res. 43, 85-95 (1976).

5. Mulvihill, D.M., Fox, P.: J. Dairy Res. 44, 319-324 (1977).

6. Koning, P.J. de, Rooijen, P.J. van, Kok, A.: Neth. Milk Dairy J. 23, 55-70 (1969).

7. Righetti. P.G., Molinari, B.M., Molinari, G.: J. Dairy Res. 44, 69-72 (1977).

8. Raymond, M.N., Bricas, E. Salesse, R., Garnier, J., Garnot, P., Ribadeau Dumas, B.: J. Dairy Sci. 56, 419-422 (1973).

9. Collin, J.C., Muset de Retta, G., Martin, P.: J. Dairy Res. 49, 221-230 (1982).

10. Rothe, G.A.L., Harboe, M.K., Martiny, S.C.: J. Dairy Res. 44, 73-77 (1977).

11. Harboe, M.K.: Neth. Milk Dairy J. 35, 367-369 (1981).

12. Garnot, P., Thapon, J.L., Mathieu, C.M., Maubois, J.L. Ribadeau Dumas, B.: J. Dairy Sci. 55, 1641-1650 (1972).

13. Collin, J.C., Martin, P., Garnot, P., Ribadeau Dumas, B., Mocquot, G.: Milchwissenschaft 36, 32-35 (1981).

14. Martin, P., Collin, J.C., Garnot, P., Ribadeau Dumas, B., Mocquot, G.: J. Dairy Res. 48, 447-456 (1981).

15. Provisional International IDF Standard 110 : 1982.

16. Axelsen, N.H., Krøll, J., Weeke, B. (eds.): A Manual of quantitative Immunoelectrophoresis. Scandinavian J. of Immunol. 2., Suppl. No. 1 (1973).

17. Andrén, A., Koning, P.J. de, Björck, L.: Neth. Milk Dairy J. 37, 11-20 (1983).

18. Hansen, P.: North European Dairy J. 9, 290-294 (1982).

19. Branner-Jörgensen, S., Schneider, P., Eigtved, P.: United States Patens no. 4,357,357 (1982).

20. Cornelius, D.A.: United States Patent no. 4,348,482,(1982).

21. Mesrob, K., Stoeva, P.: Milchwissenschaft 33, 483-486 (1978).

22. Reps, A., Poznanski, S., Zelazowska, H., Jederychowski, Z., Chojnowski, W.: Milchwissenschaft 36, 733-740 (1981).

23. Fox, P.F.: Neth. Milk Dairy J. 35, 233-253 (1981).

24. Lynggaard - Jensen, B., Peitersen, N.: Procedings from the International Symposium on Use of Enzymes in Food Technology, Versailles, (ed. Dupuy, P), 309-313 (1982).

25. Stadhouders, J., Hup, G., Waals, C.B. van.: Neth. Milk Dairy J. 31, 3-15 (1977).

26. Mehaia, M.A., Cheryan, M.: Milchwissenschaft 38, 708-710 (1983).

27. Law, B.A., Wigmore, A.: J. Dairy Res. 49, 137-146 (1982).

28. Dalgleish, D.G.: J. Dairy Res. 48, 65-69 (1981).

29. Garnot, P., Corre, C.: J. Dairy Res. 47, 103-111 (1980).

30. Nishimori, K., Kawaguchi, Y., Hidaka, M., Uozumi, T., Beppu, T.: J. Biochem. 91, 1085-1088 (1982).

31. Moir, D., Mao, J., Schumm, J.W., Vovis, G.F., Alford, B.L., Taunton-Rigby, A.: Gene, 19, 127-138 (1982).

32. Unilever N.V.: Dutch patent application 4537 (1982).

33. Emtage, J.S., Angal, S., Doel, M.T., Harris, T.J.R., Jenkins, B., Lilley, G., Lowe, P.A.: Proc. Natl. Acad. Sci., U.S.A., 80, 3671-3675 (1983).

34. Celltech limited: PCT patent application GB83/00152 (1983).

35. Genex Corporation: UK patent application 2123005 (1983).

36. Genencor.: US Patent application 452,227 (1982).

37. Hicks, C.L., O'Leary, J., Bucy, J.: Abstract presented at the Am. Dairy Sci. Association meeting, June 24-27 (1984).

mRNA´S FOR CHYMOSIN AND PEPSIN, TWO MAIN ASPARTIC PROTEINASES OF BOVINE STOMACH AND ANALYSIS OF THEIR TRANSLATION PRODUCTS

Marie Lipoldová, Jiřina Černá, Mirko Takáč, Stanislav Zadražil and Ivan Rychlík

Institute of Molecular Genetics, Czechoslovak Academy of Sciences, 166 37 Prague, Czechoslovakia

Introduction

Both chymosin and pepsin represent mammalian degradative enzymes whose role is to digest nutrient proteins and which may be regarded as major types of gastric proteinases. These enzymes are secreted as inactive precursors, prochymosin and pepsinogen, which are irreversibly converted into the active enzymes through limited proteolysis resulting in a final loss of the 42 amino-terminal residues of the zymogens (1). Both zymogens being secretory eukaryotic proteins are already processed primary products of gene expression (2,3) coded for as preprochymosin and prepepsinogen. For the important role of these enzymes in metabolism as well as for their uses in the food industry, namely in cheese making (chymosin as the main proteolytic enzyme in calf cheese rennet and pepsins of different origin as its frequent substitutes), and in molecular biology as a possible model for gene structure research and its expression and evolution (4), they are of great interest for those working in gene manipulation and gene engineering fields.

mRNA´S are often the only starting material for cloning of eukaryotic genes and suitable and simple analytical methods for determination of their translation products are important prerequisites of functional characterization of the isolated messengers as well as of expression products of the cloned

Aspartic Proteinases and their Inhibitors

genes. In this paper we present comparative characteristics of mRNA´S for chymosin and pepsin and of their translation products.

Results

Isolation of poly(A)$^+$RNA

The preparation of poly(A)$^+$RNA from the fundal region of the fourth calf or bovine stomach was performed by two methods giving identical results: either by hot phenol extraction (5) or using guanidine thiocyanate (6), followed in both cases by separation on a poly(U)-Sepharose column (7). Poly(A)$^+$RNA was then fractionated by sucrose gradient centrifugation and the fractions obtained were examined for their translation activity in cell-free systems (8). Maximum translation activity for preprochymosin and prepepsinogen, determined in the reticulocyte lysate, was present in fractions of the respective mRNA corresponding in both cases to 15S. Using urea-polyacrylamide gel electrophoresis, the molecular weights of both mRNA´s calculated from the relative migration distances (Fig. 1) corresponded to 16S. A slight difference in the molecular weights estimated by these methods has also been described by other authors, who worked with different systems; this difference can be accounted for by various levels of secondary structure of the ribosomal RNA´s used as size markers and mRNA´s themselves (9). The estimated average sizes of the mRNA´s for preprochymosin and prepepsinogen are in a good agreement with the chain length of the coded molecules of the corresponding zymogens and in the case of preprochymosin they have also been confirmed by others (10,11). As shown in Fig. 2, the differences between the mRNA´s from two parts of the stomach (fundal and pyloric) of one individual are greater than those between the mRNA preparations from the fundal parts of the stomach of the adult animal and calf. So far, nothing is known about

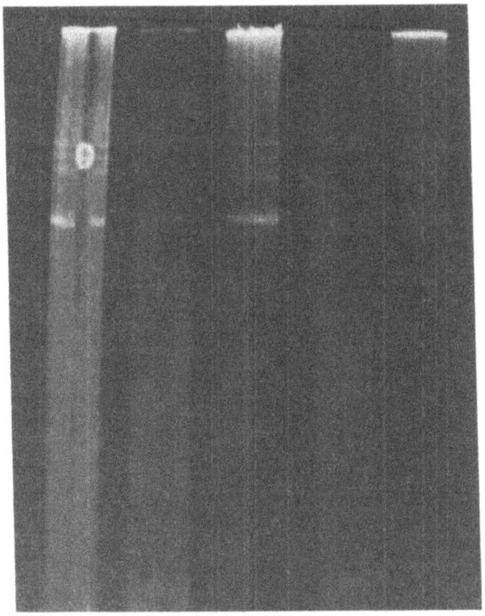

1 2 3 4 5

Figure 1. Comparison of poly(A)$^{+}$RNA´s from calf and bovine stomach by urea-polyacrylamide gel electrophoresis (8M urea in 3.5 % gel). 1 - mRNA from bovine stomach, 2 - fraction of 2 with maximum translation activ ity for prepepsinogen, 3 - mRNA from stomach of calf fed milk and hay, 4 - fraction of 5 with maximum translation activity for prechymosin, 5 - mRNA from stomach of calf fed milk only

further products, if any, which could be coded for by mRNA fractions found on the gels (Figs.1 and 2).

Translation products synthesized in cell-free systems and in oocytes

Cell-free systems, when programmed with mRNA for a secretory protein, synthesize precursors extended at their amino termini by the signal peptide usually without any processing (12-15). Xenopus oocytes, on the other hand, after being microinjected mRNA coding for a secretory protein perform translation, processing and secretion of the correct translation products - zymogens (16-18). Hence, the translation systems performing synthesis of proteins under direction of exogenous mRNA in

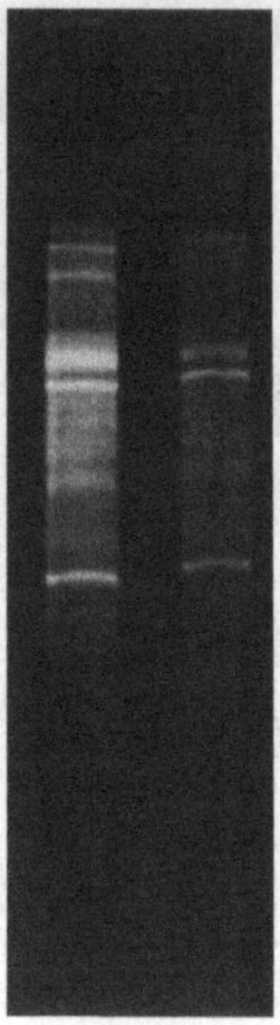

Figure 2. Urea-polyacrylamide gel electrophoresis (as in Fig. 1) of poly(A)$^+$mRNA from fundal (left) and pyloric (right) part of bovine stomach

vitro (reticulocyte lysate and wheat germ system) (19) and in vivo (Xenopus laevis oocytes) were used for further characterization of the isolated mRNA´s and of their products (Fig. 3).

Milk-clotting activity of translation products

It has been reported that mRNA´s injected into oocytes were translated into biologically active products (17,23-25). We tried, therefore, to demonstrate the enzymatic activity of chymosin by using the milk-clotting assay. We tested the milk-clotting activity of translation products accumulated in the incubation medium using the simple microplate assay of Emtage (26). As shown in Fig. 4 the activated incubation medium (acidification to pH 3.0 followed by neutralization to pH 6.3), surrounding oocytes microinjected with preprochymosin mRNA, contained proteins with milk-clotting activity (Fig. 4A, lanes

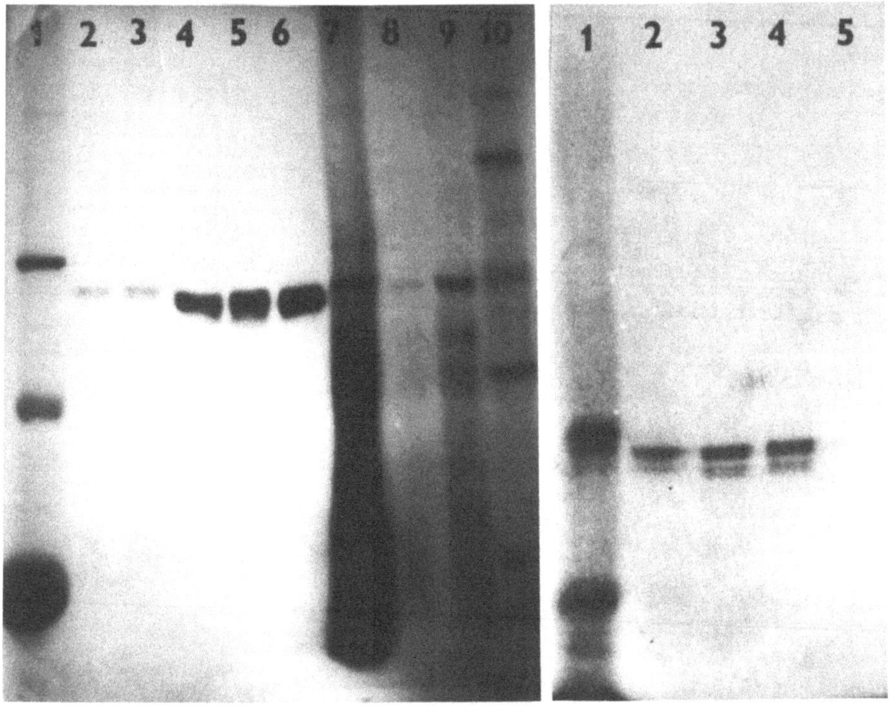

A B

Figure 3. Fluorograms (20) of translation products synthesized in reticulocyte lysate, wheat germ system, and in oocytes, immunoprecipitated (21) and separated by SDS-PAGE in gradient gel (7-15 %). A - translation of preprochymosin mRNA: 1 - reticulocyte lysate control, 2 and 3 - reticulocyte lysate, 4 and 5 - oocyte extract, 6 - oocyte incubation medium, 7 - wheat germ system (no immunoprecipitation), 8 and 9 - wheat germ system, 10 - marker proteins of M_r90 000,69 000, 46 000,30 000 and 14 000, respectively (from top to bottom); B - translation of prepepsinogen mRNA: 1 - reticulocyte lysate, 2 - oocyte extract, 3 and 4 - oocyte incubation medium, 5 - oocyte extract control

4 and 5), whereas no such activity was detected in the medium after incubation of oocytes injected with water only (Fig. 4A, lane 3). Similarly, oocytes injected with prepepsinogen mRNA exported proteins clotting milk (Fig. 4B, lane 4) and oocytes

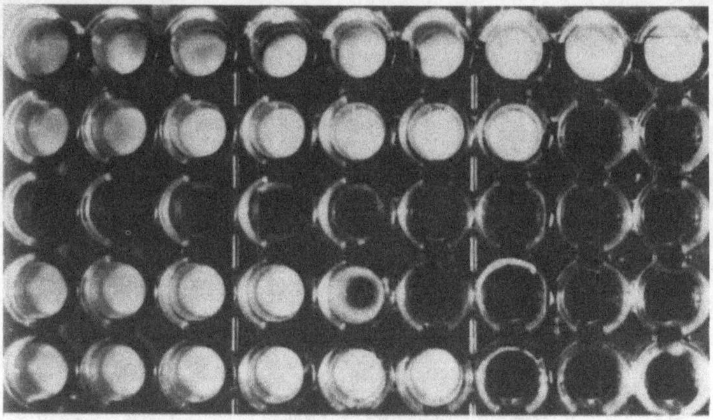

A

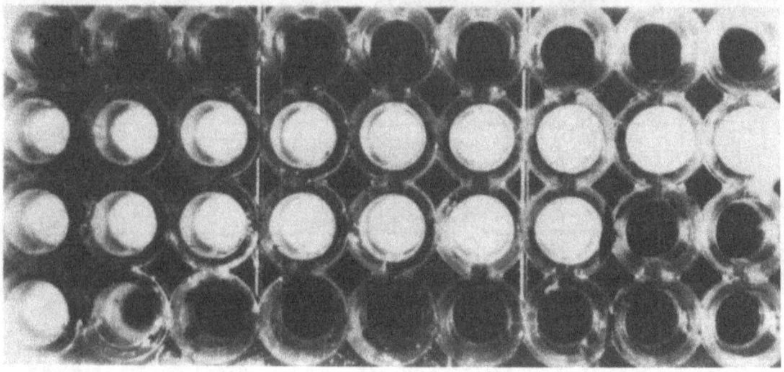

B

Figure 4. Milk-clotting assay of prochymosin and pepsinogen. A microplate with clotted milk in the wells is shown. From left to right the samples in the wells were diluted gradually 1:1 with water. A - prochymosin:1 - calf chymosin, 2 - calf prochymosin after activation, 3 - oocyte incubation medium control (48 h), 4 - oocyte incubation medium 48 h after injection of mRNA, 5 - oocyte incubation medium 96 h after injection of mRNA; B - pepsinogen: 1 - oocyte incubation medium control (48 h), 2 - chymosin, 3 - swine pepsinogen (after activation), 4 - oocyte incubation medium 48 h after injection of mRNA.

injected with water were deficient in this activity (Fig. 4B, lane 1).

Proteolytic activity of products synthesized in oocytes

Another experimental approach to the detection of enzymatic activity of proteinases synthesized in oocytes is the demonstration of their proteolytic activity directly on the separating gel using our specific modification of the Miskin and Soreq procedure (27). Casein (0.1 %) is embedded as the specific protein substrate in the polyacrylamide gel which is used for SDS-PAGE separation of the oocyte translation products. After renaturation of the separated proteins by washing off SDS from the gel (27) the latter is incubated at pH 3.0 for specific activation of acid proteinases and proteolysis of the present substrate. Staining of the treated gel with Coomassie brilliant blue then visualizes white spots in every place of the gel where the acidic proteinase has split the embedded casein. As demonstrated in Fig. 5 the in situ estimated proteolytic activity corresponds to enzymatic functions of the protein activated products which were synthesized in oocytes under specific direction by the microinjected preprochymosin mRNA isolated from calf stomach. This analytical method could similarly be used for the detection of the proteolytic activity of the product whose synthesis is dependent on the corresponding gene cloned in a heterologous host.

All the data presented in this paper show that the isolated poly(A)$^{+}$RNA´s represent fully functional mRNA´s for preprochymosin and prepepsinogen and can be translated and processed into the biologically active enzymes.

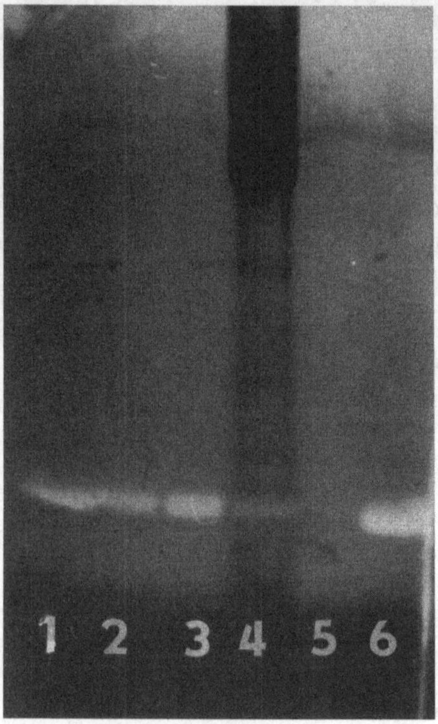

Figure 5. Detection of proteolytic activity of chymosin in gels after SDS-PAGE of oocyte translation products. The procedure is explained in the text. 1 - prochymosin standard, 2 - oocyte incubation medium 48 h after injection of mRNA, 3 - oocyte incubation medium 96 h after injection of mRNA, 4 - oocyte extract 48 h after injection of mRNA, 5 - oocyte incubation medium control (48 h), 6 - prochymosin standard.

References

1. Pedersen, V.B., Christensen, K.A., Foltmann, B.: Eur. J. Biochem. 94, 573-580 (1979).
2. Blobel, G., Dobberstein, B.: J. Cell Biol. 67, 835-851 (1975).
3. Davis, B.D., Tai, P.C.: Nature 283, 433-438 (1980).
4. Neurath, H.: Science 224, 350-357 (1984).

5. Scherrer, K.: In Fundamental Techniques in Virology (Habel, K. and Salzman, P., eds), pp. 413-432, Academic Press, New York and London (1969).

6. Chirgwin, J.M., Przybyla, A.E., MacDonald, R.J. Rutter, W.J.: Biochemistry 24, 5294-5299 (1979).

7. Molloy, G.R., Jelinek, W., Salditt-Georgieff, M., Darnell, J.E.: Cell 1, 43-53 (1974).

8. Pelham H.R.B., Jackson, R.J.: Eur. J. Biochem. 67, 247-256 (1976).

9. Woo, S.L.C., Rosen, J.M., Liarakos, C.D., Choi, Y.C., Busch, H., Means, A.R., O'Malley, B.W.: J. Biol. Chem. 250, 7027-7039 (1975).

10. Nishimori, K., Kawagushi, Y., Hidaka, M., Uozumi, T., Beppu, I.: J. Biochem. (Tokyo) 91, 1085-1088 (1982).

11. Harris, T.J.R., Lowe, P.A., Eaton, M.A.W., Millican, T.A., Patel, T.B., Bose, C.C., Carey, N.H., Doe, M.S.: Nucleic Acids Res. 10, 2177-2187 (1982).

12. Boime, I., Boguslawski, S., Caine, J.: Biochem. Biophys. Res. Commun. 62, 103-109 (1975).

13. Devillers-Thiery, A., Kindt, T., Scheele, G., Blobel, G.: Proc. Natl. Acad. Sci. USA 72, 5016-5020 (1975).

14. Chan, S.J., Keim, P., Steiner, D.F.: Proc. Natl. Acad. Sci. USA 73, 1964-1968 (1976).

15. Cox, G.S., Weintraub, B.D., Rosen, S.W., Maxwell, E.S.: J. Biol. Chem. 251, 1723-1730 (1976).

16. Colman, A., Morser, J.: Cell 17, 517-526 (1979).

17. Soreq, H., Sagar, A.D., Seghal, P.B.: Proc. Natl. Acad. Sci. USA 78, 1741-1745 (1981).

18. Valle, G., Besley,J.,Colman, A.: Nature 291, 338-340 (1981).

19. Roberts, B.E., Paterson, B.M.: Proc. Natl. Acad. Sci. USA 70, 2330-2334 (1973).

20. Bonner, W.M., Laskey, R.A.: Eur. J. Biochem. 46, 83-88 (1974).

21. Kessler, G.: J. Immunol. 115, 1617-1624 (1975).

22. Laemmli, U.K.: Nature 227, 680-685 (1970).

23. Labarca, C., Paigen, K.: Proc. Natl. Acad. Sci. USA 74, 4462-4465 (1977).

24. Shulman, L., Revel, M.: Nature 288, 98-100 (1980).

25. Miskin, R., Soreq, H.: Nucleic Acids Res. 9, 3355-3365 (1981).

26. Emtage, J.S., Angal, S., Doel, M.T., Harris, T.J.R., Jenkins, B., Lilley, G., Lowe, P.A.: Proc. Natl. Acad. Sci. USA 80, 3671-3675 (1983).

27. Miskin, R., Soreq, H.: Anal. Biochem. 118, 252-258 (1981).

PROTEOLYTIC DEGRADATION OF MUSCLE DURING THE SALT-CURING PROCESS OF HERRING.

Sigríður Ólafsdóttir, Sigurður Magnússon and Jón B. Bjarnason.
Science Institute, University of Iceland, Dunhaga 3,
107 Reykjavík, Iceland.

Introduction

Salt-cured Icelandic herring is a fishery product made from herring in the feeding stage of its annual cycle. The ripening, resulting in a particular taste and a soft consistency, is poorly understood, but it has been suggested that it results largely from a partial degradation of the fish muscle by proteolytic enzymes. The origin of the proteolytic enzymes and the nature of the degradation of the herring muscle has been a source of some investigative efforts and considerable debate and speculation.

The absence of the characteristic organoleptic properties in cured herring from which the intestines were completely removed has been demonstrated (1). Luijpen made a number of observations on salt-cured maatjes herring, from which he concluded that the appendices pyloricae indeed were responsible for the sensoric properties of the ripened product (2).

Since the content of proteolytic enzymes in the fish varies with the seasons, the fish is suitable for the manufacture of salt-cured herring only for a few months. Other variables like age and size of the fish or its source may also influence the suitability of herring for the salt-curing process. For these reasons there is a demand for a manufacturing method which is less dependent on the protease activity of the herring.

The aim of our investigation was to further our understanding of the origin of the proteolytic activities responsible for the

Aspartic Proteinases and their Inhibitors

curing process and to obtain a fuller insight into the subsequent degradation of the fish muscle.

Methods

Before the herring was put in the barrels the head was cut off and the stomach and a part of the gut torn out. About 90 kg of herring were put in each barrel and sugar, potassium sorbate and 13 kg of salt added. A sample was taken from herring muscle on the first day of the curing process and then muscle and brine samples 6, 13, 21, 28, 42, 57, 83 and 118 days after the process was initiated and by that time the herring was fully cured. The brine (16 - 18% salt) was filtered and an aliquot was set aside for nitrogen content determinations and electrophoresis investigations. The rest was stored at 2 - 4°C until measurements of proteolytic activity were made. The herring was washed, flitched, the skin was cut off and then it was minced. Some of the mince was homogenized for total nitrogen and TCA-soluble nitrogen content determinations. For gel filtration and electrophoresis experiments small portions of the mince were homogenized in distilled water and centrifuged (ca. 1000xg for 20 min.). The supernatant was used for the experiments but the sediment discarded. The rest of the mince was stored at -18°C until used for measurments of proteolytic activity.
Electrophoresis experiments. Laemmli's electrophoresis buffer system was used on acrylamide slab gels. The acrylamide concentration was 12.5% for the brine samples but linear acrylamide gradient, 7.5 - 15.0%, was used for the muscle samples.
Gel filtration experiments. For analysis of muscle samples a Sephacryl S-200 superfine (Pharmacia Fine Chemicals) gel was used in a 2.5 x 100 cm column that was equilibrated with 10 mM borate, 100 mM NaCl, 2 mM $CaCl_2$, pH 9.0.
Proteolytic activity was measured against three different protein substrates, bovine hemoglobin, myofibrillar (water insoluble), and sarcoplasmic (water soluble) protein fractions

from herring muscle. Water extraction was made for each muscle sample to use as source of proteolytic enzymes but the brine was just filtered. The incubation mixture consisted of 0.25 ml of muscle extract or brine, 0.25 ml of 8% (w/v) protein substrate and 0.50 ml 0.25 M citrate or phosphate buffer. The buffer solutions were one of six different pH values ranging from 3.5 to 7.5. The mixture was incubated for 48 hrs. at 25°C. All measurments were done in triplicate and the pH was measured in one of the tubes at the beginning, in the middle and at the end of the incubation time. The mean pH was used to express the pH of that incubation. The reaction was stopped by adding 1.0 ml 10% TCA. The mixture was centrifuged and Folin-positive substance was measured in the supernatant. Protein hydrolysis was compared with standards and expressed as tyrosin equivalents. Blank mixtures were made by adding the TCA before all protein solutions. Nitrogen content was determined by using the Kjeldahl hydrolysis method and subsequent spectrophotometric analysis. Total nitrogen was multiplied by factor 6.25 and expressed as grammes of protein pr. 100 ml of sample.

Results and discussion

Electrophoresis experiments. Results from the electrophoresis of muscle samples are shown on figure 1. Some of the changes in the banding pattern are marked with letters from a to k. Changes that most likely result from protein degradation can be seen on the first six days og the curing process, e.g. bands labeled a, b, c, g and i. Other changes that could result from protein degradation later in the process are visible in bands d, h and j. Some of the bands increase in strenght during the course of the process. This can be explaned either by proteins being released into the water soluble fraction or these are products from degradation of larger proteins. Some of the bands that appear late in the process are degraded, e.g. bands labeled h and j. None of the protein bands have yet been identified nor the sites of cleavage determined. Therefore no conclusions can be drawn from these results about the origin of the enzymes that cause the degradation of the fish muscle proteins at this stage.

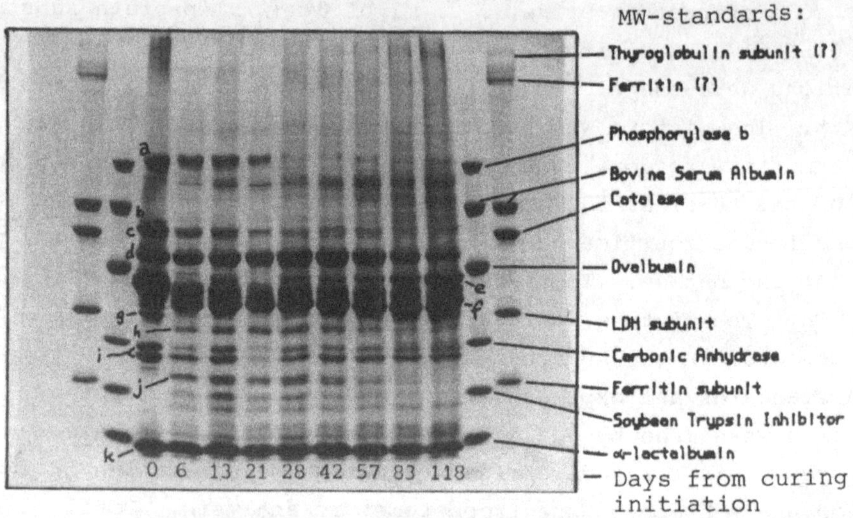

Fig. 1. Electrophoresis of muscle proteins from herring at various stages in the curing process. Laemmli's buffersystem. Linear acrylamidegradient 7.5-15%.

Results from electrophoresis of brine samples demonstrated increasing concentration of three major bands of MW. ca. 45500, 17500 and 7000. The protein bands have not been identified so nothing can be asserted about their origin. It is clear from these results that protein concentration increases in brine during the course of the curing process.

Gel filtration experiments. The results from gel filtration experiments of muscle extracts from different stages of curing demonstrated that the amount of large proteins is very little in the beginning. It increases in the first week, then decreases until the sixth week and increases again until the end of the process. These changes could be linked to changes in the proteolytic activity in the muscle. Proteins seem to be released from the tissues constantly but the decrease of proteins larger than MW. ca. 270000 is coincident with increase in proteolytic activity in the neutral pH range in the muscle, and in the end the amount of large proteins increases as the proteolytic activity decreases. The ever increasing amount of peptides of MW ca. 2500 also suggest proteolytic degradation.

Proteolytic activity. Figure 2a shows the proteolytic degradation of myofibrillar proteins at pH 5.8, 6.4 and 7.0 caused by enzymes in the brine at various stages of the curing process. The proteolytic activity towards myofibrillar proteins increased rapidly in the brine at the highest pH's the first three weeks and then decreased until about the 60th day and stayed about constant from that time until sampling was ceased. In figure 2b the proteolytic degradation of myofibrillar proteins at pH 6.2, 6.8 and 7.5 caused by enzymes extracted from muscle at various stages of the curing process are shown. A rise and fall of activity as in the brine can be seen in the muscle but it occurs later in the process. The pH optimum for the degradation seems to be about 7.5 or higher. It is known that some remains of the gut are necessary for the curing process to take place (2). It is therefore tempting to suggest that the activity shown in fig.2 is caused by digestive enzymes, diffusing from the gut-remains into the brine in the first 3-4 weeks, and in about 5 weeks of curing appearing in the muscle. Degradation of hemoglobin and sarcoplasmic proteins by enzymes in the muscle extracts was also observed at the same time as the myofibrillar protein degradation but the activity was not as extensive.

At pH 3.8, 4.5 and 5.4 hemoglobin was by far the best substrate for proteolytic enzymes. The measurements of protein degradation in brine at the more acidic pH's showed increasing activity all through the curing process that suggests constant diffusion of acidic proteases from gut and/or muscle into the brine. This increase of activity in the brine could be seen in the degradation of all of the three protein substrates. In the muscle extracts the proteolytic activity at pH 3.8, 4.7 and 5.8 was about the same towards the muscle protein substrates from the first sample to the last. The degradation of hemoglobin by the first muscle sample extract was considerable, then increased somewhat in the beginning of the process, decreased again in the fourth to sixth

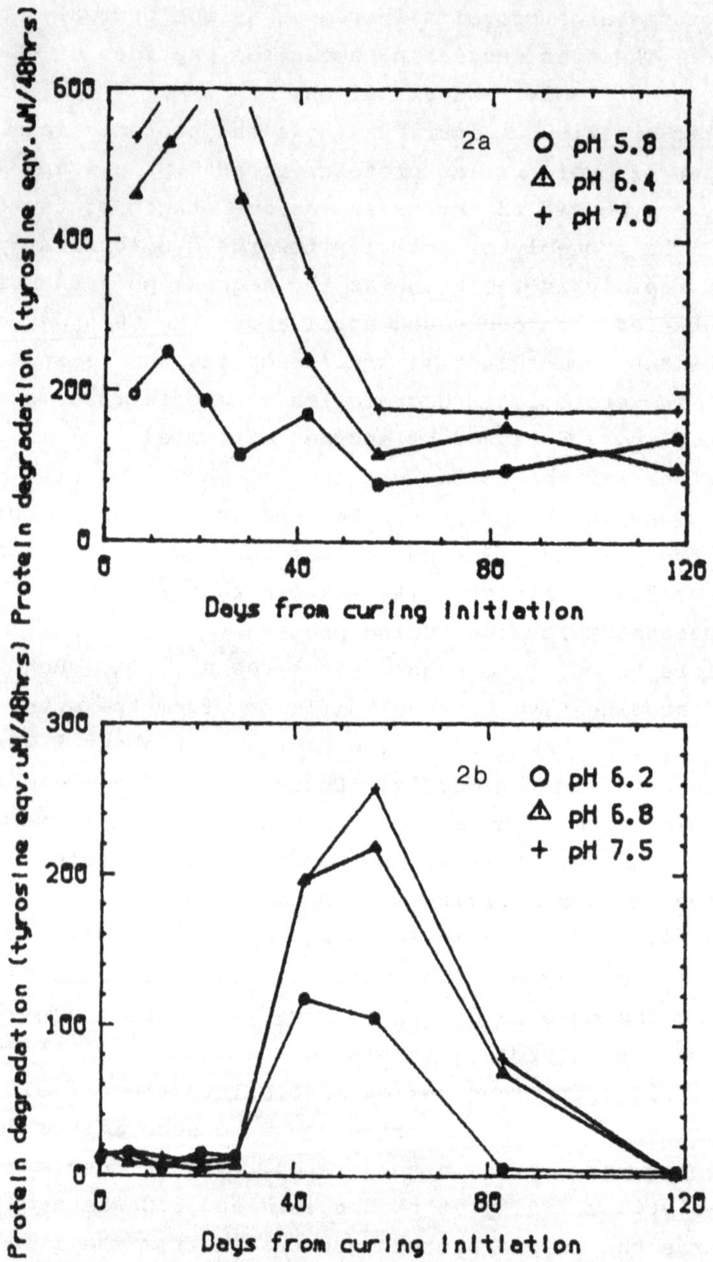

Fig.2. Degradation of myofibrillar proteins by brine (fig.2a) and muscle extracts (fig.2b) during the course of the salt-curing of herring.

weeks and was about constant after that. Most likely the protein degradation at the more acidic pH's is primarily caused by cathepsin-like enzymes from the muscle. Although catheptic activity can be detected in the degradation activity assays, the cathepsins are probably not active in the high salt concentration that persists in the muscle after the first week of the curing process (3). Therefore the protein degradation that is observed in the electrophoresis and gel filtration experiments is most likely not a result of catheptic activity. Proteases from the intestines are much less inhibited by high salt concentrations and could therefore be active during the curing process (3).

Nitrogen content determinations. In the brine the total nitrogen and the TCA-soluble nitrogen concentration increased constantly during the course of the curing process. The proteins probably originate both from the intestines and muscle, and the peptides (TCA-soluble nitrogen) diffuse out of the herring muscle and intestines or are degradation products from proteins in the brine. Total nitrogen concentration of the muscle increased the first week of curing and decreased until an equilibrium state was reached in about 6 weeks. These results reflect the osmotic flow of water out of the muscle the first week and then flow of salt and water into the muscle until equilibrium is reached. The TCA-soluble nitrogen fraction showed almost no changes at all. Perhaps the changes in TCA-soluble nitrogen concentration are not observed because of the increased amount of water flowing into the muscle diluting the peptides as soon as they are made.

The major findings of this investigation is the demonstration of protein degradation in the muscle seen in gel electrophoresis, gel filtration and nitrogen concentration determinations. Of major importance are also the findings of active proteases with a pH optimum in the weakly alkaline pH range that appear in the muscle and disappear about the time of termination of the curing process. These proteases are also seen to degrade myofibrillar proteins from herring which are probably the natural substrates in the process. The weakly alkaline proteases seem to appear first in the brine and then diffuse into the muscle which suggests that their origin might be in the herrings intestines.

References

(1) Liebert, F.: Rapp. verkand. Rijksinst. Visserijondera, 1 (I), 124-132 (1913).

(2) Luijpen, A.F.M.G.: De invloed van het kapen op de rijping van gezouten maatjesharing. Thesis. Utrecht.

(3) Kiesvaara, M.: On the soluble nitrogen fraction of barrel-salted herring and semi-preserves during ripening. Techn. Res. Centre of Finland, Helsinki (1975).

List of Participants

ANDERSEN, K.E.	Institute of Fisheries, University of Tromsö, POB 3083, Guleng, N-9001 Tromsö (Norway)
ANDREATTA, R.H.	Ciba-Geigy AG, K-122.127, Basel (Switzerland)
ANDREEVA, N.	Institute of Molecular Biology, Academy of Sciences of the USSR, Vavilova 32, Moscow 117984 (USSR)
ANTONOV, V.K.	Shemyakin Institute of Bioorganic Chemistry, Academy of Sciences of the USSR, Vavilova 32, Moscow 117312 (USSR)
BAUDYŠ, M.	Institute of Organic Chemistry and Biochemistry, Czechoslov. Acad. Sci., Flemingovo 2, 166 10 Prague 6 (Czechoslovakia)
BERÁNKOVÁ, E.	Katedra biochemie a mikrobiologie VŠCHT, Suchbátarova 5, 166 28 Prague 6 (Czechoslovakia)
BJELLAND, S.	Institute of Fisheries, University of Tromsö, POB 3093, Guleng, N-9001 Tromsö (Norway)
BLUNDELL, T.L.	Laboratory of Molecular Biology, Dept. of Crystallography, University of London, Birkbeck College, Malet St., London WC1E 7HX (United Kingdom)
BOGER, J.S.	Merck Sharp and Dohme Research Laboratories, West Point, PA 19486 (USA)
BREIPOHL, G.	Hoechst AG, Pharma Synthese, 3033 Frankfurt/Main (FRG)
BROWN, D.	ICI Pharmaceuticals Division, Alderley Park, Macclesfield, Cheshire (United Kingdom)
BULANTOVÁ, E.	Institute of Organic Chemistry and Biochemistry, Czechoslov. Acad. Sci., Flemingovo 2, 166 10 Prague 6 (Czechoslovakia)

BURT, D.W.	Department of Biochemistry, University of Leicester, University Road, Leicester LE1 7RH (United Kingdom)
CHARLTON, P.A.	Mammalian Biochemistry Dept., Glaxo Group Research Ltd., Greenford Road, Greenford, Middlesex UB6OHE (United Kingdom)
CHERRY, P.C.	Medicinal Chemistry Department, Glaxo Group Research Ltd., Greenford (United Kingdom)
DAMBMANN, C.	NOVO Industri AS, NOVO Allé, DK-2880 Bagsvaerd (Denmark)
DELANEY, R.	Department of Biochemistry and Molecular Biology, University of Oklahoma Health Sciences, POB 26901, Oklahoma City, OK 73190 (USA)
DREYER, T.	Carlsberg Laboratory, Department of Chemistry, Gl. Carlsbergvej 10, 2500 Valby (Denmark)
DUNN, B.M.	Department of Biochemistry, University of Florida, College of Medicine, Box J-245, Gainesville, FL 32610 (USA)
ERKELENS, J.	Unilever Research Laboratory, POB 114, 3130 AC Vlaardingen (The Netherlands)
FÁBRY, M.	Institute of Molecular Genetics, Czechoslov. Acad. Sci., Flemingovo 2, 166 37 Prague 6 (Czechoslovakia)
FOLTMANN, B.	Institute of Biochem. Genetics, University of Copenhagen, O. Farimagsgade 2A, 1353 Copenhagen K (Denmark)
FOUNDLING, S.I.	Department of Crystallography, Birkbeck College, University of London, Malet St. London WC1E 7HX (United Kingdom)
GILDBERG, L.A.	Institute of Fishery Technology Research, POB 677, 9001 Tromsö (Norway)
HALLETT, A.	Department of Chemical Pathology, Royal Postgraduate Medical School, London W12 OHS (United Kingdom)
HALÁT, G.	Neurobiolog. ústav CFV SAV, 040 01 Košice (Czechoslovakia)

HARBOE, M.	Chr. Hansen's Laboratorium A/S, 3 Sanct Annae Plads, 1250 Copenhagen K (Denmark)
HARRIS, J.C.	Wellcome Foundation, Langley Court, Beckenham, Kent, BR3 3BS (United Kingdom)
HAUSDORF, G.	Institute Physiol. Biol. Chemistry, Humboldt University, Hessische Str. 3-4, 1040 Berlin (GDR)
HEMMINGS, A.M.	Department of Crystallography, Birkbeck College, University of London, Malet St., London WC1E 7HX (United Kingdom)
INAGAMI, T.	Department of Biochemistry, Vanderbilt University, Nashville, TN 37232 (USA)
JABULKOVÁ, D.	Institute of Organic Chemistry and Biochemistry, Czechoslov. Acad. Sci., Flemingovo 2, 166 10 Prague 6 (Czechoslovakia)
JAMES, M.N.G.	Department of Biochemistry, University of Alberta, Edmonton, Alberta T6G 2H7 (Canada)
JINDRA, A.	II. interní klinika FVL UK, U nemocnice 2, 128 08 Prague 2 (Czechoslovakia)
JONES, D.M.	Department of Chemical Pathology, Royal Postgraduate Medical School, London W12 OHS (United Kingdom)
KAY, J.	Department of Biochemistry, University College Cardiff, POB 78, Cardiff CF1 1XL, Wales (United Kingdom)
KIRCHHÜBEL, W.	VEB Wissenschaftlich-technisch-ökonom. Zentrum der Milchindustrie, 1400 Oranienburg (GDR)
KLEINE, R.	Institute of Physiological Chemistry, University of Halle/S. (GDR)
KOHOUT, J.	I. chirurgická klinika FVL UK, U nemocnice 2, 128 00 Prague 2 (Czechoslovakia)
KORBOVÁ, L.	Ústav patologické fyziologie, U nemocnice 5, 128 53 Prague 2 (Czechoslovakia)
KOSTKA, V.	Institute of Organic Chemistry and Biochemistry, Czechoslov. Acad. Sci., Flemingovo 2, 166 10 Prague 6 (Czechoslovakia)

KOTACZKOWSKA, M.	Institute of Biochemistry, University of Wroclaw, Tamka 2, 50-137 Wroclaw (Poland)
LAH, T.	Biochemical Department, J. Stefan Institute, Jamova 39, 61000 Ljubljana (Yugoslavia)
LECKIE, B.	Blood Pressure Unit, Western Infirmary, Glasgow G11 6NT (United Kingdom)
LIPOLDOVÁ, M.	Institute of Molecular Genetics, Czechoslov. Acad. Sci., Flemingovo 2, 166 37 Prague 6 (Czechoslovakia)
MELOUN, B.	Institute of Organic Chemistry and Biochemistry, Czechoslov. Acad. Sci., Flemingovo 2, 166 10 Prague 6 (Czechoslovakia)
MIKEŠ, O.	Institute of Organic Chemistry and Biochemistry, Czechoslov. Acad. Sci., Flemingovo 2, 166 10 Prague 6 (Czechoslovakia)
ODA, K.	Department of Agricultural Chemistry, University of Osaka Prefecture, Sakai, Osaka 591 (Japan)
ÓLAFSDÓTTIR, S.	Science Institute, University of Iceland, Dunhaga 3, 107 Reykjavík (Iceland)
PATESTOS, N.	Department of Biochemistry, University College, Belfield, Dublin 4, Ireland
PEARL, L.H.	Laboratory of Molecular Biology, Dept. of Crystallography, Birkbeck College, University of London, Malet St., London WC1E 7HX (United Kingdom)
PFENNINGER, E.G.	SANDOZ Ltd., Preclinical Research, CH-4002 Basle (Switzerland)
PICHOVÁ, I.	Institute of Organic Chemistry and Biochemistry, Czechoslov. Acad. Sci., Flemingovo 2, 166 10 Prague 6 (Czechoslovakia)
POHL, J.	Institute of Organic Chemistry and Biochemistry, Czechoslov. Acad. Sci., Flemingovo 2, 166 10 Prague 6 (Czechoslovakia)

POLANOWSKI, A.J.	Institute of Biochemistry, University of Wroclaw, Tamka 2, 50-137 Wroclaw (Poland)
RADDATZ, P.	E. MERCK, Postfach 4119, Frankfurterstr. 250, D-6100 Darmstadt (FRG)
REID, W.A.	Department of Pathology, University of Leeds, Leeds LSZ 9JT (United Kingdom)
RIECHMANN, L.	Universität Bremen, Fachbereich Biologie, Postfach 330440, D-28 Bremen 33 (FRG)
ROHOŽKOVÁ, D.	Výzkumný ústav chorob revmatických, Na slupi 4, 128 50 Prague 2 (Czechoslovakia)
ROLPH, C.	University College Cardiff, Department of Biochemistry, Cardiff CF1 1XL, Wales (United Kingdom)
RYLE, A.P.	Department of Biochemistry, E.U. Medical School, George Sq., Edinburgh, Scotland (United Kingdom)
SAFRO, M.G.	Institute of Molecular Biology, Acad. Sci. of the USSR, Vavilova 32, Moscow 117984 (USSR)
SAMLOFF, I.M.	Veterans Administration Medical Center, 16111 Plummer Street, Sepulveda, CA 91343 (USA)
SCHIRLIN, D.G.	Merrell Dow Research Institute, 16 Rue d'Ankara, 67084 Strasbourg Cedex (France)
SCHMITGES, C.J.	E. MERCK, Exp. Med. Forschung, Frankfurter Str. 250, D-6100 Darmstadt (FRG)
SCHNABEL, E.	BAYER AG, Institut f. Biochemie, D-56 Wuppertal 1 (FRG)
SCHNABEL, I.	Schimmelweg 6, D-56 Wuppertal 11 (FRG)
SEDLÁČKOVÁ, J.	Institute of Organic Chemistry and Biochemistry, Czechoslov. Acad. Sci., Flemingovo 2, 166 10 Prague 6 (Czechoslovakia)

SIBANDA, B.L.	Department of Crystallography, Birkbeck College, University of London, Malet St., London WC1E 7HX (United Kingdom)
SIELECKI, A.R.	8347-120 Street, Edmonton, Alberta (Canada)
SIMONARSON, B.	Institute for Experimental Pathology, Keldur, University of Iceland, POB 110, IS-121 Reykjavík (Iceland)
STEPANOV, V.M.	Institute of Genetics and Selection of Industrial Microorg., 113545 Moscow (USSR)
SZÉCSI, P.B.	Department of Clinical Chemistry, Bispebjerg Hospital, DK-2200 Copenhagen N (Denmark)
TAKAHASHI, K.	Department of Biochemistry, Primate Research Institute, Kyoto University, Inuyama, Aichi 484 (Japan)
TANG, J.J.N.	Oklahoma Medical Research Foundation, 825 N.E. 13th Street, Oklahoma City, OK 73104 (USA)
TOMÁŠEK, V.	Institute of Molecular Genetics, Czechoslov. Acad. Sci., Flemingovo 2, 166 37 Prague 6 (Czechoslovakia)
TURK, V.	Department of Biochemistry, J. Stefan Institute, Jamova 39, Ljubljana 61000 (Yugoslavia)
TURKOVÁ, J.	Institute of Organic Chemistry and Biochemistry, Czechoslov. Acad. Sci., Flemingovo 2, 166 10 Prague 6 (Czechoslovakia)
VALLER, M.J.	Department of Biochemistry, University College Cardiff, POB 78, Cardiff, Wales (United Kingdom)
WATSON, F.E.	Department of Crystallography, Birkbeck College, University of London, Malet St., London WC1E 7HX (United Kingdom)
WIEDERANDERS, B.	Institute of Physiological Chemistry, M.-Luther University Halle, DDR-4020 Halle, Hollystr. 1 (GDR)

WILIMOWSKA-PELC, A.	Institute of Biochemistry, University of Wroclaw, Tamka 2, 50-137 Wroclaw (Poland)
WILUSZ, T.J.	Institute of Biochemistry, University of Wroclaw, Tamka 2, 50-137 Wroclaw (Poland)
WOOD, J.M.	Ciba-Geigy Ltd., K-125.1045, CH-4002 Basel (Switzerland)
ZIPSER, J.S.	Biochemistry Department, Agricultural Academy, Lubartowska 58a, Lublin (Poland)

Author Index

ABBREVIATIONS

aa:	amino acid
Abu:	L-α-aminobutyryl
Ac:	acetyl
AC:	chicken pepsin - activation peptide (p1-p42) complex, artificial
ACE:	angiotensin converting enzyme
α_2M:	alpha$_2$-macroglobulin
Amp:	γ-morpholinopropylamide
AP:	chicken activation peptide (p1-p42)
BAL:	2,3-dimercapto-1-propanol
BAPNA:	N-α-benzoyl-L-arginine-p-nitroanilide
Boc:	tert-butyloxycarbonyl
Bom:	π-benzyloxymethyl
BP:	blood pressure
BPTI:	bovine pancreatic trypsin inhibitor
CHU:	Christian Hansen unit
CP:	chicken pepsin
CPC:	calf prochymosin
CPG:	chicken pepsinogen
DAN:	diazoacetyl-D,L-norleucine methyl ester
DFP:	diisopropyl fluorophosphate
DNS:	5-dimethylamino-1-naphthalenesulfonyl, (dansyl)
DPPA:	diphenylphosphoryl azide
DTT:	dithiothreitol
EDTA:	ethylenediaminetetraacetic acid
EEO:	electroendosmosis
EPNP:	1,2-epoxy-3-(p-nitrophenoxy)propane
Et:	ethyl

FMOC: N-α-fluorenylmethoxycarbonyl
FOR: formyl
FP: fast pepsin
Fuc: fucose

GdmCl: guanidinium chloride
GlcNac: N-acetyl-α-D-glucosamine
Glu: L-2-pyrrolidone-5-carboxylic acid

Hb: hemoglobin
HC: heavy chain
Hcy: L-homocysteinyl
HMW: high molecular weight

ibu: isobutyl (i.e. 2-methyl-1-propyl)
Ibu: isobutyryl
IDF: International Dairy Federation
Iva: Isovaleryl

LC: light chain
LystaOEt: (3S,4S)-4,8-diamino-3-hydroxyloctanoic acid ethyl ester

M.A.: mixed anhydride
Man: mannose
MAP: mean arterial pressure
Me: methyl
Ms: methanesulfonyl

Nal: 3-(1´-naphthyl)alanine
NC: chicken pepsin-activation peptide (p1-p42)complex, natural
Nph: 4-nitrophenylalanine

PAD: pyridine-2-azo-p-dimethylaniline
PAGE: polyacrylamide gel electrophoresis

PCMB:	4-chloromercuribenzoic acid
pCMPSA:	p-chloromercuriphenylsulfonic acid
Pg:	pepsinogen
PG:	pepsinogen
Pht:	phthaloyl
PMSF:	phenylmethanesulfonyl fluoride
Pn:	pepsin
PRA:	plasma renin activity
RAS:	renin angiotensin system
RIE:	rocket immunoelectrophoresis
SDS:	sodium dodecyl sulfate
SMG:	submandibular gland
SMP:	slow moving proteinase
S-PI:	acetyl-pepstatin
Sta:	statine/(3S,4S)-4-amino-3-hydroxyl-6-methyl-heptanoic acid/
StaOEt:	statine ethyl ester
Suc:	succinyl
TCA:	trichloroacetic acid
Tce:	2,2,2-trichloroethyl
TFA:	trifluoroacetic acid, trifluoroacetyl
TNBS:	trinitrobenzenesulfonic acid
Z:	benzyloxycarbonyl (carbobenzoxy)
ZCl_2:	3,4-dichlorobenzyloxycarbonyl

SUBJECT INDEX

FSC
www.fsc.org
MIX
Papier | Fördert
gute Waldnutzung
FSC® C083411

Kolibri 360 GmbH
Ferdinand-Jühlke-Straße 7
99095 Erfurt, Deutschland
produktsicherheit@kolibri360.de